DAW 10/05 £44.99

Industrial Safety and Health Management

Fifth Edition

Industrial Safety and Health Management

Fifth Edition

C. Ray Asfahl
University of Arkansas

PEARSON

Prentice
Hall

Upper Saddle River, NJ 07458

Library of Congress Cataloging-in-Publication Data

Asfahl, C. Ray
  Industrial safety and health management/ C. Ray Asfahl -- 5th ed.
    p. cm.—(Prentice Hall International Series in industrial and systems engineering)
  Includes bibliographical references and index.
  ISBN 0-13-142392-4
  1. Industrial Safety 2. Industrial hygiene I. Title II. Series

T55.A83 2003
613.6'2--dc21

2003046968
CIP

Vice President and Editorial Director, ECS: *Marcia J. Horton*
Publisher: *Tom Robbins*
Acquisitions Editor: *Dorothy Marrero*
Vice President and Director of Production and Manufacturing, ESM: *David W. Riccardi*
Executive Managing Editor: *Vince O'Brien*
Managing Editor: *David A. George*
Production Editor: *Craig Little*
Director of Creative Services: *Paul Belfanti*
Creative Director: *Carole Anson*
Art Director: *Jayne Conte*
Cover Designer: *Bruce Kenselaar*
Art Editor: *Greg Dulles*
Manufacturing Manager: *Trudy Pisciotti*
Manufacturing Buyer: *Lisa McDowell*
Marketing Manager: *Holly Stark*

© 2004 Pearson Education, Inc.
Pearson Prentice Hall
Pearson Education, Inc.
Upper Saddle River, NJ 07458

Printed in the United States of America

10 9 8 7 6 5 4 3 2 1

ISBN 0-13-142392-4

Pearson Education Ltd., *London*
Pearson Education Australia Pty. Ltd., *Sydney*
Pearson Education Singapore, Pte. Ltd.
Pearson Education North Asia Ltd., *Hong Kong*
Pearson Education Canada, Inc., *Toronto*
Pearson Educación de Mexico, S.A. de C.V.
Pearson Education—Japan, *Tokyo*
Pearson Education Malaysia, Pte. Ltd.
Pearson Education, Inc., *Upper Saddle River, New Jersey*

To my partner Kela

Contents

Preface

The fifth edition of this book continues to bind together the traditional examination of time-tested concepts and techniques of safety and health management with a moden perspective on compliance with mandatory standards for workplace safety and health. It is intended to add reason, explanation, and illustration of the hazard mechanisms that form the underlying basis for the volumes of detailed standards for workplace safety and health. Industrial managers know the value of finding relevant and essential material to understand and develop a strategy for bringing their organizations into compliance with standards, reducing Workers' Compensation claims due to injuries and illnesses, improving productivity, and enhancing the overall well-being of their employees and their workstations.

Since the inception of OSHA three decades ago, one of the principal problems facing a safety and health manager has been how to find the relevant and essential information to do the job. OSHA published large volumes of detailed, mandatory standards. What was needed, though, was a strategy or guidance for coming into compliance with the volumes' standards. Also needed was understanding of the hazards so that the rationale behind the standards could justify effective actions to come into compliance. Compliance with the standards often requires analysis, planning, capital investment, and follow-through training. Such a process is better justified by a comprehensive understanding of the benefits of compliance beyond the mere avoidance of OSHA fines. One of these benefits is the reduction in workers' compensation costs—costs that have recently captured more attention from management as a significant component of direct labor costs.

OSHA has made a sincere effort to make information about their standards available to the public by posting it on their website. The website contains the full text of standards, along with public announcements, new standards-development activities, and enforcement summaries. This text further enhances OSHA's efforts by providing readers with a CD-ROM that has detailed enforcement statistics. The CD-ROM contains a proprietary, searchable database from New Century Media (NCM) Corporation that provides readers with the power to determine (1) which standards are actually being cited, (2) the numbers of these citations designated as "serious," (3) the levels of penalties proposed, (4) statistics on contested citations, and (5) other useful details. At the end of some chapters, students using this text are given exercises requiring them to use the NCM database to obtain answers to questions about standards relevant to the reading material. The objective is not to merely teach information on the various topics of industrial safety and health, but to empower students to readily find their own answers to questions relevant to their mission.

Besides providing access to the powerful NCM database, the accompanying CD-ROM gives answers to selected end-of-chapter exercises.

The 21st century has brought on new challenges and concerns for the safety and health manager. Perhaps the most significant development of such challenges has been ergonomics. Ergonomics is such a controversial issue for a safety and health manager that an entirely new chapter entitled "Ergonomics" has been added to this text, and all remaining chapters have been changed. Reinforcing these changes are new case studies, illustrations, and end-of-chapter exercises. OSHA has changed its recordkeeping forms and procedures, and these new forms and new rules for reporting are included herein. Students can download a hard copy of the forms by using the CD-ROM provided. This fifth edition also has added some humor and historical perspectives that may round out the reader's knowledge and appreciation of the culture of the field of industrial safety and health management.

Instructors using this text are advised to contact their Prentice Hall representatives to obtain the free, comprehensive *Instructor's Manual* and computer disk available to instructors only. These teaching aids include answers to the end-of-chapter exercises, audiovisual suggestions, additional supporting information, and a quiz bank of more than 1000 questions for convenient use in composing comprehensive examinations covering the subject matter. Also available are answer keys for every quiz question and page references to the text to help in settling any debates after examinations. Most of the quiz bank and almost all of the end-of-chapter exercises have been classroom tested. Also on the disk are lecture outlines for each chapter.

ACKNOWLEDGMENTS

Many persons deserve special recognition for their contributions to this edition. Mark Friend should be recognized for his ideas about dealing with disasters, such as the September 11, 2001, terrorist attacks. Jim Collier provided important suggestions, especially with respect to Chapter 2, "Recordkeeping." Mike Belote, Merrisa Purnomo, Yun Han Long, Nicholas Jackson, Curtis Jackson, Ashlea Bennett, Wynell Farrish, Carolye Asfahl, and Greg Tumbush provided assistance in the development of the NCM database. Reviewers Robert Feyen, Larry McDonald, and Fereydoun Aghazadeh provided many valuable suggestions, while industrial realism and case studies came from my students. Especially noteworthy were the contributions of Donna Reams, Paul Angelis, and Richard LaScala. Cary Pollack and Bill Bell, President of Envision New Media Design, provided exciting insights into the potential of media techniques for the delivery of safety and health training. Kristin Landers and Kela Asfahl cheerfully assisted with the manuscript.

CHAPTER 1

The Safety and Health Manager

Everyone wants a safe and healthful workplace, but what each person is willing to do to achieve this worthwhile objective can vary a great deal. As a result, the management of each firm must decide at what level, along a broad spectrum, the safety and health effort will be aimed. Some managers deny this responsibility and attempt to leave the decision to employees. This strategy seems to square with hallowed principles of personal freedom and individual responsibility. But such a denial of responsibility by management results in a decision by default, and usually the result is a relatively low level of safety and health in the workplace.

Is the foregoing an indictment on the judgment of the individual worker? Not really, because without a commitment on the part of management, the worker usually is unable singlehandedly to build safety into his or her job station. The behavior of the worker is the most important determinant for his or her safety, but behavior alone cannot make a dangerous job safe. Furthermore, even if a given worker has a strong inclination to be careful and to guard his or her health, there are plenty of production motivations and other quite natural incentives to undermine safe attitudes when management has not made a commitment to safety and health.

One person, usually designated as safety director or industrial hygienist, sets the tone of the safety and health program within a firm. In fact, right at the start, it says something about the commitment of management when a firm decides to designate a person by title to the responsibility of safety and health. But naming someone safety director or manager of safety and health is just a beginning step. Many such persons have little authority and have been largely ignored by management and worker alike, especially in the past. It was not unusual for a safety director's work to be typified by public relations activities, such as posting motivational signs and compiling statistics. These are still important functions, but much more responsibility for this function is now recognized.

Something happened in the 1970s to change dramatically the role of the typical safety director in industrial firms throughout the country. The passage of the Occupational

Safety and Health Act of 1970 created the Occupational Safety and Health Administration (OSHA), a federal agency whose regulations would have a large impact on the role of the typical safety director. Chapter 4 discusses this impact in detail, but the balance of this chapter discusses the enlarged role of the person within a firm charged with industrial safety and health.

There is little doubt that OSHA enhanced the authority of the safety manager in the typical industrial plant in the United States. Prior to OSHA, few safety managers dared to interfere with production schedules to alleviate a safety or health problem. But prominent OSHA cases in the news media have brought to the attention of top management personnel the dire consequences that can ensue when serious safety or health problems are not adequately addressed.

The field of occupational health has probably benefited even more from OSHA than has the field of occupational safety. Prior to OSHA, occupational health seemed to be a matter too remote to really concern anyone, except perhaps the plant nurse. And the plant nurse had little authority to influence policy or even to take action to prevent hazards. Prior to OSHA, the plant nurse was chiefly concerned with first aid, after the fact, and physical examinations, not with hazard abatement and prevention.

In describing the functions of today's executive charged with the safety and health responsibility, this text will use the designation *safety and health manager*, recognizing the dual nature of the job. Also, the term *manager* envisions the enlarged scope of responsibility, which includes analysis of hazards, compliance with standards, and capital investment planning, in addition to the conventional functions described earlier. The purpose of this book is to provide tools and guidelines to safety and health managers to help them execute their increased duties.

Dealing with applicable standards is one of the greatest challenges facing today's safety and health manager, and to meet this challenge is a primary purpose of this book. Since only 10% of the standards generate 90% of the activity, safety and health managers need guides to the important parts of the standards. Frequently cited standards should receive prime attention because they indicate areas in which industries are having difficulty complying or areas to which enforcement agencies are giving a great deal of attention. In either event, safety and health managers have a need to know these frequently cited standards so that they can bring their facilities within compliance. Information technology has made this task easier, and today's safety and health manager can take advantage of the full text of OSHA standards on the Internet by accessing the OSHA website. In addition, national inspection statistics are available by accessing the New Century Media database available on the CD-ROM that accompanies this book. Besides the frequency of citation, safety and health managers need to know the "why" behind the standards. Until the safety and health manager learns what hazards a particular standard is intended to prevent, he or she will have a difficult time persuading either management or employees that a given situation needs correction.

A REASONABLE OBJECTIVE

Top management sometimes turns a deaf ear to the pleas of the safety and health manager for plant improvements. But the safety and health manager is sometimes a crusader with a one-track mind. Any safety and health manager who feels that elimination of workplace hazards is an indisputable goal is naive. In the real world, we must choose among the following:

1. Hazards that are physically infeasible to correct
2. Hazards that are physically feasible, but are economically infeasible to correct
3. Hazards that are economically and physically feasible to correct

Until the safety and health manager comes to grips with this reality, he or she cannot expect to enjoy the approval of top management. Some safety and health managers have faced this reality on the surface, but within their hearts they may resent the attitudes of top executives who are unwilling to support their efforts to eliminate all workplace hazards. But this resentment is unjustified because it is an unrealistic and naive strategy to attempt to eliminate all hazards.

It may be surprising to some readers to discover that this book, which is supposed to be about safety and health, does not really advocate the elimination of all workplace hazards. Such a goal is unattainable, and to reach for it is poor strategy because it ignores the need for discriminating among the hazards that *can* be corrected. To see how such a naive strategy is not even in the interest of safety or health, consider Case Study 1.1.

CASE STUDY 1.1

A safety and health manager receives three suggestions from three different operating personnel as follows:

1. Install a drain to remove water that occasionally collects around the die-casting area.
2. Post a warning sign that advises forklift truck drivers to slow down.
3. Improve sanitation by cleaning the rest rooms more frequently.

There is a safety or health rationale for the correction of all three of these problems. Should they be corrected?

Some managers would accept the safety and health rationale as *all they need* to begin action to correct the problems listed in Case Study 1.1. But this would be a naive response. More data are needed to decide what to do. While busying the plant maintenance department to correct the foregoing three problems (which may or may not be consequential), a serious electrocution or respiratory hazard may be going unchecked or maybe even unnoticed. By reacting to every hazard that happens to arise, the safety and health manager may be missing opportunities to have a really significant impact on worker safety and health. At the same time, such overreaction may also be deteriorating the safety and health manager's credibility with top management. Even the law does not call for the elimination of all hazards, just the ones that are "recognized." Therefore, let it be clearly understood that our objective is to eliminate unreasonable hazards, *not all hazards* in the workplace. The goal of this book, then, is to assist the safety and health manager in (1) detecting hazards and (2) deciding which ones are worth correcting. The goal is an ambitious one, and this book certainly claims no break-throughs for solving this difficult problem. But any light that can be shed on the mystery of which hazards are most significant and which standards, among the thousands, are most important is sorely needed by safety and health managers everywhere.

SAFETY VERSUS HEALTH

This chapter has already implied that early "safety directors" did not emphasize health problems. It is essential that today's safety and health manager give sufficient attention not only to safety hazards, but also to health hazards, which are steadily gaining in importance as new data about industrial disease are being uncovered.

What really is the difference between safety and health? The words are so common that almost everyone has a firm image of the concept of safety versus the concept of health. There is no question that machine guarding is a safety consideration, and airborne asbestos is a health hazard. But some hazards—such as those associated with paint spray areas and welding operations—are not so easy to classify. Some situations may be both a health *and* a safety hazard. This book will draw the following line between safety and health:

Safety deals with acute effects of hazards, whereas health deals with chronic effects of hazards.

An acute effect is a sudden reaction to a severe condition; a chronic effect is a long-term deterioration due to a prolonged exposure to a milder adverse condition. Everyday concepts of health and safety fit this definition, which separates the two. Industrial noise, for instance, is usually a health hazard because it is usually the long-term exposure to noise levels in the range 90 to 100 decibels that does the permanent damage. But noise can also be a safety hazard because a sudden acute exposure to impact noise can *injure* the hearing system. Many chemical exposures have both acute and chronic effects and thus are both safety and health hazards.

Industrial hygienists, those who concentrate on health hazards, are known by their sophisticated instruments and scientific expertise. These tools are necessary to the industrial hygienist because of the tiny effects they must measure in order to determine whether a chronic hazard exists. By contrast, the safety specialist, instead of being an expert with precise scientific instruments, usually has more industrial process experience and practical on-the-job knowledge. This difference in backgrounds may generate some confrontation between safety professionals and health professionals—though they should be partners, they often compete.

The bases of competition between safety and health professionals are classics: young versus old, new versus old, and education versus experience. Safety professionals are usually older and have more industrial experience; their career field is more traditional and more entrenched in industrial organizations. Health professionals are typically younger, have more college education, and occupy newer job positions. In recent years, however, the distinctions between the career safety professional and the career health professional are disappearing.

Degree of hazard is another point of contention between safety and health. Both sides think their hazards are more grave. Safety professionals can point to fatalities on the job and feel an urgency in protecting the worker from imminent danger from accidents. To them, the industrial hygienists with their meters, pumps, and test tubes are checking for microscopic hazards that really are not urgent. But to the industrial hygienist, the safety professional's work sometimes seems superficial. The industrial hygienist is fighting those insidious unseen hazards on the job that can be just as lethal as a falling crane. There are probably more occupational health fatalities than safety fatalities, but the statistics will not reflect this difference because the health fatalities are delayed and often are never diagnosed.

Another problem with identifying health hazards is that the signs of occupational illness are often identical to common symptoms arising from normally occurring illnesses encountered off the job. For instance, a common cold causes respiratory congestion, headaches, and perhaps fever. These same symptoms also could be the result of a dangerous exposure to a toxic chemical or other occupational hazard. The industrial hygienist is tasked with sorting out these symptoms and identifying occupational hazards to be controlled. Considerable expertise is required to do this, and the problem is much more subtle than that of the person charged with controlling common safety hazards alone.

Because of age and experience, a safety professional will often have managerial control over health functions, and this can rankle the industrial hygienist. The safety and health manager should be aware of this sensitivity and be careful to recognize the importance of both fields of interest. A growing number of well-educated industrial hygienists are being given overall responsibility for both safety and health.

ROLE IN THE CORPORATE STRUCTURE

Most safety and health managers wear several hats, especially in smaller firms. Often, they are responsible for security; some are also personnel managers, and even more frequently they report to the personnel manager. This is a fairly natural arrangement in that it emphasizes the importance of worker training, statistics, job placement, and the industrial-relations aspect of safety and health. The growing importance of engineering to workplace safety and health, however, strains the placement of the safety and health manager within the personnel department, which traditionally has little interaction with engineering.

The safety and health manager is virtually never associated with the purchasing function, but one of the first goals of the safety and health manager should be to obtain some input into the purchasing process. Used-equipment dealers, even new-equipment dealers, often have bargain-priced machines, compressors, tractors, forklifts, and other pieces of equipment that fail in some way or another to meet minimum safety standards. The purchasing agent usually is not knowledgeable in safety and health standards and is easy prey for these dealers because the price is right. What is needed is a knowledgeable person to check specifications and prevent the costly purchasing error of buying equipment that does not meet current safety and health standards. When standards change, another category of equipment sometimes becomes obsolete, and the safety and health manager should warn the purchasing department when these changes occur.

A recent role of the safety and health manager is as a liaison with government agencies, a condition brought about by the arrival of OSHA. Some safety and health managers have a dual responsibility for environmental protection activities. Sometimes the safety and health manager is considered a staff member in the legal department. This arrangement stresses an adversarial role and is not recommended because it tends to detract from both workplace safety and health and from constructive relations with enforcement agencies.

A related field is consumer product safety. The Consumer Product Safety Commission (CPSC) is a federal agency whose enabling legislation is obviously patterned after OSHA's. The Product Safety and Liability Act was passed the year after the passage of the OSHA Act, and the wording of the two laws is remarkably similar. Although both fields consider the safety of machines and equipment, the CPSC concentrates on the responsibility of the manufacturers of products, whereas OSHA concentrates on the responsibility of the employer who places products into use in the workplace.

The decade of the 1980s saw renewed interest in environmental protection and especially in the prevention and emergency cleanup of disastrous accidental releases of toxic substances. In 1984, in a shocking disaster in Bhopal, India, at least 2500 civilians were killed in a single industrial accidental release of deadly methyl isocyanate gas. Without doubt, this incident had its impact on public policy in the United States. Because of the close relationship to worker safety and health, the responsibility for compliance with Environmental Protection Agency (EPA) requirements is often made a part of the duties of the safety and health manager. Chapter 5 reveals more of this relationship between safety inside and safety outside the plant.

RESOURCES AT HAND

The demands for training materials, the need for ideas for hazard correction, and the importance of the latest interpretations of standards guarantee that the successful safety and health manager will not try to work alone, insulated from the career field. A variety of resources have arisen to meet these needs of the safety and health professional.

Professional Certification

Safety and health managers can establish themselves with their peers as well as with their employers by achieving professional certification. This requires relevant work experience, letters of recommendation, and an examination. Education is also required, but there is partial trade-off between education and experience. Safety professionals should apply to

Board of Certified Safety Professionals of America
208 Burwash Avenue
Savoy, IL 61874
(217) 359-9263
www.bcsp.org

Health professionals should apply to

> American Board of Industrial Hygiene
> 6015 W. St. Joseph, Suite 102
> Lansing, MI 48917
> (517) 321-2638
> www.abih.org

The examinations are quite difficult and the background screening process is stringent. Few people qualify for certification in both fields. Industry, government, and the public have come to recognize the professional designations CSP for Certified Safety Professional and CIH for Certified Industrial Hygienist. To an increasing extent, professional certification, such as CSP or CIH, is mandatory for filling certain required positions. Many states are reforming their Workers' Compensation laws, and some are including requirements for accident-prevention plans for "extrahazardous" workplaces (ref. Zumar). These plans must be formulated by personnel or consultants who have such qualifications as the CSP or CIH credential. One objective of this book is to prepare students to complete either the CSP or CIH examination successfully.

Professional Societies

Two professional societies are foremost in the career field of occupational safety and health:

> American Society of Safety Engineers (ASSE)
> 1800 East Oakton St.
> Des Plaines, IL 60018
> (847) 768-3434
> www.asse.org

> American Industrial Hygiene Association (AIHA)
> 2700 Prosperity Avenue
> Suite 250
> Fairfax, VA 22031
> (703) 849-8888
> www.aiha.org

Leading journals in the field, *Professional Safety* and the *American Industrial Hygiene Association Journal*, are published by these two professional societies, respectively. Each holds annual conferences, and local chapters often conduct seminars and workshops on topics of current interest.

A somewhat narrower but influential organization is

> American Conference of Government Industrial Hygienists (ACGIH)
> 1330 Kemper Meadow Drive
> Cincinnati, OH 45240
> (513) 742-2020
> www.acgih.org

An important committee of this organization is

Committee on Industrial Ventilation
1330 Kemper Meadow Drive
Cincinnati, OH 45240

This committee's publication *Industrial Ventilation* (ref. Indus Vent) is probably the best recognized manual of recommended practice in the field of ventilation.

National Safety Council

Although not a professional society in the strict sense of the term, the National Safety Council (NSC) is one of the most important organizations in the field. The Council's headquarters are

National Safety Council
1121 Spring Lake Drive
Itasca, IL 60143
(630) 285-1121
www.nsc.org

The NSC is broad in scope and encompasses all kinds of safety, not just occupational safety. The membership of the NSC consists principally of organizations and businesses. It has been in existence for over 85 years and was recognized by the U.S. Congress in 1953.

Corporate membership in the NSC affords many benefits for the safety and health manager. The Council is the principal focal point for information about safety hazards. The library at the Council's national headquarters contains a wealth of information, and although it is open to the public, companies who are members of the Council have special privileges for copying services and research. Each year, the Council publishes comprehensive summaries of accident statistics in its book *Accident Facts* (ref. Accident). The NSC's *Accident Prevention Manual for Industrial Operations* (ref. McElroy), now in its 10th edition and in two volumes, is the most comprehensive general reference in the field.

The NSC makes use of its current statistical data to continually recognize and award member firms with excellent-safety awards. The annual edition of *Work Injury and Illness Rates* (ref. Work) lists company names by Standard Industrial Classification (SIC) Code that have the "best records known in industry."

Standards Institutes

The age of OSHA enforcement has brought increased recognition of the national standards-producing organizations. The following are the most prominent among these organizations:

American National Standards Institute (ANSI)
25 West 43rd Street
New York, NY 10036
(212) 642-4900
www.ansi.org

National Fire Protection Association (NFPA)
1 Batterymarch Park
P.O. Box 9101
Quincy, MA 02269
(617) 770-3000
www.nfpa.org

American Society of Mechanical Engineers (ASME)
Three Park Avenue
New York, NY 10016–5990
(800) 843-2763
www.asme.org

American Society for Testing and Materials (ASTM)
100 Barr Harbor Drive
P.O. Box C 700
West Conshohocken, PA 19428–2959
(610) 832-9585
www.astm.org

These organizations have standing committees that invite public comment and prepare voluntary standards for occupational safety and health. At the outset, OSHA issued many existing standards produced by these organizations, designating such existing standards as representing the "national consensus." Most of OSHA's national-consensus standards are derived from either ANSI or NFPA standards.

Trade Associations

If a problem pertains to a specific industry or type of equipment, an association of manufacturers may be called upon to furnish safety and health data. Some people decry such use of trade-association data as biased, but many careful studies of safety and health problems have been performed by trade associations. The following are useful associations in this regard:

1. American Foundrymen's Society (AFS)
2. American Iron and Steel Institute (AISI)
3. American Metal Stamping Association (AMSA)
4. American Petroleum Institute (API)
5. American Welding Society (AWS)
6. Associated General Contractors of America (AGCA)
7. Compressed Gas Association (CGA)
8. Industrial Safety Equipment Association (ISEA)
9. Institute of Makers of Explosives (IME)
10. National Electrical Manufacturers Association (NEMA)
11. National LP-Gas Association (NLPGA)
12. National Machine Tool Builders Association (NMTBA)
13. Scaffolding, Shoring, and Forming Institute (SSFI)

National trade associations are especially useful in providing audiovisual training materials if one is willing to accept the fact that such materials also serve to promote the industry's products.

Government Agencies

A program of free consultation is usually provided by state agencies and in some states by private consulting firms. There is an understandable reluctance to call a government agency for help with a safety-standards problem, but the purpose of these consultation agencies is to assist in increasing workplace safety, not to write citations. In most states, the consultation function is performed by a completely separate agency from the enforcement agency. In states that have comprehensive state plans, with one agency responsible for both enforcement and consultation, care is taken to maintain confidentiality of records.

The National Institute for Occupational Safety and Health (NIOSH) has a wealth of research data on the hazards of specific materials and processes. NIOSH uses these data to write criteria for recommended new standards. In addition to its research function, NIOSH acts as a source of technical information for questions about safety and health. NIOSH has a telephone hot line for this purpose: (800) 35–NIOSH.

OSHA itself can be of value to the safety and health manager seeking information. Some safety and health managers would never consider calling OSHA to discuss a problem for fear of precipitating an inspection. However, problems can be posed in a hypothetical setting, and OSHA personnel can understand the need for keeping such questions hypothetical. Most OSHA personnel will be glad to furnish whatever answers are available in the interest of encouraging employers to keep their facilities safe and healthful. OSHA has also opened the doors to its national training institute in Des Plaines, Illinois, for training the general public in "voluntary compliance." A wealth of information is also available through OSHA's website at www.osha.gov

SUMMARY

Safety and health management is a career field that has been both enhanced and made more challenging by the institution of the federal agency OSHA. The current chapter has provided a brief introduction to this career field and to some of the organizations and boards that have given the field both identity and assistance in fulfilling its mission.

Chapter 2 proceeds to describe how safety and health managers can go about performing their responsibilities within their organizations. Chapter 3 addresses the principal goal of the safety and health manager—the reduction of workplace hazards—describing four basic approaches to solving this problem. OSHA has had such an impact on the field that it deserves a chapter of its own; Chapter 4 describes OSHA in general, including both the positive and the negative aspects of this controversial agency. The remaining chapters address specific hazard categories, advising safety and health managers on what to do to eliminate hazards while complying with established standards. Any factual data or approach strategies that can assist the manager in understanding the hazard mechanisms and in tackling the most significant problems first will be helpful, even if many questions are left unanswered.

EXERCISES AND STUDY QUESTIONS

1.1 Why are some safety and health standards cited more frequently than others?

1.2 What else, besides frequency of citation, does a safety and health manager need to know about safety and health standards?

1.3 Should safety and health managers attempt to eliminate all workplace hazards? Why or why not?

1.4 Identify three categories of hazards with respect to feasibility of correction.

1.5 Describe at least two disadvantages of overreacting to minor hazards in the workplace.

1.6 How does a safety hazard differ from a health hazard?

1.7 Name three safety hazards and three health hazards.

1.8 Name some chemicals that are both health and safety hazards.

1.9 Name some physical agents that can be both health and safety hazards.

1.10 Why does an industrial hygienist need more scientific instruments to evaluate hazards than does a safety specialist?

1.11 Which type of hazard appears to be more grave: safety or health?

1.12 What aspects of the safety and health manager's job are related to the personnel department?

1.13 What disadvantage is associated with placing the safety and health manager within the personnel department?

1.14 What disadvantage can be seen in placing the safety and health manager within a firm's legal department?

1.15 Compare the missions of the following two federal agencies: OSHA and CPSC.

1.16 What national, but nongovernmental, safety organization is officially recognized by an act of the U.S. Congress?

1.17 What is ANSI, and what relationship does it have to the field of safety and health?

1.18 Compare the missions of OSHA and EPA. Why might the same person within a given industrial plant have the responsibility for dealing with both agencies?

1.19 What is the federal government's telephone hot line for technical information on safety and health? What agency responds to this number?

1.20 What event occurred in 1970 that enhanced the authority of the safety director in a typical industrial firm?

1.21 Compare attitudes before and after the passage of the OSHA Act with regard to occupational health and the job of the plant nurse.

RESEARCH EXERCISES

1.22 Check the Internet for web pages for the National Safety Council, the American Society of Safety Engineers, and the American Industrial Hygiene Association. What resources to help the safety and health manager can be found through these websites?

1.23 Browse the Internet to learn about current requirements to become certified as a CSP. Also check requirements to become a CIH.

1.24 Find the home pages of several CIHs on the Internet. Check to see whether any job openings are listed on the Internet for CIHs.

1.25 Find the home pages of several CSPs on the Internet. Check to see whether any job openings are listed on the Internet for CSPs.

1.26 Do research to determine the percentage of CSPs working in various industries. What percentage are consultants? What percentage are government employees? What percentage are employed by insurance companies?

1.27 Go to the OSHA website and check out the Voluntary Protection Program (VPP).

1.28 Check the latest news releases from OSHA as posted on their website.

1.29 Explain the three principal reasons that health hazards are more difficult to identify than safety hazards.

Development of the Safety and Health Function

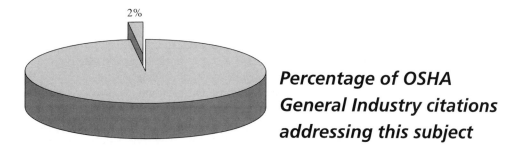

2%

*Percentage of OSHA
General Industry citations
addressing this subject*

The safety and health function has both line and staff characteristics, and the safety and health manager needs to recognize which elements of the function belong to which category. The physical accomplishment of workplace safety and health is a line function. For example, operator work practices are the responsibility of the workers themselves, as directed by their line supervisor. In industries where maintenance departments are recognized as another line function, the correction of facilities problems is again the direct responsibility of the maintenance operators and their line supervisors. The safety and health manager then performs a staff function by acting as a "facilitator" in assisting, motivating, and advising the line function in achieving worker safety and health.

The interest of line personnel in receiving this advice and assistance from the safety and health manager depends on how important the goal of safety and health is to top management. The successful safety and health manager will be keenly aware of this need for top management support. Respect and approval of top management must be won by responsible decisions and actions by the safety and health manager. A necessary ingredient of such decisions and actions is a recognition of the important principle stated in Chapter 1—that is, the goal is to eliminate unreasonable hazards, *not all hazards.* The respect and approval of top management may be difficult to win because safety and health managers are too often such emotional crusaders for the cause that they lose their credibility and with it their eligibility to be considered "managers." On the other hand, federal regulation has added a measure of urgency and credibility to those whose efforts make industries safe and healthful, and this has substantially strengthened the position of the safety and health manager in the management hierarchy.

Similarities can be drawn between the safety function and other staff functions, such as quality control and production control. Like safety and health, quality and production goals must be achieved by line personnel, facilitated by the staff function. This principle is recognized in such clichés as "You can't inspect quality into a product" and "Safety is everybody's business."

Once the approval of top management has been won, the safety and health manager is advised to document this approval in a written safety and health policy statement, issued by top management. This written policy becomes the documented authority to line personnel that top management does have safety and health goals and wants these goals met. The everyday actions of management then reinforce the written policy, as discussed in Chapter 1. If, however, top management fails to practice what it preaches in the policy statement, it is the duty of the safety and health manager to go back to management and redetermine its level of commitment to safety and health.

Having established by word and deed management's commitment to safety and health, the safety and health manager is ready to proceed with the staff function of facilitating the safety and health program throughout the plant. To do this, plant-operating personnel will have needs that the safety and health manager can satisfy. To make workers aware of hazards, supervisors and the workers themselves need regular training in hazard recognition and correction. Statistics and accident records are needed to keep management and operating personnel advised of how well the company and its departments are doing in achieving their safety and health goals. Sometimes the safety and health manager can provide a plantwide stimulus for worker safety and health by using contests and awards for safety performance. Finally, of increasing importance is the safety and health manager's role in dealing with safety and health standards and in assisting operating personnel in achieving compliance with these standards. The remainder of this chapter itemizes staff functions of the safety and health manager's office with guidance for the development of each.

WORKERS' COMPENSATION

Workers' compensation laws (formerly known as workmen's compensation laws) provided an initial structure for industrial safety. The first such laws were introduced in state legislatures in 1909, and now all states have comparable legislation. It is invariably the duty of the safety and health manager to implement the workers' compensation system within the plant, so it is considered here as the first staff function to be discussed.

Workers' compensation legislation has the ostensible purpose of protecting the worker by providing statutory compensation levels to be paid by the employer for various injuries that may be incurred by the worker. There is an ulterior feature, however, that provokes labor to be dissatisfied with the workers' compensation system. This feature is the immunity from additional liability that the workers' compensation system grants to the employer, except in cases in which "gross negligence" can be proved.

Table 2.1 lists examples of statutory compensation levels for various types of permanent injury. To most people, the various levels seem too low to compensate adequately

TABLE 2.1   Sample Statutory Compensation Levels for Permanent Injuries
(Compensation at $66\frac{2}{3}$ % of Average Weekly Pay)

Type of permanent injury	Compensation level[a] (weeks)
Arm amputated	
At or above the elbow	210
Below the elbow	158
Leg amputated	
At or above the knee	184
Below the knee	131
Hand amputated	158
Thumb amputated	63
Finger(s) amputated	
First	37
Second	32
Third	21
Fourth	16
Foot amputated	131
Toe amputated	
Great toe	32
Other toes, per toe	11
Loss of sight in one eye	105
Loss of hearing in one ear	42
Loss of hearing in both ears	158
Loss of testicle	53
Loss of both testicles	158

[a] These compensation levels are in addition to any compensation paid for the healing period. The sample compensation levels were obtained from workers' compensation levels for Arkansas and are intended only as an approximate guide. Exceptions and special cases exist, and tables vary somewhat from state to state.

Source: Arkansas Workers' Compensation Commission (ref. Arkansas Workers').

for the permanent injury to the worker. Historical evolution of the rates is slow, and it can be seen that public sensitivity to worker injuries has increased over the years. This sensitivity has resulted in outcries for reform of the workers' compensation system. On the other side of the issue is management's position that industry can never fully compensate monetarily for everything that may happen to workers in the course of their duties. Since some risk is inescapable in any line of work, management's general position is that in consideration of the salary and wages that workers receive, part of the normal risk of injury must be borne by the worker.

Typically, the firm does not pay the workers' compensation payments directly; rather, it carries insurance against compensation claims. The insurance company is vitally interested in the safety and health within the plant, and this provides a major impetus to the development of the safety and health program. The accident experience of the firm is reflected in the levels of the insurance premium, which can be adjusted up or down depending on plant safety experience. The insurance industry applies an "experience rating" expressed as a decimal fraction to be multiplied by the standard premium rate. The experience rating is based on a three-year average of the firm's actual claims experience and can be less than or greater than 1.00. An experience rating of 1.00 would represent no modification at all and would be applied to a firm that is estimated by the insurance company to have a standard, typical risk. A large company with an experience rating of perhaps 0.80 can save thousands of dollars in annual workers' compensation insurance premiums. Even in the first year of coverage under workers' compensation insurance, a company can profit from an effective safety and health program because the insurance underwriter depends upon prior loss data and an initial assessment of the firm's hazards before setting the initial annual premium. A good insurance company will make regular inspections of facilities to be sure that installations and practices are safe. This is a direct and measurable monetary stimulus to the safety program.

Some companies choose to self-insure against workers' compensation claims. This may make economic sense when claims experience and premium levels are compared. But the intangible benefits provided by the insurance company must also be considered in order to make a rational decision. Besides the regular inspections mentioned earlier, the insurance companies are valuable sources of technical advice to their clients. Many insurance companies provide training films and other valuable aids regarding the conduct of the safety and health program. Insurance companies even have research centers for the purpose of reducing workers' compensation claims by studying such hazards as cumulative trauma, low back pain, biomechanics, acoustics, and work physiology (ref. Lorenzi). If the safety and health manager is not receiving these aids and services from the insurance company, he or she should request them and also consider alternative vendors when policy renewal time presents itself.

The number of companies that have elected to self-insure has led to a new type of consultant called a *loss-control representative*. This consultant's objective is to keep workers' compensation claims low by supplying the type of services normally provided by an insurance carrier. A significant part of these services is maintaining close relationships with employees who do file claims. This serves the purpose of showing an interest in the well-being of, and providing encouragement to, the honest claimant, while at the same time uncovering evidence of fraudulent claims in the case of the dishonest worker who is either malingering or truly injured, but whose injury occurred off the job.

At the dawn of the 21st century, there is evidence that state workers' compensation programs are on the threshold of significant change. Added to labor's dissatisfaction with the system is management's alarm over the sharply rising costs of workers' compensation insurance premiums. Historical norms for workers' compensation premiums were around 4–5% of payroll until near the end of the 20th century. Recent figures are much higher, with some industries in the area of 20–30% of payroll. Such high costs are focusing attention on the importance of safety and health in the overall equation of industry in general and manufacturing in particular. At the same time, states are groping for ways to reduce the costs of workers' compensation premiums while improving worker protection against injury and illness.

The various state reforms for workers' compensation can be roughly divided into three categories:

1. Managed care programs
2. Reduction of false claims
3. Prevention of injury and illness

The concept behind "managed care" is to reduce the cost of claims by close supervision of each claim, with the objective of getting the worker back on the job as soon as is practical. The objective of the second approach is to detect instances in which workers are attempting to take advantage of the system by malingering or by blaming their employers for injuries or illnesses they actually incurred off the job. The incidence of workers' compensation fraud is significant. Insurance fraud is now considered the second-largest white-collar crime in the United States, behind income tax evasion (ref. Fraud). Therefore, it is appropriate that the workers' compensation system focus on this problem.

The third approach, prevention, constitutes the most dramatic change in the concept of workers' compensation programs. The original purpose of workers' compensation— to ensure financial compensation for injured employees—is seen to be broadened to include general regulation and enforcement of workplace safety and health. The experience modifier adjustment for workers' compensation premiums, described earlier in this section, is being replaced with more forceful incentives for employers, including mandatory inspections and establishment of safety improvement programs. Since such an approach pertains more to enforcement and regulation, the new workers' compensation developments will be discussed later in Chapter 4.

RECORDKEEPING

Forms, reports, and recordkeeping make up no small part of the safety and health manager's job. According to the National Safety Council, "Just one OSHA data sheet takes U.S. safety managers a cumulative 54 million hours a year to complete. And that's just one of dozens of forms (the safety manager) may be responsible for" (ref. National Safety). In addition, the safety manager must keep up with the latest developments, including changes to rules and procedures. The official public notification device of the federal government is the *Federal Register*, which is printed daily by the U.S. Government Printing Office. In this document alone, the government prints approximately 70,000 pages every year (ref. National Safety).

The National Safety Council established the first national system of industrial safety recordkeeping. This system was standardized and designated the Z16.1 system by the American National Standards Institute. In the 1970s, the federal agency OSHA set mandatory recordkeeping requirements very similar to the Z16.1 system, which was voluntary. There were some differences, however, that made year-by-year comparisons of safety and health records infeasible when one year was based on traditional Z16.1 records and another on the federal system. This is particularly unfortunate in that it confounds attempts to determine from statistical records whether the federal agency has had any beneficial impact on worker safety and health. Some specific industries and hazard categories, such as trenching and excavation cave-ins in construction, have shown visible improvements since OSHA's inception, but other gains have been obscured by the change in the statistical records system. Other variations in conditions, such as employment levels and recession cycles, have also acted to blur statistical comparisons. Statistical studies on this subject will be examined in Chapter 4 in the section titled "Public Uproar."

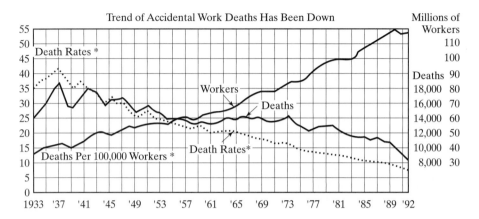

FIGURE 2.1

Trends in workers, deaths, and death rates, 1933–1992 (*Source*: National Safety Council, *Accident Facts, 1993 Edition.* Chicago: NSC; used with permission).

The recording of worker fatalities is more consistent than that of injuries and illnesses; thus, fatality statistics can be used to observe trends both before and after federal regulation. Figure 2.1 shows that the long-range trend of industrial fatalities is downward. Note that since the implementation of the OSHA law in the early 1970s there has been very little visible impact on this trend when all industries are considered together.

The National Safety Council continues to publish data on workplace fatalities in its revised-format Injury Facts, the successor to its former publication, Accident Facts. Although no longer reported in the same format as that shown in Figure 2.1, the trend in workplace fatalities is still downward (ref. Injury). The latest available statistics at this printing were for the year 2000, in which the total number of workplace fatalities was 5935 for a workforce of approximately 138 million workers. The fatality rate per 100,000 workers was 4.3, the lowest in history.

Traditional Indexes

Familiar statistical measures are *frequency* and *severity*, which were defined by the old Z16.1 system. Frequency measured the numbers of cases per standard quantity of workhours, and severity measured the total impact of these cases in terms of "lost workdays" per standard quantity of workhours.

Some injuries, such as amputations, are quite severe, but might result in few or no lost workdays. To avoid a distortion in severity rates in such cases, standard lost-workday charges were arbitrarily set for permanent injuries, such as amputations or loss of eyesight. The greatest need for arbitrary severity charges was for fatalities, because when you think about it, a fatality is not really a lost-workday case in the literal sense of the term; neither is a permanent total disability, because the worker never works again.

Another obsolete term is *seriousness*, which is the ratio of severity to frequency. This produced a measure of the relative average importance of injuries and illnesses without regard to the number of hours worked during the study period.

Incidence Rates

The current system of recordkeeping represents an enlargement of the old Z16.1 system. The total injury–illness incidence rate includes all injuries or illnesses that require medical treatment, plus fatalities. Compare this with the traditional *frequency rate*, which included only those cases in which the worker missed at least a day of work.[1] *Medical treatment* does not include simple first aid, preventive medicine (such as tetanus shots), or medical diagnostic procedures with negative results. *First aid* is described as "one-time treatment and subsequent observation of minor scratches, cuts, burns, splinters, and so forth, which do not ordinarily require medical care." First aid is not considered medical treatment even if it is administered by a physician or registered professional personnel. Regardless of treatment, if an injury involves loss of consciousness, restriction of work or motion, or transfer to another job, the injury is required to be recorded. Certainly regulating agencies, by their recordkeeping criteria, would not want to discourage medical treatment for an injury that should receive attention, so the U.S. Bureau of Labor Statistics (ref. Recordkeeping) has listed sample types of medical treatment. (See Appendix B.) Any injury that receives or should have received one or more of these types of treatment is almost always considered recordable. Appendix C gives examples of first aid given for injuries that are not normally recordable unless they qualify for recording for another reason, such as loss of consciousness or transfer to another job.

To compute the incidence rate, the number of injuries is divided by the number of hours worked during the period covered by the study. The value obtained is then multiplied by a standard factor to make the rate more understandable. Specifically,

$$\text{total injury–illness incidence rate} = \frac{\text{number of injuries and illnesses including fatalities}}{\text{total hours worked by all employees during the period covered}} \times 200{,}000$$

$$(2.1)$$

Without the factor of 200,000, the incidence rate would be a very small fraction indeed, as it should be. One should expect a very small number of recordable injuries and illnesses per single hour worked. The choice of the number 200,000 is not entirely arbitrary. A full-time worker typically works approximately 50 weeks per year at 40 hours per week. Thus, the number of hours worked per year per worker is approximately

$$40 \text{ hours/week} \times 50 \text{ weeks/year} = 2000 \text{ hours/year}$$

So 200,000 hours represents the number of workhours spent by 100 workers in a year:

$$100 \text{ workers} \times 2000 \text{ hours/year/worker} = 200{,}000 \text{ hours/year}$$

Thus, the total injury–illness incidence rate represents the number of injuries expected by a 100-employee firm in a full year if injuries and illnesses during the year follow the

---

[1]The ANSI Z16.1 system labeled such lost-workday cases as "disabling injuries" whether the disability was temporary or permanent.

same frequency as observed during the study period. Note from Equation (2.1) that the actual period for gathering the incidence-rate data need not be a year or any other specific time period. A fairly long period is needed, however, to obtain a representative number of cases, especially when the incidence is low. A typical data collection period is one year.

Sometimes the safety and health manager will want to relate current total injury–illness incidence rates to the traditional frequency rate. The old frequency rate used a factor of 1,000,000 hours instead of 200,000. Thus, rates were standard as "per million man-hours," as they were called in those days. Note that such a factor related to a standard year for a firm employing 500 employees, not 100 employees. Thus, the old frequency rates should be higher than the current total injury–illness incidence rates—but they generally are not because one must remember that the current rate includes all cases involving medical treatment, not just lost-workday cases. Also, days in which the worker was still on the job, but was unable to perform his or her regular job due to an injury or illness, are now taken into account. These days are called *restricted work activity* days and may be lumped together with lost workdays or considered separately, depending on the statistic desired. Unless specifically stated otherwise, today's interpretation of *lost workdays* includes days in restricted work activity as well as days away from work.

The term *incidence rate* is really a general term and, in addition to the total injury–illness incidence rate, includes the following:

1. Injury incidence rate
2. Illness incidence rate
3. Fatality incidence rate
4. Lost-workday-cases incidence rate (LWDI)
5. Number-of-lost-workdays rate
6. Specific-hazard incidence rate

All of the foregoing rates use the standard 200,000 factor. Note the difference between rates 4 and 5 in the foregoing list. Rate 4 counts *cases* in which one or more workdays were lost or in which the worker was transferred to another job. Rate 5 counts the total *number* of workdays lost or days in which the worker was transferred to another job.

In counting the number of lost workdays, the date of the injury or onset of illness should not be counted, even though the employee may leave work for most of that day. Thus, if the employee returns to his or her *regular* job and is able to perform all regular duties full time on the day after the injury or illness, no lost workdays are counted. Count all calendar days lost, not just regular workdays lost. The reader may note that counting calendar days, instead of just workdays, represents a policy change on the part of OSHA. This change was instituted in the 1990s at the same time that the record-keeping forms were revised. The number-of-lost-workdays rate compares to the old severity rate, except that no arbitrary charges are assessed for permanent partial disabilities and except for the 200,000 factor.

The specific-hazard incidence rate is useful in observing a small part of the total hazards picture. For specific hazards, injury incidence, illness incidence, fatality incidence, and all of the other rates can be computed. Care must be taken in selecting the corresponding total hours worked to be used in the denominator in calculating specific-hazard incidence rates. Since specific hazards are narrower and fewer workers are exposed, data should be collected over several years to achieve meaningful results for specific-hazard incidence rates.

The most widely recognized standard incidence rate is the *lost-workday-cases incidence rate*, commonly known as the LWDI. A somewhat surprising characteristic of the LWDI is that it considers injuries only—not illnesses. Illnesses are more difficult to track than injuries because there are often time delays in their diagnosis. Also, it is more difficult to prove work-relatedness for chronic exposures, which may have a variety of concurrent causes. The LWDI, because it is based on clear evidence, is considered a more precise and robust measure of the effectiveness of the firm's overall safety and health program. Also, perhaps for the same reasons, the LWDI considers only lost-time injuries, not all injuries. Remember, though, that restricted work activity cases are considered to be lost-time cases. Finally, the LWDI does not include fatalities, whether they be by illness or injury. Fatalities should always be considered a rare occurrence of grave importance and as such should not be averaged among the more common injury statistics on which the LWDI is based.

The prominence of the LWDI is largely historical, because it was once used by OSHA as a criterion to determine whether to conduct a general inspection of a randomly selected establishment. Now, the LWDI is applied to entire industries, as designated by the four-digit Standard Industrial Classification (SIC) number, rather than to individual firms. Thus, on a national basis, if an industry SIC is identified as having an LWDI higher than the national average for all industries, that entire industry is identified as a priority for inspection. Whether an individual company actually receives an inspection, however, is subject to several additional factors, such as in which OSHA region and area it is located, the available inspection resources in that region or area, how recently the firm has received an inspection, the number of high-priority requests (such as major accident investigations or employee complaints) that arise in that region or area, and the number of resources already committed for named target areas (such as construction). The priorities for OSHA inspections will be examined in more detail in Chapter 4.

Every year the National Safety Council gathers incidence statistics from surveys of its member companies and publishes them in *Injury Facts* (ref. Injury). Since the surveys are voluntary, they cannot be relied on to represent all member companies of the National Safety Council or the general population of industries nationwide. However, the NSC reports are frequently used as benchmarks for comparison. Figure 2.2 is a reprint of the NSC's report for 2000 (ref. Injury).

Recordkeeping Forms

The format for keeping injury and illness records has been standardized. The basic form is the *Log of Work-Related Injuries and Illnesses*, displayed in Figure 2.3. Figure 2.4 shows a summary to post annually, so that employees can see what injuries and illnesses have been recorded for the year. The summary is required to be posted in a prominent place in the workplace on February 1 each year and to remain posted until April 30. It is the employer's responsibility to enter data correctly into the log and summary. General records are required to be saved for a period of at least five years.

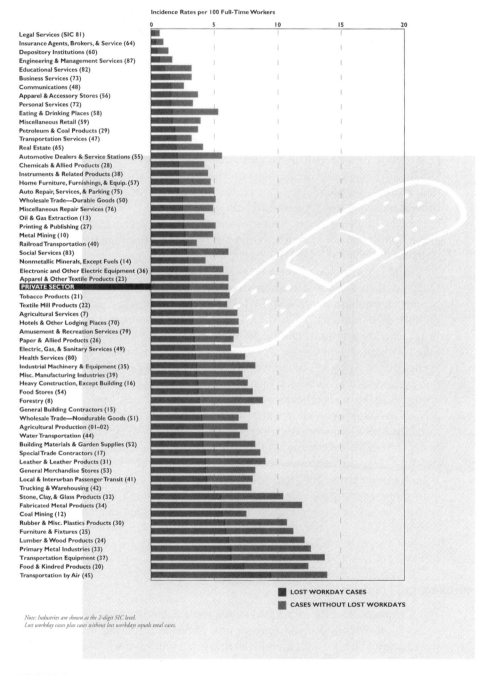

FIGURE 2.2

Comparison of incidence rates for various industries by Standard Industrial Classification (SIC) Code. (*Source*: National Safety Council. *Injury Facts, 2002 Edition*. Itasca, IL.)

OSHA's Form 300

Log of Work-Related Injuries and Illnesses

Year 20 ___

U.S. Department of Labor
Occupational Safety and Health Administration

Form approved OMB no. 1218-0176

Attention: This form contains information relating to employee health and must be used in a manner that protects the confidentiality of employees to the extent possible while the information is being used for occupational safety and health purposes.

You must record information about every work-related death and about every work-related injury or illness that involves loss of consciousness, restricted work activity or job transfer, days away from work, or medical treatment beyond first aid. You must also record significant work-related injuries and illnesses that are diagnosed by a physician or licensed health care professional. You must also record work-related injuries and illnesses that meet any of the specific recording criteria listed in 29 CFR Part 1904.8 through 1904.12. Feel free to use two lines for a single case if you need to. You must complete an injury and illness incident Report (OSHA Form 301) or equivalent form for each injury or illness recorded on this form. If you're not sure whether a case is recordable, call your local OSHA office for help.

Establishment name _____

City _____ State _____

Identify the person

(A) Case no.	(B) Employee's name	(C) Job title (e.g., Welder)

Describe the case

(D) Date of injury or onset of illness	(E) Where the event occurred (e.g., Loading dock north end)	(F) Describe injury or illness, parts of body affected, and object/substance that directly injured or made person ill (e.g., Second degree burns on right forearm from acetylene torch)

Classify the case

Using these four categories, check ONLY the most serious result for each case:

Death (G)	Days away from work (H)	Remained at work — Job transfer or restriction (J)	Remained at work — Other recordable cases (K)

Enter the number of days the injured or ill worker was:

On job transfer or restriction (K)	Away from work (L)
___ days	___ days

Check the "Injury" column or choose one type of illness:

(M) Injury (1)	Skin disorder (2)	Respiratory condition (3)	Poisoning (4)	Hearing loss / All other illnesses (5)

Page totals ▶

Be sure to transfer these totals to the Summary page (Form 300A) before you post it.

Page ___ of ___

Public reporting burden for this collection of information is estimated to average 14 minutes per response, including time to review the instructions, search and gather the data needed, and complete and review the collection of information. Persons are not required to respond to the collection of information unless it displays a currently valid OMB control number. If you have any comments about these estimates or any other aspects of this data collection, contact: US Department of Labor, OSHA Office of Statistics, Room N-3644, 200 Constitution Avenue, NW, Washington, DC 20210. Do not send the completed forms to this office.

FIGURE 2.3

OSHA Form 300: Log of Work-Related Injuries and Illnesses. (This form is available in larger format on the CD-ROM that accompanies this book.)

23

OSHA's Form 300A

Summary of Work-Related Injuries and Illnesses

Year 20___

U.S. Department of Labor
Occupational Safety and Health Administration

Form approved OMB no. 1218-0176

All establishments covered by Part 1904 must complete this Summary page, even if no work-related injuries or illnesses occurred during the year. Remember to review the Log to verify that the entries are complete and accurate before completing this summary.

Using the Log, count the individual entries you made for each category. Then write the totals below, making sure you've added the entries from every page of the Log. If you had no cases, write "0."

Employees, former employees, and their representatives have the right to review the OSHA Form 300 in its entirety. They also have limited access to the OSHA Form 301 or its equivalent. See 29 CFR Part 1904.35, in OSHA's recordkeeping rule, for further details on the access provisions for these forms.

Number of Cases

Total number of deaths	Total number of cases with days away from work	Total number of cases with job transfer or restriction	Total number of other recordable cases
_____	_____	_____	_____
(G)	(H)	(I)	(J)

Number of Days

Total number of days of job transfer or restriction	Total number of days away from work
_____	_____
(K)	(L)

Injury and Illness Types

Total number of . . .
(M)

(1) Injuries _____
(2) Skin disorders _____
(3) Respiratory conditions _____

(4) Poisonings _____
(5) All other illnesses _____

Post this Summary page from February 1 to April 30 of the year following the year covered by the form.

Public reporting burden for this collection of information is estimated to average 50 minutes per response, including time to review the instructions, search and gather the data needed, and complete and review the collection of information. Persons are not required to respond to the collection of information unless it displays a currently valid OMB control number. If you have any comments about these estimates or any other aspects of this data collection, contact: US Department of Labor, OSHA's Office of Statistics, Room N-3644, 200 Constitution Avenue, NW, Washington, DC 20210. Do not send the completed forms to this office.

Establishment information

Your establishment name _____

Street _____

City _____ State _____ ZIP _____

Industry description (e.g., Manufacture of motor truck trailers) _____

Standard Industrial Classification (SIC), if known (e.g., SIC 3715) _____

Employment information *(If you don't have these figures, see the Worksheet on the back of this page to estimate.)*

Annual average number of employees _____

Total hours worked by all employees last year _____

Sign here

Knowingly falsifying this document may result in a fine.

I certify that I have examined this document and that to the best of my knowledge the entries are true, accurate, and complete.

_____ Title: _____
Company executive

_____ (___) ___ - ____ Date: __/__/__
Phone

FIGURE 2.4

OSHA Form 300A: Summary of Work-Related Injuries and Illnesses. (This form is available in larger format on the CD-ROM that accompanies this book.)

The person responsible for completing the log and summary may need some guidance in distinguishing between occupational injuries and illnesses. Examples of occupational injuries include lacerations, fractures, sprains, and amputations that are the result of a work accident or an exposure involving a single incident in the work environment. Animal bites—such as insect or snake bites—are considered injuries. Even chemical exposures can be considered injuries if they result from a one-time exposure.

An illness is any abnormal condition or disorder not classified as an injury and caused by exposure to environmental factors associated with employment. Illnesses are usually associated with chronic exposures, but some acute exposures can be considered illnesses if the exposure is the result of more than a single incident or accident. More detailed classification in the log and summary is required for illnesses than for injuries. Some guidance in classifying illnesses can be found in Appendix D.

Besides the log and summary, there is the *Injury and Illness Report*. (See Figure 2.5.) Each page of the report corresponds to a single-line entry in the log.

To illustrate the calculation of the various incidence rates and to demonstrate the use of the standard forms, Case Study 2.1 will now be analyzed.

CASE STUDY 2.1

A metal products fabrication and assembly plant employs 250 workers and has the following injury–illness experience for the year (workers are employed on a regular 40-hour-workweek basis):

File 1 January 31: Press-blanking operator lacerates hand on strip stock scrap from punch press; first aid received, no medical treatment; worker remains on the job.

File 2 February 19: Maintenance worker, not wearing eye protection, operating grinding machine in tool room, incurs eye injury from flying chip; medical treatment required; injury occurs on Tuesday, employee returns to regular job at regular time on Thursday.

File 3 February 27: Assembly worker becomes "ill" due to noxious odors from remodeling operation in the assembly area; receives permission from supervisor to take the rest of the day off; does not go to a doctor or clinic; reports to regular work on time the next day.

File 4 March 2: Sewing machine operator's right ring finger pulled into unguarded drive belt pulley on sewing machine; small fracture revealed by X ray; splint applied; worker returns to regular work at regular time the next day.

File 5 March 19: Dockworker sprains ankle on loading dock; moved to office job for two workweeks.

File 6 May 2: Maintenance worker entangles finger in rope as winch is released; taken to clinic for X ray; no fractures found; no treatment; worker returns to regular work the next day.

File 7 June 7: Yard worker exposed to poison ivy while clearing weeds in tank-farm area behind plant; Rash develops; treated with prednisone adrenocortical steroid drug by prescription; no time lost.

OSHA's Form 301
Injury and Illness Incident Report

U.S. Department of Labor
Occupational Safety and Health Administration

Form approved OMB no. 1218-0176

This *Injury and Illness Incident Report* is one of the first forms you must fill out when a recordable work-related injury or illness has occurred. Together with the *Log of Work-Related Injuries and Illnesses* and the accompanying *Summary*, these forms help the employer and OSHA develop a picture of the extent and severity of work-related incidents.

Within 7 calendar days after you receive information that a recordable work-related injury or illness has occurred, you must fill out this form or an equivalent. Some state workers' compensation, insurance, or other reports may be acceptable substitutes. To be considered an equivalent form, any substitute must contain all the information asked for on this form.

According to Public Law 91-596 and 29 CFR 1904, OSHA's recordkeeping rule, you must keep this form on file for 5 years following the year to which it pertains.

If you need additional copies of this form, you may photocopy and use as many as you need.

Completed by _____

Title _____

Phone (_____) _____ Date ___/___/___

Attention: This form contains information relating to employee health and must be used in a manner that protects the confidentiality of employees to the extent possible while the information is being used for occupational safety and health purposes.

Information about the employee

1) Full name _____

2) Street _____
   City _____ State _____ ZIP _____

3) Date of birth ___/___/___

4) Date hired ___/___/___

5) ☐ Male
   ☐ Female

Information about the physician or other health care professional

6) Name of physician or other health care professional _____

7) If treatment was given away from the worksite, where was it given?
   Facility _____
   Street _____
   City _____ State _____ ZIP _____

8) Was employee treated in an emergency room?
   ☐ Yes
   ☐ No

9) Was employee hospitalized overnight as an in-patient?
   ☐ Yes
   ☐ No

Information about the case

10) Case number from the *Log* _____ *(Transfer the case number from the Log after you record the case.)*

11) Date of injury or illness ___/___/___

12) Time employee began work _____ AM / PM

13) Time of event _____ AM / PM ☐ Check if time cannot be determined

14) **What was the employee doing just before the incident occurred?** Describe the activity, as well as the tools, equipment, or material the employee was using. Be specific. *Examples:* "climbing a ladder while carrying roofing materials"; "spraying chlorine from hand sprayer"; "daily computer key-entry."

15) **What happened?** Tell us how the injury occurred. *Examples:* "When ladder slipped on wet floor, worker fell 20 feet"; "Worker was sprayed with chlorine when gasket broke during replacement"; "Worker developed soreness in wrist over time."

16) **What was the Injury or Illness?** Tell us the part of the body that was affected and how it was affected; be more specific than "hurt," "pain," or sore." *Examples:* "strained back"; "chemical burn, hand"; "carpal tunnel syndrome."

17) **What object or substance directly harmed the employee?** *Examples:* "concrete floor"; "chlorine"; "radial arm saw." *If this question does not apply to the incident, leave it blank.*

18) **If the employee died, when did death occur?** Date of death ___/___/___

Public reporting burden for this collection of information is estimated to average 22 minutes per response, including time for reviewing instructions, searching existing data sources, gathering and maintaining the data needed, and completing and reviewing the collection of information. Persons are not required to respond to the collection of information unless it displays a current valid OMB control number. If you have any comments about this estimate or any other aspects of this data collection, including suggestions for reducing this burden, contact: US Department of Labor, OSHA Office of Statistics, Room N-3644, 200 Constitution Avenue, NW, Washington, DC 20210. Do not send the completed forms to this office.

FIGURE 2.5

OSHA Form 301: Injury and Illness Report. (This form is available in larger format on the CD-ROM that accompanies this book.)

File 8 July 6: Assembly worker loses two workdays recuperating from severe allergic reaction to wasp stings incurred while cleaning out his attic at home; medical treatment with prescription drugs.

File 9 August 4: Maintenance worker using ungrounded portable electric drill to repair equipment in assembly area is electrocuted. Date of death: August 4.

File 10 August 7: Loaded pallet in loading dock area falls from forklift on dockworker's left foot; worker was not wearing steel-toed shoes; worker examined in hospital emergency room and X ray revealed no fractures or other injuries; worker receives whirlpool therapy and goes home; worker reports back to his regular job on time the next day and wears his company-issued safety shoes.

File 11 August 9: Maintenance worker in the tool room incurs injury from foreign object in the eye; irrigation method used to remove foreign object, which was not embedded in the eye; worker returns to regular job.

File 12 September 11: Worker in final assembly diagnosed with carpal tunnel syndrome (CTS) from repetitive work; surgery prescribed; worker misses three weeks of work before returning to regular job with engineering improvements to the workstation.

ANALYSIS

The first step is to complete the OSHA 300 Log of Occupational Injuries and Illnesses, generating one line on the log for each incident file. Figure 2.6 displays the completed log and the rationale for each entry is as follows:

File 1   The key word is "first aid." This case is not recordable.

File 2   This is a lost-workdays injury case. Do not count the date of injury (Tuesday). Do not count Thursday either because the worker returned to work at the regular time. Only one day was lost. Mark columns H, L, and M(1).

File 3   This case is not recordable. The worker felt "ill," but there was no medical treatment, and although the worker left work one afternoon, the date of the onset is not counted. The worker returned to work on time the next day, so no lost time is counted.

File 4   This is a recordable injury, as the X ray was positive, revealing a fracture, which is always recordable. The worker returned to her regular work at the regular time the next day, however, so no time was lost. Mark columns J and M(1).

File 5   This is a lost-time injury. Even though the worker returned to work, he was assigned to a different job, so OSHA's position is that the days at the restricted work activity count as lost workdays. Record in columns I, K, and M(1).

File 6   Unlike File 4, the X ray in this case was negative. Since there was no fracture and no medical treatment and the worker returned to the same job the next day on time, this case is not recordable.

OSHA's Form 300

Log of Work-Related Injuries and Illnesses

Year 20__

U.S. Department of Labor
Occupational Safety and Health Administration

Form approved OMB no. 1218-0176

Attention: This form contains information relating to employee health and must be used in a manner that protects the confidentiality of employees to the extent possible while the information is being used for occupational safety and health purposes.

You must record information about every work-related death and about every work-related injury or illness that involves loss of consciousness, restricted work activity or job transfer, days away from work, or medical treatment beyond first aid. You must also record significant work-related injuries and illnesses that are diagnosed by a physician or licensed health care professional. You must also record work-related injuries and illnesses that meet any of the specific recording criteria listed in 29 CFR Part 1904.8 through 1904.12. Feel free to use two lines for a single case if you need to. You must complete an injury and illness incident Report (OSHA Form 301) or equivalent form for each injury or illness recorded on this form. If you're not sure whether a case is recordable, call your local OSHA office for help.

Establishment name _____
City _____ State _____

Identify the person

(A) Case no.	(B) Employee's name	(C) Job title (e.g. Welder)	(D) Date of injury or onset of illness	(E) Where the event occurred (e.g. Loading dock north end)	(F) Describe injury or illness, parts of body affected, and object/substance that directly injured or made person ill (e.g. Second degree burns on right forearm from acetylene torch)
1					
2	worker #2	maint.	2/19	toolroom	NOT RECORDABLE
3					
4	worker #4	seamst	3/2	sewing room	eye injury (flying chip) RECORDABLE
5	worker #5	dockwkr	3/19	loading dock	finger fracture - belt/pulley
6					sprained ankle NOT RECORDABLE
7	worker #7	yardwkr	6/7	tank farm	skin rash (acem) - poison ivy
8					RECORDABLE
9	worker #9	maint	8/4	assembly	electrocution - portable drill
10					NOT RECORDABLE
11					NOT RECORDABLE
12	worker #12	assy	9/11	final assy	CTS - repetitive motion RECORDABLE

Classify the case

Using these four categories, check ONLY the most serious result for each case:

(G) Death	(H) Days away from work	Remained at work — (I) Job transfer or restriction	Remained at work — (J) Other recordable cases	Enter the number of days the injured or ill worker was: (K) On job transfer or restriction	(L) Away from work
	✓				___ days
			✓		___ 1 days
					___ days
			✓		___ days
		✓		14	___ days
					___ days
			✓		___ days
					___ days
	✓				___ days
					___ days
					___ days
	✓			21	___ days

Check the "Injury" column or choose one type of illness:

(M) (1) Injury	(2) Skin disorder	(3) Respiratory condition	(4) Poisoning	(5) All other illnesses
✓				
✓				
✓				
	✓			
✓				
			✓	

Page totals ▶ 14 22

Be sure to transfer these totals to the Summary page (Form 300A) before you post it.

Page 1 of 1

Public reporting burden for this collection of information is estimated to average 14 minutes per response, including time to review the instructions, search and gather the data needed, and complete and review the collection of information. Persons are not required to respond to the collection of information unless it displays a currently valid OMB control number. If you have any comments about these estimates or any other aspects of this data collection, contact: US Department of Labor, OSHA Office of Statistics, Room N-3644, 200 Constitution Avenue, NW, Washington, DC 20210. Do not send the completed forms to this office.

FIGURE 2.6

OSHA 300 Log for Case Study 2.1. (This form is available in larger format on the CD-ROM that accompanies this book.)

28

File 7   Poison ivy from on-the-job exposure is classified as an occupational illness and is identified in column M(2) as "Skin Disorder." (See Appendix D.) No time was lost, so a check also goes in column J.

File 8   Incidents occurring off the job are not recordable.

File 9   This is an injury-type fatality and should be recorded in columns G and M(1).

File 10   The negative X ray and whirlpool therapy during the first visit to medical personnel are both considered first aid, not medical treatment. (See Appendix C.) This case is not recordable.

File 11   Since the irrigation method was used and the object was not embedded in the eye, this eye injury is considered a first-aid case and is thus not recordable. (See Appendix C.)

File 12   Because CTS is due to "repeated motion," it is classified as a column M(5) illness. (See Appendix D.) This is a recordable lost-time illness. The lost time is in the days-away-from-work category, so it is recorded in columns H and L.

Calculation of Incidence Rates

$$\text{LWDI (injuries only)} = \frac{2 \times 2000,000}{250 \times 2000} = 0.8$$

$$\text{Injury incidence rate} = \frac{3 \times 200,000}{250 \times 2000} = 1.2$$

$$\text{Illness incidence rate} = \frac{2 \times 200,000}{250 \times 2000} = 0.8$$

$$\text{Fatality incidence rate} = \frac{1 \times 200,000}{250 \times 2000} = 0.4$$

$$\text{Number-of-lost-workdays rate} = \frac{26 \times 200,000}{250 \times 2000} = 10.4$$

$$\text{Specific-hazard incidence rate (eye injuries)} = \frac{1 \times 200,000}{250 \times 2000} = 0.4$$

Some explanation of the calculations for Case Study 2.1 may be helpful. The LWDI is calculated in a prescribed way that excludes all fatalities and all illnesses, regardless of whether time was or was not lost. Remember that the LWDI is an *incidence rate* and should not be confused with the number-of-lost-workdays rate. Also remember that OSHA considers both days away from work and restricted work activity days as lost workdays. Note that injuries that do not result in lost time (column J of the OSHA 300 form) do not count toward the LWDI. In the specific-hazard incidence rate calculation, only one eye injury (File 2) was included in the calculation. The File 11 eye injury satisfied the Appendix C definition of first aid and thus, as a nonrecordable injury, was excluded from the calculation.

The 250-employee firm in Case Study 2.1 provides ample data to show meaningful calculations for the various incidence rates. But many firms are much smaller. For very small firms, the calculations are obviously inappropriate. It is not uncommon for small businesses to operate for several years without a single injury or illness. Recognizing that the general injury–illness recordkeeping system was designed for larger firms, Congress exempted small firms with 10 or fewer employees from general recordkeeping requirements.

However, Congress granted only a partial exemption. The federal Bureau of Labor Statistics (BLS) conducts annual surveys of occupational injuries and illnesses based on a random, stratified sample of industries. If a firm is to be included in a sample, it will be notified by the BLS. To keep the statistics representative of all industries, small firms are *not* excluded. Therefore, if a firm receives notice that it has been selected for participation in the survey for a given year, the firm must respond to the BLS. Thus, selected sample firms must keep the OSHA log and injury–illness statistics even though they normally would be exempt due to the firm's small size.

The current general recordkeeping system is based on federal standards and has remained relatively static since the early 1970s. As was stated earlier in this chapter, the required retention period for these general records is five years. However, in the early 1980s, special recordkeeping requirements were established for toxic chemicals in the movement that became known as "right-to-know." It will be seen in Chapter 5 that the recordkeeping requirements for toxic chemicals are much more comprehensive and have led to the development of computer information systems for safety and health. The required retention period for hazardous chemical exposure records and medical records under the "right-to-know" standards is 30 years.

ACCIDENT CAUSE ANALYSIS

So far, this chapter has discussed the more visible, busy functions of the safety and health manager, many of which are required by state or federal agencies. But even more important to the health and safety of workers are some of the jobs that the safety and health manager is not required to, but should, do. One of these voluntary, but important, tasks is a thorough analysis of the potential causes of injuries and illnesses that have already occurred in the plant. Even accidents or incidents that may not actually have caused injuries or illnesses, but which could have, should be studied to prevent their recurrence. Any occurrence of an unplanned, unwanted event is a datapoint to consider in the prevention of future illnesses and injuries. Accident cause analysis and subsequent dissemination of this information to personnel who will be exposed to the hazards in the future is believed to be the most effective way of preventing injuries and illnesses. The literature of injury case histories is filled with accounts of cases in which workers are killed by conditions that had previously caused accidents or injuries to others. Case Study 2.2 will be used to illustrate this point.

CASE STUDY 2.2

A worker was struck in the head and killed by the sudden movement of a large wrench used for releasing gates on the bottom of railroad hopper cars. The worker used a powerful, 3- to 4-foot-long wrench to trip a mechanical latch on the bottom gate of the car. The wrench was supposed to be of a ratchet type, so that when the gate was tripped, the tremendous weight of the bulk material in the hopper car would not suddenly force the wrench back on the worker. But for some reason the ratchet wrench was not available, and workers had been using an ordinary rigid wrench to release the latch. Only a week before the fatality occurred, another worker had narrowly escaped the same injury when he lost control of the same wrench in the same operation.

Sometimes the accident analysis leads to a design change in a product or process. In other cases, work procedures are changed to prevent future occurrences, or at least to minimize the adverse effects of these occurrences. Even when nothing can be changed to prevent a future occurrence, at least workers can be informed of what happened, what caused the accident, under what conditions the accident might occur again, and how to protect themselves in such an event. Informing workers of the facts and causes of accidents that have already happened to their coworkers is the single most effective method of training workers to avoid injury and illness. Thus, accident cause analysis is the foundation on which safety and health engineering, capital investment planning, training, motivation, and other functions are based. There are other types of accident analysis; statistical frequency analysis was discussed earlier, and cost analysis will be discussed later in this chapter. But no type of analysis is as important as the determination of causes of accidents that have already occurred and might occur again.

Accident cause analysis, as essential as it is, does have some disadvantages. The main disadvantage is the obvious one: It is after the fact—that is, it is too late to prevent any injury or loss that occurred as a result of the accident under analysis. Another disadvantage is that the focus of the analysis can easily degenerate into an exercise in assigning blame or allocating legal liability. Recognizing these disadvantages, the analyst should strive to stay focused on the objective of identifying what processes, procedures, or management practices need to be changed to prevent future occurrences of the same or similar accidents.

ORGANIZATION OF COMMITTEES

The value of using safety and health committees has long been recognized. Committees are appointed from the ranks of the operating personnel of the regular line organization. The appointments are temporary, so that workers throughout the organization rotate on and off a committee periodically. The committees then make facilities inspections, evaluate safety and health suggestions, analyze accident causes, and make recommendations.

Several natural advantages of the committee approach make it a winning strategy. In general, operating personnel know a lot more about their processes and machines than does the safety and health manager. Many valuable and practical ideas can

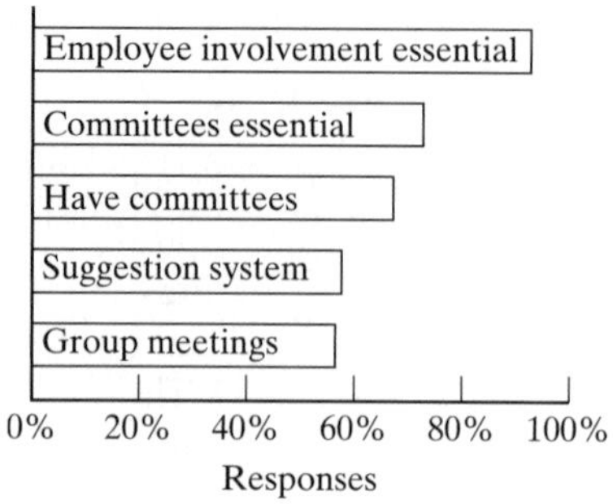

FIGURE 2.7

Employee involvement in safety and health [*Source:* 1993 National Safety Council study (ref. OSHA, 1993)].

come from operating personnel if staff persons will listen. Also, operating personnel may more readily accept new policies and procedures if these procedures arise from other operating personnel like themselves. Then there is the advantage of exposure. Sooner or later, nearly everyone has his or her turn on a safety committee, which means that the direct activity of the safety and health program is a product of plantwide participation. Some workers have no appreciation for or sensitivity to safety and health hazards until they take their turn on the committee. Indirectly, then, the committee becomes a vehicle for safety and health training.

The early 1990s was the stage for general public debate on the issue of whether companies should have employer–employee committees for safety and health. At the congressional level, a key element of reform legislation was a provision for mandatory safety and health programs with employer–employee committees. The idea was particularly attractive to labor unions. To test the popularity of the idea nationwide, the National Safety Council conducted a study in late 1993, obtaining 249 responses from a sampling of 2500 nonagricultural companies. If those responses can be assumed to be representative, the study suggests that a general consensus favors employee involvement and safety and health committees, as Figure 2.7 seems to bear out (ref. OSHA, 1993).

Despite its advantages, there are pitfalls to the committee approach. The safety and health manager should provide resources and guidance to the committee so that it will have the necessary tools and knowledge to function effectively. Otherwise, the committee may make ridiculous suggestions and be disappointed when management does not approve them or does not follow through with capital support. Also, committees must be conditioned not to expect miracles. Some orientation or training is necessary so that committee members will comprehend the goal of targeting recognized hazards, but not all hazards. Finally, committees should not be allowed to degenerate into spy parties with the objective of discrediting the processes or procedures of other departments.

SAFETY AND HEALTH ECONOMICS

Safety and health managers are sometimes dismayed to discover that top management bases safety and health decisions on dollars and cents. But the cold reality is that business exists to make profits, and everything a business does is either directly or indirectly related to economics. Safety and health managers who are naive enough to think that the humanitarian objective of worker safety and health transcends the more crude issues of profit and loss should ask themselves the following question: How *much* safety and health staff activity is justified by the humanitarian objective?

The prevention of employee injuries and illnesses can be formulated as an economic objective; such a formulation is more meaningful to management than vague humanitarian aspirations. Accidents, injuries, and illnesses have undeniable costs that contribute nothing to the value of products manufactured or services performed by the firm. Occupational injuries alone have been estimated to total over $132 billion annually (ref. Injury). The annual cost of injuries and illnesses in many industries dwarfs the total profits picture. This is a reality that almost any top manager will want to consider. Although it is true that many of these costs are subtle and difficult to estimate, the existence of these costs is in no way diminished by this fact.

One obvious and direct category of costs from injuries and illnesses is the payment of workers' compensation insurance premiums, which are based on a firm's injury and illness experience. Self-insured firms have the actual claims data on which to calculate these direct costs. In addition to these claims are medical costs that may be covered by insurance. Since these costs are directly identified to injuries and illnesses in accounting records, they are sometimes called "direct costs" of injuries and illnesses. Workers' compensation premiums have recently been increasing sharply. Historically, premiums have been in the range of 1–2% of total payroll. In recent years, however, rates have been much higher, as Table 2.2 attests.

Despite the significantly higher premium rates for workers' compensation insurance, these "direct costs" of injuries and illnesses have been referred to by some analysts as the "tip of the iceberg." (See Figure 2.8.) The intangible costs of accidents, although hidden, appear to be much greater than the so-called "direct costs." It is the job of the safety and health manager to attempt to estimate these costs and to keep management apprised so that rational investment decisions can be made.

The National Safety Council, in its *Accident Prevention Manual for Industrial Operations*,[2] lists the following categories of hidden costs of accidents:

1. Cost of wages paid for time lost by workers who were not injured. These are employees who stopped work to watch or assist after the accident or to talk about it, or

TABLE 2.2 Sample Workers' Compensation Insurance Premium Rates

SIC Code	Description	W.C. rate (% of payroll)
8039	Department stores	2.91
2003	Bakeries	4.40
2883	Cabinet manufacturing	6.92
8829	Nursing homes	4.25
5022	Bricklayers	10.14
5645	Carpentry (light residential)	18.86
5551	Roofers	29.53

Source: Arkansas, 2002.

---

[2]*Accident Prevention Manual for Industrial Operations: Administration and Programs Volume*, 8th ed. Chicago: National Safety Council, 1981, pp. 214–215 (used with permission).

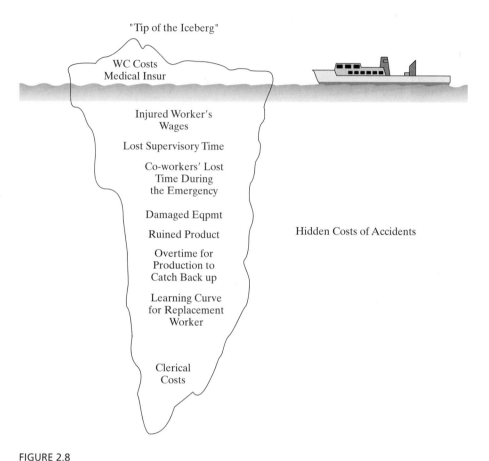

"Tip of the Iceberg"

WC Costs
Medical Insur

Injured Worker's
Wages

Lost Supervisory Time

Co-workers' Lost
Time During
the Emergency

Damaged Eqpmt

Ruined Product   Hidden Costs of Accidents

Overtime for
Production to
Catch Back up

Learning Curve
for Replacement
Worker

Clerical
Costs

FIGURE 2.8

Direct costs of accidents (workers' compensation premiums and medical insurance) represent only the "tip of the iceberg."

who lost time because they needed equipment damaged in the accident or because they needed the output or the aid of the injured worker.

2. Cost of damage to material or equipment. The validity of property damage as a cost can scarcely be questioned. Occasionally, there is no property damage, but a substantial cost is incurred in putting back in order material or equipment that has been thrown into a state of disorder. The charge should, however, be confined to the net cost of repairing or putting in order material or equipment that has been damaged or displaced, or to the current worth of the equipment less salvage value if it is damaged beyond repair.

An estimate of property damage should have the approval of the cost accountant, particularly if the current worth of the damaged property used in the cost estimate differs from the depreciated value established by the accounting department.

3. Cost of wages paid for time lost by the injured worker, other than worker's compensation payments. Payments made under workers' compensation laws for time lost after the waiting period are not included in this element of cost.

4. Extra cost of overtime work necessitated by the accident. The charge against an accident for overtime work necessitated by the accident is the difference between normal wages and overtime wages for the time needed to make up lost production, and the cost of extra supervision, heat, light, cleaning, and other extra services.

5. Cost of wages paid supervisors for time required for activities necessitated by the accident. The most satisfactory way of estimating this cost is to charge the wages paid to the foreman for the time spent away from normal activities as a result of the accident.

6. Wage cost caused by decreased output of injured worker after return to work. If the injured worker's previous wage payments are continued despite a 40% reduction in his output, the accident should be charged with 40% of his wages during the period of such low output.

7. Cost of learning period of new worker. If a replacement worker produces only half as much in his first two weeks as the injured worker would have produced for the same pay, then half of the new worker's wages for the two-week period should be considered part of the cost of the accident that made it necessary to hire him. A wage cost for time spent by supervisors or others in training the new worker also should be attributed to the accident.

8. Uninsured medical cost borne by the company. This cost is usually that of medical services provided at the plant dispensary. There is no great difficulty in estimating an average cost per visit for this medical attention. The question may be raised, however, whether this expense may properly be considered a variable cost. That is, would a reduction in accidents result in lower expenses for operating the dispensary?

9. Cost of time spent by higher supervision and clerical workers on investigations or in the processing of compensation application forms. Time spent by supervision (other than the foreman or supervisor covered in Item 5) and by clerical employees in investigating an accident, or settling claims arising from it, is chargeable to the accident.

10. Miscellaneous usual costs. This category includes the less typical costs, the validity of which must be clearly shown by the investigator on individual accident reports. Among such possible costs are public liability claims, cost of renting equipment, loss of profit on contracts canceled or orders lost if the accident causes a net long-run reduction in total sales, loss of bonuses by the company, cost of hiring new employees if the additional hiring expense is significant, cost of *excess* spoilage (above normal) by new employees, and demurrage. These cost factors and any others not suggested here would need to be well substantiated.

Every firm is different, and if time and staff resources permit, the best way to estimate hidden costs of accidents is to survey and analyze the individual company's recent accident data. When performing such an analysis, it must be remembered that noninjury accidents can also be costly and are generally caused by the same types of conditions and practices that result in injury accidents. Therefore, noninjury accidents should also be included when one attempts to assess the total cost of accidents.

Most firms cannot afford the luxury of a comprehensive, statistically reliable, in-house study of hidden accident costs. An alternative is to turn to national studies of average costs of various accident categories, and apply these survey averages as estimates of in-house costs. Two well-known studies of uninsured costs of accidents were reported by Grimaldi and Simonds (ref. Grimaldi) and Imre (ref. Imre) in 1975. Although the data were gathered over a span of several years, when the dollar figures were adjusted for inflation to a common representative year, the results of the two studies were shown to corroborate each other, recognizing that rough approximations are all that can be hoped for in such studies.

Though the Grimaldi and Simonds and Imre studies are classics in the field of estimating the costs of accidents, many safety professionals consider them too old and too conservative to be relevant to today's costs. Even when adjusted by the Consumer Price Index (CPI), the classic estimates are often seen as too low to be realistic. Another difficulty with the classic studies is that the classification of accidents is not clear. The four general classifications are "lost time," "first aid," "doctors' cases," and "no-injury cases." These four general classes seem to overlap for some accidents. Furthermore, the classification of fatalities does not seem to be adequately addressed.

In the mid-1990s, the National Safety Council modernized its approach to estimating the cost of accidents as published in *Accident Facts* (ref. Accident). The NSC estimate of the average total cost per worker fatality is $790,000. For worker injuries, the corresponding figure is $28,000. These figures are much higher than would be derived using the classical methods employed by Grimaldi, Simonds, and Imre. Even so, the NSC estimates do not include any estimate of property damage costs.

The U.S. Air Force has compiled estimates of accident cost categories for use in their investigations of aircraft accidents and other loss incidents (ref. AFI 91-204). Despite the difficulty associated with estimating human loss, the Air Force has even attempted to place some kind of cost figure on the loss of life. For a rated[3] officer fatality, the estimated cost to the Air Force is $1,100,000 as of the publication of AFI 91-204 (1995).

For permanent total disability, the estimated cost is slightly higher ($1,300,000, including lost-workday and hospitalization-day costs). The average estimated cost for permanent partial disability is $210,000. For temporary disabilities, the lost-workday cost estimate is $425/day, or $466/day during hospitalization. For an injury that does not result in lost workdays, the estimated average cost is $120. The corresponding costs for civilian employees are generally less, probably due to the lower government investment in training for these individuals. Civilian fatalities are estimated at $460,000 each, permanent total disabilities at $385,000 each, permanent partial disabilities at $250,000 each, and lost workdays at $350 per day. Hospitalization-day costs are $466 per day, and the no-lost-time cases (at $120 each) are estimated to be the same as for the military-rated officer.

Turning to the nonmilitary environment, the U.S. Department of Energy (DoE) places a similar dollar value on human life ($1 million per fatality) and reportable injuries ($2000 per case) in cost studies and annual reports. In addition, for cases involving lost workdays, DoE estimates $1000 loss per workday lost (refs. crites, Briscoe). Another estimate (ref. Barciela) places the hidden costs of accidents anywhere from $5 to $50 per dollar of workers' compensation claims.

---

[3]The term *rated* means that the officer is in flight-duty status, which means that the officer has been trained for and receives extra compensation for flight duties.

At the beginning of this section, it was stated that the direct cost of worker injuries and illnesses, workers' compensation insurance costs, represented the "tip of the iceberg" compared to the total costs incurred. Recent National Safety Council and U.S. Air Force estimates seem to bear out this theory.

TRAINING

Training or training support may be the most important staff function to be performed by the safety and health manager. Despite a recent trend toward concentration on unsafe conditions, experts still attribute most worker injuries and illnesses to unsafe acts. Unsafe work habits are deeply rooted, even in new, young workers. Our society and its standards of status, as influenced by the media (especially television), places a premium on high-risk activity. From an early age, children learn that heroes are people who are daring, lucky, and risk their lives, especially in their life work. In some jobs—such as space exploration, the military, law enforcement, and firefighting—it is occasionally both necessary and rational to take big risks. And those who take these risks indeed deserve to be called heroes. Unfortunately, though, the desire for recognition, status, and the esteem of their peers causes people to take unnecessary risks in activities that do not warrant such risk. A good example of this phenomenon is exhibited in the automobile driving habits of people of all ages. Deep-rooted, unsafe habits and lack of knowledge about specific job hazards are major barriers to worker safety and health. It is on these two problems that the training program should be focused, and training is perhaps the most challenging and important function of the safety and health manager.

One of the biggest mistakes a safety and health manager can make is to assume that he or she is the principal trainer in safety and health. The principal trainers in safety or health or in any other aspect of the job are the first-line supervisors. Their direct contact with the workers will determine how the job will be done. A corollary to this principle is that most training in safety and health is informal and is conducted on the job. In fact, training by example is a very important delivery mode, and new workers are more influenced by what the supervisor and experienced workers *do* than by what they *say*.

Recognizing that most training takes place between supervisor and worker, there is still a need for classroom training in safety and health principles, standards, and hazards recognition, especially for supervisors. The safety and health manager can provide this training directly or can act as a facilitator in bringing useful information and training aids into the plant. Commercially available videos are especially helpful in addressing basic principles applicable to all workplaces. An example principle addresses the inherent hazard in workers' tendencies to save time by taking shortcuts for comfort or convenience. The video titled "It's About Time," produced by Envision, Inc., addresses this general hazard principle.

Safety and health managers should not "try to reinvent the wheel" in their development of training materials. Audiovisual packages and outlines are available, and when the safety and health manager's time, including overhead, is considered, it is usually much more reasonable to purchase or rent training materials than to attempt to create original material in-house. Chapter 1 enumerated some sources to assist safety and health managers in developing this aspect of their job. Chapter 3 also will address the subject of training and how it becomes a part of the overall objective of hazard avoidance.

Drug and Alcohol Abuse

Safety and health managers are taking a more assertive posture to control the effects of drug and alcohol abuse in the workplace. Drug and alcohol abuse has been shown to be a greater problem than was once thought. Consider the experience of the Aluminum Company of America's Vancouver, Washington, plant (ref. Houston). Following the lead of a sister ALCOA plant in Davenport, Iowa, the company decided to try a preemployment drug screening test for all applicants. To the surprise of management, in three months of testing, half of the 750 applicants failed the test. The test was a urinalysis designed to indicate whether drugs had been used in the preceding two or three days and was conducted by a hospital laboratory service. The test results indicated that the use of marijuana was the most prevalent problem. ALCOA hired 130 of the applicants who passed the test and, according to the personnel manager, found the people hired to be better workers than the company had had before the drug screening program had been added to the hiring process.

It is not difficult to justify a carefully planned and executed drug and alcohol abuse plan for any company, and the safety and health manager should take the lead in establishing one. Indeed, there is no choice in certain sectors of the transportation industry subject to mandatory testing for marijuana, cocaine, opiates, amphetamines, and PCP, under rules issued by the U.S. Department of Transportation. The program prescribes random, preemployment, periodic, reasonable cause, and postaccident testing (ref. Drug). Drug and alcohol testing may make even more sense in other industries. And when a treatment program becomes necessary, few parties have as much influence over an employee's decision to enter treatment as does his or her employer, whether the treatment is for drug abuse or alcohol abuse.

A key question to ask management is whether they can imagine a situation in which the firm might some day need to terminate an employee because excessive drug or alcohol abuse has affected his or her job. If the answer is yes, the firm is exposed to litigation risks if a policy on drug and alcohol abuse is not in place. If your company has a rule against drug abuse, employees should be informed, and new employees should be required to consent in writing that they are working under the company's drug abuse policy as a condition of employment (ref. Wilkinson). By making preemployment drug screening a requisite for employment, safety and health hazards can be prevented, but to be consistent, the company should apply the same rules against drug abuse to existing employees. Otherwise, the firm may face a discrimination charge from an applicant who has been refused employment.

Besides screening programs to detect a problem, both in the cases of new applicants and existing employees, many employers are instituting *employee-assistance programs* to deal with the difficulties of employees who have a recognized drug or alcohol abuse problem. The rationale is that some well-trained and competent employees are too valuable to lose because of a drug or alcohol problem. Therefore, instead of viewing the problem as a matter of discipline, the condition is viewed as a sickness requiring treatment and therapy to restore the worker to full usefulness. Such programs have the intangible benefit of conveying to workers in general that the company cares about the well-being of its employees and would rather see them cured than terminated.

As pervasive as the drug and alcohol abuse problem is, there is no doubt that safety and health managers will continue to encounter increased responsibilities to establish and maintain programs to control the hazards these problems present.

JOB PLACEMENT TESTING

The success with laboratory tests for drug and alcohol abuse suggests that perhaps other tests can be used to screen applicants so that safer and more dependable employees can be recruited. Such tests have been used, with quantifiable results that have been extensively validated. One such test, developed by Behavioral Science Technology, Inc., is the Job Candidate Profile (JCP) (ref. Kamp) Dramatic reductions in workers' compensation claims have been demonstrated after the implementation of the JCP testing system.

A program of placement testing should not be entered into lightly, as it is possible to run afoul of Title VII of the Civil Rights Act of 1964. Preemployment screening tests must not be discriminatory against females or racial minorities. Of course, it is possible that more individuals representing racial minorities or more females will happen to fail a given test than do male Caucasians. The Equal Employment Opportunity Commission (EEOC) has published guidelines that prescribe limits for failure rates for any selection test or procedure. If the failure rate for racial minorities or females is less than 80% of the failure rate for white males, the test is considered to have an adverse impact on those racial minorities or females.

Another concern is the Americans with Disabilities Act of 1990 (ADA), which protects disabled persons from job discrimination. However, Congress did not intend for the ADA to prohibit job screening for alcohol or drug abuse. Therefore, despite the fact that alcohol and drug abuse can sometimes be considered "diseases" or "handicaps," alcohol or illegal-drug users cannot use this as a defense to force an employer to consider them for employment without regard to their existing alcohol or drug problem. Chapter 4 will explore the ADA in more detail. As can be seen, the preemployment screening problem can be legally complicated, but if these problems can be overcome, the policy of using preemployment tests of proven reliability and validity can be a sound and effective method of reducing workplace injuries.

THE SMOKE-FREE WORKPLACE

Public opinion has shifted in the direction of sober respect for the serious hazards of tobacco smoke. In the past, the general health concern was the deleterious effects of smoking on the smokers themselves. More recently, the concern has shifted to the *nonsmoker*, who is the victim of what has become known as "passive smoking." Wells estimates (ref. Wells) that some 46,000 nonsmoking Americans die each year from exposure to tobacco smoke. Glantz (ref. Glantz) cites cigarette smoke as a source of more than 4000 chemical air contaminants, including 43 known carcinogens. OSHA, in its Advanced Notice of Proposed Rulemaking, notes the presence of chemicals such as acrylonitrile, arsenic, benzene, lead, cadmium, formaldehyde, and vinyl chloride in tobacco smoke. For all of these chemicals, there is "sufficient evidence" of carcinogenicity in humans or animals. The concentrations of these dangerous substances may be tiny in tobacco smoke, but it should be noted that each of these chemicals is the subject of a separate OSHA air contaminant standard. These and other air contaminants will be addressed in Chapter 9. It is evident that the increasing concerns of nonsmoking American workers about passive exposure to tobacco smoke cannot be ignored by the safety and health manager, nor by Congress and federal agencies. OSHA has already taken steps to deal with smoking in the workplace in advance of any workplace standards that deal specifically with this problem. OSHA officials have testified before congressional subcommittees studying this problem. (See refs. Douglass, Scannell.)

In 1994, OSHA published a "Proposed Rule on Indoor Air Quality" in the *Federal Register*. Although other indoor air contaminants are addressed, it is clear that tobacco smoke is the primary target of this proposed standard. In the case of tobacco smoke, employers would be required either to prohibit smoking in the entire building or to establish designated smoking areas. The proposed standard requires direct exhaust ventilation for the designated smoking area and maintenance of continuous negative pressure in the area so that tobacco smoke will be contained within that area. If the exhaust system breaks down, smoking must be prohibited even in the designated smoking area until repairs are made. Even cleaning and maintenance activities in the smoking area are restricted to those times in which personnel are not smoking in the area.

It should not be difficult to understand that OSHA would have problems with promulgating a sweeping new standard as pervasive as the Proposed Rule on Indoor Air Quality. At press time for this book, the new standard was still under review. Political pressure exists on both sides of the issue, however. A national antismoking organization named "ASH" (Action on Smoking and Health) has been pressing OSHA for the past 20 years to promulgate a rule to regulate smoking in the workplace (ref. Work Practices). ASH has sought court action to force OSHA to act on conclusions reached by several federal agencies that tobacco smoke is a "Group A carcinogen." OSHA, by its own rules, gives priority to promulgating standards that respond to carcinogen hazards. ASH insists that OSHA adhere to these priorities with respect to tobacco smoke, now that tobacco smoke has been found to be a carcinogen. In March 1997, the U.S. Court of Appeals refused to set a time limit for a final OSHA rule, as ASH would have liked, but warned OSHA to "proceed with due speed even when statutory and regulatory deadlines are not mandatory."

A milestone in the mounting sentiment against public smoking occurred on June 20, 1997, when legislation was introduced to Congress to delineate a comprehensive settlement between the tobacco industry and suits brought by attorneys general of 40 of the 50 states. The settlement contained sweeping features, including a confirmation of the authority of the federal Food and Drug Administration (FDA) to regulate tobacco products, a monetary settlement to be paid by the tobacco industry having a 25-year face value of $358.5 billion, and comprehensive bans on outdoor advertising that is believed to be focused on youthful smokers (ref. Tobacco).

BLOODBORNE PATHOGENS

In the words of Warner Green in *Scientific American* (ref. Greene), "AIDS is the defining immunologic problem of our time. The HIV pathogen stands out as the preeminent threat to human health and therefore is the most intensely studied virus in history." According to the Global AIDS Policy Coalition, the estimated number of people infected by HIV as of the end of 1992 was 19.5 million (ref. Felsenthal). By 2002 that figure may have doubled. The alarming AIDS crisis has captivated the attention of not only the medical profession, but also the military, elected officials, and the general public. Although workplace exposures are rare, some occupational exposures have resulted in bloodborne disease incidence and subsequent death. Workers in some industries have become sensitized to the threat, and it is not surprising that OSHA responded with the promulgation of a standard for bloodborne pathogens, made effective March 6, 1992.

The HIV virus has the spotlight because of the alarming growth of the epidemic, the lack of any cure, the lack of any preventive immunization, and because it eventually leads to AIDS and certain death. But despite these sinister aspects of AIDS, in the occupational arena the hepatitis B virus (HBV) actually kills more victims than does HIV.

It is well known that the health professions are the primary at-risk occupations for bloodborne pathogens, and these jobs are the primary focus of the OSHA standard. Indeed, hospitals have long known of and dealt with the risk of hepatitis B outbreaks among their staff. Although the medical professions are the primary focus, the OSHA standard is not limited to these workplaces. The question to be asked is whether the worker will be exposed to blood or other potentially infectious materials, which includes some wastes and tissues of infected animals. The precautions to be taken to defend against HIV infection are basically the same as for HBV, so the OSHA standard addresses them together.

For workplaces that have one or more employees who may encounter occupational exposures, OSHA expects the employer to have a written exposure control plan. This plan must be accessible to employees and is subject to update at least annually. The employer must identify and list those jobs that are subject to exposure.

As with other health hazards, OSHA looks first to elimination of HBV and HIV hazards by engineering and work practice control measures. A large percentage of the occupational incidence of HIV infection is from accidental contact with "sharps," such as needles and broken glass vessels for human blood. A simple and reasonable system for sharps disposal is a practical first step toward control of the hazard and compliance with the OSHA standard.

An orderly and effective system of housekeeping, laundry, and waste disposal is another significant step in controlling the hazard and complying with the standard. Washing, cleaning, and disinfecting exposed surfaces are particularly effective in destroying HIV and HBV. Provision for consumption and storage of food must consider the need for separation from potential exposures. Applying cosmetics or lip balm and handling contact lenses are prohibited in work areas with a reasonable likelihood of occupational exposure. The eyes are suspected to be a somewhat vulnerable path for contracting HIV or HBV.

Besides engineering and work-practice controls, there still is a need for personal protective equipment. The employer's duty is to supply the necessary equipment and, further, to require employees to use it, unless, under unusual circumstances, the employee elects, for professional reasons, to refuse to use the equipment. One can picture an emergency medical scenario in which a medical professional might elect to forgo personal protective equipment in order to immediately render lifesaving aid to the victim of a medical crisis.

Bloodborne pathogens, especially AIDS, are more a concern of society and the medical profession in general than to the typical workplace. This vital topic, though, is too important to ignore and will continue to receive attention from OSHA and those safety and health managers whose workers are potentially exposed. OSHA has published guides, fact sheets, and even a *Sample Bloodborne Pathogens Exposure Control Plan* booklet (refs. Bloodborne, Most, Sample).

WORKPLACE VIOLENCE

Ask the average person what the leading cause of workplace fatalities is and you will probably hear "falls," "electrocutions," or perhaps "asphyxiation." But, according to recent statistics (ref. DeGroff), *workplace violence* is the leading cause of occupational fatalities for working women and the second leading cause for working men. Prevention of workplace violence is usually considered someone else's responsibility, but, increasingly, the safety and health manager is taking the initiative to control this significant hazard. The federal agency OSHA, too, is taking a look at this problem, and though no standards had been set as of this writing, guidelines have been issued by OSHA for comment. The first-draft guideline was issued in April 1996.

Not surprisingly, the most hazardous exposure to workplace violence is the check-out clerk in a night retail establishment—that is, the convenience store clerk. OSHA guidelines address six risk factors usually present in such environments (ref. DeGroff):

1. Exchange of money with the public
2. Working alone or in small numbers
3. Working late at night or during early morning hours
4. Working in high-crime areas
5. Guarding valuable property or possessions
6. Working in community settings

Some of the preceding risk factors are unavoidable in night retail store operations. However, the OSHA guidelines recommend measures that are intended to control or reduce the severity of the hazards. The following elements are included:

1. Management commitment and employee involvement
2. Worksite analysis
3. Hazard prevention and control
4. Training and education

The OSHA recommendations are necessarily general and are obviously intended to cause employers and employees to think about this hazard category and to concentrate on methods to reduce its impact. At the same time, technology is advancing the means of detection and apprehension of criminals who attempt violent crimes in the workplace.

The night retail establishment workplace is irrelevant to the jobs of most safety and health managers. However, even in the manufacturing, processing, construction, and service industries, workplace violence is on the increase. In fact, according to Moore (ref. Moore), homicide in the workplace is the fastest-growing violent crime in the United States. Unlike homicide off the job, it is not appropriate to blame alcohol or drugs in workplace homicides. Workplace homicide is often related to despair over downsizing or a termination notice for some other reason. There is evidence that homicide in the workplace is committed in a methodical and selective way (ref. Psychological).

The safety and health manager needs to be alert to this hazard and take necessary steps to keep it under control. Managers need plans and procedures for dealing with incidents when they do occur and for trying to prevent violence before it happens. A logical first step is to train supervisors in conflict management and in the importance of basic fairness in dealing with their subordinates. Managers also need to be ready with ideas for investment to start making a difference if it becomes apparent that workplace-violence hazards are present. Some possibilities include cellular phones for workers in dangerous zones, more intensive maintenance and replacement of motor vehicles so as to prevent breakdown exposure, assignment of tasks to pairs of workers instead of to single workers, and closer supervision of schedule with interim reporting at scheduled intervals.

SUMMARY

Safety and health on the job, like production quality or any other desirable factory characteristic, is achieved by the workers themselves. Thus, the actual achievement of safety and health is a line function. The safety and health manager, then, has a staff function in facilitating the line organization, especially first-line supervisors, in achieving the goal of safety and health.

It cannot be assumed that safety and health is really a "goal" of line management. More or less everyone wants safety and health in the workplace, but the degree of management's commitment to this goal must be assessed and documented by the safety and health manager.

Once management's commitment to the goal of safety and health is ascertained, the safety and health manager can get to the important functions of dealing with workers' compensation, collecting and analyzing statistical records, economic analysis, safety and health training, and dealing with both hazards and violations of safety and health standards.

In the 1990s, safety and health managers were saddled with responsibility for managing new programs dealing with changes in our society, government, and the environment. Drug and alcohol abuse are problems to both society and the workplace. The safety and health manager must administer programs that protect workers and their companies, while sidestepping discrimination pitfalls. The discrimination issue also is the focus of the Americans with Disabilities Act and touches on the social issue of the smoke-free workplace. Finally, the AIDS crisis and other diseases attributed to bloodborne pathogens are headline concerns addressed by OSHA and by the modern safety and health manager as well.

EXERCISES AND STUDY QUESTIONS

2.1 Is worker safety a line or a staff function?

2.2 What is the safety and health manager's relationship to the line organization?

2.3 Why is it often difficult for safety and health managers to win the respect and approval of top management?

2.4 What principle is embodied in the statement "Safety is everybody's business"?

2.5 What should the safety and health manager do if top management does not "practice what it preaches" in safety and health policies?

2.6 Is the workers' compensation system a federal or a state program? How long has it been in existence?

2.7 What are the ostensible and ulterior purposes of the workers' compensation system?

2.8 Should employers bear all the risk of injury to their workers?

2.9 Who pays the workers' compensation claims of injured workers?

2.10 What is a loss-control representative?

2.11 What is the Z16.1 system?

2.12 Why is it difficult to prove by statistics whether OSHA has been of any benefit?

2.13 What is the disadvantage of using lost workdays as a measure of the severity of injury or illness?

2.14 What constitutes a recordable injury or illness?

2.15 A firm employing 300 workers has 25 recordable injuries or illnesses in a year. What is its total injury–illness incidence rate?

2.16 How does the injury–illness incidence rate compare with the traditional frequency rate?

2.17 Compare the terms *frequency*, *severity*, and *seriousness*.

2.18 What is important about February 1 in the life of a safety and health manager?

2.19 How long are files of records required to be kept?

2.20 Are there advantages and pitfalls to the use of safety and health committees?

2.21 Compare the importance of direct and hidden costs of accidents.

2.22 Name some categories of hidden costs of accidents.

2.23 How can records of noninjury accidents be important to the safety and health manager?

2.24 Who are the principal trainers in safety and health in industry?

2.25 During a 6-month period, a firm employing 50 employees has 18 injuries and illnesses requiring medical treatment, in 4 of which cases the employee lost at least one day from work.

(a) Calculate the general injury–illness incidence rate.

(b) Calculate the traditional frequency rate.

(c) Is this a very dangerous industry?

2.26 For the year 2001, a firm with 25 employees has two medical-treatment injuries, plus one injury in which the worker lost three days of work. Calculate the injury incidence rate and the LWDI.

2.27 A firm has 62 employees. During the year, there are seven first-aid cases, three medical-treatment injuries, an accident in which an injured employee was required to work one week in restricted work activity, a work-related illness in which the employee lost one week of work, a work-related illness in which the employee lost six weeks of work, and a fatality resulting from an electrocution. Calculate the total incidence rate, the number-of-lost-workdays rate, and the LWDI.

2.28 Top management has set a safety and health objective for the year for a plant employing 135 employees. The objective is to reduce the LWDI for the firm to a level lower than the national average: 3.6. By May 1, the safety and health manager has logged 12 first-aid cases, three lost-time injuries, and two illnesses, both of which resulted in hospitalization. Based on these preliminary results, does it appear that the firm will meet top management's objective for the year? Show calculations to justify your conclusion.

2.29 A chemical plant employing 900 employees (employees work a regular 40-hour work-week) has the following safety and health record for the year 2003:

File 1   Forklift truck drops pallet load of packaged raw material; no injuries; some material wasted; pallet destroyed; extensive cleanup required.

File 2   Worker suffers heat cramps (illness) from continuous exposure to hot process; admitted to hospital for treatment: 2 weeks off.

File 3   Worker burns hand on steam pipe; first aid received and worker returns to workstation.

File 4   Worker suffers dermatitis from repeated contact with solvent; one week of work lost; another four weeks of work restricted to an assembly job.

File 5   Worker fractures finger in packaging machine; worker sent to hospital for treatment; back on the job the next day.

File 6   Maintenance worker lacerates hand when screwdriver slips; five sutures given; worker back on the job the next day.

File 7   Pressure vessel explodes; extensive damage to processing area; miraculously, no one is injured.

File 8   Worker gets poison ivy from exposure a week earlier while removing weeds around the plant perimeter fence; worker receives doctor's treatment, but no workdays are lost.

File 9   Worker becomes ill from continuous exposure to hydrogen sulfide leaks from furnace area; misses 2 weeks' work; leaks are repaired.

File 10 Worker gets severe poison ivy from weekend outing with Boy Scout troop; misses 2 days of work.

File 11 Maintenance worker falls from fractionating tower and is killed.

File 12 Worker fractures arm in transmission system that powers pulverizer mill; loses three days of work and an additional six weeks of work is in the production scheduling office before returning to regular job.

(a) Calculate the following incidence rates:

   1. LWDI
   2. Total injury incidence rate
   3. Total illness incidence rate
   4. Fatality incidence rate
   5. Number-of-lost-workdays rate (injuries and illnesses)
   6. Specific-hazard incidence rate (fractures)

(b) How does the safety and health record of this firm compare with that of other manufacturing companies and with industries in general?

2.30 On February 1, a 50-employee firm posts its annual OSHA log for the previous year, as shown in Figure 2.9. Complete the table and calculate the following:

(a) Total injury incidence rate

(b) Total illness incidence rate

      (c) Number-of-lost-workdays rate

      (d) LWDI

2.31 A certain large firm with 1400 employees in 1998 pays an annual Workers' Compensation insurance premium of $120,000. The experience modifier for this firm for 1998 was 1.05. By the year 2001, the firm has shown a dramatic improvement in worker safety and health, and accordingly, the insurance rating bureau revises the experience modifier for this firm to 0.80. What is the actual and percent savings in Workers' Compensation premium for this firm in 2001 compared with the premium in 1998?

2.32 The National Safety Council provides annual reports of incidence statistics from surveys gathered from member companies (ref. Accident). Identify which combinations of column totals on the OSHA 300 log correspond to each of the NSC reporting categories named as follows:

      (a) Lost-workday cases

      (b) Cases involving days away from work & deaths

      (c) Nonfatal cases without lost workdays[4]

      (d) Total cases

      (e) Lost workdays

      (f) Days away from work

2.33 Complete the column totals in the OSHA 300 log in Figure 2.9 for a firm that has 165 employees. In your school library, check current annual National Safety Council summaries (ref. Injury) to compare the incidence statistics of this firm with corresponding reporters to the National Safety Council for the most recent year available. If your library does not have *Injury Facts*, published by the National Safety Council, compare Figure 2.9 statistics with 2000 rates shown in Figure 2.2.

2.34 In cost analysis studies, what dollar cost does the U.S. Air Force associate with a human fatality? How does the cost vary between military-rated officers and civilian personnel? Why is there a difference?

2.35 What is the National Safety Council estimate of the cost of a human fatality? How much is the corresponding estimate for a worker injury?

2.36 What is the Air Force's estimate of the cost of an injury that does not result in one or more lost workdays?

2.38 What is the organization called "ASH," and for the promulgation of what OSHA standard is it attempting to apply political pressure?

2.39 What is the leading cause of occupational fatalities among working women?

2.40 Discuss the findings of the ALCOA Company with regard to drug testing.

2.41 What litigation risks should a firm consider if it terminates employees because of alcohol or drug use? Explain how this risk is affected by a drug and alcohol abuse policy within the company.

2.42 Is alcohol and drug abuse a significant factor in workplace homicides? Why or why not?

2.43 How has preemployment testing become a controversial issue?

---

[4]The NSC now uses the OSHA definition of "lost workday" to include days away from work, plus days of restricted work activity.

OSHA's Form 300

Log of Work-Related Injuries and Illnesses

Year 20 ____

U.S. Department of Labor
Occupational Safety and Health Administration
Form approved OMB no. 1218-0176

You must record information about every work-related death and about every work-related injury or illness that involves loss of consciousness, restricted work activity or job transfer, days away from work, or medical treatment beyond first aid. You must also record significant work-related injuries and illnesses that are diagnosed by a physician or licensed health care professional. You must also record work-related injuries and illnesses that meet any of the specific recording criteria listed in 29 CFR Part 1904.8 through 1904.12. Feel free to use two lines for a single case if you need to. You must complete an Injury and Illness Incident Report (OSHA Form 301) or equivalent form for each injury or illness recorded on this form. If you're not sure whether a case is recordable, call your local OSHA office for help.

Establishment name **B & D Castings**
City **Buffalo** State **TX**

Identify the person / Describe the case

(A) Case no.	(B) Employee's name	(C) Job title (e.g., Welder)	(D) Date of injury or onset of illness	(E) Where the event occurred (e.g., Loading dock north end)	(F) Describe injury or illness, parts of body affected, and object/substance that directly injured or made person ill (e.g., Second degree burns on right forearm from acetylene torch)
1	Empl A	Maint	month/day	Warehse C	Ankle sprain
2	Empl B	Assy	month/day	Final Assy	CTS – wrist discomfort
3	Empl C	Plating	month/day	Dip tank area	Nauseous from vapors
4	Empl D	Maint.	month/day	Receiving	Fatality – fall from ladder
5	Empl E	Press Oper	month/day	Fabrication	Hand laceration – strip stock
6	Empl F	Yard Wkr	month/day	Tank farm	Poison ivy – arms & face
7	Empl G	Grinder	month/day	Fabrication	Grinding dust – pneumoconiosis
8	Empl H	Plating	month/day	Dip tank area	Skin rash on hands – solvent
9	Empl I	Operator	month/day	Foundry	Burned hand – hot casting

Classify the case

CHECK ONLY ONE box for each case based on the most serious outcome for that case:

Columns: (G) Death, (H) Days away from work, (I) Job transfer or restriction, (J) Other recordable cases

Case	(G) Death	(H) Days away from work	(I) Remained at work – Job transfer or restriction	(J) Remained at work – Other record-able cases
1	☐	☐	☐	☑
2	☐	☐	☑	☑
3	☐	☑	☑	☐
4	☑	☐	☐	☐
5	☐	☐	☐	☑
6	☐	☐	☐	☑
7	☐	☑	☑	☐
8	☐	☐	☐	☑
9	☐	☑	☑	☐

Enter the number of days the injured or ill worker was:

Case	(K) On job transfer or restriction	(L) Away from work
1	8 days	___ days
2	56 days	___ days
3	___ days	14 days
4	___ days	___ days
5	___ days	___ days
6	___ days	___ days
7	___ days	2 days
8	___ days	___ days
9	3 days	___ days

Check the "Injury" column or choose one type of illness:

Columns: (M)(1) Injury, (2) Skin disorder, (3) Respiratory condition, (4) Poisoning, (5) Hearing loss, (6) All other illnesses

Case	(1) Injury	(2) Skin disorder	(3) Respiratory condition	(4) Poisoning	(5) Hearing loss	(6) All other illnesses
1	☑	☐	☐	☐	☐	☐
2	☑	☐	☐	☐	☐	☐
3	☐	☐	☑	☐	☐	☐
4	☑	☐	☐	☐	☐	☐
5	☑	☐	☐	☐	☐	☐
6	☐	☑	☐	☐	☐	☐
7	☐	☐	☑	☐	☐	☐
8	☐	☑	☐	☐	☐	☐
9	☑	☐	☐	☐	☐	☐

Page totals ▶

Be sure to transfer these totals to the Summary page (Form 300A) before you post it.

Page ____ of ____

FIGURE 2.9

OSHA form 300 for Exercises 2.30 and 2.33. (This form is available in larger format on the CD-ROM that accompanies this book.)

2.44 What basic plant procedures are considered particularly effective in controlling blood-borne pathogens such as HIV and HBV?

2.45 What part of the body is considered a particularly vulnerable path of entry for HIV or HBV in occupational scenarios?

RESEARCH EXERCISES

2.46 Examine significant citations disclosed in the news media to find an outstanding example of large OSHA fines for recordkeeping violations.

2.47 This chapter has stated that workplace violence is increasing as a cause of workplace fatalities. Examine current research to determine whether this trend is continuing.

2.48 Besides homicide, what other acts of violence occur in the workplace? Obtain annual estimates in each category if available.

2.49 Examine current developments in the comprehensive settlement between the tobacco industry and the states. What monetary damages has the tobacco industry been required to pay?

2.50 Find recent reports on the cumulative death toll from the AIDS epidemic. If possible, determine what percentage of the victims can be attributed to workplace exposure.

CHAPTER 3

Concepts of Hazard Avoidance

Hazards involve risk or chance, and these words deal with the unknown. As soon as the unknown element is eliminated, the problem is no longer one of safety or health. For example, everyone knows what would happen if he or she jumped off a 10-story building. Immediate death would be virtually a certainty, and such an act is not properly described as unsafe; it is described as suicidal. But to work on the roof of a 10-story building with no intention of falling off becomes a matter of safety. Workers without fall protection on an unguarded rooftop of a building are exposed to a recognized hazard. This is not to say that such workers will be killed or even to say that they will come to any harm whatsoever, but there is that chance, that unknown element.

Dealing with the unknown makes the job of the safety and health manager a difficult one. If the safety and health manager pushes for a capital investment to enhance safety or health, who is able to *prove* later that the investment was worthwhile? Improved injury and illness statistics help and may look impressive, but they do not actually prove that the capital investment was worthwhile because no one really knows what the statistics would have shown had the investment *not* been made. It is in the realm of the unknown.

Since safety and health deal with the unknown, there is no step-by-step recipe for eliminating hazards within the workplace. Instead, there are merely concepts or approaches to take to whittle away at the problem. All of the approaches have merit, but none is a panacea. Drawing on their own strengths, different safety and health managers tend to concentrate on certain favorite approaches familiar to them. The objective of this chapter is to present various approaches so that the safety and health manager will have a variety of tools, not just one or two, to deal with the unknown elements of worker safety and health. Both the good and the bad will be discussed for each approach. The good is often obvious or taken for granted. But the drawbacks of each approach must be squarely faced, too, so that safety and health managers can see the limitations and draw on the strengths of each approach to accomplish their missions.

THE ENFORCEMENT APPROACH

This is the approach initially taken by OSHA, but OSHA was certainly not the first to use it. Safety rules with penalties for breaking them have existed almost since people first began to deal with risks. The pure enforcement approach says that, since people neither assess hazards properly nor make prudent precautions, they should be given rules to follow and be subject to penalties for breaking those rules.

The enforcement approach is simple and direct; there is no question that it has an impact. The enforcement must be swift and sure and the penalties sufficiently severe. If these conditions are met, people will follow rules to some extent. Using the enforcement approach, OSHA has without doubt forced thousands of industries to comply with regulations that have changed workplaces and made millions of jobs safer and more healthful. The preceding statement sounds like a glowing success story for OSHA, but the reader should know that the enforcement approach has failed to do the whole job. It is difficult to see any general improvement in injury and illness statistics as a result of enforcement, although some categories—such as trenching and excavation cave-ins—have shown marked improvement. Despite its advantages, there are some basic weaknesses in the enforcement approach, as the statistics suggest. These weaknesses will now be examined.

At the foundation of any enforcement approach lies a set of mandatory standards. Mandatory standards must be worded in absolutes, such as "always do this" or "never do that." The wording of complicated exceptions can alleviate the problem somewhat, but requires the anticipation of every circumstance to be encountered. Within the framework of the stated scope of the standard, recognizing all exempt situations, each rule must be absolutely mandatory to be enforceable. But mandatory language employing the words *always* and *never* is really inappropriate when dealing with the uncertainties of safety and health hazards. To see how this is true, consider Case Study 3.1.

CASE STUDY 3.1

Suppose that a properly grounded electrical appliance used for the resuscitation of injured employees is equipped with a three-prong plug. But in the midst of an emergency, it is discovered that the wall receptacle is the old, ungrounded, two-hole variety. With no adapter in sight and an employee in desperate need of the appliance, who would not bend back or cut off the grounding plug and proceed to save the employee's life?

Of course, this example states an extreme case, and we must be "reasonable" and use our "professional judgment," but in the arena of enforcement and mandatory standards, who is going to say what is "reasonable"? Everyone knows what is reasonable in such an extreme case, but countless borderline cases occur every day in which it is not certain whether the proper course of action is to violate or not to violate the rule. Consider Case Study 3.2.

CASE STUDY 3.2

A dangerous fire was in progress as flammable liquids were burning in tanks. To shut off the source of fuel, a thinking employee quickly turned off the adjacent tank valves to avert a more dangerous fire that could have cost many lives, not to speak of property damage. Did the employee receive a medal for his meritorious act? The answer is no. Instead, the company received an OSHA citation because the employee was not wearing gloves! The valves were hot, and because the employee went ahead and closed the valves, burning his hands, the company was issued a citation.

If a government agency will issue a citation for failure to wear gloves while closing a valve in an emergency, who will have the courage to "be reasonable" and to go ahead and act even if a violation is the consequence? A strikingly similar case is described in Case Study 3.3.

CASE STUDY 3.3

In a trench cave-in accident in Boise, Idaho, a worker was buried and coworkers, "Good Samaritans," bravely jumped into the trench in the emergency to attempt to free the buried worker. OSHA responded by fining the company $8000 because of the humanitarian response of the rescue workers to the emergency. This action was ridiculed by some U.S. Senators, who awarded OSHA the infamous "Red Tape Award" for issuing the citation (ref. OSHA, 1993).

Although OSHA later rescinded the fines in the Idaho trench rescue case, one can see that the enforcement approach leads to problems when it is the only response to dealing with a safety or health hazard. Sometimes a fine is a negative and inappropriate response in a vain attempt to place blame after the fact when an accident has occurred. In the face of a pure enforcement approach, many industry employees and employers alike will gradually retreat to a defensive position, failing to produce, and blaming their lack of productivity on the government.

As stated earlier, OSHA did not invent the enforcement approach for dealing with hazards. Other mandatory safety rules and laws are familiar to everyone. Sometimes overzealous and oppressive rules can destroy themselves by alienating the very persons they are intended to protect. A notorious example is the mandatory helmet law for motorcycle riders. The helmet manufacturers can present impressive statistics that show how helmets save lives, at least in some accidents. Such statistics should be a strong motivation to motorcyclists to wear helmets. But in certain situations, the use of a helmet has disadvantages that may cause motorcyclists to hate the law that requires them *always* to wear a helmet. Thus, it is illegal to give a friend a trial ride around the block without obtaining an extra helmet for the passenger for that one ride. If a temporary skin rash or scalp treatment prevents an operator from wearing the helmet for a day or two, he or she must not ride, even if the motorcycle is the sole means of transportation. Where to put the helmet during a brief stop can also become extremely awkward in some situations. One motorcyclist in Houston became so frustrated that he complied with the letter of the law while he defied its spirit by riding his motorcycle around and about the city wearing his helmet on his elbow!

THE PSYCHOLOGICAL APPROACH

Contrasted with the enforcement approach is an approach that attempts to reward safe behaviors. This is the approach employed by many safety and health managers and may be identified as the psychological approach. The familiar elements of the psychological approach are posters and signs reminding employees to work safely. A large sign may be at the front gate of the plant displaying the number of days since a lost-time injury. Safety meetings, departmental awards, drawings, prizes, and picnics can be used to recognize and reward safe behaviors.

Religion versus Science

The psychological approach emphasizes the religion versus the science of safety and health. Safety meetings at which the psychological approach is used are typified by attempts at persuasion, sometimes called "pep talks." The idea is that employees can be rewarded into wanting to have safe work habits. Peer pressure can be brought to bear on an employee when the entire department may suffer if one person has an injury or an illness.

Top Management Support

The psychological approach is very sensitive to the support of top management. If such support is absent, the approach is very vulnerable. Recognition pins, certificates, and even monetary prizes are small reward if workers feel that to win these rewards they are not pursuing the real goals of top management.

Workers are able to sense the extent of management's commitment to safety by the day-to-day decisions it makes, not by written proclamations that everyone should "be safe." A rule requiring safety glasses to be worn in the production area is undermined when top management does not wear safety glasses when visiting the production floor. If safe practices are ordered to be cast aside when production must be expedited to fill an order on time, workers find out just how much worker safety and health mean to top management. Most safety and health managers will want to get a written endorsement of the plant safety program from top management. However, unless top management really understands and believes in the safety and health program, the written endorsement is not very valuable. The true orientation of top management will soon show through. Safety and health managers should beware of this pitfall when seeking such a written endorsement.

Worker Age

New workers, especially new young workers, are particularly influenced by the psychological approach to safety and health. Workers in their late teens or early twenties enter the workplace having recently emerged from a social structure that places a great deal of importance on daring and risk taking. These new workers are watching supervisors and more experienced coworkers to determine what kind of behavior or work habits earn respect in the industrial setting. If their older, more experienced peers wear respirators or ear protection, the young workers may also adopt safe habits. If highly respected coworkers laugh at or ignore safety principles, young workers may get off to a very bad start, never taking safety and health seriously.

In fairness to young workers, there is a complacency factor in older, more experienced workers that sometimes leads to tragic accidents at the end of the older worker's career. Case Study 3.4 tells the sad story of a worker who took an extra shift before a vacation prior to his retirement.

CASE STUDY 3.4

EXPERIENCED WORKER KILLED

On an extra weekend shift, a steel mill worker was removing a five-ton piece of equipment using a crane. The equipment was attached to the overhead crane, but did not lift properly because one of the equipment "hold-downs" was still attached. This caused the equipment to cock to one side. The worker saw the problem and went into the mill to detach the hold-down. Since the lift was under crane tension, the release of the hold-down caused the load to swing unexpectedly. The worker was crushed in a pinch point between the mill stand and the hold-down. The employee was 62 years old and had been employed in the industry for 33 years. Tragically, he did not quite make it to retirement.

Accident reports confirm that a large percentage of injuries are caused by unsafe acts by workers. This fact emphasizes the importance of the psychological approach in developing good worker attitudes toward safety and health. This approach can be bolstered by training in the hazards of specific operations. Once the subtle hazards are made known to workers who otherwise would not know about these hazards, the development of safe attitudes becomes less difficult.

THE ENGINEERING APPROACH

For decades, safety engineers have attributed most workplace injuries to unsafe worker acts, not unsafe conditions. The origin of this thinking has been traced to the great pioneering work in the field by the late H. W. Heinrich (ref. Heinrich), the first safety engineer so recognized. Heinrich's studies resulted in the widely known ratio 88:10:2.

Unsafe acts	88%
Unsafe conditions	10%
Unsafe causes	2%
Total causes of workplace accidents	100%

Recently, Heinrich's ratios have been questioned, and efforts to recover Heinrich's original research data have produced sketchy results. The current trend is to give increasing emphasis to the workplace machinery, environment, guards, and protective systems (i.e., the *conditions* of the workplace). Accident analyses are probing more deeply to determine whether incidents that at first appear to be caused by "worker carelessness" could have been prevented by a process redesign. This development has greatly enhanced the importance of the "engineering approach" to dealing with workplace hazards.

Three Lines of Defense

From the profession has emerged a definite preference for the engineering approach when dealing with health hazards. When a process is noisy or presents airborne exposures to toxic materials, the firm should first try to redesign or revise the process to "engineer out" the hazard. Thus, engineering controls receive first preference in what might be called *three lines of defense* against health hazards. These are identified as follows:

1. Engineering controls
2. Administrative or work-practice controls
3. Personal protective equipment

The advantages of the engineering approach are obvious. Engineering controls deal directly with the hazard by removing it, ventilating it, suppressing it, or otherwise rendering the workplace safe and healthful. This removes the necessity of living with the hazard and minimizing its effects as contrasted with the strategies of administrative controls and the use of personal protective equipment. This preference of strategies will be considered again in Chapters 10 and 12.

Safety Factors

Engineers have long recognized the chance element in safety and know that margins for variation must be provided. This basic principle of engineering design appears at various places in safety standards. For example, the safety factor for the design of scaffold components is 4:1. For overhead crane hoists, the factor is 5:1, and for scaffold *ropes*, the factor is 6:1 (i.e., scaffold ropes are designed to withstand six times the intended load).

The selection of safety factors is an important responsibility. It would be nice if all safety factors could be 10:1, but there are trade-offs that make such large safety factors unreasonable, even infeasible, in some situations. Cost is the obvious, but not the only, trade-off. Weight, supporting structure, speed, horsepower, and size are all factors that may be affected by selecting too large a safety factor. The drawbacks of large safety factors must be weighed against the consequences of system failure in order to arrive at a rational decision. There are many degrees of difference between situations when evaluating the consequences of system failures. Compare the importance of safety factors in the Kansas City hotel disaster[1] of 1981 to a failure in which the only loss is some

---

[1] Two skywalks collapsed in the crowded multistory lobby of the Hyatt Regency Hotel in Kansas City, Missouri, on July 17, 1981, killing 113 persons.

material or damaged equipment. Obviously, the former situation should employ a larger safety factor than the latter. Selection of safety factors depends on the evaluation or classification of degree of hazard, a subject that will be treated in depth later in this chapter.

Fail-Safe Principles

Besides the engineering principle of safety factors, there are additional principles of engineering design that consider the consequences of component failure within the system. These principles are labeled here as *fail-safe principles*, and three are identified:

1. General fail-safe principle
2. Fail-safe principle of redundancy
3. Principle of worst case

Each of these principles will be considered here, and their applications will appear again and again in subsequent chapters dealing with specific hazards.

General Fail-Safe Principle

The resulting status of a system, in event of failure of one of its components, shall be in a safe mode.

Systems or subsystems generally have two modes: active and inert. With most machines, the inert mode is the safer of the two. Thus, product safety engineering is usually quite simple: If you "pull the plug" on the machine, it cannot hurt you. But the inert mode is not always the safer mode. Suppose that the system is a complicated one, with subsystems built in to protect the operator and others in the area in the event of a failure within the system. In this case, pulling the plug to disconnect the machine might deactivate the safety subsystems so essential to protecting the operator and others in the area. In the case of such a system, disconnecting the power might render the system more unsafe than it was when the power was on. Design engineers need to consider the general fail-safe principle so as to ensure that a failure within the system will result in a safe mode. Thus, it may be necessary to provide backup power for the proper functioning of safety subsystems. Case Studies 3.5 and 3.6 will illustrate this concept.

CASE STUDY 3.5

An electric drill has a trigger switch that might be continuously depressed to operate the drill. The trigger switch is loaded with a spring, so that if some failure (on the part of the operator in this case) results in the release of the trigger, the machine will return to a safe mode (off, in this case). Such a switch is often called a *deadman control*. This example illustrates the common situation in which the inert state of the system is the safer one.

CASE STUDY 3.6

This example illustrates the more uncommon situation in which the inert state of the system is the more dangerous state. Consider an automobile with power steering and power brakes. When the engine dies, both steering and braking become very difficult; so at least as far as these subsystems are concerned, the inert state is more dangerous than the active one.

As will be seen in later chapters, industrial systems have characteristics similar to those described in the foregoing case studies. Sometimes safety standards specify that system failures be accommodated in such a way that subsystems for safety will continue to be operative.

The general fail-safe principle is the one that embodies the literal meaning of the term *fail safe*. However, industry and technology often associate another concept with the term *fail safe*, and that is the concept of *redundancy*.

Fail-Safe Principle of Redundancy

A critically important function of a system, subsystem, or components can be preserved by alternate parallel or standby units.

The design principle of redundancy has been widely used in the aerospace industry. When systems are so complicated and of such critical importance as a large aircraft or a space vehicle, the function is too important to allow the failure of a tiny component to bring down the entire system. Therefore, engineers back up primary subsystems with standby units. Sometimes, dual units can be specified right down at the component level. For extremely critical functions, three or four backup systems can be specified. In the field of occupational safety and health, some systems are seen as so vital as to require design redundancy. Mechanical power presses are an example.

Still another fail-safe design principle is the principle of *worst case*.

Principle of Worst Case

The design of a system should consider the worst situation to which it may be subjected in use.

This principle is really a recognition of *Murphy's law*, which states, "If anything can go wrong, it will." Murphy's law is no joke; it is a simple observation of the result of chance occurrences over a long period of time. Random events that have a constant hazard of occurrence are called *Poisson processes*. The design of a system must consider the possibility of the occurrence of some chance event that can have an adverse effect on safety and health.

An application of the principle of worst case is seen in the specification of explosionproof motors in ventilation systems for rooms in which flammable liquids are handled. Explosionproof motors are much more expensive than ordinary motors, and industries may resist the requirement to install explosionproof motors, especially in those processes in which the vapor levels of the substances mixed never even get close to the explosive range. But consider the scenario presented by a hot summer day on which a spill happens to occur. The hot weather raises the vapor level of the flammable liquid being handled. A spill at such an unfortunate time dramatically increases the liquid surface exposure, which makes the problem many times worse. At no other time would the ventilation system be more important. But if the motor is not explosionproof and is exposed to the critical concentration of vapors, a catastrophic explosion would occur as soon as the ventilation system switched on.

The concept of *defensive driving* is well known to all drivers and serves to explain the principle of worst case. Defensive drivers control their vehicles to the extent that they are prepared for the worst random event they can reasonably imagine.

Design Principles

Engineers have come to rely on a variety of approaches or "engineering design principles" to reduce or eliminate hazards. Some of these principles are listed here to stimulate thinking of the various paths that can be taken in dealing with hazards.

1. *Eliminate* the process or cause of the hazard. Often a process has been performed over so long a period that it is erroneously considered essential to the operation of the plant. After many years of operation, a process becomes institutional and plant personnel tend to accept it without question. However, it is the duty of safety and health professionals to question old and accepted ways of doing things if these ways are hazardous. Hazards that may have been considered acceptable in the days when the process was originally designed may now be considered unacceptable. New thinking may reach a different conclusion on the question of just how critical the need is for a particular process.

2. *Substitute* an alternate process or material. If a process is essential and must be retained, perhaps it can be substituted with another method or material that is not so dangerous. A good example is the substitution of less hazardous solvents for benzene, which has been found to cause leukemia. Another example is changing a machining process to perform the machining dry—that is, without the benefit of cutting fluid. Certainly many machine tool cutting operations require cutting fluid, but for some materials and processes, cutting fluid may not be absolutely necessary and the drawbacks might outweigh the benefits.

3. *Guard* personnel from exposure to the hazard. Perhaps a process is absolutely essential to the operation of the plant, and there is no substitute for it or for the hazardous materials that must be dealt with by it. In these cases, it is sometimes possible to control exposure to the hazard by guarding personnel from exposure to the hazard.

4. Install *barriers* to keep personnel out of the area. Contrasted with guards, which are attached to the machine or process, are independent barriers that are installed around the process or machine to keep personnel out of danger. Such barriers may seem more like an administrative function or an operational procedure, but the engineer who designs the process can specify in particular what barriers are needed around a process and where to place them.

5. *Warn* personnel with visible or audible alarms. In the absence of other protective design features of the system, the engineer can sometimes design the machine or process in such a way that the system warns the operator or other personnel when exposure to a significant hazard is imminent or likely. To be effective, the alarm should be used sparingly so that the personnel do not ignore the flashing light or the beeping alarm and continue operating the process despite the exposure.

6. Use *warning labels* to caution personnel to avoid the hazard. Sometimes an essential hazardous operation cannot be eliminated, substituted with a less hazardous process or material, or adequately guarded from personnel exposure. In these situations, at least it is often possible to attach a warning label to the process or device that reminds personnel of hazards that are not controlled by the machine or process itself. This design approach is not as effective as the preceding approaches, because personnel may not read or heed the warning labels. Despite the limited effectiveness of warning labels, they are better than complete disregard for the hazard in the design process.

7. Use *filters* to remove exposure to hazardous effluents. Certain hazards require a different perspective on the part of the design engineer. The exhaust of dangerous effluents is an example. The engineer can sometimes design filter systems in the machine or process itself to deal with gases or dusts that may be undesirable products of the process.

8. Design *exhaust ventilation* systems to deal with process effluents. Sometimes the undesirable products of a process are too hazardous or are impractical to filter out of the breathing air in the environment of a process. In these cases, sometimes the process or machine design itself can include features that exhaust the harmful agents as they are being produced. Again, these features may seem to be within the purview of someone else, such as a ventilation expert or plant maintenance engineer. Nevertheless, the designer of the process itself should not overlook opportunities to incorporate these features into the original design of the process or machine.

9. Consider the *human interface*. After the more straightforward engineering principles of dealing with hazards are included in the design process, it is a good idea to once again review and identify all the interfaces of the process or machine with personnel. At what points does it become necessary for persons to interact with the machine? At these points, are personnel exposed to hazards? The human interfaces so identified should include both the equipment interfaces and the material interfaces. Each interface so identified should be checked again for possible design features that can further control hazards using the other engineering design principles enumerated in this section.

Engineering Pitfalls

It is easy to get caught up in the idea that technology can solve our problems, including elimination of workplace hazards. Certainly, an inventor of a new gadget to prevent injuries or illnesses can become quickly enamored with it and present a convincing argument that the new invention should be installed in workplaces everywhere. When standards writers are persuaded by these arguments, they often require all appropriate industries to install the new device. Several things can go wrong, however.

Recalling the case against the enforcement approach, certain unusual circumstances can make the engineering solution inappropriate or even unsafe. A good example is the use of spring-loaded cutoff valves in air lines for compressed-air tools. The purpose of the cutoff valve is to prevent hose whip action by stopping the flow of air if the tool accidentally separates from the hose. The sudden flow of air overcomes the spring-loaded valve and closes it, stopping the flow. The problem comes when several tools are operated from the same main hose and flow reaches maximum even during normal use. The cutoff then becomes a nuisance and impedes production.

A second problem with the engineering approach is related to the first: Workers remove or defeat the purpose of engineering controls or safety devices. The most obvious example is the removal of guards from machines. Before faulting the worker for such behavior, take a close look at the guard design. Some guards are so awkward that they make the work nearly impossible. Some machine guards are so impractical that they conjure up doubts about the motives of the equipment manufacturer. There is a legal motivation to install an impractical guard on a new machine so that users will take the guard off before putting the machine in service. When the user modifies a machine by removing a guard, the manufacturer is absolved of guilt for any accident that later occurs, but that might reasonably have been prevented by the guard.

An irony of the engineering approach is that, if the engineered system does not do the job for which it was intended, it can do more harm than good by engendering a false sense of security.

CASE STUDY 3.7

A FALSE SENSE OF SECURITY

A printing press operator was proudly demonstrating a new printing press to his family at an open house intended to display the state-of-the-art safety devices engineered into the new equipment. One of the high-tech features was a photoelectric sensing device that was engineered to detect any object (such as the operator's hands) that had violated the danger zone at the inrunning nip point to the printing rolls. The system was designed to immediately stop the rolls whenever an object was sensed. So proud of the design feature of the system was the operator that he demonstrated by thrusting his hand repeatedly into the danger zone. Ultimately, he succeeded in beating the system, and the printing press did indeed amputate the end of one of his fingers. Incredible as this case study may seem, this incident actually happened. One could question the judgment of the operator in tempting the machine in such a foolish way, but the tendency exists to trust the engineering implicitly. Thus, workers are exposed to hazards due to the false sense of security that engineering systems sometimes engender.

Such false sense of security can even lead to new operator procedures that depend on the safety device to control the operation so that the work can be hastened. The best example that comes to mind is the hoist limit switch on an overhead crane. If the hoist load block approaches too close to the bridge, the hoist limit switch is tripped, shutting off the hoist motor. The idea sounds good, but the operator can take advantage of the device by *depending* on the switch to stop the load during normal operation. The hoist limit switch is not intended as an operating control, but workers can and do use it that way. The only defense against such use appears to be proper training and safe attitudes on the part of the operator—that is, the psychological approach.

Finally, the engineered system can sometimes cause a hazard, as illustrated in the example that follows in which a pneumatic press ram pinned an operator's hand on the *upstroke*. (See Figure 3.1.) The press was equipped with a two-hand control that was designed for safety's sake not to allow the press to be actuated except by both hands of the operator. Ironically, the two-hand control created a hazard. This press was later redesigned to place a shield in front of the ram so that the operator would be unable to reach into the area above the ram.

Foot controls for machines provide a good example of the conflicts that arise between the hazards that engineering controls are designed to prevent and the hazards that they create. Accidental tripping is a problem with foot controls, so engineers have fashioned enclosures into which the operator must insert his or her foot before stepping on the control itself. The problem with these enclosures is that they make the process of activating the foot trip more complicated. More operator attention is required to move the foot in the right ways to get it inside the enclosure and then operate the pedal. Supposedly, this is good, because then a careless motion will not accidentally actuate the machine. However, because of the sometimes awkward additional motions that the enclosure requires, some operators position their feet so that they can keep a foot on the pedal at all times, so-called "riding the pedal." Unfortunately, riding the pedal increases the likelihood that the operator will accidentally trip the machine, the very hazard that the foot pedal enclosure was intended to prevent. This problem has been studied extensively by Triodyne, Inc. (ref. Barnett).

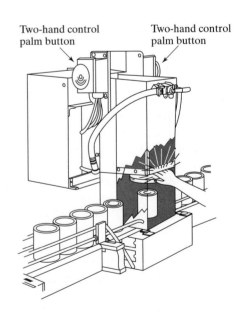

Two-hand control palm button   Two-hand control palm button

FIGURE 3.1

Pneumatic press ram pins operator's hand on the upstroke; two-hand control safety device prevents the operator from reactivating the press to release her hand.

As another example, robots are being used to work in hot, noisy environments, to lift heavy objects, and to otherwise serve in places where humans might be injured or suffer health hazards. Most industrial robots are simply mechanical arms programmed by computer to feed material to machines or to do welding. But the mindless swinging of these mechanical arms can cause injury to workers who get in the path of the robot. The irony is that a hazard is created by the robot, the very purpose of which was to reduce hazards. One solution is to make the robot more sophisticated, giving it sensors to detect when a foreign object or person is in its path. Another solution is simply to install guardrails around the robot or otherwise keep personnel out of the danger zone.

In summary, the engineering approach is a good one and deserves the recent emphasis it is receiving. However, there are pitfalls, and the safety and health manager needs a certain sophistication to see both the advantages and disadvantages in proposed capital equipment investments in safety and health systems. Upon review of the preceding examples of engineering pitfalls, it can be seen that almost every problem can be dealt with if some additional thought is given to the design of the equipment or its intended operation. The conclusion to be reached is that engineering can solve safety and health problems, but the safety and health manager should not naively assume that the solutions will be simple.

THE ANALYTICAL APPROACH

The analytical approach deals with hazards by studying their mechanisms, analyzing statistical histories, computing probabilities of accidents, conducting epidemiological and toxicological studies, and weighing costs and benefits of hazard elimination. Many, but not all, analytical methods involve computations.

Accident Analysis

Accident and incident (near miss) analysis is so important that it already has been discussed extensively in Chapter 2. No safety and health program within an industrial plant is complete without some form of review of mishaps that have actually occurred. The subject is mentioned again here to classify it as within the analytical approach and to show its relationship to other methods of hazard avoidance. Its only drawback is that it is *a posteriori*—that is, the analysis is performed after the fact, too late to prevent the consequences of the accident that has already happened. But the value of the analysis for future accident prevention is critical.

Accident analysis is not used nearly enough to assist in the other approaches to hazard avoidance. The enforcement approach would be much more palatable to the public if the enforcing agency would spend more time analyzing accident histories. That way, citations would be written only for the most important violations. The psychological approach could also be strengthened a great deal by substantiating persuasive appeals with actual results of accidents. The engineering approach needs accident analysis to know where the problems are and to design a solution to deal with all of the accident mechanisms.

Failure Modes and Effects Analysis

Sometimes a hazard has several origins, and a detailed analysis of potential causes must be made. Reliability engineers use a method called *failure modes and effects analysis* (FMEA) to trace the effects of individual component failures on the overall, or "catastrophic," failure of equipment. Such an analysis is equipment oriented instead of hazard oriented. In their own interest, equipment manufacturers sometimes perform an FMEA before a new product is released. Users of these products can sometimes benefit from an examination of the manufacturer's FMEA in determining what caused a particular piece of equipment to fail in use.

The FMEA becomes important to the safety and health manager when the failure of a piece of equipment can result in an industrial injury or illness. If a piece of equipment is critical to the health or safety of employees, the safety and health manager may desire to request a report of an FMEA performed by the manufacturer of, or potential bidder for, the equipment. In practice, however, FMEA is usually ignored by safety and health managers until *after* an accident has occurred. Certainly, safety and health managers should at least know what the initials FMEA represent so that they are not dazzled by the term in the courtroom should equipment manufacturers use it to defend the safety of their products.

One beneficial way of using FMEA *before* an accident occurs is in preventive maintenance. Every component of equipment has some feasible mechanism for eventual failure. To simply use equipment until it eventually fails can sometimes have tragic consequences. Consider, for example, the wire rope on a crane, or the chain links in a sling, or the brakes on a forklift truck. The FMEA can direct attention to critical components that should be set up on a preventive maintenance schedule that permits parts to be inspected and replaced *before* failure.

Fault-Tree Analysis

A very similar, but more general, method of analysis than FMEA is *fault-tree analysis*. Whereas FMEA focuses on component reliability, fault-tree analysis concentrates on the end result, which is usually an accident or some other adverse consequence. Accidents are caused at least as often by procedural errors as by equipment failures, and fault-tree analysis considers all causes—procedural or equipment. The method was developed by Bell Laboratories under contract with the U.S. Air Force in the early 1960s. The objective was to avoid a potential missile system disaster.

The term *fault tree* arises from the appearance of the logic diagram that is used to analyze the probabilities associated with the various causes and their effects. The leaves and branches of the fault tree are the myriad individual circumstances or events that can contribute to an accident. The base or trunk of the tree is the catastrophic accident or other undesirable result being studied. Figure 3.2 shows a sample fault-tree diagram of the network of causal relationships that can contribute to the electrocution of a worker using a portable electric drill.

The diagram in Figure 3.2 reveals the use of two symbols in coding causal relationships. Figure 3.3 deciphers this code. It is essential that the analyst be able to distinguish between the AND and the OR relationship for event conditions. All of the causal event conditions are required to be present to cause a result to occur when the

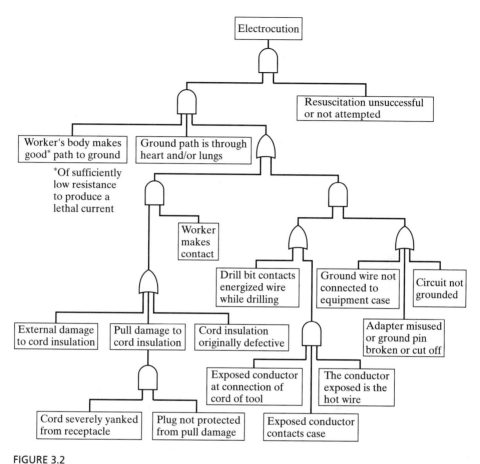

FIGURE 3.2

Fault-tree analysis of hazard origins in the electrocution of workers using portable electronic drills (not double insulated).

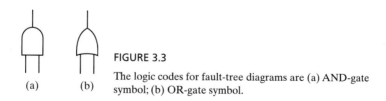

FIGURE 3.3

The logic codes for fault-tree diagrams are (a) AND-gate symbol; (b) OR-gate symbol.

conditions are connected by an AND gate. However, a single condition is sufficient to cause a result to occur when conditions are connected by an OR gate. For example, oxygen, heat, and fuel are all required to produce fire, so they are connected by an AND gate, as shown in Figure 3.4. But *either* an open flame or a static spark may be sufficient to produce ignition heat for a given substance, so these conditions are connected by an OR gate, as shown in Figure 3.5. Note that Figures 3.4 and 3.5 could be combined to start the buildup of a fault-tree branch.

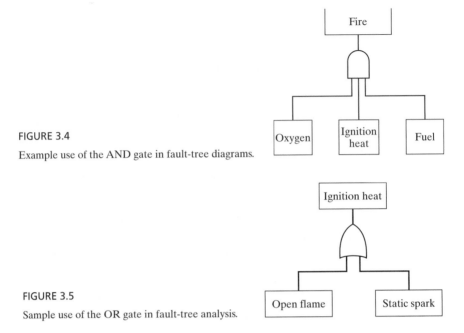

FIGURE 3.4

Example use of the AND gate in fault-tree diagrams.

FIGURE 3.5

Sample use of the OR gate in fault-tree analysis.

One difficulty with fault-tree analysis is that it requires each condition to be stated in absolute "yes/no" or "go/no-go" language. The analysis breaks down if a condition as stated may or may not cause a specified result. When the analyst is confronted with a "maybe" situation, it usually means that the cause has not been sufficiently analyzed to determine what additional conditions are also necessary to effect the result. Therefore, the difficulty in dealing with a "maybe" situation forces the analyst to look more deeply into the fault relationships, so the "difficulty" may be a benefit after all.

Fault-tree analysis permits the analyst to compute quantitative measures of the probabilities of accident occurrence. The computation is tricky, however, and is at best only as good as the estimates of the probabilities of occurrence of the causal conditions. It seems intuitive that one should add the probabilities of events leading into an OR gate and multiply the probabilities of events leading to an AND gate. This intuition is wrong in both cases and is shown for the OR gate as follows: Suppose, in the example shown in Figure 3.5, that the probabilities of occurrence of the two event causes were as given in the following table:

Event cause	Probability of occurrence (%)
Open flame	50
Static spark	50

To add these two probabilities would result in a 100% chance of reaching "ignition heat," an obviously erroneous result. A 50–50 chance of either of two possible causes is insufficient basis to find a 100% certainty of any resulting event. To make the logic even more convincing, the reader should rework the example using the following probabilities:

Event cause	Probability of occurrence (%)
Open flame	60
Static spark	70

Obviously, no result can have a probability of 130%! The correct probability computation is to subtract the "intersection," or probability that both independent causes would occur:

$$P[\text{ignition heat}] = P[\text{open flame}] + P[\text{static spark}]$$
$$- P[\text{open flame}] \times P[\text{static spark}]$$
$$= 0.6 + 0.7 - 0.6 \times 0.7$$
$$= 1.3 - 0.42 = 0.88 = 88\%$$

For OR gates in which there are several event causes, the computation becomes very complex. Adding to the complexity is the question of whether the event causes are *independent*—that is, the occurrence of one event does not affect the likelihood of the occurrence of the other event cause(s). A special case of *dependent* events are events that are *mutually exclusive*—that is, the occurrence of one event *precludes* the occurrence of the other(s). The mutually exclusive condition simplifies the computation, but event causes in fault-tree diagrams are typically not mutually exclusive.

Fully resolving the problem of fault-tree computations would require a treatise on probability theory, which is beyond the scope of this book. Suffice it to say that fault-tree probability computations are more complex than most people think they are and therefore are usually done incorrectly.

Despite the computational problems associated with fault-tree computations, the fault-tree diagram itself is a useful analytical tool. The diagramming process itself forces the analyst to think about various event causes and their relationship to the overall problem. The completed diagram permits certain logical conclusions to be reached without computation. For instance, in Figure 3.2, the event "Worker makes contact" is key because, from the diagram, it can be seen that the prevention of this event would preclude any of the five events to the lower left in the diagram from causing electrocution. Even more important is the event "Worker's body makes good path to ground." Prevention of this one single event is sufficient to prevent electrocution, according to the diagram. The reader can no doubt draw other interesting conclusions from Figure 3.2. These conclusions can result in a more sophisticated understanding of the hazard and in turn may lead to revisions to make the diagram more realistic. Such a developmental process leads to the overall goal of hazard avoidance.

Loss Incident Causation Models

Closely related to both fault-tree analysis and failure modes and effects analysis is a model that emphasizes the causes of *loss incidents*, whether or not the incident results in personal injury, as reported by Robert E. McClay (ref. McClay). McClay's model attempts to take a universal perspective in which the entire causal system is examined, including primary causes—labeled *proximal* causes—and secondary causes—labeled *distal* causes. A proximal cause would be a direct hazard in the conventional sense of the word—for example, a missing guard on a punch press. By contrast, an example of a distal cause would include a management attitude or policy that is deficient in allocating resources and attention to the elimination of hazards. Distal causes are as important as proximal causes, because, although the effect of distal causes is less direct and immediate, distal causes create and shape proximal causes.

A critical point in the progression of the loss incident causation model is the *point of irreversibility*. McClay identifies this as the point at which the various interacting proximal causes will result in a loss incident. Despite the number and variety of proximal causes, only a few select cases will result in a sequence of events in which the point of irreversibility is reached. Once this point is reached, a loss incident is unavoidable. This is still not to say that an injury will occur. A loss incident can occur, and still no personnel might be exposed, or perhaps whatever exposure does occur is not injurious. Factors such as personnel exposure in the event of a loss incident affect the severity of the effects of the incident after the point of irreversibility has been exceeded. Such factors affecting the outcome can be either negative or positive—that is, they can be *aggravating factors*, which make the outcome more severe, or they can be *mitigating factors*, which make the outcome less severe.

A diagramming convention has been proposed by McClay to assist the analyst in visualizing the universal model of causation. Figure 3.6 defines the symbols to be used in depicting the model. Note in Figure 3.6 that proximal causes are represented by three different symbols, each denoting one of three categories of hazards: physical condition,

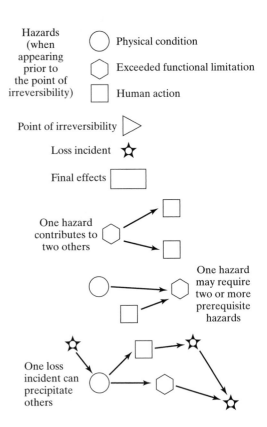

FIGURE 3.6

Symbols used in diagramming loss incidents.
(*Source*: McClay, reprinted by permission of Professional Safety, Des Plaines, IL.).

exceeded functional limitation and human action. Each of these three types of hazards can have a causal relationship with either of the other two. Further, a loss incident can in turn have a causal relationship on other proximal causes or other loss incidents, as illustrated in the examples in the diagram. Figure 3.7 summarizes the universal model of loss incident causation, revealing the relationship between distal and proximal causes and representing the region prior to the point of irreversibility as the *sphere of control*. The value in visualizing a loss incident causation system in a diagram such as the one shown in Figure 3.7 is twofold: It permits the analyst or the safety and health manager to distinguish the factors and conditions that can be controlled and allows him or her to perceive the consequences, good and bad, to be derived from applying resources to eliminate hazards or to mitigate their effects. As such, the analytical approach can be useful in assisting the safety and health manager to define and accomplish *reasonable objectives*, as suggested in Chapter 1.

Toxicology

Toxicology is the study of the nature and effects of poisons. Industrial toxicology is concerned especially with identifying what industrial materials or contaminants can harm

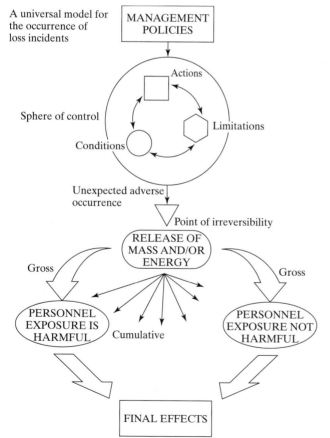

FIGURE 3.7

A universal model for the occurrence of loss incidents. (*Source*: McClay, reprinted by permission of Professional Safety, Des Plaines, IL.)

workers and what should be done to control these materials. This is really a broad statement because virtually every material is harmful to living organisms if the exposure rate or quantity is great enough.

Many toxicological studies are performed on animals to provide a basis for conclusions about the hazards to humans. These animal studies are essential because most toxicological experiments would cause death or serious harm to human subjects. The disadvantage is that animal defenses to various toxic substances vary between species. The more the field of toxicology advances, however, the more various toxic materials can be classified and their effects predicted even before experiment. Rabbit–man, monkey–man, mouse–man, and guinea-pig–man comparisons of species' defenses to various agents are becoming fairly well known.

A field that relates to both pharmacology and toxicology is *pharmacokinetics*. Bischoff and Lutz (ref. Bischoff) define pharmacokinetics as follows: "Pharmacokinetics is a description of the absorption, disposition, metabolism, and elimination of chemicals in the body, and is useful in both pharmacology and toxicology." It is important to the pharmacologist to understand how medical chemicals are handled within the body. In a similar way, it is important to the industrial toxicologist to understand how toxic chemicals are handled within the body.

Epidemiological Studies

Epidemiology is contrasted with toxicology in that epidemiology studies are strictly of people, not animals. The word obviously derives from the word *epidemic*, and in the literal sense, epidemiology is the study of epidemics. The epidemiological approach examines populations of people to associate various patterns of possible disease causes with the occurrence of the disease. It draws heavily on the analytical tools of mathematical statistics.

A classic epidemiological study was the association of the disease rubella (German measles) in pregnant women to birth defects in the infants born of those pregnancies. The study began with a curiosity observed by N. McAlister Gregg, an Australian ophthalmologist, in 1941. He observed eye cataracts among infants born of mothers who had had German measles during pregnancy in 1939 and 1940. The phenomenon might have been unnoticed except for the Australian epidemic of rubella during the World War II buildup. Years of statistical epidemiological studies later confirmed a strong relationship between rubella during pregnancy and a wide variety of birth defects in the infants born of those pregnancies.

Epidemics are usually thought of as attacking a general population at a specific time in a specific geographical area. Examples are the bubonic plague in Europe in the mid-14th century and the rubella epidemic in Australia in 1939–1940. But more subtle epidemics attack a specific subgroup of people who may be spread out over time and place. In other words, the victims of a particular epidemic may not live in one place or at one time, but instead have some other common characteristic, such as what they *do*. This aspect of epidemiology is what makes it important to occupational safety and health. Thus, lung fibrosis may not be a very common disease in any location or at any time. But when the population of only those persons who have worked with asbestos is examined, it can be seen that, after a long latency period, lung fibrosis can be considered an epidemic. Epidemiological studies linking lung fibrosis to asbestos have led to the identification of a type of lung fibrosis known as *asbestosis*. Other epidemiological links are *brown lung* to textile workers, *black lung* to coal miners, and angiosarcoma to vinyl chloride workers. A recent epidemiological study is examined in Case Study 3.8.

CASE STUDY 3.8

EPIDEMIOLOGICAL STUDY

This study was performed in the early 1990s by researchers at Johns Hopkins University in an investigation funded by the IBM Corporation (ref. Computer). The population studied was pregnant women who worked with diethylene glycol dimethyl ether and ethylene glycol monethyl ether acetate—chemicals used in the making of computer chips. Only 30 women were studied in the target population, but the results were seen to be significant because of the high percentage of miscarriages that occurred in this group. Of the 30 women studied, 10 had miscarriages—a rate of 33.3%! This compares with a miscarriage rate of 15.6% among women not exposed to the chemicals.

From Case Study 3.8, it can be seen that an epidemiological study can be a powerful tool to linking a potential hazard to observed occupational diseases. It becomes an excellent preliminary step to in-depth studies that pinpoint the causal relationship that underlies the observed link.

Both epidemiology and toxicology are important elements in the analytical approach to avoiding hazards, but the safety and health manager does not typically perform such studies. The studies provide the basis for mandatory standards that are subsequently used in the enforcement approach. Safety and health managers may also use the results of such analytical studies to substantiate the psychological approach or as a justification for an engineering approach to a health problem.

Cost–Benefit Analysis

Chapter 1 set the record straight on the importance of hazard costs. Like it or not, people do make cost judgments on occupational safety and health—not just management, but workers, too. In the world of reality, funds do have limitations, and cost–benefit analyses must be used to compare capital investment alternatives. Safety and health managers who feel that they can justify at any cost capital investment proposals that can be shown to have the possibility of preventing injuries and illnesses can easily appear naive. There is always more than one opportunity to improve safety and health, and cost–benefit analyses provide the basis for deciding which ones to undertake first.

The biggest difficulty with cost–benefit analysis is the estimation of the benefit side of the picture. Benefits to safety and health consist of hazard reduction, and to make the cost–benefit analysis computation, some quantitative assessment of hazard must be made. Such probabilities of injury or illness are very difficult to determine. Statistical data are being compiled at the state level, as discussed in Chapter 2. But these data are still usually of insufficient detail to permit a quantitative determination of existing risk. In addition to this determination, an estimate of expected risk *after* the improvement must also be made, since it cannot be assumed in general that every safety and health improvement will completely eliminate hazards. Case Study 3.9 illustrates the method of cost–benefit analysis.

In Case Study 3.9, the reader will perceive the large degree of speculation in the estimates of hazard costs. Such speculation casts doubt on the entire analysis, but it is better to speculate and calculate than to completely ignore the cost–benefit trade-off. This is an opportune point to lead into the next section of this chapter—hazards classification. There is a great deal to be gained from a subjective analysis of hazard costs, without resorting to overly sophisticated quantitative analysis, the credibility of which "hard" data may not support.

CASE STUDY 3.9

COST–BENEFIT ANALYSIS OF INSTALLING A PARTICULAR MACHINE GUARD

Cost

Amortization of initial investment	
Initial cost	$4000
Expected useful life	8 years
Salvage value	0
Interest cost on invested capital	20%
Annual cost = $4000 × (20% interest factor for 8 years)	
= $4000 × 0.26061 =	$1042
Expected cost of annual maintenance (if any)	0
Expected annual cost due to reduction in production rate (if any)	800
Total expected annual cost	**$1842**

Benefit

Estimated tangible costs per injury of this type	$350
Estimated intangible costs per injury of this type	2400
Total costs per injury	$2750
Average number of injuries per year on this machine due to this hazard	1.2
Expected number of injuries of this type after guarding	0.1
Expected reduction in injuries per year	1.1
Expected benefit = $2750 × 1.1 =	**$3025**

Since the total expected benefit of $3025 is more than the total expected cost of $1842, the conclusion is that it would be worthwhile to install this machine guard.

HAZARDS CLASSIFICATION SCALE

The absence of hard data to support quantitative cost–benefit analyses leaves a void of tools or benchmarks for use by the safety and health manager, safety committee, or other party on whom the decision responsibility for safety and health improvement may be thrust. Some sort of ranking or scale is needed to distinguish between serious hazards and minor ones so that rational decisions can be made to eliminate hazards on a worst-first basis.

OSHA does recognize four categories of hazards or standards violations as follows:

- Imminent danger
- Serious violations
- Nonserious violations
- De minimis violations

Chapter 4 explains these categories in more detail. However, the categories are rather loosely defined and are distinguished chiefly by the extent of the penalty authorized for each type. The imminent danger category qualifies OSHA to seek a U.S. District Court injunction to force the employer to remove the hazard or face a court-ordered shutdown of the operation. The de minimis violations, by contrast, are merely technical violations that bear little relationship to safety or health; they typically do not carry a monetary penalty. But the designations of which violations will be recognized as falling into which category of hazard is especially unclear.

It is perhaps impossible to define clear-cut categories in every instance, but much is to be gained by some type of subjective ranking of workplace hazards. The thesis of this book is that a scale of 1 to 10 should be attempted, crude as that scale might be. Until people start talking about degrees of hazards on such a quantitative scale, little progress can be made toward establishing an effective and orderly strategy for hazard elimination. On a 10-point scale, a "10" is characterized as the worst hazard imaginable, while a "1" is the least significant or mildest of hazards.

The 10-point scale is recommended because such a scale has become very popular in everyday speech. Facilitated by the media, especially television, the public has come to understand a statement such as "on a scale of 1 to 10, that item (tennis match, ski slope, kiss, etc.) was at least a 9." The familiarity of this popular jargon can be employed to characterize workplace hazards.

Table 3.1 is a first attempt to describe subjectively each of 10 levels of hazards. The definitions address basically four types of hazards: fatalities, health hazards, industrial noise hazards, and safety (injury) hazards. A clear-cut delineation is obviously difficult, and some readers no doubt will disagree with the wording of the definitions. Criticisms of the scale will reflect both the shortcomings of the definitions and the biases of the critics themselves. Acoustical experts, for instance, may want to place a high degree of emphasis on excessive noise hazards. Other specialists will want to emphasize other areas.

One critical test is met by the proposed scale. Within each hazard type, each successive level of the scale describes a progressively more severe hazard. Individual industries may devise more suitable definitions, but the idea is to start talking and thinking about hazards in terms of a 10-point scale.

TABLE 3.1   Category Descriptions for a 10-Point Scale for Workplace Hazards

1. "Technical violations;" OSHA standards may be violated, but no real occupational health or safety hazard exists.

2. No real fatality hazard

   Health hazards minor or unverified

   Even minor injuries unlikely

3. Fatality hazard not of real concern

   Health hazards have exceeded designated action levels

   or

   Sound exposure action levels exceeded (e.g., continuous exposure in the range 85–90 dBA)

   or

   Minor injury risks exist, but major injury hazard is very unlikely

4. Fatality hazard either remote or nonexistent

   Health hazards characterized by illnesses that are usually temporary; controls or personal protective equipment may not be required

   or

   Temporary hearing damage will result without controls or protection, and a few workers may incur partial permanent damage

   or

   Minor injuries likely, such as cuts and abrasions, but major injury risk is low

5. Fatality hazard either remote or not applicable

   Long-range health *may* be at risk; controls or personal protective equipment *advisable* or *required* by OSHA

   or

   Hearing damage may be permanent without controls or protection (e.g., continuous 8-hour exposure in the range of 95–100 dBA)

   Major injuries such as amputation not very likely

6. Fatality hazard unlikely

   Long-range health definitely at risk; controls or personal protective equipment *required* by OSHA

   or

   Hearing damage likely to be permanent without controls or protection (e.g., continuous 8-hour exposure in the range 100–105 dBA)

   or

   Major injury such as amputation not very likely, but definitely *could* occur

7. Fatality not very likely, but still a consideration

   or

   Serious long-range health hazards are proven; controls or personal protective equipment essential to prevent *serious* occupational illness

   or

   Hearing damage obviously would be *severe* and permanent without protection (e.g., continuous 8-hour exposure in excess of 105 dBA)

   or

   Major injury such as amputation could easily occur

8. Fatality possible; this operation has never produced a fatality, but a fatality easily could occur at any time

   or

   Severe long-range health hazards are *obvious*; controls or personal protective equipment essential to prevent *fatal* occupational illness

   or

   Major injury is likely; amputations or other major injuries *already* have occurred in this operation in the past

TABLE 3.1   *(Continued)*

9. Fatality likely; similar conditions have produced fatalities in the past; conditions too risky for normal operation; rescue operations are undertaken for injured workers with rescuers using personal protective equipment
10. Fatality imminent; risks are grave; some employees earlier in the day have died or are dying; conditions are too risky even for daring rescue operations except perhaps with exotic rescue protection

One criterion omitted from the scale definitions is cost of compliance, or cost of correcting a given hazard. Cost is a completely different criterion and is almost independent of level of hazard. That is, it can easily cost just as much or more to correct a Category 2 hazard as it does to correct a Category 9 hazard. Cost is an important criterion in the decision-making algorithm, but is omitted from the scale definitions to permit a clear ranking of hazard priority first. Once the hazards have been sorted out, costs of hazard correction can be estimated and capital allocated for such correction according to a rational capital investment policy.

Another missing criterion, perhaps conspicuously so, is any mention of OSHA legal definitions of *imminent danger, serious violation*, and so on. To place these legal designations into fixed positions in the hazard scale would undermine the objectivity of the classification. Many persons have a preconceived notion of what legal penalty should or should not be imposed for a given hazardous situation. This bias is found in OSHA officers as well as their industrial counterparts. For instance, if a plant safety and health manager happens to believe that a given situation is not serious enough to warrant a plant shutdown, he or she would tend to prohibit selection of any designation that would bear the legal OSHA designation "imminent danger." This is despite the fact that reason would place the Category 10 definition into the imminent danger category.

Three credible profiles of the OSHA legal classification superimposed on the proposed 10-point scale are displayed in Figure 3.8. Profile A has an industry flavor and represents the viewpoints of some business executives. Profile A cannot be considered an extreme position because thousands of American businesses are headed by persons who believe that no government official should have the right to step in and close their businesses with or without a court order, regardless of hazards. Therefore, some business executives do not recognize the legal category "imminent danger." Profile A does at least recognize the imminent danger category, although the profile is

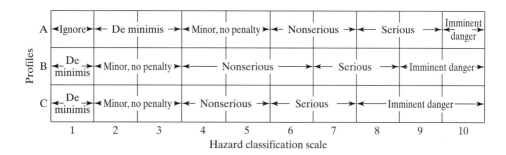

FIGURE 3.8

Three profiles of legal designations for hazards superimposed on the 10-point hazard classification scale.

skewed to the right. Profile C shows a contrasting position, being skewed to the left. Some OSHA officers have demonstrated that their positions are very close to Profile C. Profile C likewise is not the ultimate in the left extreme because some people believe that any fatality or amputation hazard should be classified as serious regardless of the remoteness of the hazard. Profile B represents a middle-of-the-road position.

A final consideration in the classification of hazards would be the industry environment in which the hazard is portrayed. What seems hazardous to a structural steelworker working 100 feet above the ground on a narrow beam is in an entirely different category from what seems hazardous to an accountant. A similar comparison could be drawn between a coal miner and a computer analyst. It is only hoped that as a first step at least, coal miners should know what other coal miners mean when they talk about Category 4 hazards. Similarly, accountants should be uniform in their conception of Category 9 hazards, for example.

Except for some very broad categories, such as "serious" and "nonserious," federal law has little to say about degree of hazard. No consistent criterion is available to safety and health managers, who must decide which problems must be tackled first. The 10-point scale recommended by this book for ranking safety and health hazards offers an opportunity to bring some order to this perplexing problem. The scale is a vehicle for encouraging all parties to focus on the degree of each hazard so that rational decisions can be made to correct those problems that are indeed significant.

It is more useful to rank hazards if some weight is placed on the likelihood of occurrence of the accident or loss incident. A fatality hazard is of course severe in terms of results, but if the likelihood of occurrence is extremely remote—as in air transport, for example—one cannot say that the hazard itself is severe. Risk analysis is the study that deals with this problem, and the U.S. Air Force has devised a "Risk-Assessment Code (RAC)" (ref. AFI 91-202). The RAC system considers four levels of "severity" and four levels of "mishap probability," as shown in Table 3.2.

TABLE 3.2   Risk-Assessment Codes

| | | Mishap | Probability | |
		A	B	C	D
Severity					
	I	1	1	2	3
	II	1	2	3	4
	III	2	3	4	5
	IV	3	4	5	5

Mishap severity:

1. Death or permanent total disability, resource loss or fire damage more than $1,000,000.

2. Permanent partial disability, temporary total disability in excess of three months, resource loss or fire damage $200,000 or more, but less than $1,000,000.

3. Lost workday mishap, resource loss or fire damage $10,000 or more, but less than $200,000.

4. First aid or minor medical treatment, resource loss or fire damage less than $10,000, or a violation of a requirement in a standard.

TABLE 3.2   (*Continued*)

Mishap probability

(a) Likely to occur immediately or within a short period of time

(b) Probably will occur in time

(c) Possibly will occur in time

(d) Unlikely to occur

RAC Designations

1. "Imminent danger"
2. "Serious"
3. "Moderate"
4. "Minor"
5. "Negligible"

From Table 3.2, it can be seen that the Air Force RAC system results in a scale of 1 to 5 after considering both severity and mishap probability. The scale may be rather arbitrary, but it does make sense. Also, the RAC codes do create some order out of both severity and risk of occurrence in a single code. Case Study 3.10 will demonstrate the generation of a risk-assessment code for a given case of severity and mishap probability.

CASE STUDY 3.10

RISK-ASSESSMENT CODE

A defective condition has been noted in the instrument panel of a military aircraft. The panel often falsely indicates a fault in the oxygen system when in fact no fault exists. Pilots have sometimes ignored this fault indication because they have believed it to be a fault in the instrument itself, not a genuine fault in the oxygen system. The result, although highly unlikely, could be that a real oxygen hazard could go undetected, resulting in possible loss of the aircraft in flight and loss of life. The resulting assessment of the severity of this hazard is Category I and the assessment of mishap probability is Category D. From Table 3.2, the appropriate RAC to assign to this risk is Code 3.

A British standard titled "Standard Code of Practise for Safety of Machinery" (BS5304:1988) sets up a classification system that assigns points for three hazard criteria—severity, potential for injury, and frequency of access—as follows:

Severity

Fatal/LTD	Loss of life, or long-term disability requiring hospitalization or treatment	6 points
Major	Permanent disability, loss of limb or sight, etc.	4 points

| Serious | Loss of consciousness, burns, laceration, broken bones; anything requiring hospital treatment | 3 points |
| Minor | Bruise, small cuts, light abrasion; anything which may require no more than local medical assistance | 1 point |

Potential for injury

Certain	6 points
Probable	4 points
Possible	2 points
Unlikely	1 point

Frequency of access

Frequent	Many times per day	4 points
Occasional	Once or twice a day	2 points
Seldom	Weekly or less	1 point

The procedure is to add the total point score of the three categories to arrive at an overall hazard level or score to be used in decision making to alleviate the hazard (ref. standard) The procedure is demonstrated in Case Study 3.11.

CASE STUDY 3.11

BRITISH STANDARD HAZARD CLASSIFICATION

A punch press is manually operated in a high-production operation at a standard production rate of 720 cycles per hour. In this setup, the operator feeds the press every cycle. If ever the die closes with the operator's hand in the danger zone, an amputation is virtually a certainty. Obviously, every time the operator feeds the press, the hazard is present, but effective engineering controls using two-hand actuating devices makes the possibility of injury each cycle very remote. The problem is to examine severity, potential for injury, and frequency of access to obtain an overall assessment of the hazard level. By referring to the category definitions for each, the following assessment is made: The severity is "Major" (Code 4), because the hazard is an amputation; the potential for injury is very low because of the effective engineering control, "Unlikely" (Code 1); the frequency of access is thousands of cycles every day, so it is assigned a Code 4, "Frequent." The overall assessment is obtained by adding the three codes together to result in an overall assessment of $4 + 1 + 4 = 9$.

The concept of hazard classification can be carried a step further and applied to national decisions to invest billions in risk abatement. There is growing concern for the need for a risk-assessment system that recognizes some level of risk associated with various national priorities for risk abatement. John C. Nemeth (ref. Nemeth) writes, "I am convinced that well-founded, consistent risk assessment is the only way to proceed rationally. We need to get down to business and standardize." Jeremy Main (ref. Main) raises the question of whether our nation is allocating our hazardous cleanup expenditures in a rational way. Main compares $8 billion spent per year dealing with asbestos hazards, which are thought to cause from zero to eight cancer deaths per year, while $0.1 billion spent per year in dealing with radon hazards, which are believed to cause as many as 20,000 cancer deaths per year. Perhaps the nation's policymakers should use some sort of hazard classification scale to determine where the most money should be spent, rather than yielding to whatever thrust is politically popular at the time.

SUMMARY

The sophisticated safety and health manager is not content with one approach to dealing with workplace hazards. There is too much uncertainty to solve the enormous problems neatly with a simple approach, such as "awards for no lost-time accidents" or "fines for anyone who breaks the rules." These two approaches, together with all the others, have their place, but an integrated program using the strengths of all approaches has the greatest potential for success.

While becoming more sophisticated, safety and health managers may fall into the trap of becoming more enamored with their impressive-looking analyses, scientific formulas, and statistics. Some of the best analyses may be subjective, not quantitative. The 10-point hazard classification scale suggested in this chapter represents an opportunity for safety and health managers to talk and think about workplace hazards in terms of *degree*. Corporate top management has been looking for years for this breed of safety and health manager to emerge, who can discern between the really significant problems, the ordinary problems, and the trivial ones.

EXERCISES AND STUDY QUESTIONS

3.1 Name the four categories of seriousness of hazards as recognized by OSHA.

3.2 What is the OSHA designation for minor standards violations that bear little or no relationship to safety and health?

3.3 Consider the list of hazards that follows, and rank each on a scale of 1 to 10 (10 being the worst). Also, classify each into the four OSHA categories according to your opinion.

(a) Ground plug (third prong) is cut off on a power cord for an office computer.

(b) Ground plug is cut off on a power cord for a shop wet-vac vacuum cleaner.

(c) An electric drill with faulty wiring causes an employee to receive a severe shock, after which he refuses to use it. Another employee scoffs at the hazard, claims that he is "too tough for 110 volts," and picks up the tool to continue the job.

3.4 A well-known safety rule is not to pull an electrical plug out of a wall socket by means of the cord. Have you broken this rule? Do you regularly do so? What is the reason for this rule? Do you think the rule is an effective one?

3.5 Do you believe that safety or health rules are "made to be broken"? Why or why not?

3.6 Name four basic approaches to hazard avoidance.

3.7 Recall your first supervisor in your first full-time job. Did he or she even mention safety or health on the job? Was your supervisor's influence on your safety habits positive, negative, or neutral?

3.8 What is the Heinrich ratio?

3.9 What are the three lines of defense against health hazards?

3.10 What is the standard safety factor for crane hoists? for scaffolds?

3.11 Name three general fail-safe principles. Can you think of a real-world example of any such principles?

3.12 What does Murphy's law have to do with occupational safety and health?

3.13 What is the purpose of FMEA?

3.14 Construct a fault-tree diagram describing potential causes of fire in paint spray areas. Compare your analysis with others in the class.

3.15 Construct a fault-tree diagram that describes the ways in which a pair of dice can be thrown to "roll seven." Calculate the "risk" or chances of rolling seven.

3.16 Explain the difference between toxicology and epidemiology.

3.17 Alter the fault-tree diagram of Figure 3.2 to consider the possibility that the portable electric drill might be double insulated (i.e., sheathed in an approved plastic case to prevent electrical contact with the metal tool case).

3.18 In the universal loss incident causation model, what is the difference between proximal factors and distal factors? To which category does management policy belong?

3.19 Explain the concept of point of irreversibility. Does this point guarantee that a personal injury will occur? What are the roles of aggravating and mitigating factors?

3.20 Accident Causes A, B, and C each has a probability of occurrence of about 1 in 1000, but the causes are mutually exclusive. Suppose that Cause B does in fact occur in a given situation. What are the chances that Cause A will occur in this situation?

3.21 Using a pair of dice for a simulation device, suppose that an outcome of 11 represents an industrial accident.

  (a) Draw a fault-tree diagram to illustrate the ways in which this accident could happen.

  (b) Calculate the probability that this accident will happen in a given throw of the dice.

  (c) Are the causes of this accident mutually exclusive?

3.22 The following are some causes of slips and falls:

  (a) oil leaks from forklift trucks

  (b) water or wax on floor during cleaning operations

  (c) ice on walkways or loading docks during winter weather

  (d) slippery heels on footwear

From the perspective of a total plant safety program, are these causes mutually exclusive? Why or why not? For a given single accident, are these causes mutually exclusive? Why or why not?

3.23 The following are three among many causes of injury to an operator of a punch press:

  (a) barrier guard of adequate size but placed too high—worker can reach under the guard.

  (b) barrier guard of adequate size but placed too low—worker can reach over the guard.

  (c) openings in the barrier guard are too large—worker can reach through the guard.

Which of these causes are mutually exclusive?

3.24 A certain type of injury has tangible costs of $15,000 per occurrence and intangible costs estimated to be $250,000 per occurrence. Injury frequency is 0.01 per year, but would be reduced by half with the installation of a new engineering control system. What annual benefit would the new system provide?

3.25 A certain ventilation system would cost a firm approximately $60,000, which can be amortized over its useful life at a cost of $15,000 per year. Annual maintenance costs are expected to be $600 per year and monthly operating costs (utilities) $150 per month. The system is expected to facilitate production by reducing the amount of machine cleaning for an expected annual savings of $1200 per year. The primary benefit of the proposed ventilation system is expected to be the elimination of the need for respirators, which cost the company $4000 per year in equipment, maintenance, employee training, and respiratory system management. The ventilation system is expected to reduce short-term-illness complaints by an average of 6 per year and long-term illnesses by an average of 0.2 per year. Short-term illnesses result in a total cost of $600 per occurrence, including intangibles. Long-term illnesses are expected to result in a total cost of $30,000 per occurrence, including intangibles. Use a cost–benefit analysis to determine whether the ventilation system should be installed. What is the primary benefit of the ventilation system?

3.26 On a scale of 1 to 10 (10 being the worst), how would you rate each of the following hazards?

(a) A 10-foot balcony does not have a guardrail. Workers regularly work close to the edge every day without fall protection.

(b) Same as part (a), except that the working surface is outdoors and is very slippery in rainy weather.

(c) No guardrail on a 10-foot balcony accessed only twice per year by a maintenance worker to service an air conditioner.

(d) An unguarded flat rooftop accessed only by air conditioner maintenance personnel. The closest necessary approach to the edge of the roof is 25 feet.

(e) Broken ladder rung (middle rung of a 12-foot ladder).

(f) Overfilled waste receptacles in lunchroom.

(g) Two-ton hoist rope with dangerously frayed and broken wires in several strands.

(h) Mushroomed head on a cold chisel.

3.27 What limits the region of sphere of control, and what factors belong to this region?

3.28 Under what circumstances is it incorrect to use the simple sum of causal event probabilities to calculate the probability of an event that results from either of two sufficient causal events?

3.29 Event A has probability of occurrence p_a, Event B has probability of occurrence p_b, and A and B are independent. Either A or B is sufficient cause for loss incident Event C to occur. Write a formula to calculate the probability of occurrence of loss incident Event C.

3.30 Event A has probability of occurrence 0.3, Event B has probability of occurrence 0.2, and A and B are independent. Either A or B is sufficient cause for loss incident Event C to occur. Calculate the probability of occurrence of loss incident Event C.

3.31 Study the OSHA standards to find examples of the application of each of the three fail-safe principles.

3.32 What engineering concept appears to have been misapplied in the Kansas City hotel sky-walk disaster?

3.33 What is a "deadman control?" Describe an example not given in this book.

3.34 Describe an example of the use of "redundancy" in engineering design not given in this book.

3.35 Defensive driving is an example application of which of the three fail-safe principles?

3.36 Describe how FMEA can be a benefit to a preventive maintenance program.

3.37 It can be said that almost any substance is poison to humans. Explain how this can be true by citing an example of a seemingly harmless substance.

3.38 Explain the term "pharmacokinetics" and how it applies to occupational safety and health.

3.39 In what way is the field of epidemiology useful to occupational safety and health?

3.40 What persons or institutions perform toxicology and epidemiology studies and why? Would safety and health managers usually be expected to perform such studies?

3.41 Which do you think is a more serious hazard, radon or asbestos? Why?

RESEARCH EXERCISES

3.42 From your own experience, from library research, or from interviews with others, construct an actual case history of a fatal accident that was blamed on carelessness, but could have been prevented by a better engineering design.

3.43 Select a real hazard and gather information on the possible causes that could lead to an accident relevant to this hazard. Construct a fault-tree diagram to relate the causes to the loss event.

3.44 Search the Internet for details on the Kansas City hotel skywalk disaster.

3.45 Consider the following accident case history:

On July 4, 1980, three workers, ages 14, 16, and 17, were installing a sign at a bait shop along a state highway. They were using a truck-mounted extensible metal ladder to unload and position an upright steel support for the sign. Two of the workers were holding and guiding the steel support while the third stood on the flatbed truck operating the extensible-ladder controls. The metal ladder came into contact with a 13,200-volt overhead power line. The two workers guiding the steel support were standing on the ground and were subjected to immediate electrocution. The third worker attempted to break contact with the power line by operating the controls, but the controls had become useless, probably because the control wiring had become burned by the high-voltage contact. The worker then leaped from the truck and ran around to the front to try to drive the truck away to break contact. As he grasped for the door handle to the truck cab, he was still standing on the ground, which provided a path for the current through his body. The utility company had equipped the line with a "recloser" that normally would have broken the circuit under these conditions, but for a variety of reasons it failed to do so in this case. Therefore, the power remained on for a rather lengthy period. The high voltage and burning current finally broke down the dielectric strength of the rubber tires and they blew out. This shifted the truck's position, and contact with the power line was broken, but not before one worker had been burned in half and another's legs were burned off; all three workers died. How could future accidents of this type be prevented? Compare the four basic approaches to hazard avoidance in preventing accidents of this type.

STANDARDS RESEARCH QUESTIONS

3.46 Use the OSHA website or the NCM database to search for OSHA General Industry standards that relate to training. Determine the percentage of citations that are classified as "serious."

3.47 Search the OSHA standards for the use of the word *engineering* in any of the General Industry standards.

3.48 From what you have learned in this chapter, compose a list of five words that you associate with the engineering approach to safety and health hazards. Do a search of the OSHA standards to determine whether any of these words appear in the General Industry standards.

CHAPTER 4

Impact of Federal Regulation

The field of industrial safety and health took a giant leap in the early 1970s. There is some question, however, whether the leap was forward or backward. On December 29, 1970, Congress passed the Williams–Steiger Occupational Safety and Health Act of 1970, creating OSHA, the Occupational Safety and Health Administration of the U.S. Department of Labor. OSHA got off to a bad start and immediately became the target of sharp public criticism. But at the same time, the agency created an immediate focus of attention on the field of industrial safety and health. Changes in federal administrations have changed OSHA's methods, and these methods need to be examined here, along with the basic approaches on which OSHA was founded. Regardless of how OSHA may change, its impact on the field is a permanent one, and this book would not be complete without examining this impact.

Most people think of OSHA when the subject of federal safety and health regulations is mentioned. But there are also MSHA, the Mine Safety and Health Administration; TOSCA, the Toxic Substances Control Act; and CPSC, the Consumer Product Safety Commission. Most of these other agencies and legislation governing safety and health are patterned after the OSHA agency and legislation. Accordingly, this chapter examines the basic approach of OSHA and the impact it has had on the field of industrial safety and health.

STANDARDS

The most significant change that OSHA brought to industry was a book of federal standards. Never before had virtually all general industries been subjected to a book of federally prescribed, mandatory rules for worker safety and health. This set of rules formed the basis for inspection, citations, penalties, and virtually every activity in which OSHA was engaged. One rule, however, was not in the book; it is the most important rule of all and will be discussed first.

General Duty Clause

Congress decided to set up one general rule for all to follow, and this rule was included in its entirety in the text of the law that created OSHA. This rule, called the General Duty Clause, might be called the First Commandment of OSHA. It is quoted as follows:

Public Law 91–596

> Section 5(a) Each employer . . .
>
> (1) shall furnish to each of his employees employment and a place of employment which are free from recognized hazards that are causing or are likely to cause death or serious physical harm to his employees. . . .

The General Duty Clause is cited by OSHA whenever a serious safety or health violation is alleged for which no specific rule seems to apply. This has drawn some criticism from industry because after a serious accident has happened, it is easy to conclude that a situation was unsafe. However, before the accident occurred, it might not be clear what should have been done, especially in the absence of any specific rule to provide guidance. Sometimes a specific rule will also apply somewhat, and OSHA has cited both the General Duty Clause and the specific standard. When a situation is clearly in violation of a specific standard, OSHA usually does not bother to cite the General Duty Clause.

If the General Duty Clause is the First Commandment of OSHA, the second is like unto it:

Public Law 91–596

> Section 5(b) Each employee shall comply with occupational safety and health standards and all rules, regulations, and orders issued pursuant to this Act which are applicable to his own actions and conduct.

Note that Section 5(a)(1) describes a responsibility for employers, whereas Section 5(b) pertains to employees. Penalties are prescribed for employer violations, but no penalty is prescribed for employee violations. Section 5(a)(1) has been cited quite frequently, but as far as this author knows, there has never been an OSHA citation of Section 5(b).

Promulgation

In addition to the General Duty Clause, the law that created OSHA also set up the machinery for OSHA to issue new standards too technical and detailed to be included one by one in the text of the law passed by Congress. The law provides safeguards intended to ensure that OSHA will be fair and give all interested parties an opportunity to support or object to proposed new standards. In addition to promulgating new standards, OSHA is also permitted to revise old standards or even revoke them, following the same rulemaking procedure with prescribed safeguards.

National Consensus

The OSHA law recognized the existence of national consensus standards already in use prior to the enactment of the OSHA law. This is a very important part of the law because it gave OSHA the authority to bypass the procedural safeguards just mentioned and issue standards without consulting the public. The principle was that the standards by their prior existence had already been accepted by the public. OSHA's authority to issue national consensus standards expired in early 1973, two years after the effective date of the Act. Thus, after two years, OSHA was no longer permitted to "grandfather in" safety and health standards. The largest number of national consensus standards were developed by the two leading national standards-producing organizations, the American National Standards Institute (ANSI) and the National Fire Protection Association (NFPA).

Besides the national consensus standards, any established federal standard was also permitted to be adopted as a general standard by OSHA. The established federal standards had formerly applied only to limited groups such as construction industries or government contracts. With the OSHA law, Congress permitted OSHA to extend these standards to virtually all industries.

The biggest issue with the national consensus standards was whether they ever really represented the national consensus. Virtually none of these standards had ever been enforced, and the language used made it obvious that the original drafters of many of the standards had never intended for them to be mandatory rules to be enforced with monetary penalties for offending employers. This issue will be examined further in the section entitled "Public Uproar."

Structure

When discussing standards, this book will sometimes use the terms *horizontal standard* or *vertical standard*. The safety and health manager needs to understand these terms. Prior to OSHA, states enforced codes for safety and health by industry, issuing separate handbooks of rules for each industry to follow. These standards have come to be called vertical standards. OSHA's basic approach, by contrast, is to generalize and organize standards by hazard sources, regardless of industry. These are called horizontal standards. There are certain standards that OSHA limits in scope to a particular industry, but these are the exceptions; the basic structure of the standards is horizontal.

OSHA almost had to use the horizontal approach. The nature of the OSHA law itself was to authorize the new agency to generalize on specific standards. Imagine for a moment the morass of books that would have been generated if OSHA had attempted to write a separate standards book for each industry, sifting through the national consensus standards to determine which standards should be included in each. Unusual industry categories would inevitably "fall through the cracks" under a vertical scheme, characterizing themselves as being under the scope of neither this nor that standard.

By using the horizontal approach, OSHA left the job of sifting through the standards to the public. Thus, a general book of standards entitled *General Industry, Part 1910*, was published to cover virtually all industries. A manufacturer of teacups must

Figure 4.1 Venn diagram to represent the relationship of two different overlapping types of equipment.

Abrasive wheels and tools

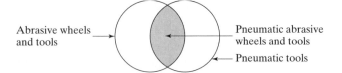

Pneumatic abrasive wheels and tools

Pneumatic tools

read out of the same book as does an airline, and both must determine whether they need to be concerned with a provision entitled "horse scaffolds."

The construction industry is one for which OSHA published a vertical standard entitled *Construction Standards, Part 1926*, but even these special categories of industries are also covered by the more general Part 1910 with respect to any hazard for which no standard exists in the narrower vertical standard. OSHA did make the job of compliance a little easier by subsequently publishing a larger construction standard that included those parts of the general industry standard that OSHA might cite at construction sites. However, it did not abdicate the right to turn to general industry standards, which were not included in even the larger version of the construction standard.

The concept of vertical versus horizontal standards extends down within the subdivisions and individual paragraphs of the OSHA standard. For instance, OSHA standard 1910.106 pertains to flammable and combustible liquids and is basically a horizontal standard to be applied to general industry. However, within the standard there is paragraph 1910.106(i), which is entitled "Refineries, chemical plants, and distilleries." Therefore, that portion of 1910.106 should be considered a vertical standard applying only to those industries named.

One problem in determining the scope of all standards and of vertical standards in particular is the infeasibility of an exclusive classification of industries and equipment. Consider the diagram in Figure 4.1 (called a Venn diagram), which relates two different classes of equipment. The circles represent the two classes, and the shaded area is the overlap between the two groups. The circles could also be considered safety standards, and thus equipment represented by the shaded area would be subject to both standards.

Another classification for standards divides them into *specification standards* and *performance standards*. The specification type is easier to enforce because it spells out in detail exactly what the employer must do and how to do it. The performance type has its advantages, however, in that it permits the employer latitude in devising innovative ways to eliminate or reduce hazards. In short, the specification standard emphasizes methods, whereas the performance standard emphasizes results. The comparison is best shown by an example.

Example 4.1

(a) Example of a *specification* standard:

OSHA standard 1910.110   Storage and handling of liquefied petroleum gases

(c) Cylinder systems

(5) Containers and equipment used inside buildings or structures
    (vi) Containers are permitted to be used in industrial occupancies for processing, research, or experimental purposes as follows:
        (a) ...
        (b) Containers connected to a manifold shall have a total water capacity not greater than 735 pounds (nominal 300 pounds LP-gas capacity) and not more than one such manifold may be located in the same room unless separated at least 20 feet from a similar unit.

(b) Example of a *performance* standard:

OSHA standard 1910.36   General requirements (under Subpart E—Means of Egress)
    (b) Fundamental requirements
        (2) Every building or structure shall be so constructed, arranged, equipped, maintained, and operated as to avoid danger to the lives and safety of its occupants from fire, smoke, fumes, or resulting panic during the period of time reasonably necessary for escape from the building or structure in case of fire or other emergency.

The standard in part (a) leaves no doubt in the reader's mind as to what must be done and specifies exactly the limitations of containers and manifolds within the room. There is no leeway for the employer to devise a better way—perhaps safer and cheaper—to minimize hazards. However, once the exact standards have been met, the employer can feel safe from citation.

In the performance standard stated in part (b), it can be seen that employers have all kinds of latitude to set up their buildings in ways to avoid "undue danger." If a new type of effective smoke alarm is developed and placed on the market, the employer has the choice of installing the new alarm or adopting a different strategy to enhance safety in the event of "fire or other emergency." But with the performance standard, a disagreement may develop between an enforcement inspector and the employer over the effectiveness of the methods selected by the employer to protect against "undue hazards."

Most standards do not exactly fit either type, but do rather roughly fit one or the other category. Employers tend to prefer performance standards, whereas enforcement officials generally prefer specification standards. However, this issue does not clearly divide the two groups. Sometimes employers prefer specification standards because it is easier to determine whether a given facility or equipment meets a specification standard. Also, a few enforcement officials prefer the performance standard because if an accident does occur, it tends to strengthen the case for a citation regardless of the specific steps the employer may have taken to prevent the hazard.

NIOSH

The National Institute for Occupational Safety and Health (NIOSH) was established by the OSHA law to carry on research and training. NIOSH is mentioned here because of its important role in recommending new standards to OSHA. OSHA has the sole authority to promulgate new standards, but NIOSH develops the criteria for new standards and conducts the research necessary to justify the need for new standards.

NIOSH is often thought of as being concerned primarily with health and toxic materials, but it is important to remember that NIOSH also has responsibility for safety-standards research and development.

Although the OSHA law infused new significance into the agency, NIOSH actually predated OSHA in existence. As early as 1914, NIOSH was part of the Department of Industrial Hygiene and Sanitation in Pennsylvania. In 1937, it became the Division of Industrial Hygiene as a part of the National Institutes of Health. Note the exclusion of the word safety in its title during this period. This orientation persisted for decades as the agency was made a part of the Bureau of State Services in 1944 and of the Bureau of Occupational Health in 1968. Two years later, the OSHA law elevated the agency to include safety and made it a part of the Department of Health, Education, and Welfare (HEW), which was subsequently changed to the Department of Health and Human Services (HHS).

ENFORCEMENT

Occupational safety and health ought to be topics of vital interest and importance regardless of OSHA, but most of us would not have realized this if OSHA had not been given the authority in the 1970s to inspect industries and issue citations with monetary penalties.

Inspections

The original OSHA law gave the OSHA official the right to enter a factory or other workplace without delay (at reasonable times) upon presenting credentials. Such credentials consisted of basic identification of the officer, but no court-issued search warrant. The right of a government official to do this was later challenged by an Idaho businessman, and the U.S. Supreme Court ruled in his favor in the famous Barlow decision in 1978. Employers may now invoke the Fourth Amendment to the U.S. Constitution and require OSHA to obtain a search warrant to conduct an inspection.

Some company managers take the position that no time is a "reasonable time" to be visited by an OSHA inspector. They argue that their plant has such proprietary processes that a visit by an inspector would jeopardize their trade secrets. Congress anticipated this excuse and provided that any information gathered that might reveal a trade secret be kept confidential. Actually, this provision was essential to prevent OSHA from conflicting with existing laws to protect trade secrets.

OSHA inspections are generated according to the following priorities:

1. Imminent danger
2. Fatalities and major accidents
3. Employee complaints
4. High-hazard industries

Imminent danger would be a situation in which death or serious physical harm could be expected to occur immediately. Time is of the essence in these situations, and ordinary enforcement procedures may be too late to protect workers. OSHA may go to a U.S. district court and get a temporary injunction to remove employees from the work area in an imminent-danger situation. Such an action would be rare and must be reasonable in allowing some employees to remain to correct the condition and to permit a safe and orderly shutdown. If the process is continuous, a sudden shutdown might cripple the equipment and might even be unsafe. OSHA realizes this, and in the case of continuous process operations, the process does not have to be completely closed down; a pilot crew is permitted to remain and maintain the capacity of the process to resume normal operations.

OSHA requires a telephone call or other notification within eight hours on occurrence of fatal accidents or accidents in which three or more persons are hospitalized. This notification will invariably trigger an OSHA inspection because this category of inspections is second in priority only to imminent-danger inspections. In fact, at first, fatalities were considered top priority; later, it was decided that it would be better for OSHA to assign top priority to serious, perhaps fatal, hazards before the accident occurs rather than after the fact. OSHA has a policy of investigating both categories, however, within 24 hours of notification.

An employee can request that OSHA investigate a hazard by filing with OSHA a complaint describing the hazard believed to exist in the workplace. To be valid, the complaint must be signed by the employee. Many workers fear recrimination or other reprisal if they are subsequently identified by their signatures. But OSHA is bound by law to keep the origins of an employee complaint confidential if the employee so requests. However, the employee has additional safeguards against discrimination, as will be seen later in this chapter.

Next in priority to complaints come the inspections of industries that are shown by statistical records to be particularly hazardous. Early in the implementation of OSHA's enforcement program, a collection of industries were named target industries. Subsequently, a target health standards group was named. Focus on the Target Industries Program (TIP) shifted later to a special-emphasis program, which named trenching and excavation cave-ins as the principal concern. Next came the National Emphasis Program (NEP), emphasizing foundries and metal stamping.

An enduring emphasis since the inception of OSHA has been the construction industry, generally recognized as hazardous, especially from a safety standpoint. Other industries are identified as hazardous on the basis of statistical evidence that the national average LWDI for the four-digit SIC classification for the given industry is higher than the national average LWDI for all industries. The LWDI is a moving target, generally improving by moving lower every year.

Citations

After the inspection, OSHA may issue a citation for alleged violations of specific standards or of the General Duty Clause. The statutory limit is six months, so if no citation has been received within six months, the employer can be assured that a citation will not be forthcoming. A growing percentage of firms do not receive a citation upon inspection. OSHA has on occasion pointed to this percentage and designated these firms as "in compliance." The designation may be a misnomer, however, because the depth of inspections varies considerably. A really thorough wall-to-wall inspection of a comprehensive manufacturing plant of any size except the very smallest is almost certain to reveal some OSHA violations. The OSHA compliance officer may be reluctant to issue a citation and may overlook some minor items if it is felt that the employer has shown a good-faith effort to comply with the standards.

OSHA recognizes that some conditions, even though they represent a violation of the letter of the law, have no direct or immediate relationship to safety or health. In these instances, OSHA may issue a de minimis notice in lieu of a citation. De minimis notices do not carry a monetary penalty.

If a citation is issued, it is not a very private matter. To the chagrin of the safety and health manager, the citation must be prominently posted near where the violation occurred. Employees and management alike have the opportunity to read the citation and note the alleged violation. This increases the overall awareness of OSHA's enforcement powers and may spawn complaints by employees for other potential violations. As far as management is concerned, the safety and health manager can prepare top management by explaining that many companies do receive a citation when inspected by OSHA; there is no disgrace to being found out of compliance with a few detailed provisions of the standards.

Table 4.1 arrays the statutory maximum penalty structure for OSHA violations. Note that many of the penalties curiously begin with the number "7." There is some history behind this somewhat peculiar penalty structure. Congress originally set penalty maximums in round numbers such as $1000 (for a nonserious violation) and $10,000 (for willful or repeat violations). However, the Omnibus Budget Reconciliation Act of 1990 authorized a sevenfold increase in the amount of OSHA penalties (ref. Foulke). The provision that sets penalties for "willful violation resulting in death" was omitted from the revision which increased penalty levels sevenfold. This omission resulted in the odd situation that statutory maximum penalties for willful violations resulting in death are actually lower ($10,000 and $20,000 for first and second offenses, respectively) than are the statutory maximum penalties for willful violations *not* resulting in death ($70,000).

OSHA penalties can be quite severe, as Table 4.1 indicates, but actual fines are usually quite inconsequential. Penalties for nonserious violations are usually less than $100. OSHA has a formula for computing a reduced penalty, taking into consideration

TABLE 4.1   OSHA Penalties (Statutory Maximums)

Offense	Maximum penalty	Offender
Nonserious violations	$7000	Employer
Serious violations	$7000	Employer
Willful violations (min. $5000)	$70,000	Employer
Repeat violations	$70,000	Employer
Failure to abate a violation	$7000 *per day*	Employer
Willful violation resulting in death		
(min. $5000) to employee	$10,000 and/or 6 months in prison	Employer
Second offense: willful violation resulting in death	$20,000 and/or 1 year in prison	Employer
Failure to post notices, citations, etc.	$7000	Employer
Giving advance notice of an inspection	$7000 and/or 6 months in prison	Anyone
Falsifying records or reports	$10,000 and/or 6 months in prison	Anyone
Killing an OSHA inspector	Life imprisonment	Anyone

the size of the company (small businesses tend to be assessed smaller fines), the history of previous violations, and "good faith" shown by the employer. By the time a given citation is issued, the formula for computing a possible reduction in penalty will have already been applied.

Although most OSHA penalties are small, it is possible for fines to become quite severe. Note that the failure-to-abate penalty is assessed for every day a violation remains uncorrected. Some willful and repeat violations have resulted in fines of hundreds of thousands of dollars assessed on a single firm, although these cases are rare.

In addition to the statutory penalty categories listed in Table 4.1, an additional category, egregious violation, was established administratively by OSHA in the 1980s. Worse than a willful violation, an egregious violation is a glaring or flagrant violation, which may invoke even higher penalties. Citation of egregious violations requires clearance from OSHA's national headquarters in Washington, DC.

The following questions are frequently asked in seminars about OSHA:

1. Who in the organization goes to prison when the "employer" is found to be guilty of a willful violation resulting in death?
2. Where does the money collected in OSHA fines go—to OSHA's budget to pay inspectors?

In answer to the first question, "employer in a complex organization" can be construed to represent the entire chain of supervision, from the supervisor of the employee victim to the chief executive officer of the firm. However, what few cases have been prosecuted indicate that OSHA will usually attempt to zero in on only one guilty person to go to prison. The rationale is that OSHA attempts to prosecute the manager with the highest authority who knew about the violation and had the authority to correct it, but failed to do so, resulting in the employee's death. A notable exception will be examined in Chapter 7 in a case in which three members of top management in a single company, including the owner, were indicted for criminal violations.

The second question, "Where does the money go?" apparently has its origins in some state and local enforcement schemes. It would hardly seem reasonable to deposit fine money into the coffers of those agency officials who assess the fines. But the misconception that OSHA is permitted to keep the money from fines collected seems to persist in the minds of the public. The origins of this idea appear to be the tradition of state boiler inspection agency regulations, which provide for money collected to be deposited into the boiler inspection division accounts. OSHA does not use this strategy, however, and all fines are deposited directly into the U.S. Treasury and not earmarked for OSHA's use. OSHA fines would be insufficient to operate OSHA anyway, since the total amount collected annually is generally very low compared to OSHA's annual budget. It must be admitted that the 1990 budgetary move by Congress to increase OSHA's penalty structure sevenfold most certainly had the effect of adding credence to the opinion that OSHA fines were expected to fund OSHA's budget.

A big mistake made by some managers is to pay OSHA fines and then consider the matter closed. Such a strategy ignores the far more important aspect of the citation—the prescribed abatement period. The OSHA fine itself may be inconsequential; the big impact of the OSHA citation, if it is accepted, is that each item listed in the citation must be corrected, regardless of cost. The cost of correcting violations is usually much greater than the amount of the OSHA penalty.

Indeed, a recent estimate of OSHA's cost to business exceeds $33 billion per year (ref. Mukherjee). It should be acknowledged that such an estimate does not take into consideration the benefit to the company from decreased direct and indirect costs of injuries and illnesses prevented.

Before accepting a citation, the safety and health manager should stop and think about whether it is really feasible to correct the violation. There is only a 15-day period (working days) after receipt of the citation for a decision to be made on whether to contest the citation. Appeals can even be taken through the judicial processes all the way to the Supreme Court.

Appeals of citations are not to be confused with variances. Appeals are for employers who have already been cited. Variances are for employers who need time to comply or have an alternative to compliance that is more practical and still protects employees.

Prescribed abatement periods can be seen to be very important. It is easy to overlook abatement periods and concentrate on the nature of the violations and the proposed penalties. The abatement period can seem to pass quickly, and then the firm is subject to a possible reinspection and potentially more severe penalties. OSHA is known to be very reasonable in extending abatement periods if the employer contacts OSHA and gives reasons for a request to extend. But if the employer takes no action and the OSHA compliance officer returns, a severe penalty is likely. It is easy for OSHA to be too optimistic about timetables for changes in facilities, allowing insufficient time for administrative approvals, delayed equipment deliveries, installation schedules, and, of course, Murphy's law. It is up to the safety and health manager to bring up the subject of abatement and to ensure that OSHA sets reasonable periods.

Employee Discrimination

One penalty that is sometimes costly does not appear in Table 4.1. This is the penalty paid by the employer who is found to have discriminated against an employee for filing an OSHA complaint, answering the OSHA compliance officer's questions during an inspection, or exercising any other right afforded employees under the OSHA law. Company management should be very careful to document reasons for discharging any employee, especially if that employee has a history of complaining about safety and health hazards. This type of complaining is no basis for dismissal, and the company that dismisses such an employee may find itself later reinstating the employee with back pay. No fine must be paid to the government for such an infraction, but if the case has lain idle for many months, the cost of reinstatement with back pay can be substantial.

Besides dismissal of the employee, OSHA recognizes other, more subtle means of discrimination. Any of the following employer actions is considered by OSHA to be illegal discrimination if it is used as punishment for the employee's exercise of rights under OSHA:

- termination
- demotion

- assignment to an undesirable job or shift
- denial of promotion
- threats or harassment
- blacklisting the employee with other employers

OSHA has even told employees that the employer may be in violation of the law for suddenly punishing an employee for doing something else wrong after the employee has protested a hazardous condition. It would be especially incriminating to single out a complaining employee for punishment for some unrelated action that goes unpunished for other employees. As can be seen, the whole matter can be extremely delicate, and the safety and health manager should ensure that all persons in management, from first-line supervisors on up, are aware and careful not to discriminate either directly or indirectly against any employee for complaining about safety and health violations, either during the worker's employment with the company or in the future if the worker seeks employment elsewhere.

One questionable worker "right" is whether an employee can walk off the job because of unsafe or unhealthy conditions and still expect to be paid. The courts have ruled against employees who sue to be paid by their employer for not working when they have walked off the job due to unsafe or unhealthful conditions.

PUBLIC UPROAR

OSHA survived a very stormy first decade. Despite the validity of its purposes, OSHA quickly grew to be one of the most hated agencies the federal government had ever created. At times, its demise seemed imminent, but it continued to survive.

At the root of public criticism of OSHA lie the OSHA standards. Much has been said about "nitpicking inspectors," "unjustified fines," and "gestapo techniques," but these criticisms would probably never have arisen if the standards themselves had been set up differently.

The original standards had a few obsolete provisions that have subsequently been deleted. Another problem with the standards was that advisory provisions containing language such as "should" were incorporated as mandatory rules with language such as "shall." The courts frowned on this, and so did the public. The government has since worked hard to eliminate advisory provisions from the standards. Federal standards have also been criticized for their level of detail, vagueness, redundancy, and irrelevance.

Some federal standards have seemed to do more for industries manufacturing safety equipment than they have done to protect the worker. A good example is the original standards for fire extinguishers. OSHA now permits alternatives to fire extinguishers for many applications.

A favorite weapon of OSHA's critics is the old question "Has it done any good?" They usually feel comfortable with the question without any examination of the record because they believe that OSHA could not have any measurable beneficial effect. It is difficult to assess the impact of federal regulation on worker safety and health because even the statistical recordkeeping has changed somewhat since OSHA was created. It is not even agreed on whether the changes in recordkeeping have made injury and illness rates look better or worse. Some believe that the presence of OSHA enforcement with OSHA compliance officers examining injury and illness records has tempted management to hide injuries and illnesses, making the overall record of injuries and illnesses under OSHA look better than it really is. Others believe that because injuries and illnesses requiring medical treatment are now required by law to be recorded, statistical summaries will show more injuries and illnesses and thus make OSHA look worse. There is no question, however, that, in some areas, OSHA has had a positive impact on worker safety and health. Dramatic results have been claimed in the reduction of fatalities from trench and excavation cave-ins. The overall picture regarding OSHA's effect on fatalities is not impressive, as Figure 4.2 reveals. Although the overall trend in fatalities in the workplace has shown an impressive downward trend throughout the 1900s, no improvement can be seen in this downward trend as a result of OSHA. Griffin (ref. Griffin) studied the statistical significance of OSHA's effect on fatalities and found no statistically significant effect. In fact, he found the trend to follow a straight line very closely with a statistical correlation coefficient (r^2) of 0.99 when 38 years of data were analyzed (from 1954 to 1991). Figure 4.2 shows this very strong linear relationship, with no apparent changes in the trend anywhere near the beginning of OSHA enforcement in 1971 or in any year thereafter.

The concept of cost effectiveness continues to gain in importance. Thus, the old question "Does a federal regulation do any good?" is changing to "Does the regulation do enough good to warrant the cost of compliance?" The cost of compliance is a much greater consideration than is the cost of monetary penalties assessed by regulatory agencies.

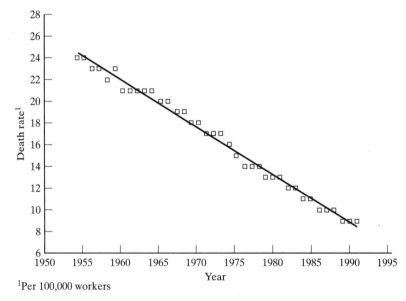

[1]Per 100,000 workers

Figure 4.2   The workplace fatality rate follows a very linear downward trend, with no apparent change in this trend associated with the arrival of the OSHA era (*Source:* Ref. Griffin).

ROLE OF THE STATES

Prior to OSHA, occupational safety and health were generally considered to be concerns of the states, not of the federal government. The general feeling among supporters of OSHA was that the states historically had not been doing a sufficient job in establishing and enforcing adequate standards for occupational safety and health. Recognizing, however, that some states might develop effective occupational safety and health standards and enforcement programs, the OSHA law provides for state plans to be submitted to OSHA for approval.

Enforcement

It must first be said about state enforcement that OSHA itself has been given no authority to regulate state agencies, counties, or municipalities. Even federal agencies are exempt from regular OSHA enforcement—a sore point with some private employers—but the OSHA law does provide for federal agency coverage as a part of the responsibilities of the heads of the various federal agencies. Obviously, it would be impractical for the federal government to cite and penalize itself, so OSHA does make inspections but does not issue citations to its sister agencies. As far as states are concerned, if the federal government were to issue citations to, and assess penalties on, state and local governments, there would be sovereignty problems, so that is prohibited.

But if a state submits a plan for occupational safety and health standards and enforcement, to be approved by OSHA the plan must contain a program applicable to employees of state agencies and political subdivisions of the state. In addition, the state standards and enforcement must be at least as effective as the corresponding federal standards and enforcement. About half of the states have achieved "as effective as" status and have had their state plans approved. (See Appendix G.) Three such states (Connecticut, New Jersey, and New York) formerly had comprehensive state plans, but withdrew from enforcement of the private sector while retaining their authority to enforce standards for state agencies. These states now have federal enforcement of the private sector within their borders. Safety and health managers operating in any of the state-plan states should consult with appropriate state officials for copies of standards and enforcement procedures. In most state-plan states, the standards are virtually identical to the federal OSHA standards.

The status of state programs was immediately brought into question with the occurrence of one of the worst fire disasters in the nation's history in Hamlet, North Carolina, the morning of September 3, 1991 (ref. LaBar). The causes and details of this tragedy are discussed more fully in Chapter 7; in summary, 25 people were killed and another 56 were injured. Ironically, the catastrophe occurred in North Carolina, the state that was selected by OSHA as the nation's first to be approved to replace federal OSHA inspectors with state inspectors for enforcement of standards that were supposed to be "as effective as" the federal OSHA standards. Not surprisingly, the tragedy precipitated a thorough review of the North Carolina state plan for standards and enforcement. The final result was that North Carolina was permitted to continue to operate its state plan, but several changes were mandated to assure its effectiveness. The evaluation of the North Carolina plan became the first step in an extended program to evaluate all state plans. Approximately one year after the accident, OSHA announced that all state plans had been evaluated, that significant progress had been made, and that all approved state plans would be permitted to continue (ref. OSHA, 1992).

Consultation

Besides standards and enforcement programs, OSHA also delegates to states authority and responsibility for consultative assistance in occupational safety and health to employers on request of the employers seeking such assistance. OSHA has authority to make federal grants to states to support enforcement, consultation, and other purposes of the OSHA law. Recent political tides have been in the direction of transferring federal agency authority and activity to the jurisdiction of the states. OSHA is pursuing this path with the state consultation programs. Cooperation with a state agency can even result in temporary immunity from OSHA citation in some cases. The possibility is worth checking into and safety and health managers should consider state consultation as a part of their overall strategies. There is no charge for state consultation, and as of this writing, it was available in some form in every state.

Employer Reluctance

The biggest problem with the OSHA consultation program is a reluctance on the part of the employer to call a government official, either federal or state, to "ask for an inspection." Many company managers, realizing that the standards are so complex that some fault can be found if one digs deep enough, simply do not want to involve the government inspectors if they can avoid it. This thinking is reminiscent of the joke known as "The Three Biggest Lies" about OSHA inspections:

1. The company management welcomes the OSHA inspector with the greeting, "We're glad to see you!" (Lie #1)
2. The OSHA inspector says, "Well, we're just here to help you." (Lie #2)
3. The union steward says, "Well, I sure didn't call him." (Lie #3)

Despite the joke, it is true that OSHA has made genuine efforts to help employers voluntarily comply with standards. The agency must walk a tightrope, however, to avoid alienating labor interests by granting immunity to cooperating employers.

Inspection Immunity

It was stated earlier that cooperation with a state agency can result in "temporary immunity from OSHA citation in some cases." It should come as no surprise that this is a sensitive issue, and as such, OSHA policy has evolved somewhat over the years. At first, temporary immunity from enforcement inspections was granted to employers who requested consultations, with one important exception: If serious violations were found, the consultation inspector was obliged to notify OSHA. This exception chilled employers to the whole idea of requesting an inspection of any kind. Picture the awkward position of the safety and health manager who convinces management that a consultation inspection should be requested and then later finds the company embroiled in an enforcement inspection with stiff penalties to be paid. The policy was later modified slightly to state that the cooperating employer was temporarily immune from inspection provided that serious violations were corrected within an agreed-upon time period. Thus, if the employer cooperates, enforcement inspections will not be triggered by the consultation visit, even if serious violations are found. OSHA agency officials, of course, understand employer reluctance to request an inspection, and at the same time understand labor pressures to not exempt any employer who fails to comply with standards. This dilemma and others are addressed in the next section.

POLITICAL TRENDS

From this chapter, it can be easily seen that OSHA is a politically charged subject, so much so that it has been often thought that the agency was on the brink of repeal. But despite its problems, OSHA appears to be quite durable and has survived repeated attempts to repeal or modify it.

The Barlow decision was an important one, but it really did not affect OSHA's enforcement operations as much as the public first thought it would. Those firms insisting on inspection warrants are usually inspected later; the procedure merely delays the process somewhat.

An attempt to exempt small businesses and family farms merely shifted enforcement emphasis. The same can be said of a strategy to inspect only those firms that have a lost-workday rate greater than the national average. An exemption of one category means emphasis on another, provided that the staffing and organization of the agency remain intact.

One indicator of OSHA's future is the congressional acts that have been patterned after the OSHA law and have been passed despite criticism directed at OSHA. The laws governing product safety and liability and mine safety were approved in the 1970s and were constructed and worded in a manner similar to the OSHA law. This illustrates that Congress has continued to endorse the concept of OSHA.

The evolution of OSHA standards is very slow. Once a standard has been adopted as the national consensus, it becomes very difficult to revoke the standard at a later date by saying that the standard did not represent the "national consensus" after all. Such a step repudiates or at least reflects adversely on the thousands of citations enforcing the given standard in inspections prior to the revocation. Any revision of the standards that can be considered subjective or controversial can be challenged in the courts. All of this tends to preserve the status quo despite widespread controversy. OSHA has been able to overcome the inertia of existing standards by making some rather broad changes to some of the standards, the fire protection standard being the most notable example.

Returning to the subject of consultation, OSHA can be seen vacillating between trying to satisfy labor and keeping peace with management. In December 2000, OSHA revised its rules on the consultation program to stiffen the program somewhat. Always aware of its labor constituency, OSHA authorized employees to take a larger role in the consultation inspection, with union or other employee representation during the consultation inspection. In addition, OSHA required that the employer post a list of serious violations found, so that employees can see the results of the consultation. Employers were required to leave the posting up for three days or until the hazards were corrected, whichever was *later*. Then, less than a year later, in 2001, OSHA finally began to recognize that it just was not reaching some employers who were shy about calling either OSHA or a state agency supported by OSHA to perform a consultation inspection. Therefore, OSHA initiated a new program in which specialists provide training and compliance assistance without the threat of inspection. The compliance assistance specialists were kept separate and distinct from OSHA's enforcement program. One can see that, politically, OSHA has a great deal of difficulty in balancing industry and labor interests when it comes to defining its position on consultation.

A final question to be addressed in this section is "Which political party, Democrat or Republican, defends or condemns OSHA?" Most would call this question a "no-brainer," because Democrats are thought to be more liberal and pro labor and, therefore, pro OSHA. Republicans are thought to be the protectors of business interests and personal and entrepreneurial liberty and, therefore, anti OSHA. However, it is politically naïve to buy into this oversimplification of the political dynamics. The real differences are more subtle. If we examine the political history of OSHA and the controversial issues related to the agency, the distinctions between the two major political parties begins to blur. Consider, for instance, that it was under a Republican presidential administration (but a Democratic Congress) that OSHA was first conceived. Richard Nixon signed the OSHA law into existence. Despite this fact, the Republican sweep of both the White House *and* Congress in 1980 was thought to spell doom for OSHA. It did not turn out that way. Indeed, toward the end of the decade, OSHA enforcement appeared to intensify, with the assessment of some exceptionally large penalties. This period also saw the inception of egregious violations. The Democrats were back with the election of Bill Clinton in 1992, but OSHA was kept low-key, apparently with the intent of not arousing the ire of business interests. By the year 2000, with the return of the Republicans to power in both Congress and the White House, OSHA detractors had already given up hope that they would bring about outright repeal of the law. Analogous to the Democrats' fear of alienating businesses, the Republicans did not want to alienate labor, so the distinction between the two remains blurred. One can see a difference between the parties, however, in specific issues, such as ergonomics. This controversial subject will be discussed later. But first, some positive developments will be identified in which both management and labor have cooperated to further the objectives of the agency.

Positive Developments

Gilbert J. Saulter (ref. Saulter) summarized his assessment of public sentiment by saying that OSHA was reaching "maturity." He cited successes in achieving public awareness by observing that the number of university academic programs in safety and health had tripled since 1970. Saulter emphasized the success of the program of exemption through consultation, whereby employers seek free consultation, usually provided by the states, and thus achieve limited immunity from citation.

SHARP Program

OSHA's "Safety and Health Achievement Recognition Program" (SHARP) is designed to provide recognition to small companies that cooperate and participate in OSHA's consultation program. To obtain this recognition, the employer must be proactive and go beyond simply requesting a consultation visit. Specifically, the employer must (1) receive a comprehensive safety and health consultation visit; (2) correct all workplace safety and health hazards; (3) adopt and implement effective safety and health management systems; and (4) agree to request further consultative visits if major changes in working conditions or processes occur that may introduce new hazards (ref. OSHA Consultation). The reward to employers for agreeing to cooperate in these ways is exemption from OSHA programmed inspections for a period of one year. Note carefully the use of the word "programmed." This means that OSHA will exempt the employer from the random selection of the facility for general inspections. However, if there is a serious accident or an employee complaint during the year of the "exemption," OSHA will be there.

VPP Program

Another positive program is the Voluntary Protection Program (VPP). This program is a comprehensive one that involves a serious commitment on the part of the employer to maintain an exemplary safety and health program for its employees. Such a program can be compared to the corresponding Malcolm Baldridge award for manufacturing quality. The first step is called "Demonstration," which identifies the company as seeking recognition from the VPP program. After a lengthy series of steps, the firm qualifies for the next rating, entitled "Merit." The highest designation, reserved for outstanding safety and health programs, carries the designation "Star."

The Voluntary Protection Program (VPP) is a procedure for giving recognition to noteworthy safety and health programs in companies that are voluntarily doing a good job. There are three levels of recognition as follows, in increasing levels of achievement: Demonstration, Merit, and Star. (ref. Lee).

Recent OSHA emphasis is on the management of safety and health, an approach that is heralded in this book. OSHA has become increasingly interested in both training and the effectiveness of employee committees for safety and health. The stress is on hazards analysis and feedback to correct safety and health problems so that the same injuries and illnesses do not recur.

Like the emphasis on management is the attempt by OSHA to use a systems approach to dealing with hazards. This view of safety and health as a complex systems problem that impacts many other facets of the production system is a further sign that the agency is maturing. Sophisticated OSHA technical personnel now look beyond the citation and abatement of individual violations and seek solutions more basic to the problem(s) at hand.

Ergonomics

In no arena is the systems approach more evident than in that of ergonomics. Ergonomics is the study of human capability in relation to the work environment, and solutions to ergonomics problems usually require a sophisticated analysis involving perhaps a redesign of the workstation to fit the process. One of the hottest issues of the 1990s was a proposed OSHA standard for ergonomics. The subject of ergonomics is mentioned here because it is one of the most controversial political issues related to occupational safety and health. Due to the importance of this subject, Chapter 8 is devoted entirely to it.

OSHA's emphasis on ergonomics has been highly visible to the public in the form of severe penalties, especially in the meat-packing industry. An example is OSHA's inspection and citation of a snack food manufacturer in which a total of $1,384,000 in penalties was proposed for alleged violations, including $875,000 for repetitive motion illnesses at $5000 each (ref. OSHA Proposes). The trend is expected to continue and will have an impact on production that goes beyond the issue of safety and health.

"Extra-Hazardous Employer" Status

This chapter has emphasized OSHA as the primary regulator and enforcer of safety and health standards, but there are other agencies and laws to consider. In Chapter 2, the workers' compensation system was examined, and it was stated there that the system is being expanded from its traditional role of compensation of the injured to a more general role that includes regulation and enforcement. One of the prominent aspects of this new development is the creation of the designation "Extra-Hazardous Employer" (ref. Hazardous). The management of a company unhappily discovers that it has received this designation when notified by the Workers' Compensation Commission of the state of the company's residence. The designation is determined by a calculation of the company's lost-time injuries incidence rate and comparison with the "expected rate" for that company's industry (SIC). When the company's rate exceeds the level that can "reasonably be expected" for that industry, the company is designated an "Extra-Hazardous Employer."

Being dubbed "extra hazardous" carries with it certain responsibilities for management. The first of these is to have a safety consultation within a short time period, usually 30 days. Options for this consultation may include the state labor department, the company's insurance carrier, or a professional consultant approved by the Workers' Compensation Commission of the state of residence.

The written report of the safety consultant is then used by the company as a basis for the development of a mandatory "accident-prevention plan." The plan must be more than a general document; it must address each of the hazards or unsafe practices identified in the safety consultation report. The plan must also include provision for the following:

1. statement of company safety policy
2. analysis of hazards
3. recordkeeping
4. education and training
5. in-house audits or inspections
6. accident investigations
7. periodic review of abatement effectiveness
8. implementation schedule

The plan must be more than a "paper plan." Six months after submittal of the plan by the company, the Workers' Compensation Commission may return to the worksite and conduct a follow-up inspection to determine whether the plan was actually implemented. Penalties are an option if the company has failed to comply.

The foregoing description of the Extra-Hazardous Employer program was based on specific states' programs—in particular, Texas and Arkansas. Other states may use different approaches. As was stated in Chapter 2, the principal purpose of workers' compensation reform measures is to reduce the enormous increases in workers' compensation claims, with benefit to employers and employees alike. The Extra-Hazardous Employer program is seen as a preventive or proactive approach that attempts to reduce the number of injuries or illnesses that arise on the job before they become a claim in the workers' compensation system.

The Extra-Hazardous Employer program of state workers' compensation agencies can also be seen as a means of requiring employers to develop safety and health programs and plans, even though OSHA's own attempts to require such programs have not met with political approval. The more aggressive workers' compensation programs can also be seen as a threat to OSHA's authority to enforce safety and health standards. This is bound to create political controversy over who has authority to enforce safety and health in the workplace and ultimately may lead to a constitutional showdown on the issue of states' versus federal rights to enforce workplace safety and health.

Americans with Disabilities Act (ADA)

On July 26, 1990, the Americans with Disabilities Act of 1990 was signed into law and immediately became the focus of attention of employers and institutions that deal with the public. The law was in response to findings that an estimated 43,000,000 Americans have one or more physical or mental disabilities (ref. Public Law 101–336). This number is expected to increase as the population as a whole is growing older.

The American public and businesses alike were already familiar with the term *discrimination* as it applied to civil rights legislation forbidding discrimination based on race, color, religion, sex, and national origin. The ADA extended this concept to the handicapped with regard to employment and access to public facilities and services. Specifically, the ADA prohibits discrimination against the disabled in regard to job application procedures; the hiring, advancement, or discharge of employees; employee compensation; job training; and other terms, conditions, and privileges of employment.

The ADA has had a significant impact on the field of worker safety and health because it has been common practice to discriminate against job candidates whose safety or the safety of their coworkers may be at stake if these candidates have physical or mental impairments that might affect their performance of the particular job for which they are being hired. The ADA does not forbid medical examinations and screening tests for employment, but it has regulated such practices to assure that all candidates are tested fairly, not just the impaired, and that all characteristics being screened truly are significant to the health and safety of the worker or coworkers. In addition, the ADA requires that the results of medical preemployment exams be kept confidential.

A controversial issue with respect to medical tests and preemployment screening is the subject of drug testing and alcohol abuse. The relevant question is this: Is the use of drugs or alcohol considered to be a handicap and thus protected from discrimination by the ADA? Congress anticipated this question and did not forbid drug testing in its passage of the ADA law. Current users of illegal drugs or alcohol may be barred from the workplace by the employer, but former users who have completed a supervised drug rehabilitation program are again eligible to work. Indeed, former drug users are protected from general discrimination in the hiring process. Companies should avoid a blanket policy that excludes former drug users. Each former drug user should be evaluated to determine whether he or she poses a "direct threat" to himself or herself or to the health or safety of others that cannot be eliminated or reduced by reasonable accommodation, as is demonstrated by Case Study 4.1.

CASE STUDY 4.1
EEOC SUES EXXON

The Exxon Corporation was sued in 1995 by the Equal Employment Opportunity Commission (EEOC) for issuing a blanket policy excluding all employees with a record of past substance abuse from certain jobs (ref. Gonzales). The job description involved in the case was "aircraft flight engineer," a "designated" safety position. Although it was recognized that the job "aircraft flight engineer" had a direct relation to safety, the subject issue was whether a former drug user was still a risk in this particular job. Even though the job impacted safety, Exxon was sued because it was allegedly not clear that former drug use would contribute to the hazard in this job.

As can be seen from Case Study 4.1, employers must be careful to avoid blanket policies that bar the hiring of former drug users. The ADA law requires that employers determine on a case-by-case basis whether the former drug user poses a direct threat through individualized assessment based on medical analysis or other objective factual evidence before excluding the former drug user from a given job.

A related controversy is over the definition of an impairment. If a mental or physical condition qualifies as an "impairment" under the ADA law, a person having the impairment is protected from discrimination in employment—that is, the employer must make reasonable accommodations to accept the individual for employment. The law excludes such behavioral characteristics as transvestitism and transsexualism from protection, but it does not exclude diseases such as AIDS or HIV positivity. The law also excludes compulsive gambling, kleptomania, and pyromania from protection under ADA as an impairment.

Another impact the ADA has had on the field of industrial safety and health is in the construction and remodeling of buildings and facilities. The law requires that the employer make reasonable accommodation to permit handicapped employees to perform jobs as much as possible like the jobs performed by nonhandicapped employees. Recognition is given to those situations that would result in undue hardship to the employer to provide such reasonable accommodation. Undue hardship is judged on an individual basis considering the level of difficulty or expense to comply. Factors to be considered are the nature and cost of the accommodation and the overall financial resources and size of the company.

SUMMARY

This chapter has provided a glimpse of the effect of government regulation on the field of industrial safety and health and the controversy this regulation has spawned. The primary agency that affects the field is OSHA, but other agencies and regulatory laws, such as the ADA, are also having an impact. Despite the controversy over government regulation, it is very difficult to repeal federal regulation once it is in place. It cannot be questioned that OSHA and associated government regulation have had a profound effect on the field of industrial safety and health management.

EXERCISES AND STUDY QUESTIONS

4.1 What is NIOSH? What is its role?

4.2 By what procedure can an employer request time to comply with an OSHA standard before an enforcement inspection?

4.3 Describe the procedure established for employers who believe that OSHA has issued an unfair citation.

4.4 Identify at least three legal rights extended to employees by OSHA.

4.5 What is the General Duty Clause?

4.6 What is a national-consensus standard as defined by the OSHA law? Have any such standards been adopted in the 1990s? Why or why not?

4.7 Compare performance versus specification standards.

4.8 Compare horizontal versus vertical standards.

4.9 Explain the significance of the Barlow decision.

4.10 Explain the difference between appeal and variance as far as OSHA is concerned.

4.11 Suppose that you are a writer of new standards. Pick a familiar hazard and write a two- or three-sentence paragraph for a possible standard to protect against this hazard. Write the standard first in the style of a specification standard and then in the language of a performance standard.

4.12 Compare advantages of specification standards versus performance standards from the viewpoint of the employer and then from the viewpoint of an enforcement agency.

4.13 List in order of priority four OSHA inspection categories.

4.14 What is the difference between a repeat violation and failure to correct a violation? How do the penalties differ?

4.15 Name some public criticisms of OSHA standards.

4.16 Describe some employee impairments for which the Americans with Disabilities Act explicitly prohibits discrimination.

4.17 Name several employee behavioral characteristics that do not qualify as impairments under the Americans with Disabilities Act.

4.18 Is it prohibited to discriminate against kleptomaniacs by refusing employment to them? Why or why not?

4.19 This chapter has stated that OSHA's highest inspection priority is the "imminent danger" category. However, very few imminent danger inspections are actually performed. Explain this anomaly.

4.20 How much does industry spend each year on OSHA compliance?

4.21 When was the origin of the original agency that eventually became NIOSH? When was safety added to NIOSH's mission?

4.22 Under what circumstances is it permissible for a company to leave a pilot crew on duty when OSHA obtains a court order to prohibit workers from entering an "imminent-danger" area?

4.23 When a fatality or major accident occurs, an employer is required to quickly report the incident to OSHA. Within what time limit must the employer report? According to OSHA policy, how soon after the report will OSHA respond with an inspection?

4.24 What industry has received OSHA emphasis since the inception of the law?

4.25 How has state "sovereignty" become an issue with regard to OSHA enforcement?

4.26 Under what circumstances are state agencies and political subdivisions covered by safety and health standards and enforcement?

4.27 What tragic accident in 1991 precipitated a general review of the effectiveness of state plans for safety and health standards and enforcement?

4.28 Why is it especially difficult to revoke an existing standard once it has been put into effect?

4.29 The ADA law was passed in response to a finding that millions of Americans have one or more physical or mental disabilities. In this finding, how many millions of Americans were estimated to be in this category?

RESEARCH EXERCISES

4.30 Do research to provide an estimate of the current number of Americans who have one or more mental or physical disabilities.

4.31 Use the OSHA website (www.osha.gov) to study the impact of OSHA's General Duty Clause by examining the number of citations of this clause in industries having more than 250 employees. What industry group is cited most? What percentage of all citations for the General Duty Clause is for firms employing 250 employees or more?

4.32 Use the OSHA website (www.osha.gov) to check OSHA statistics to see whether the OSHA General Duty Clause for employees [Section 5(b)] has been cited at all for the current fiscal year.

4.33 Find the names of five consulting firms that could assist managers with compliance to the Americans with Disabilities Act (ADA).

4.34 Check government statistics to estimate the total number of disabled workers employed by the federal government. What percentage of the total federal workforce is disabled?

4.35 Check out "What's New at NIOSH" using the website http://www.cdc.gov/niosh/. What do you suppose the letters "cdc" stand for in this site address?

4.36 This chapter discussed the ADA and requirements for making reasonable accommodations to employ the handicapped. Search the Internet and other possible sources for actual case histories in which the employer was sued or fined for failure to make reasonable accommodation to employ the handicapped. What is the largest fine you can find on record for this type of violation?

4.37 Search out state workers' compensation programs for policies with regard to employers designated as "Extra Hazardous." What states have such programs?

4.38 In Exercise 4.37, your answer should have included the state of Texas. What is the monetary fine in Texas for failure to implement an "accident-prevention plan?"

4.39 Is it possible for OSHA to cite a firm for an "egregious" violation if the offense is only in recordkeeping? Search for an example of egregious recordkeeping violation in the news media.

4.40 Examine the account of the citation of Samsung Guam, Inc. (SGI), one of the largest OSHA fines on record. What dollar penalty did OSHA propose? In the final settlement, how much did SGI agree to pay?

STANDARDS RESEARCH QUESTIONS

4.41 Use the NCM database search tool to estimate the percentage of annual OSHA citations that are designated as "repeat" violations. Do the same for "willful" violations.

4.42 Use the NCM database search tool to estimate the average OSHA proposed penalty for a "General Duty Clause" violation.

4.43 Take a sampling of standards from general industry and determine the percentage of citations in your sample that have arisen from an employee complaint.

CHAPTER 5

Information Systems

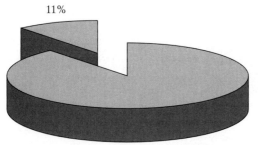

11%

**Percentage of OSHA
General Industry citations
addressing this subject**

There are two schools of thought regarding responsibility for risk in the workplace. The more ambitious of these two factions places full responsibility on the employer not only to identify hazards but also to eliminate them so that the employee is assured a safe and healthful workplace regardless of the nature of the hazard. For the most part, this is the approach used by the drafters of the OSHA law. It is true that the law contains a general-duty clause for employees as well as employers, but there is no question that the enforcement provisions of the law are to confirm compliance on the part of the employer, not the employee.

The second of these schools of thought is more conservative in that it recognizes the inability of the employer to completely eliminate some hazards and accordingly shifts some of the responsibility to the employee by requiring the presence of information systems that provide the employee with data specifying the nature and degree of hazard associated with the job. The theory of this school of thought is that the employee is thus given the necessary data with which to evaluate the risks and take action accordingly.

As OSHA entered its second decade of existence, it increasingly shifted toward this more conservative approach, reflecting the more conservative political climate that was ushered in by the change in government administrations in the United States in 1980. Even OSHA's critics acknowledged the fairness of a system of disclosing knowledge to employees of hazards to which he or she would be exposed and of which the employer had knowledge. Thus began the movement that became known as the *right-to-know* movement, along with regulations requiring *Material Safety Data Sheets* (MSDSs) and labeling for hazardous materials to which employees or the public might be exposed.

Although more conservative in concept, the right-to-know movement should not be interpreted as a weakening in the protection of the rights of the worker to a safe and healthful work environment. To the contrary, knowledge of the hazard can be a potent weapon in the employee's fight for improved safety and health—often, the employer is aware of this power in the hands of the employee. The specter of future litigation for today's hazards is a powerful motivator for careful concern on the part of the employer, especially if the company is large and can be shown by attorneys to have a "deep pocket." Despite the immunities provided by workers' compensation laws, employers are increasingly exposed to litigation risk as a result of exposing employees and others to hazards. Further, information systems that proliferate knowledge of these hazards heighten that risk.

HAZARD COMMUNICATION

Action in the right-to-know movement was precipitated by OSHA in late 1983 with the promulgation of the hazard communication standard (29 CFR 1910.1200). A significant provision of this standard is the requirement that manufacturers and importers must label containers they ship and provide an MSDS for each hazardous chemical they produce or import. Employers in industries that make use of the hazardous substances also have responsibilities to maintain hazard communication programs to protect their employees.

Container Labeling

The hazard communication standard placed the labeling responsibility on the manufacturer or importer of the substance. Just about any container one can think of is included, except pipes; the labeling of pipes is the concern of other OSHA standards. The labeling of some substances is governed by rules administered by another regulatory agency. For this and other reasons, some substances are excluded from the OSHA requirement, as follows:

- pesticides
- food, drugs, or cosmetics
- alcoholic beverages
- substances covered by the labeling requirements of the Consumer Product Safety Commission (CPSC)
- hazardous wastes
- tobacco or tobacco products
- wood or wood products
- "articles"

The term *articles* is to be interpreted as manufactured items formed to a specific shape or design during manufacture, having end use dependent on that shape or design, and that do not result in hazardous chemical exposures during normal use. The distinction really is whether the item is a material or a manufactured object. In some cases, it may be difficult to distinguish between articles and materials. One example given is that of a desk, which is an article, versus a piece of lumber, which is considered a material, even though both the desk and the piece of lumber might be objects manufactured from wood.

Material Safety Data Sheets

In addition to labeling, chemical manufacturers or importers must provide MSDSs for hazardous substances. The hazard communication standard lists specific categories of information that must be included in the MSDS. Figure 5.1 is an exhibit of a blank form that has been used as a generic format to comply with the standard.

A delicate issue with MSDSs is the problem of dealing with trade secrets. There is likely no industry more sensitive about trade secrets than the chemical industry, and the OSHA standard addresses this issue in detail. Basically, the manufacturer or importer, and in turn the employer, can withhold specific chemical identities from the MSDS, but only if they can justify this position according to specific criteria contained in the standard. Even then they must disclose the chemical identities to health professionals on request in accordance with criteria specified in the standard. In nonemergency situations, the manufacturer, importer, or employer may require a confidentiality agreement. Disputes are anticipated as an inevitable consequence of conflicting interests between the manufacturers and persons claiming a need to know chemical identities. Accordingly, the standards provide for referrals to OSHA to weigh the evidence on both sides, and, if appropriate, subject the manufacturer, importer, or employer to OSHA citation.

Another problem to consider is how to deal with mixtures in which a hazardous substance is only an ingredient. What must be done depends on circumstances best illustrated by a decision diagram. (See Figure 5.2.)

Employee Hazard Communication Program

After the manufactured or imported chemical is distributed to others, responsibility for protection of employees from potential exposure becomes the responsibility of the employer in those firms that "use" (i.e., "package, handle, react, or transfer") the hazardous substances. A principal requirement for such employers is that they have a written hazard communication program. The safety and health manager should ensure that employees know about this program, because the workers themselves may be asked about it by federal enforcement inspectors. One required component in the written program is a list of the hazardous chemicals known to be present in the workplace. For each of the hazardous chemicals listed, an MSDS must be on hand and available to employees. If the substance was purchased prior to the right-to-know era and no MSDS is on hand for a given substance, the employer is required to obtain or generate one. Safety and health managers can turn to current manufacturers or distributors, or perhaps write their own MSDSs. Sometimes a state consultation agency, as described in Chapter 4, can be of assistance in identifying hazardous substances and in preparing the MSDS.

Federal standards permit the MSDS to be kept in any form, even within operating procedures. Sometimes it is more practical to address multiple hazards as a process,

Material Safety Data Sheet May be used to comply with OSHA's Hazard Communication Standard, 29 CFR 1910.1200. Standard must be consulted for specific requirements.	**U.S. Department of Labor** Occupational Safety and Health Administration (Non-Mandatory Form) Form Approved OMB No. 1218–0072
IDENTITY *(As Used on Label and List)*	*Note: Blank spaces are not permitted. If any item is not applicable, or no information is available, the space must be marked to indicate that.*

Section I

Manufacturer's Name	Emergency Telephone Number
Address *(Number, Street, City, State, and ZIP Code)*	Telephone Number for Information
	Date Prepared
	Signature of Preparer *(optional)*

Section II — Hazardous Ingredients/Identity Information

Hazardous Components (Specific Chemical Identity; Common Name(s))	OSHA PEL	ACGIH TLV	Other Limits Recommended	% *(optional)*

Section III — Physical/Chemical Characteristics

Boiling Point		Specific Gravity (H$_2$O = 1)	
Vapor Pressure (mm Hg.)		Melting Point	
Vapor Density (AIR = 1)		Evaporation Rate (Butyl Acetate = 1)	
Solubility in Water			
Appearance and Odor			

Section IV — Fire and Explosion Hazard Data

Flash Point (Method Used)	Flammable Limits	LEL	UEL
Extinguishing Media			
Special Fire Fighting Procedures			
Unusual Fire and Explosion Hazards			

(Reproduce locally)	OSHA 174, Sept. 1985

FIGURE 5.1

Material Safety Data Sheet (MSDS). (This form is available in larger format on the CD-ROM that accompanies this book.)

Section V — Reactivity Data

Stability	Unstable		Conditions to Avoid
	Stable		

Incompatibility (*Materials to Avoid*)

Hazardous Decomposition or Byproducts

Hazardous Polymerization	May Occur		Conditions to Avoid
	Will Not Occur		

Section VI — Health Hazard Data

Route(s) of Entry:	Inhalation?	Skin?	Ingestion?

Health Hazards (*Acute and Chronic*)

Carcinogenicity:	NTP?	IARC Monographs?	OSHA Regulated?

Signs and Symptoms of Exposure

Medical Conditions
Generally Aggravated by Exposure

Emergency and First Aid Procedures

Section VII — Precautions for Safe Handling and Use

Steps to Be Taken in Case Material Is Released or Spilled

Waste Disposal Method

Precautions to Be Taken in Handling and Storing

Other Precautions

Section VIII — Control Measures

Respiratory Protection (*Specify Type*)

Ventilation	Local Exhaust		Special
	Mechanical (*General*)		Other

Protective Gloves	Eye Protection

Other Protective Clothing or Equipment

Work/Hygienic Practices

☆ U.S.G.P.O. 1986–491–529/45775

FIGURE 5.1 (*Continued*)

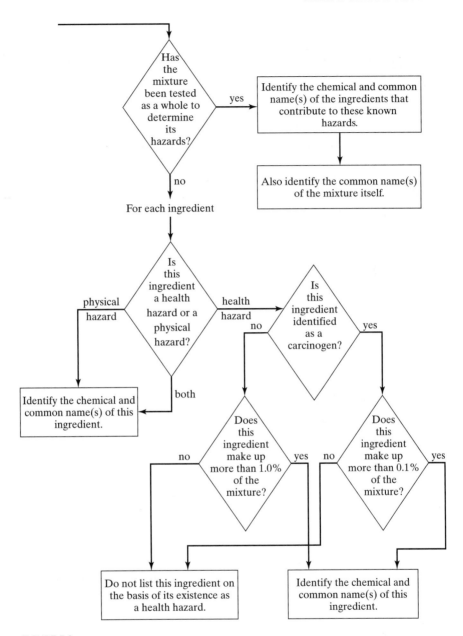

FIGURE 5.2

Decision diagram for reporting the content of mixtures in a Material Safety Data Sheet (MSDS).

not as separate hazardous chemicals. In designing the right-to-know system, however, the safety and health manager should ensure that the required information is available for each hazardous chemical present among a group of materials in a single process and that the information is readily accessible to employees of each work shift.

In addition to retaining and maintaining MSDSs, the employer must maintain the labels provided by the manufacturer or importer of the substance. Note, however, that labels are not required for portable, in-house containers intended for immediate use.

Record Retention

Records are of ever-increasing importance, and the safety and health manager should set up information systems that trace the identities, location of use, and time of use for hazardous substances, along with each employee exposure, for a retention period of *at least 30 years*. Employee medical records—except records of health insurance claims, which are maintained separately—must be preserved and maintained for the duration of employment *plus* 30 years. The reason for the very long retention period is to permit tracing of the cause of illnesses that may have extremely long latency periods after exposure to hazardous substances. Such a responsibility, of necessity, presupposes a sophisticated, often computerized, information system.

Sale of the business or closing the business does not relieve the employer of the record-retention responsibility. Upon sale of the business, the successor employer is required to receive and maintain the records. If the business is closed permanently, the employer may be required to transfer the records to NIOSH, depending on the requirements of specific standards pertaining to the hazardous substances in question.

ENVIRONMENTAL PROTECTION AGENCY

It was stated in Chapter 1 that the safety and health manager often holds responsibility for compliance with Environmental Protection Agency (EPA) regulations. In this section, we examine this important additional responsibility that is often placed on the safety and health manager. The decades of the 1970s and 1980s saw more than $850 billion in expenditures by the American people to clean up the environment (ref. cleaning). This is a serious investment on the part of our society, an investment that has been paid largely by industry and in turn by our society in increased costs of services and especially goods produced by these industries. How effectively and efficiently the company manages this investment in large part determines its overall competitiveness and vitality as a firm. Thus, an enterprising safety and health manager will see environmental protection as an opportunity to have a significant impact on the management of the company.

The dual roles of OSHA and the EPA were recognized in the enactment of the "Superfund Amendments and Reauthorization Act of 1986" (SARA) on October 17, 1986. OSHA responded with its standard 29 CFR 1910.120—hazardous waste operations and emergency response, sometimes called HAZWOPER. This OSHA standard covers hazardous substance response operations under the Comprehensive Environmental Response, Compensation, and Liability Act (CERCLA) of 1980 and major corrective actions taken in cleanup operations under the Resource Conservation and Recovery Act (RCRA) of 1976.

One part of the SARA provisions, Title III, is the Emergency Planning and Community Right-to-Know Act of 1986 (ref. Title III). Title III establishes requirements for federal, state, and local governments and industry regarding emergency planning and reporting on hazardous and toxic materials.

At the outset of enforcement of the SARA law, the EPA published the initial list of 402 extremely hazardous substances. The list is subject to revision by the EPA at any time, but the latest available list at the time of this book's printing appears in Appendix E. It is essential that the safety and health manager keep this important list on hand and up to date for reference so that the firm can be in compliance if it uses any of the materials on the list. There are reporting requirements for these substances, and there are federal standards for protecting workers who may be exposed to them.

Medical Surveillance

Medical surveillance requirements make up one of the elements of the OSHA standards for dealing with the EPA-listed hazardous substances. And wherever there is a medical surveillance program, there are medical records. Fundamental to the right-to-know concept is the employees' right of access to their personal medical records held by the company that employs them. A medical surveillance program is required in the following situations:

1. all employees who may be exposed to health hazards at or above the established permissible exposure limits for 30 days or more a year, whether or not the employee uses a respirator for protection against the hazards.
2. all employees who wear a respirator for 30 or more days a year.
3. employees designated by the employer to plug, patch, or otherwise temporarily control or stop leaks from containers that hold hazardous substances or health hazards (i.e., members of hazardous material (HAZMAT) teams).

The medical surveillance program serves both to determine whether the person is fit to work with the hazardous materials present and to recognize any adverse effects on the worker arising from the exposure. It will be seen in Chapter 12 that various conditions, such as heart problems, the wearing of a beard, or even a perforated eardrum, could disqualify the employee from working in jobs requiring the use of a respirator.

Monitoring the adverse effects of exposure is the other, and perhaps more important, purpose of the medical surveillance program and the purpose that relates most to the right-to-know principle. The worker wants to know everything the employer knows about his or her health and its possible deterioration due to hazardous exposures on the job.

Federal standards prescribe intervals at which medical examinations and consultations shall be made. The regular times are as follows:

1. prior to assignment to duties that may require hazardous material exposure.
2. at least every 12 months during assignment to such duty.
3. upon termination of such duty, unless the employee has had an examination within the last six months.

In addition to these regular times, an examination is required as soon as possible if an employee develops signs or symptoms indicating possible overexposure or if an unprotected employee becomes exposed in an emergency situation. Also, the examining physician might advise that increased frequency of examination is medically necessary, and in such instances, the employer is obliged to comply. All medical examinations and procedures are required to be performed by or under the supervision of a licensed physician, at a reasonable time and place, without cost to the employee, and without loss of pay for employee time lost.

It is the business of the safety and health manager to be aware of the circumstances that require a medical surveillance program and to ensure that company management implements such a program if necessary. In presenting such a recommendation to management, the safety and health manager can point out certain legal benefits to the company over and above the benefits of increased safety and health of employees and avoidance of OSHA fines. A record of the initial medical examination can be proven to be a valuable piece of evidence of preexisting conditions or symptoms in the event such conditions or symptoms surface during worker exposure. Management may already be aware of that, but what many management personnel have overlooked is the importance of the medical examination at the end of employment also. This serves the purpose of documenting conditions, symptoms, or lack of symptoms at the end of the period of risk. In this age of right to know, employees are usually aware that they can take legal action at a later date if symptoms related to the hazardous material exposure reveal themselves after the fact. At such time, the content of the medical examination at employee termination will have obvious value to the employer as well as to the employee.

Full disclosure in the spirit of a person's right to know extends to the physician also. The employer is required to provide the examining physician with a copy of the federal standard covering medical examinations and must provide data pertaining to the employee's job, anticipated exposure levels, personal protective equipment to be used, and information from previous examinations.

It is the responsibility of the employer to obtain and furnish the employee with a copy of a written opinion from the examining physician containing the results of the examination, including any opinions or recommendations by the physician regarding increased risk or limitations on the employee's assigned work. In the interest of privacy, however, the written opinion obtained by the employer is prohibited from revealing specific findings or diagnoses unrelated to occupational exposure.

Reporting

Title III of the SARA act requires the EPA to establish an inventory of toxic chemical emissions from certain facilities. This requirement is in effect a reporting requirement for manufacturing facilities (SIC[1] Codes 20xx through 39xx) that have 10 or more employees and have manufactured, processed, or otherwise used a listed toxic chemical in excess of specified threshold quantities. The term *listed* refers to one of the extremely hazardous substances listed by the EPA. (See Appendix E.) Threshold quantities vary and are different, depending on whether the facility uses *or* manufactures the hazardous substance. Facilities in the manufacturing industries (SIC Codes 20xx through 39xx) that *use* listed toxic substances in quantities over 10,000 pounds in a calendar year are required to submit toxic chemical release forms (EPA Form R) by July 1 of the following year. For firms that *manufacture* or *process* these materials, the threshold

---

[1]SIC refers to Standard Industrial Classification, and codes 20 through 39 are identified with manufacturing industries. Appendix F identifies the principal manufacturing categories of the SIC Code.

quantity is 25,000 pounds per year, for amounts above which the firm is required to submit the toxic chemical release form.

For firms that both manufacture *and* use the same substance, if the threshold quantity is exceeded in either instance, then the firm must report, a point illustrated by Case Study 5.1. In Case Study 5.2, neither threshold is exceeded.

CASE STUDY 5.1

A fertilizer plant (SIC Code 2873) produces 22,000 pounds of ammonia, 16,000 pounds of which it uses within the plant. Is the firm required to report to EPA, and if so, in what fashion?

Solution

Yes, the firm has a manufacturing SIC (28xx through 39xx) and exceeds the "use" threshold; therefore, the firm must report to EPA using the Toxic Chemical Release Inventory Reporting Form R in accordance with Title III of the SARA Act of 1986. Since one of the thresholds was exceeded (the "use" threshold), the firm must complete a full report based on all activities and releases of ammonia from its facility, not just the releases from the use activity, (This study assumes that the firm had 10 or more employees.)

CASE STUDY 5.2

During a year of operations, a manufacturer of chemical coatings (SIC Code 2821) processes 20,000 pounds of cresol and uses 6000 pounds of it within the plant. Is the firm required to report to EPA, and if so, in what fashion?

Solution

No, the firm is not required to report for either its processing or its use of cresol, because neither the "use" nor the "processing" thresholds were exceeded.

One other point should be made about threshold amounts. Stockpiling a material that is neither processed nor used within the plant does not count toward the computation of whether the threshold quantity has been exceeded. The EPA may ask about total stockpiled amounts in the course of collecting data for a chemical for which a threshold has been exceeded, but stockpiling does not constitute processing or use.

Firms that manufacture or process materials in excess of the threshold quantity in a year are required to submit a toxic chemical release form. The form is in four parts and is too lengthy to be included here, but it can be obtained by requesting EPA Form R from the agency's regional or national office or by contacting the designated state

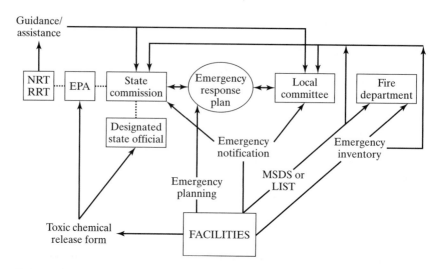

FIGURE 5.3

Major information flow requirements under Title III of the SARA act.

office (Designated Section 313 contact). The form is designed to gather data for a national computer information system and includes entries disclosing basic data on the process within the plant, including maximum amounts on site; receiving stream or water body, if applicable; quantities of release into the atmosphere; off-site disposal locations; and waste-treatment methods and efficiency.

In addition to the requirement for filing the toxic chemical release form, emergency notification is required by telephone, radio, or in person if there is a release of a listed hazardous substance that exceeds the reportable quantity for that substance. This includes substances listed by EPA as extremely hazardous (see Appendix E) and substances subject to the emergency notification requirements under CERCLA. Each such notification must be followed with a written confirmation including additional information on actual response actions taken, any known or anticipated data on chronic health risks associated with the release, and advice regarding medical attention necessary for exposed individuals. The requirements for information flow between private companies and the various governmental agencies that have a role in the control of hazards arising from toxic substances result in a complex information network that is best visualized by use of a diagram. (See Figure 5.3.)

COMPUTER INFORMATION SYSTEMS

In the preceding section, reference was made to a national computer information system for toxic chemical data. Increasingly, safety and health managers, both within and outside government, are turning to computerized databases for quick access to detailed facts about the thousands of toxic substances and other workplace hazards. This development follows on, and sometimes coordinates with, earlier computerized management information systems for administration and reporting under the record-keeping requirements of OSHA as described in Chapter 2.

Artificial Intelligence and Expert Systems

One of the technologies that shows particular promise in the management of safety and health information systems is the general field of artificial intelligence, or more specifically, expert systems. Artificial intelligence is that general field of development that is attempting to make computers "think" or respond to problem-solving situations more in the manner in which humans do. Expert systems is the branch of artificial intelligence that encompasses computer systems that give advice based on a knowledge base of logic rules provided by a human expert. Thus, the computer is able to answer questions posed by a wide variety of problem situations and provide advice or assistance much as a human expert would. A key point to understand with these systems is that an exhaustive list of answers to specific questions has not been preprogrammed into the computer. Rather, the human expert has provided the basic logic or rules of thumb, and the computer is capable of gleaning what it needs to know from this knowledge base to answer specific questions in the future. Augmenting the logic of the human expert is the computer's power to rapidly pinpoint facts accessed from massive tables contained in on-line data systems. A prominent element in the advancement of expert systems is the evolution of *natural language* interfaces that empower the computer to understand requests expressed in ordinary English, instead of in the rigid formats of computerese. Such interfaces are often called *intelligent front ends*, a reference to their appending to the front of existing computerized database management systems to make these systems more user friendly and more capable of understanding human requests for answers to questions.

Industrial Chemical Hazards Database

Safety and health research at the University of Arkansas (ref. Parker) has been conducted to develop a prototype computer system for answering questions about safety and health hazards. The system, identified as the Industrial Chemical Hazards Database (ICHD) is structured using the relational database tool R:BASE®. Queries to the database are made using the natural language access tool CLOUT®.[2]

The design of the relational database consists of a primary table with secondary tables for cross-reference. Each chemical in the database is keyed to its CAS number.[3] This identifier is used to key the chemical name to a variety of synonyms that might be employed by various workers or users that might query the database via a computer terminal. The user need not know the CAS number itself; it is a cross-reference tool, internal to the database. Thus, using natural language, a user may inquire about "coal tar naphtha," and the relational database will recognize that this is a synonym for benzene and will respond to the user's questions as if the questions were about benzene. The user can inquire about such particulars as protective eyewear required or recommended, if any, first-aid instructions, or perhaps washing instructions. If the user does not know what chemical he or she is dealing with, but has symptoms of exposure or can identify target organs affected, the relational database can use its cross-referencing capability to

---

[2]R:BASE and CLOUT are registered trademarks of Microrim, Inc.
[3]Chemical Abstracts Number, a well-known reference list for substances.

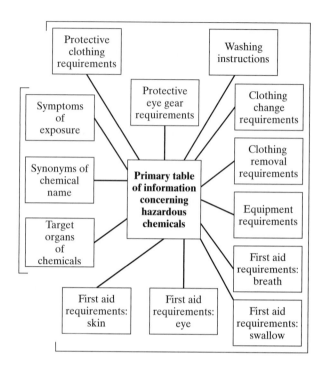

FIGURE 5.4

Industrial chemical hazards database design.

retrieve answers to inquiries, in addition to identifying the chemical or groups of chemicals that match the symptoms. Figure 5.4 diagrams the relationships of the various files in the ICHD database available for retrieval in response to a user inquiry. Case Study 5.3 exhibits an application of the ICHD relational database using an inquiry worded in a natural language format acceptable to the system.

CASE STUDY 5.3

Suppose a medical examination reveals that a worker has developed symptoms of kidney damage. The plant safety and health manager becomes curious about whether any hazardous chemicals used in the plant could be associated with kidney damage and poses the following question: "Give me a list of all the chemicals that harm the kidney." The ICHD system would automatically scan all of the relevant tables in the relational database to find the desired information, merging information using the CAS number cross-reference key to provide the requested response.

The CLOUT® natural language interface has built-in features to break down the structure of the sentence and interpret the meaning using a basic vocabulary of 300 words and capacity for the user to customize the system with another 500 buzzwords or common phrases used in the industry in which the ICHD system is being implemented.

The system is dynamic in that the language can be updated at any time to accommodate new terminology adopted by users. A single user can "teach" the system new words to be used on a temporary basis and then discarded, or the system can be instructed to remember the new vocabulary indefinitely.

Other database systems similar to the ICHD have been developed and proliferate the Internet. In the current information age, there is a wealth of data for dealing with the hazards of chemical exposures. Employers and workers alike can access this data and be "in the know" about the hazards of specific chemicals, how to protect against these hazards, what jobs are vulnerable to exposure to these chemicals, how to recognize the symptoms of exposure, what to expect in terms of long-term health hazards, and the case histories of others who have been exposed to these chemicals. This powerful knowledge has changed the ways in which people are willing (or not willing) to work in certain jobs and has changed the way that employers do business.

SUMMARY

This chapter has afforded a glimpse of current developments in managing and making accessible the large volumes of detailed data necessary to protect workers from hazards in the workplace. The evolution is the result of two developments: advancements in information technology and increased interest on the part of workers and the public who are asserting their right to know about hazardous substances to which they are, or might become, exposed. Regulated by both EPA and OSHA, chemical hazards were the first to receive attention in the right-to-know movement. But it is projected that information system needs for safety and mechanical hazards will follow.

EXERCISES AND STUDY QUESTIONS

5.1 With regard to responsibility for risk in the workplace, describe the change in emphasis that occurred in the early 1980s.

5.2 Does right to know represent a strengthening or weakening of workers' powers in the fight for improved safety and health? Explain.

5.3 What type of container is expressly exempted from the hazard communication standards for labeling?

5.4 For purposes of hazard communication, what is the difference between an article and a material?

5.5 How may chemical companies protect their trade secrets despite requirements to furnish MSDSs?

5.6 Are MSDSs required for mixtures in which some ingredients are hazardous and others are not? Explain.

5.7 Is an MSDS required for a chemical that was purchased by a plant prior to the hazard communication standard?

5.8 How long must records be kept to trace the use of hazardous substances? Why?

5.9 How long must employee medical records be preserved?

5.10 What should the safety and health manager do about records if a firm decides to go out of business?

5.11   What do the acronyms SARA, CERCLA, and RCRA represent, and how do they relate to OSHA?

5.12   What is the function of a HAZMAT team?

5.13   Under what circumstances is a medical surveillance program required for a given employee? How frequently are medical examinations required for such an employee?

5.14   If an emergency notification of an accidental release of a toxic chemical is required, who should be notified?

5.15   What are expert systems?

5.16   What commercially available computer tools for relational databases are used by the Industrial Chemical Hazards Database (ICHD) system?

5.17   What does the term *HAZWOPER* represent?

5.18   How many billions of dollars are estimated to have been spent by Americans to clean up the environment in the 1970s and 1980s?

RESEARCH EXERCISES

5.19   Use your favorite search engine on the Internet to find how many Internet sites are related to the term "*HAZWOPER.*"

5.20   Determine the criteria for how much HAZWOPER training is required for various employees. (*Hint:* What determines whether an employee needs 8, 24, or 40 hours of training?)

5.21   Find the names of five firms or institutions that offer 40-hour courses in HAZWOPER training for client companies.

5.22   Check the Internet to find an MSDS for hydrazine sulfate.

STANDARDS RESEARCH QUESTIONS

5.23   Search the OSHA general industry standards for the term *hazard communication*. Use the NCM database to examine the annual number of citations related to "hazard communication." What percentage of these citations are "serious?"

5.24   Determine which general industry OSHA standards relate to the term *MSDS*. How many citations are found on the NCM database that pertain to "MSDS"?

5.25   Determine whether any OSHA general industry standards are being cited for "medical surveillance" violations.

CHAPTER 6

Process Safety and Disaster Preparedness

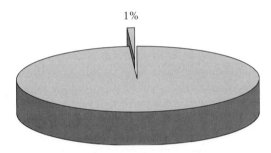

1%

Percentage of OSHA General Industry citations addressing this subject

Of major impact on the field of safety and health management in the early 1990s was the promulgation by OSHA of the standard for Process Safety Management of Highly Hazardous Chemicals. The decade of the 1980s was witness to major tragedies involving explosions and catastrophic release of hazardous chemicals that resulted in numerous fatalities both to employees and to the general public. These tragedies were of such significance that they attracted worldwide attention.

The most prominent tragedy was the Bhopal, India, disaster, which killed 2500 civilians in a single chemical company accident, mentioned in Chapter 1. There is no doubt that this catastrophe affected national policy here in the United States and influenced the development of the process safety standard. Another major tragedy was the Phillips Petrochemical Plant explosion in October 1989 in which an explosion and fire at a plant near Houston, Texas, killed 24 and injured another 128 workers. As might be expected, a substantial OSHA inspection was conducted at Phillips after this incident and a large fine was levied. But this disaster prompted OSHA to seek more than an after-the-fact approach with inspections and fines. The resulting OSHA Process Safety Standard sought to forestall such catastrophes in the future.

Some readers may be inclined at this point to skip this chapter as not applicable to their operation. "Process safety" seems to apply only to chemical plants and petroleum refineries, but the reader is cautioned that, in the early years of enforcement of the process safety standard, OSHA has been shown to adopt an aggressive definition of the word "process." Thus, a poultry processing plant, for example, can be seen to be covered by the process safety standard, because its process might employ chlorine for refrigeration, a dangerous chemical. Even a discrete-items manufacturing plant, might employ dangerous acids in its plating operations and thus fall under the scope of the process safety standard because it processes or stores a dangerous chemical in excess of a threshold amount.

PROCESS INFORMATION

In Chapter 5, the growing influence of information systems on the field of industrial safety and health was emphasized. This influence was evident in the content of OSHA's process safety standard. Before any analysis of the process is to begin, OSHA requires the employer to compile information on the highly hazardous chemicals to be used or produced by the process, the equipment to be used in the process, and the technology of the process itself. It is clear that OSHA's intent is for this information to be available to the union or other employee representative at the plant.

The safety and health manager (or whomever has been designated to deal with process safety hazards and standards) should first address the problem of where to find information about the chemicals used in the process. In Chapter 5, we studied the primary document of information regarding chemicals used within an industrial plant, and that document is the MSDS. The MSDS may provide all information necessary to comply with process safety requirements, but if it does not do this, the safety and health manager can turn to standard reference volumes on the properties of hazardous chemicals. The safety and health manager can win the confidence of the committees or teams of engineers, employees, and employee representatives assigned to analyze a hazardous process by knowing these standard reference volumes and relying on them when advising the analysis team. Following are a few of the popular standard references regarding hazardous chemicals:

- Irving Sax, *Dangerous Properties of Industrial Materials* (ref. Sax)
- Robert E. Lenga, *Sigma-Aldrich Library of Chemical Safety Data*
- Gessner G. Hawley, *The Condensed Chemical Dictionary* (ref. Hawley)
- *NIOSH Registry of Toxic Effects of Chemical Substances*

These references were used in the development of Case Study 6.1.

CASE STUDY 6.1

HAZARDOUS CHEMICAL INFORMATION FOR PROCESS SAFETY ANALYSIS

Name of chemical	Phosphorus chloride (PCl_3), sometimes known as phosphorus trichloride.
Toxicity information	Poison by inhalation. Moderately toxic by ingestion A corrosive irritant to skin, eyes (at 2 parts per million), and mucous membranes.

	Lethal dose (for 50% of population):
	Rats (oral): 550 mg/kg
	Lethal concentrations for inhalation (for 50% of population):
	Rats: 104 ppm for 4 hours
	Guinea pigs: 50 ppm for 4 hours
Permissible exposure limits	OSHA PEL: 8-hour time-weighted average (TWA): 0.5 ppm
Physical data	Clear, colorless, fuming liquid
	Melting point: $-111.8°$ Celsius
	Boiling point: 76° Celsius
	Density: 1.574 at 21° Celsius
	Vapor pressure: 100 mm of mercury at 21° Celsius
	Vapor density: 4.75
Reactivity data	Highly reactive with a variety of acids, oxidizers, and even water or steam. Fire and explosion hazard.
Corrosivity data	Department of Transportation Classification: Corrosive Material.
Thermal and chemical stability	Dangerous; when heated to decomposition, it emits highly toxic fumes of chlorides and PO_x. Can react with oxidizing materials.
Hazardous mixing	Potentially explosive with nitric acid, sodium peroxide, oxygen (above 100° Celsius). Violent reaction with water evolves hydrogen chloride and diphosphane gas, which then ignite. Will react with water, steam, or acids to produce heat and toxic corrosive fumes.

Case Study 6.1 was used to show how published data in chemical reference volumes can be used to provide the data necessary for analysis of hazardous processes. Not all data contained in reference volumes are required. For instance, in Case Study 6.1 under the category "hazardous mixing," only those materials that could foreseeably be inadvertently mixed with the hazardous chemical in the process under study are included.

Beyond the properties of the chemicals used in the process, OSHA wants employers to document the technology of the process, including at least a block flow diagram (Figure 6.1) or a simplified flow process diagram (Figure 6.2). In addition, process chemistry data, maximum intended inventory, and safe upper and lower limits for temperatures, pressures, flows, or compositions must be provided. Any deviations

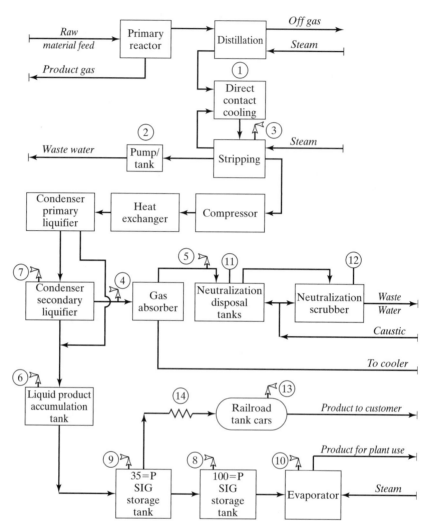

FIGURE 6.1

Example of a block flow diagram (*Source*: OSHA Std 1910.119).

from the standards of the process that might affect the safety and health of employees must be evaluated to consider consequences. These data may already be available, but if they are not, they can be developed in the process hazard analysis explained in the next section.

The process equipment must also be documented and described with such details as materials of construction and piping and instrument diagrams. Particular interest is placed on safety features, such as relief system design, ventilation, design codes and standards, material and energy balances, and safety systems, such as interlocks, detection, and suppression systems. The federal standards require that process equipment comply with "recognized and generally accepted good engineering practices." Especially with regard

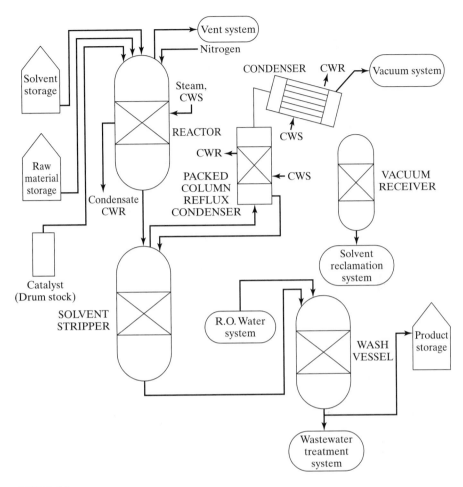

FIGURE 6.2

Example of a flow process diagram (*Source*: OSHA Std 1910.110).

to equipment purchased prior to the process safety standard, it may be advisable to uti-
lize a registered professional engineer to make a "good engineering practices" evaluation
of process equipment.

Considering the scope of the information required by OSHA to be compiled for
chemicals, processes, and process equipment, there is a need for the employer to develop a strategy for complying with the standard. It might seem straightforward to gather
all of the required information into a file drawer, but there are problems with this approach. Different departments have responsibilities for different parts of the problem.
For instance, plant maintenance might be responsible for proper functioning of process
equipment, but documentation of the properties of the chemicals processed might fall
within the responsibility of engineering or operations. It would be nice to be able to
show the OSHA inspector what he or she probably would like to see: a file drawer of
information all in one place that complies with every provision of the standard. But

this is usually not a practical solution. Especially with regard to changes and updates, keeping the central file up to date could become a nightmare. What the employer does not want the OSHA inspector to find is a nice, central file that is later shown to have incorrect or obsolete information about the process because responsible departments have made changes. A more practical solution to the problem is what Lastowka (ref. Lastowka) has referred to as the "road-map approach." This approach leaves required documents in their respective departments of responsibility. In a central file, convenient to the OSHA inspector, employee representatives, and other interested parties, is a "road map" that identifies each documentation requirement of the process safety standard and tells precisely where within the entire plant to find the pertinent detailed information required.

PROCESS ANALYSIS

The preceding section showed that federal standards require documentation of a great deal of information about a process. The main thrust of the standards, however, is in the analysis of the data. The intent of the analysis is to go beyond equipment, chemicals, and how the process works to an investigation of what can go wrong in the process and how to deal with these hazards. The analysis requirements of the process safety standard recall the methods that were studied in Chapter 3, including fault-tree analysis and failure modes and effects analysis. Some analysts refer to "what-if" analyses and "what-if" checklists, which raise questions about process interactions and outside events in addition to failure modes of the process itself. Also to be included are analyses of past incidents that have had a potential for catastrophic consequences in the workplace.

The potential value of engineering control systems must be considered. Engineering control systems might include detection and early warning of impending catastrophic events. Such systems might consist of a computer-based process monitoring system with instrumentation and alarms. Of course, the computer monitoring the process might itself fail, and the consequences of this possibility must also be considered. Even the location of the site of the facility may enter into the analysis. For instance, if the facility is located along a geologic fault, the possibility of earthquakes becomes a consideration. In addition, the human element must not be ignored. If human failure can contribute to the possibility of a catastrophe, the analysis must consider the human factor and how to mitigate any consequences of human failure. Human factors may enter the design decisions regarding the process.

Clearly, process safety analysis is an important subject, and the safety and health manager should be alert to recommend that top management take this responsibility seriously. Professional analysts with recognized credentials, assigned to the process analysis team, can go a long way in establishing the good faith of the employer in this endeavor. On the other hand, the opinions of the operators and maintenance personnel, who are intimately familiar with the process, can be even more valuable to the analysis.

Care should be taken in the documentation of the process hazard analysis to make sure that significant open-ended issues are not left unresolved—for example, without a documented strategy for alleviation of the issue. If the process has a serious hazard, the documentation of that hazard constitutes "recognition" of the hazard. Remember from Chapter 4 the importance of the word "recognized" in the wording of OSHA's General Duty Clause. Especially in the event of an accident or major incident, the OSHA inspector that makes the postaccident inspection can be expected to be looking for evidence that the company "recognized" the hazard that caused "serious physical harm" to employee(s).

Recognizing the comprehensive responsibility placed on employers to develop process hazard analyses, a phased approach was specified over a five-year period after the effective date of the standard. Then, every five years, the initial analyses are to be updated and evaluated to be sure that the analysis is consistent with the current process. Of course, such actions must be documented, and records must be kept for the life of the process.

OPERATING PROCEDURES

After the process information has been gathered and analyzed, the conclusions reached must be transformed into operating procedures that assure that anticipated hazards are actually dealt with. Procedures depend on the phase of the operation being addressed. It is in keeping with good safety and health practice to recognize a difference between *temporary operations* and *normal operations*. It is sometimes necessary to bypass certain automatic protection systems during *temporary* or *initial startup* operations, but it is still necessary in some alternative fashion to address the hazards that are thus uncovered. In an emergency, some processes must continue to be operated in an *emergency operate* mode. Of particular interest is the need to know under what conditions an emergency shutdown becomes necessary, and if it does become necessary, what must be done.

A key feature in the safe operation of a process is the capability to recognize when something has gone wrong. For this, a process needs to have preset limits for the variables under control. For instance, a centrifugal pump on a pipeline operates normally with a prespecified minimum suction pressure and maximum discharge pressure. Any time the pressure dips below the prescribed minimum on the intake side of the pump, an automatic shutdown becomes necessary to protect the pump. Pressure on the discharge side in excess of prescribed limits might exceed the design limits for the pipeline. Either of these conditions can be used to trigger emergency action to avert a more serious situation, especially when dealing with hazardous chemicals. The operating plan should tell workers what the consequences of control limit deviations are, as well as what to do to bring the process back under control. Rapid-response, on-line HELP display screens are a resource for providing such information in time to be of benefit in an emergency.

TRAINING

It is well known that operating procedures are often tucked away in some ring binder that no one reads or heeds. When catastrophic release of dangerous chemicals are at stake, however, "paper plans" for process safety are simply not good enough. There must be training for personnel who are required to execute the plan. An effective training plan has four ingredients:

1. initial training for new operators or new processes
2. refresher training at prescribed intervals, and in any event at least every three years
3. verification or testing to prove that employees understand the process and safe procedures and are current
4. documentation to confirm that the training and testing have been carried out

One method of documentation is to have employee cards that verify their currency in the process. The problem with this strategy is that employees may lose or fail to carry the cards. Federal standards do not say that the documentation must be carried on the employee's person; an employee record that verifies the training is more appropriately kept by the employer for each employee who works with the process.

CONTRACTOR PERSONNEL

This chapter began with a discussion of some major catastrophes that led to the development of the process safety standard. In the 1989 Phillips Petroleum catastrophe, some of the dead and injured workers were employed by outside contractors, not Phillips. There is no question that this catastrophe triggered a response by OSHA to include some language to protect contractor personnel in the new process safety standard, which was under preparation at that time. OSHA already had knowledge that a significant number of petrochemical and other companies were using outside contractors to perform work at their plants. In the aftermath of the Phillips catastrophe, OSHA commissioned a study by the John Gray Institute of Lamar University to study safety and health issues as they relate to contract work in the petrochemical industry.

What emerged in the OSHA standard was a requirement for the prime employer to exert a measure of control over the contract operations and conduct of contract employees. Most employers already had procedures for controlling access to their facilities by contract personnel. However, the new process safety standard required employers to check into the safety record of prospective contractor firms before contract. Information known about process hazards must be communicated to contractors as well as applicable provisions of the prime employer's emergency action plan. In addition, the prime employer is required to perform periodic evaluations to assure that the contractor is doing the job with respect to OSHA process safety standards. Perhaps the most visible requirement of all is that prime employers are required to maintain an injury and illness log for contractor employees.

Notwithstanding the new responsibilities placed on the prime employers, the contract employers are still responsible for providing safe and healthful workplaces for their own employees. Training, for instance, must still be provided by contract employers for their employees. Even for safety rules originated by the prime employer for the hazardous process, the contract employers must assure that their own employees follow the prime employer's rules.

Historically, one management strategy for dealing with difficult safety and health matters has been simply to contract away the dangerous operations or dangerous parts of an operation. The rationale behind this strategy has been to pass the responsibility for safety and health along to the contractor or subcontractor. When the OSHA inspector visits the facility, the prime contractor could thus say that OSHA did not apply because his or her own employees were not exposed to the hazard. The process safety standard, however, has removed a large proportion of the incentive for pursuing this avenue of escape from responsibility for compliance with safety and health standards.

ACTS OF TERRORISM

This chapter and the OSHA process safety standard deal with disasters and major catastrophes and how to prevent them with process planning and design. There is no doubt that the type of catastrophe represented by Bhopal, India, and the Phillips explosion and fire can be prevented, or at least mitigated, by the measures described thus far in this chapter. On September 11, 2001, however, the world was changed forever when a new type of disaster surpassed the devastation of Bhopal and Phillips combined. The deliberate attack on the twin towers of the World Trade Center in New York City and the Pentagon in Washington, D.C., on that fateful day killed more than 3000 people.

The world sees acts of terrorism daily and suicide bombings often, but September 11, 2001, brought an unprecedented new dimension to the perception of terrorism. The September 11 attacks consisted of multiple teams of terrorists, coordinated to act on the same day and strike within minutes of each other. The origins of the strike teams that morning were in different states, and each team had several members. All of the actual strike team members were committed to suicide in four separate groups, corresponding to the four commercial airline flights that were deliberately crashed that morning in the execution of the plan. The precision and coordination of the attacks and the similarity in execution made it clear that, in addition to the actual suicide teams, there were uncounted backup supporters who had some knowledge of the strategy and its execution. The September 11 attacks were obviously not isolated, impulsive strikes by a few fanatical extremists. Rather, they were a meticulously planned, broad assault on a nation by a large and well-supported organization. The reality of September 11 forced all Americans to grapple with the potential of further coordinated attacks and a general war against terrorism that would be fought primarily in the United States. In addition to the individual shock and response, employers were forced to plan for future acts of terrorism in the form of precautionary, preventive measures, as well as disaster plans to deal with actual occurrence of potential future catastrophes on their own premises.

The solution to this grave problem, which threatens every citizen whether in the workplace or elsewhere, is not an easy one. Likewise, it is difficult to conceive an appropriate governmental and regulatory response to this type of problem. The federal agency OSHA was faced with the dilemma of the threat to thousands of workers' lives, and yet there seemed to be no readily conceivable regulation or standard that could prevent recurrence of the bloodshed. For the first time in its history, OSHA was faced with a serious threat to workers' lives and had no response that seemed to have any potential to prevent or even mitigate this hazard. OSHA citation of employers seemed absurdly inappropriate. This is not to say that OSHA did not respond to this tragedy after the fact; indeed OSHA committed a tremendous amount of resources to the aftermath of the tragedy, including dispatching emergency crews of inspectors who worked around the clock to support the cleanup operations. OSHA issued over 100,000 respirators in the first year of the cleanup (ref. OSHA's Role). They also issued gloves, hard hats, and other personal protective equipment. In addition, OSHA assessed hazards to clean-up workers, taking samples and testing atmospheres, and conducted fit testing for respirators. But despite these efforts after the fact, virtually all of the casualties of this tragedy were already beyond anyone's ability to rescue them. No OSHA standard had addressed the hazard of terrorist attack, and nothing OSHA had done could prevent this most dramatic attack on the safety of the American worker. The devastation of this attack has changed the way Americans view safety and health priorities. A new Department of Homeland Security seemed more compelling than the traditional mission of OSHA.

During and immediately after the September 11 attacks, the cameras were on the emergency responders—police and firefighters—who encouraged the nation with their heroism and devotion to duty. Much analysis and aftermath investigation would use hindsight to suggest that the emergency responders were disorganized and ineffective. But the public would not tolerate the use of hindsight to blame any of the heroic responders for actions or inactions in the heat of the crisis. To the contrary, the public rallied around the police and firefighters, and there was an outpouring of moral and charitable public support from across the nation. Although the rescue workers in the trenches were honored, some blame fell upon their departments for failure to support the rank and file, saying that the separate departments had a history of political competition and reluctance to share resources.

More than a year after the disastrous attacks, ideas began to emerge from thinking safety and health professionals who have pieced together remedies and steps for preparing for the worst of nightmares that could be faced in a workplace. One such thinking professional is Dr. Mark A. Friend, CSP, professor of occupational safety in the Department of Environmental Health Sciences and Safety at East Carolina University in Greenville, North Carolina. Dr. Friend, accompanied by university student researchers, visited the disaster sites at the Pentagon and in New York, where they made observations and gathered information from witnesses. Dr. Friend summarized his findings of "Lessons Learned" in *Responder Safety* (ref. Friend). Most of the recommendations dealt with national strategies for defense against terrorist attacks, but some were steps to be taken at the employer level.

One recommendation was to "limit building access." Although many such steps have been taken around government buildings and military bases, private companies could benefit from the planning of future buildings and parking lots. Parking lots that have close proximity to buildings facilitate car and truck bombing attacks. Subtle landscaping tricks can enhance the beauty and attractiveness of company buildings and facilities while protecting the buildings from explosive blasts from car or truck bombs. The protection can take the form of restrictions to the approach of vehicles or, as in landscape berms, in limiting the blast area. A more commonly seen temporary remedy has been the grim reminder rendered by huge portable, concrete barricade structures.

In times of crisis, often the large number of volunteers overwhelms the system charged with managing the effort. The September 11 experience has shown the need for what Dr. Friend calls "incident management." Volunteers may be naïve about requirements for personal protective equipment and how to use it properly. Untrained volunteers are at risk, and when they make mistakes, sometimes they put the professionals at risk, too. In addition, some of the so-called "volunteers" are really at the site out of curiosity and can interfere with the ongoing rescue work. Preparation for major disasters includes systems for screening persons authorized to be on site and the provision of a robust system of providing and checking credentials.

There is a subtle pitfall in the plan for a state-of-the-art communication system for appropriate command and control in times of crisis. The finest computer system in the world can lose its functionality if it is not used. Therefore, it is wise to find a way to integrate sophisticated emergency information systems into daily use so that they will be operational and ready to serve in times of crisis. There is no field in which obsolescence appears more rapidly than in the computer and information systems arena. To keep a computer system operational, it must be used and updated frequently. The same can be said of the people who use the computer system.

The tragedy of September 11 was more than could be fathomed at the time of the drafting of the OSHA process safety standard. Still, some of the same steps and procedures recommended for chemical processes subject to the hazards of catastrophic release or explosion can be used to plan for dealing with terrorist attacks. A good example is the use of "what-if" analyses, discussed earlier in this chapter. In the basic planning for chemical processes themselves, a new hazard to be considered is possible sabotage of the process by terrorists.

It remains to be seen what long-range changes to the typical workplace and to society in general will emerge from the experience of September 11, 2001. Perhaps the architectural style represented by the high-rise office building will become obsolete in time. Workweek routines and styles will likely change also. Even before the September 11 attacks many workers had gone to flex schedules and work-at-home plans. The reasons for these changes in work styles include the laptop computer, cell phones, FAX, and other technology developments that were totally independent of the terrorist attacks. In short, the American workplace is in transition. The safety and health manager's job will undergo much change in this transition, and so will the role of government enforcement agencies.

SUMMARY

Major industrial catastrophes—the shocking Bhopal disaster being the worst—alerted the world to the hazards of processes dealing with chemicals dangerous not only to plant workers, but to the general public as well. OSHA responded in the early 1990s with the Process Safety Management of Highly Hazardous Chemicals standard, which has had a major impact on industrial America, especially on chemical plants. The standard prescribes a systematic approach that includes gathering information about dangerous processes and analyzing that information to anticipate catastrophes and deal with them in advance. The analysis draws on recognized principles of safety engineering and analysis and on chemical engineering tools such as block flow diagrams and flow process charts. Once the analysis is complete, operating procedures must be developed and a training program instituted to assure that the fruits of the analysis are implemented.

Recognizing the role of contractor personnel in chemical process tragedies that have already occurred, prime employers are being held to greater responsibility for contractor personnel as well. This facet of the process safety standard whittles away at the employer excuse that the safety of subcontract employees is not their responsibility.

In the aftermath of September 11, 2001, disasters from runaway, unsafe chemical processes have been overshadowed by the threat of terrorist attacks. Some of the precautionary steps in preparation for chemical process disasters are likewise recommended for dealing with the threat of terrorism.

EXERCISES AND STUDY QUESTIONS

6.1 What two major catastrophes in the 1980s affected national policy with respect to process safety?

6.2 When developing the database of information regarding hazardous chemical processes, what is the primary source of information regarding the chemicals themselves?

6.3 Name two types of diagrams that are recognized methods of documenting the technology of a hazardous process.

6.4 List some example hardware details of engineering designs that might be included in a documentation of process technology.

6.5 Explain the term *what-if analysis*.

6.6 Give an example of how plant location can affect the analysis for process safety.

6.7 How is the human element considered in process safety analysis?

6.8 How would an employer benefit from employing professional analysts to analyze the safety of a process?

6.9 What is the principal advantage of using in-house operators and maintenance personnel on the process analysis team?

6.10 How often must process analyses be updated to assure their currency?

6.11 How long must records be kept to document processes and the update of the analyses of their safety?

6.12 Is it ever legal to bypass automatic process safety systems? If so, give an example.

6.13 Is it sometimes advisable to continue to operate a dangerous process in an emergency? Does OSHA prohibit this?

6.14 Explain how a control system can recognize when danger with a process has developed.

6.15 List the four ingredients of an effective training program for process safety.

6.16 What disadvantage is associated with an employee-held card system for documenting training for process safety?

6.17 The traditional role of prime employer and subcontractor is that the prime employer is responsible for hazards to prime employees, and subcontractors are responsible for hazards to their own employees. How has the process safety standard changed this relationship?

6.18 Under what circumstances might an employer need to maintain an OSHA injury/illness log for persons other than their own employees?

6.19 What popular strategy have employers used to dodge responsibility for the safety and health of their employees? How has the process safety standard limited this strategy?

6.20 **Design Case Study**. Use library resources and the format of Case Study 6.1 to gather information from standard chemical references to document the dangers associated with hydrochloric acid.

6.21 What is the "road-map" approach to compliance with the process safety standard and what is its principal advantage?

6.22 How might a poultry processing plant fall under the scope of the process safety standard?

6.23 How might a manufacturer of wrenches and tools fall under the scope of the process safety standard?

6.24 What pitfall lies in the documentation of process hazard analyses that leave significant issues open ended?

6.25 Under what circumstances might it be especially important to employ the services of a registered professional engineer in the evaluation of process equipment?

RESEARCH EXERCISES

6.26 Check out the Internet for details regarding the Phillips Petroleum disaster in October 1989.

6.27 Study international disasters, such as Chernobyl and Bhopal. Which was worse? Are these the worst industrial disasters ever? Why or why not?

STANDARDS RESEARCH QUESTIONS

6.28 Identify (by OSHA standard number) the OSHA General Industry standard for Process Safety Management of Highly Hazardous Chemicals. Search the NCM database to determine its frequency of citation. What percentage of the citations are designated as "serious?"

6.29 Search the OSHA General Industry standard for process safety for the term *training*. Determine whether training provisions of this standard are among the more frequently cited parts of the standard.

6.30 Search the General Industry standard for process safety to determine what roles for the employee or employee representative are specified in the standard. Have these provisions been cited by OSHA? If so, are any of the citations designated as "serious?"

Buildings and Facilities

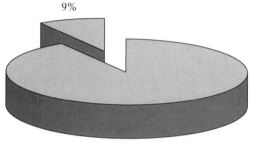

9%

Percentage of OSHA General Industry citations addressing this subject

We now turn to the business of examining hazards in various categories, highlighting applicable standards and suggesting methods of bringing about change to eliminate or reduce hazards. Most enterprises begin with a building in which to conduct operations, and this is an appropriate beginning for the examination of hazards. The enterprise is fortunate if its management considers hazards and applicable safety and health standards during its building design stages.

Safety standards for buildings, whether they be municipal, state, or federal, are usually called *codes*. For the most part, building codes apply to the construction of new buildings or to their modification. Thus, although building codes change constantly, those buildings built or remodeled before a particular change in the code are not required to be torn down and rebuilt or remodeled according to the new code. Needless to say, most buildings in existence do not meet the latest provisions of current building codes.

Some federal standards for buildings have been applied to all buildings, regardless of age. The standards have included matters of relative permanence such as floors, aisles, doors, numbers and locations of exits, and stairway lengths, widths, riser design, angle, and vertical clearance. Industry's objection has been that such standards are unfair, not only because they apply to existing buildings, but also because they are vague and generally worded. But despite these problems, industries have undertaken a large number of retrofit programs to update their buildings and facilities to satisfy federal standards.

In the defense of federal standards for buildings and facilities stands the fact that some of the most frequent categories of worker injuries and fatalities arise from improper building design, lack of guardrails, and problems with exits. Building equipment is designed and built without sufficient thought about the worker who must have access to this equipment to clean, maintain, repair, replace light bulbs, or otherwise service the buildings or facilities. Some workers are even in locations where they would be unable to escape in event of fire. Aisle widths are often set up arbitrarily, without giving thought to clearance between moving machinery and personnel.

Federal standards pertaining to buildings and facilities include the following categories:

- Walking and working surfaces
- Means of egress
- Powered platforms, manlifts, and vehicle-mounted work platforms
- General environmental controls

A few provisions of these standards generate almost all of the problems. This chapter will now single out those individual provisions and analyze each to determine what the safety and health manager should do to alleviate hazards and conform to standards.

WALKING AND WORKING SURFACES

One does not normally call a floor a *walking* or *working surface*, so why do the standards writers select such a complicated bit of jargon? This question is answered by reflecting on the hazardous locations in which people work. To be sure, many accidents, especially slips and falls, occur on floors, but consider the other walking and working surfaces—for instance, mezzanines and balconies. Then there are the even more hazardous platforms, catwalks, and scaffolds. Not to be forgotten are ramps, docks, stairways, and ladders.

Guarding Open Floors and Platforms

The most frequently cited standard in the walking-and-working-surfaces subpart is indeed one of the most frequently cited standards in all of the OSHA standards. It is repeated here in its entirety, due to its importance.

OSHA standard 1910.23—Guarding floor and wall openings and holes

(c) Protection of open-sided floors, platforms, and runways

(1) Every open-sided floor or platform 4 feet or more above adjacent floor or ground level shall be guarded by a standard railing [or the equivalent as specified in paragraph (e)(3) of this section] on all open sides, except where there is entrance to a ramp, stairway, or fixed ladder. The railing shall be provided with a toeboard wherever, beneath the open sides,

(i) Persons can pass,

(ii) There is moving machinery, or

(iii) There is equipment with which falling materials could create a hazard.

To many people, 4 feet seems an innocuous height, but the reason for this illusion is that they are thinking of jumping, not falling. Almost everyone has *jumped* without injury from elevations of more than 4 feet, but few adults experience a *fall* from that height without injury. A surprising number of *fatalities* result from falls at heights of only 8 feet. National Safety Council estimates place falls third, after workplace violence, as one of the leading causes of work-related fatalities (ref. Accident).

The illusion of safety of platforms in the range 4 to 14 feet has led to complacency on the part of building and facility designers as well as of safety and health managers. To stand on the extreme point of a precipice overlooking the Grand Canyon without a guardrail would seem foolhardy to the average person. But many "average persons" would think nothing of standing on the unguarded top of a tank 10 feet high. An unexpected event in such a situation can easily result in the worker taking a reflex action that results in a fatal fall.

The top of a tank, in the example just cited, was considered a *working surface* to be reckoned with by the facility designer. This is an important thing to remember. Even if the surface is the top of a tank, or even the top of a piece of equipment that is in the process of being manufactured, it may act as a temporary walking and working surface. But in unusual temporary situations, there are other ways to provide temporary fall protection for workers, as will be seen in Chapter 18.

Another problem area is the protection of personnel from falls from loading docks. A few loading docks are slightly less than 4 feet high, and the issue is thereby avoided. Some safety and health managers have had the area immediately adjacent to the dock built higher to comply with standards, but this is of dubious value in preventing injuries. Others have installed temporary, removable railings, and some have used chain-type gates. The chain-type gates do not qualify as "standard railings," but have sometimes been accepted as a practical substitute in an attempt to deal with the hazard in the face of a difficult situation.

A rooftop is, of course, a walking and working surface for roofing workers. This raises the question of whether guardrails, fall protection, or some other type of protection is needed for perimeters of rooftops on which persons are working. A provision of the OSHA Construction Standard specifies *catch platforms*, unless

1. The roof has a parapet.
2. The slope of the roof is flatter than 4 inches in 12 inches.
3. The workers are protected by a safety belt attached to a lifeline.
4. The roof is lower than 16 feet from the ground to the eaves.

The term *standard railings* mentioned in the platform guarding standard is further clarified by federal standards, and salient features are shown in Figure 7.1. Certain reasonable deviations from the specifications shown in Figure 7.1 are permitted. For instance, some other material in place of the midrail is permitted, provided that the alternative material provides protection equivalent to the midrail. Also, the height of the railing does not have to be exactly 42 inches. Railings of 36-inch height are acceptable if they do not present a hazard and are otherwise in compliance. This avoids the necessity of tearing down good railings and rebuilding to meet current codes, as has actually been done by some employers, as shown in Figure 7.2.

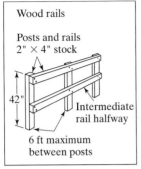

Wood rails

Posts and rails
2" × 4" stock

42"

Intermediate
rail halfway

6 ft maximum
between posts

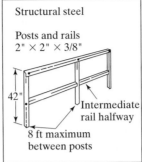

Structural steel

Posts and rails
2" × 2" × 3/8"

42"

Intermediate
rail halfway

8 ft maximum
between posts

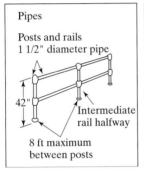

Pipes

Posts and rails
1 1/2" diameter pipe

42"

Intermediate
rail halfway

8 ft maximum
between posts

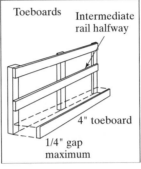

Toeboards   Intermediate
                   rail halfway

4" toeboard

1/4" gap
maximum

FIGURE 7.1

Standard railings (*Source*: NIOSH).

FIGURE 7.2

New guardrails. Some employers tore down their guardrails and rebuilt them 3 or 4 inches higher just to meet OSHA's standard in the early years of OSHA enforcement. Note the paint marks on the wall where the old guardrail had been attached. OSHA has since relaxed its rule somewhat.

One type of railing that is definitely not standard, but still can afford reasonable protection in some cases, is lines of parallel ductwork or pipes along the otherwise exposed sides. In many cases, parallel ductwork or pipes completely eliminate the hazard of the otherwise open-sided floor, platform, or runway. In fact, one might argue that such a floor, platform, or runway is not even "open-sided," provided that the parallel structure is sufficiently rigid.

In OSHA standard 1910.23(c)(1) quoted earlier, three conditions are enumerated, any one of which calls for a *toeboard*. Toeboards are vertical barriers along the exposed edges of the walking or working surface to prevent falls of *materials*. A standard toeboard is 4 inches high and leaves no more than a 1/4-inch gap between the floor and the toeboard. If the toeboard is not of solid material, its opening must not be greater than 1 inch.

Although 4 feet is the maximum height for open platforms in general, special situations require protection *regardless* of the height of the walking or working surface. Examples are open pits, tanks, vats, and ditches. When the hazardous opening size is small, it may be more practical to place a removable cover over the opening than to place a guardrail around it. Some safety and health managers have saved their firms a great deal of money by calling attention to this alternative. Another special situation is walking and working surfaces adjacent to dangerous equipment, where standard railings are appropriate.

Floors and Aisles

Judging by the level of OSHA enforcement activity, the most important consideration for floors and aisles is not how they are built, but how they are maintained. Federal standards for housekeeping require areas to "be kept clean and orderly and in a sanitary condition." The obvious question is "How is the safety and health manager to know what constitutes 'clean and orderly and in a sanitary condition'?" There is no clear-cut answer to this question, but some information can be gleaned from past cases of OSHA citations.

Example 7.1

In the railroad yard area of a steel manufacturing industry, piles of debris such as railroad ties and cables were lying about close to the track and presented tripping hazards to employees who must work in the area of the tracks. A complicating factor was that some of the debris was hidden by weeds.

Example 7.2

Dangerous accumulations of grain dust were found in many locations in a grain elevator. The dust was sufficiently concentrated so as to represent a health hazard to cleanup personnel and present a serious explosion hazard. Since the condition had been cited before, the citation was established as "serious and repeat," and the penalty was set at $10,000.

Example 7.3

A cluttered workshop had obstructions that some of the employees had to step over or go around to do their jobs.

Example 7.4

Leaking hydraulic cylinders dripped oil on the floor in a work area. No one was responsible for cleaning up the leaked oil.

Some conditions might *look* bad, but not represent a hazard. In one example, an OSHA compliance officer wrote a citation for piles (approximately 2 feet high) of sawdust around a planer and jointer in a woodworking shop. But the company resisted the citation because few employees would be in the area, and those few would be more inconvenienced than endangered. The company won its case.

It is also recognized that it is necessary in the course of many jobs for objects and materials to lie about in the work area. This is especially true of repair and disassembly work, and also of construction. What is generally considered excessive is material accumulation in the immediate work area in quantities in excess of what is actually required to do the given job, or scrap material lying hazardously about the work area in excess of a day's accumulation.

A prime measure of whether a housekeeping program is inadequate is the number of accidents such as trips and falls occurring in the area. Note that the preceding sentence was worded negatively in that too many accidents suggest an inadequate program, but "no accidents on record" does not prove that the work area is *free* of hazards. For every accident that is on file in the plant, the wise safety and health manager will have documented what steps have been taken to remove or reasonably reduce the offending hazard, whatever that hazard has been determined by analysis to be.

Finally, safety and health managers can look to their own industries for guidance. Any reasonable person knows that the definition of "clean and orderly and in a sanitary condition" is different in a foundry than in a pharmaceuticals plant. If the industry has any work practice guidelines, as put forth by trade associations, for instance, these guidelines would be very helpful in illustrating what is reasonable for a given industry.

The general housekeeping standard is targeted at a very common hazard and a very frequent source of workers' compensation claims: slips and falls. So important is this hazard to insurance companies that they sometimes have research facilities to study tribology (ref. Lorenzi). "Tribology" is the study of the mechanisms and phenomena of friction and applies to the study of trips and falls as well as to lubrication and wear of contact surfaces (ref. Lopedes).

Water on the floor is a problem in many industries, and constant surveillance becomes necessary to assure that the floor is kept clean and dry. In the design phase of buildings and facilities, attention to the problem of wet processes will suggest slopes and floor drainage systems to alleviate the problem. Another helpful device is to use a false floor or mat to provide a dry standing place for the worker who must work in a wet process.

Some buildings are not built to facilitate cleaning. The floors in others not only are poor from a cleaning standpoint, but also present trip hazards, such as protruding nails, splinters, holes, or loose boards.

Trip hazards due to uneven floors can result in serious injury. In one case in a steel mill, a difference of 1 to 3 inches existed at a point where a grating and a plate came together in the floor. The employee who worked in the area had the job of guiding red-hot steel butts from the end of a roller line. A fall could have brought the employee into contact with the red-hot butts of steel.

Aisles are important, and standards for aisles specify that permanent aisles be kept clear of hazardous obstructions and that they be appropriately marked. Ironically, the more clearly the aisles are marked, the more noticeable will be clutter or materials allowed to accumulate in the aisles. Conversely the more materials or obstructions are kept clear for forklifts or other travel, the more obvious will be the misuse of an aisle that is not appropriately marked.

There is a lesson to be learned from the irony just described. There is a tendency among safety and health managers to go out all over the plant, indiscriminately laying out aisles, and then to take pride in marking them according to appropriate standards. It is human nature for some managers to want to show everyone that they are taking positive action to enhance safety and comply with established standards. The unfortunate result is that aisles can become too regimented and so much time then spent policing them that production efficiency is lost. In fact, the effect on safety and health can thus become negative, with employer and employee alike wanting to say, "Do we really need that safety and health manager in the plant?" Aisle-marking procedures are a prime illustration of the principle that overzealous safety and health action can do more harm than good. Every time a decision is made to mark an aisle, the safety and health manager should stop and ask the question, "Can we keep this aisle free of material or other obstructions?"

Conspicuously absent from federal standards for aisles is a specific minimum width for aisles. State codes that reflect the National Fire Protection Association's Life Safety Code often specify a minimum exit access width of 28 inches. But the federal code is silent on this point except to use the language "sufficient safe clearances shall be allowed." Aisles in forging-machine areas receive special attention: "sufficient width to permit the free movement of employees." But this standard also does not specify an actual minimum width dimension. These two standards are therefore good examples of "performance standards."

The OSHA aisle-marking standard has left a mark of its own on industry. The term "appropriately marked" originally meant black or white or combinations of black and white for aisles. Untold thousands of dollars were spent by industries switching from the yellow stripes widely used for aisles to OSHA-acceptable white stripes. However, the white striping used was not as durable, and frustrated facilities superintendents would return in desperation to the more durable yellow stripe in order to maintain any stripe at all. The black-and-white rule became more and more unpopular until OSHA finally revoked the aisle color code as superfluous. In summary, the safety and health manager should be sure that aisles are well marked, but the choice of color for the marking stripe is of little importance. Thus, the aisle-marking standard was formerly a specification standard but is now a performance standard.

The discussion about floors thus far has concerned guarding of openings, surfaces, maintenance, and appropriate markings. Nothing has been said about the structural design of the floor itself or whether the floor can withstand the loads applied to it. One is reminded of the eight-story vehicle parking building in which the eighth floor collapsed on the seventh. The seventh floor was well constructed but was unable to withstand the shock load of the falling eighth floor, and from then on, all the floors came down like dominoes. An even more grim reminder was the Kansas City hotel tragedy of 1981, mentioned in Chapter 3. Almost no one really pays attention to whether a floor is overloaded. Earlier editions of this book referred to structural collapse of buildings to be so remote and rare that few people ever worry about the problem. Sadly, that statement will never again be true after the tragic events of September 11, 2001, when terrorist attacks caused the collapse of the twin World Trade Center towers in New York City.

Federal standards require marking plates for floor loads approved by the "building official." One of the chief complaints from industry safety and health managers is that the floor-loading standard does not explain its use of the term *building official*. Confusion over this term has led to phone calls to various agencies in attempts to identify some *public* official to come to the facility to make an engineering determination to use as a basis for the floor-load marking plates. Since the term "building official" is not defined, a better course of action for the safety and health manager is to secure the services of a competent professional engineer either inside or outside the company. This would show a good-faith effort to comply with the standard and would virtually eliminate the possibility of a hazard.

Floor-load marking plates are of little value if they are ignored by employees. Adherence to floor-load limits is an administrative or procedural matter and thus calls for vigilance in order to maintain compliance. A good system of records and inventories can make weights and locations part of the database of an overall management information system. If this system is computerized, the computer can monitor loads at all times and print out a warning message in the unlikely event that a distribution of inventories exceeds a floor-load limit.

Such a computerized system may seem to be an overreaction to the problem, and in fact it would be an overreaction for most companies small to medium in size. But in a large-scale computerized warehouse system, the floor-load monitoring system would represent such a small sideline to the overall computerized operating monitor that it would require only a few seconds of computer time per month. The system could be programmed to print out status reports monthly or on an exception basis whenever loads exceed or perhaps approach limits. The monthly status reports would be preferable to the exception reports because the monthly status reports would provide evidence of positive adherence to approved standards.

A word of caution is in order here. If there actually exists a violation of floor-load limits in the plant, no fancy computerized information system is going to hide it. Manual calculations can be used to determine whether a floor-load limit has been exceeded, and symptoms such as crushed, bent, or cracked floor members can be as embarrassing as they are hazardous.

The whole problem of floor and aisle design and maintenance is magnified by the use of mechanical handling equipment such as forklift trucks. Forklifts both aggravate the problems and at the same time become more of a hazard themselves whenever problems with floors exist. This problem will be revisited in Chapter 14.

Stairways

Building codes and standards for stairs are well established, but many factories and businesses have stairways that do not meet code. If the stairways have four or more risers, they need standard railings or handrails and must be kept clear of obstructions. Note the use of the words *handrail* and *railing*. **They are not the same**. A handrail, as used in this standard, is a single bar or pipe supported on brackets from a wall to furnish a handhold in case of tripping. A railing, however, is a vertical *barrier* erected along the exposed sides of stairways and platforms to prevent falls. A handrail and railing may of course be combined into the same unit, but the two terms are not interchangeable.

The OSHA standard is very flexible on the subject of the placement of stairway landings, using such language as "avoid" (long flights of stairs), and "consideration should be given to."

The placement of stairway landings is a safety consideration. Most people think that the purpose of stairway landings or platforms is to give the climber a chance to rest, and it is true that this is a supplemental purpose. However, the main purpose of the stairway landing is to shorten the distance of falls, and thus landings play an important role in building and facilities safety. Extremely long flights of stairs are obviously more dangerous than stairs interrupted by landings. To be effective, landings must be no less than the width of the stairway and a minimum of 30 inches in length measured in the direction of travel.

Ladders

Ladders are not the simple devices most people think they are. Design is critical—the construction should be neither too strong nor too weak. A weak ladder is obviously dangerous, but a ladder that is overdesigned is difficult or impossible to handle safely. A long, heavy ladder can be as much a hazard when it is carried as when it is climbed. It is easy to see why ladders must be manufactured to exacting standards.

Most firms buy ladders from reputable ladder manufacturers, and the safety and health manager can be quite sure that the ladders were constructed properly in the first place. What is more important to industrial safety is how the ladders are used and maintained. Defective ladders must be either repaired or destroyed, and while awaiting either fate, they must be tagged or marked "Dangerous; Do Not Use." Personal interviews with safety and health managers have indicated that they often saw defective ladders in half rather than risk the possibility of a defective ladder being returned to service. When a job needs to be done and there is no regular ladder in sight, the temptation is too great for maintenance or other personnel to remove the danger tag and make immediate use of a defective ladder. At best, a repair job on a portable ladder usually does not look very good and will likely arouse suspicion even if the ladder is safe. It takes a good engineer to convince anyone that the repaired ladder is just as good as new, even if the engineer has been able to convince himself or herself.

Portable metal ladders share common hazards with portable wooden ladders; a major difference, however, is the fact that metal ladders conduct electricity. Many workers are aware of the increased hazards of electrocution present when using metal ladders. Rubber or otherwise nonconducting feet are a good precaution on metal ladders, but the hazard is still present. The rubber feet should not be allowed to give rise to complacency.

Whether the portable ladder is made of wood or metal, it is the way in which the ladder is used that will chiefly determine its safety. Almost everyone knows the admonition that it is unsafe to ascend or descend a ladder with the climber facing away from the ladder. Less obvious is the fact that portable ladders are typically not designed to be used as platforms or scaffolds, and they become very weak if used at angles close to horizontal.

A common error in the use of ladders is to use ladders that are too short. For instance, when accessing a roof, the ladder needs to extend at least 3 feet above the upper point of support. For ordinary stepladders, the top should not be used as a step. Another foolish practice is to place ladders on boxes, barrels, or other unstable bases to obtain additional height. Some people even try to splice short ladders together to make longer ones.

The first consideration in the use of a portable ladder is its condition, especially the rungs. Next to the hazard of broken rungs, the greatest hazard probably is a ladder that slips or tips because it is insecurely positioned. The proper slant is 4 feet vertical to 1 foot horizontal. A safe practice is to tie off the ladder at the top so that it cannot tip or slide down. This is not always practical, however, and alternative solutions can solve the slipping problem. Sometimes a ladder can be made stable by positioning it where the structure of the wall or building limits its movement and makes the ladder safe. Another solution is to use nonslip bases, but this method may not work on some surfaces, such as oily, metal, concrete, or slippery surfaces. Metal ladders in particular may be subject to slipping and need to be equipped with good safety shoes. Safety shoes are as much to prevent slipping as they are to prevent electrocution. Even spikes may be useful on some surfaces to make the footing secure. If the ladder is used without safety shoes on a hard, slick surface, the ladder needs a footladder board to prevent slipping.

Fixed Ladders

From a safety and health standpoint, fixed ladders require a somewhat different approach than portable ladders. With portable ladders, the emphasis is on care and correct use, as just discussed. With fixed ladders, the emphasis is on design and construction. It is beyond the scope of this book to specify all of the details of fixed-ladder design. If fixed ladders need to be constructed, the designer should follow the detailed specifications laid out in established standards, not this book.

This book seeks only to alert safety and health managers to problems they may encounter with existing fixed ladders. Some fixed ladders may have been constructed long ago or perhaps without the benefit of standards. It is the duty of the safety and health manager to alert the company if fixed ladders are found to be incorrectly constructed. The safety and health manager can then recommend that design details be followed when the ladders are rebuilt.

Some knowledge of fixed-ladder design principles is necessary, however, if the safety and health manager is to know how to recognize common problems found. For brevity, this book selects some easy-to-observe problems with fixed-ladder design, illustrated in Figure 7.3. The rung offset is to prevent the foot from sliding off the end of the rung. Other means, such as siderails, would also be acceptable.

Long unbroken lengths of fixed ladders are dangerous, and ladder standards prescribe breaking the long lengths and separating successive lengths of the ladder by an offset equipped with a landing platform. Such offsets are required every 30 feet of

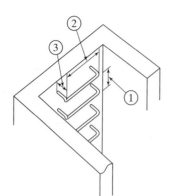

1. Distance between rungs not more than 12 inches and uniform.
2. Minimum rung length is 16 inches.
3. Back clearance minimum is 7 inches. This dimension is the most frequently violated, according to frequency of OSHA citation.

FIGURE 7.3

Easy-to-recognize problems with fixed ladders; all three are required by OSHA. Note the rung offset to prevent the foot from slipping off (based on OSHA standards).

unbroken length. When the fixed ladder is more than 20 feet long, but less than 30 feet long, protective cages are needed. But many safety and health managers do not realize that there is an alternative to installing ladder cages. On tower, water tank, and chimney ladders over 20 feet in unbroken length, ladder safety devices may be used in lieu of cage protection. Ladder safety devices usually take the form of a combination of fixed equipment on the ladder and personal equipment worn by the climber. One type uses a trolley that moves along a fixed rail. The trolley is attached to the climber's belt. The trolley moves easily up or down the rail when the climber ascends or descends, but a brake engages that stops the trolley in event of a fall. One thing to watch carefully with a rail-and-trolley device is to ensure that the device is equipped to deal with ice on the rail if the ladder is located in a climate where ice is a possibility (most outdoor locations in the United States). The safety problem with ice on the rail is somewhat indirect; the trolley simply cannot move along the ice-covered rail, making the system useless. But this is when the safety device is most needed. The problem is not without solution because there are deicing devices for trolley systems. Safety and health managers should choose their protection systems with care, however, being sure to consider all factors, including the problem of icing, in order to avoid later embarrassment and wasted investment on the part of the company. An example rail-and-trolley ladder safety device system is illustrated in Figure 7.4.

Dockboards

A dockboard, or bridge plate, provides a temporary surface over which loads can be transported, particularly during the loading or unloading of a cargo vehicle. One of the main safety hazards with these boards is that they may shift while in use. It is also possible that the surfaces connected by the dockboard can shift, such as when the cargo vehicle itself moves. Finally, the dockboard itself may not be strong enough to carry the load.

FIGURE 7.4

Example of ladder safety device system.

EXITS

Exits are usually considered doors to the outside and from a safety standpoint are considered a means of escape, especially from fire. Such thinking is accurate, but incomplete. The safety and health manager should enlarge the concept by using the more general term *means of egress* to include the following:

1. The way of exit access
2. The exit itself
3. The way of exit discharge

By considering means of egress instead of simply exits, the safety and health manager can analyze the entire building to determine whether every point in the building has a continuous and unobstructed way of travel to a public way. In this way, one must think of such building facilities as stairways, intervening rooms, locked interior doorways, and limited-access corridors. Outside the building one must think of yards, exterior storage of materials, fences, courtyards, and shrubbery. One thinks of shrubbery and landscaping as affecting neither safety nor health. However, employees might escape a burning building (or perhaps a building with a ruptured chlorine gas line) through an exit door, only to find outside that the exit empties into a courtyard tightly confined by a fence, dense shrubbery, or other obstruction.

Almost every safety and health manager at some time during his or her career encounters the embarrassment of an exit that is *locked*. This can be a two-edged problem because many safety and health managers are also responsible for plant security. In many cases, the only practical solution is to provide panic bars or other mechanisms for locking doors from outside entry while maintaining free and unobstructed egress from the inside. Where unauthorized exit can be as much a security problem as unauthorized entry, the automatic-alarm-sound type of door may be the only alternative. Facilities designers are turning more and more to the use of unlocked, automatic-alarm-sounding emergency exit doors. Even more frequently encountered than locked exits are exits that are cluttered or blocked by obstructions or impediments. Stacks of material obstructing the door or way of travel defeats the purpose of the exit.

If anyone ever doubted the importance of the OSHA requirement to keep exits unlocked and clear, that doubt was swept away by the tragedy that occurred in Hamlet, North Carolina, the morning of September 3, 1991. In one of the worst industrial accidents in U.S. history, 25 people were killed and another 56 were injured in a fire that swept through the Imperial Foods poultry processing plant (ref. Labar, 1992). The inferno was put out in only 35 minutes, but it had already done its damage. The 30,000-square-foot building, a converted 1920s vintage ice cream plant, was virtually windowless, and when the lights went out soon after the outbreak of the blaze, the 90 workers present were scrambling through a maze of production processing equipment in the dark trying to find a way out. To prevent product theft and to keep out flies, reportedly seven of the nine exit doors were routinely locked or bolted from the outside. The tragedy closed the plant permanently, and three members of top management were indicted for 25 counts each of involuntary manslaughter. The three indicted were the owner of the company, which is now bankrupt; the owner's son, who was working in the plant as operations manager; and the plant manager.

The entire nation will ever be more acutely aware of the importance of exits and building egress after the tragedy of September 11, 2001, when both colossal towers of the World Trade Center collapsed.

Americans with Disabilities Act

Attention to buildings and facilities assumed increased significance with passage of the Americans with Disabilities Act (ADA) in 1990. This law mandated that employers make reasonable accommodation for handicapped employees rather than deny them employment. This means that many changes to walking and working surfaces, exits, drinking water fountain levels, rest rooms, and other facilities became mandatory instead of voluntary. Often, the safety and health manager has been assigned the responsibility for compliance with ADA, if for no other reason than the fact that he or she has been accustomed to dealing with compliance problems with other federal regulations, such as OSHA standards.

ADA is being taken seriously by industry and public institutions as well. Expert consultants in building compliance for handicapped access can be retained to conduct an audit of facilities. Some of these consultants are themselves handicapped and derive their livelihood by visiting facilities in their wheelchairs and touring the building to check facilities for handicapped access. When a handicapped person in a wheelchair attempts to gain access to a building or its facilities, such as water fountains or rest rooms, and is unable to do so, a convincing case can be made to justify capital improvements in buildings and facilities. At the same time, the Safety and Health Manager should assure that OSHA as well as ADA standards are considered in the redesign of buildings and facilities.

ILLUMINATION

The subject of lighting was mentioned earlier. Lighting, or the lack of it, can be a safety hazard, but there is no code for minimum safe lighting except for specialized areas. For instance, if forklift trucks are operated in the plant area, the minimum general lighting level is two lumens per square foot unless the forklift trucks themselves have lights. Every exit sign should be suitably illuminated by a reliable light source giving a value of not less than five footcandles on the illuminated surface. This is not to say that the exit sign must be the kind that is *internally* lighted. An alternative to be considered is artificial lighting *external* to the sign. Also, there is nothing wrong with relying on *natural* illumination (sunlight) on the exit sign in an amount not less than five footcandles. Natural illumination can be a problem, however, if the area is accessed on second or third shifts. Incidentally, five footcandles is not very much illumination. Most plant areas are normally lit by much greater levels of illumination.

In the General Industry standard adopted as the national consensus standard in the early days of OSHA, there was no general lighting standard. However, the omission seems to be an oversight on OSHA's part. More than a decade later, OSHA promulgated a standard for "Hazardous Waste Operations and Emergency Response," commonly known as the HAZWOPER standard, discussed earlier in Chapter 5. The HAZWOPER standard focused on special emergency operations, but a little-known provision of this standard had to do with minimum illumination levels. Table 7.1 delineates these minimum illumination levels and is taken directly from the HAZWOPER standard. Review of the table clearly reveals that it was originally conceived to be a general table of illumination levels, not just a special table for HAZWOPER operations. For instance, the table specifies levels for "warehouses," "general shops," "barracks," "dining rooms," and "offices," all of which describe areas that might be associated with general employment, not only "emergency response" operations. OSHA has not seen fit to attempt to promulgate a general illumination standard for all workplaces. Instead, the agency has adopted such a standard within specific standards only. Another example standard that contains the Table 7.1 illumination minimums is the OSHA Construction standard.

TABLE 7.1   Minimum Illumination Intensities in Footcandles

Footcandles	Area or operations
5	General site areas.
3	Excavation and waste areas, accessways, active storage areas, loading platforms, refueling, and field maintenance areas.
5	Indoors: warehouses, corridors, hallways, and exitways.
5	Tunnels, shafts, and general underground work areas; (Exception: minimum of 10 footcandles is required at tunnel and shaft heading during drilling, mucking, and scaling. Mine Safety and Health Administration approved cap lights shall be acceptable for use in the tunnel heading.
10	General shops (e.g., mechanical and electrical equipment rooms, active storerooms, barracks or living quarters, locker or dressing rooms, dining areas, and indoor toilets and workrooms.
30	First-aid stations, infirmaries, and offices.

(*Source*: OSHA standard 29 CFR 1910. Subpart H, Table H-1.)

MISCELLANEOUS FACILITIES

Maintenance Platforms

The importance of planning for maintenance activities when constructing a new building was pointed out earlier in this chapter. Many modern buildings have built-in, safe suspension systems for exterior window cleaning and other exterior maintenance. Maintenance workers for buildings not so equipped are less fortunate and typically work from suspended scaffolds of the same type as construction scaffolds, which are discussed in Chapter 18. Not only are the maintenance workers less fortunate, but so are the employers of these workers and the safety and health managers who must worry about the safety of the scaffolds, the proper securing of scaffolds on the roof of the building, and other items governed by applicable standards.

Safety and health managers who do find that their buildings are equipped with powered platforms for exterior maintenance should direct most of their attention to how these platforms are being used and maintained, not how they are made. The manufacturer of such equipment would normally be very careful to adhere strictly to standards when fabricating the powered maintenance platform. Typical problems with these platforms are missing guardrails, missing toeboards, missing side mesh, disabled safety devices, and inadequate inspections or records of inspections.

Regarding the equipment itself, some companies have been tripped up for not having load-rating plates on the platform. The load rating must be stated in letters at least $\frac{1}{4}$ inch in height. The wire rope suspending the platform must also be marked with a metal tag stating its maximum breaking strength and other data, including the month and year the ropes were installed.

Workers on some types of powered platforms need to wear safety belts; on other types, they are safe without the belts. A platform supported by four or more wire ropes can be so designed that the working platform will maintain its normal position even if one rope fails. However, many powered platforms are suspended by only two wire ropes and will tip dangerously if one of the ropes fails. One of the more dangerous types of platform is known as "type T," and workers on these platforms must wear safety belts attached by lifelines. If the platform qualifies as type T, it will upset with a single wire rope failure, but it will not fall to the ground. Therefore, the lifeline may be attached either to the building structure or to the working platform. Compare this to construction industry standards (Chapter 18), which require that lifelines be secured to an anchorage or structural member instead of to the scaffold.

Public-utilities workers and tree trimmers often use platforms that are vehicle mounted, such as aerial baskets, aerial ladders, boom platforms, and platform-elevating towers. Again the majority of accidents arise from improper use of the platform rather than from equipment failure or design. This is even more true of the vehicle-mounted platforms than of the building-mounted models discussed earlier.

The most serious hazard with vehicle-mounted platforms is contact with high-voltage power lines, and this kills workers every year. This hazard is so severe that a safety distance must be maintained at all times—except, of course, in the case of electric utility companies who by the nature of their work must approach closer. For safety, the utility companies must insulate aerial devices that work closer than the standard safety distance. For nonutility companies, the accepted standard is a ten-foot distance in the case of a 40-kilovolt line, for example. Different line voltages may need higher or lower safety distances, and these distances are covered in more detail in Chapter 18 in the discussion of mobile cranes.

Sometimes special sensors called "proximity warning devices" are installed on the boom to warn the operator when the basket is too close for safety. However, these warning devices do not provide positive protection and thus should not be considered an excuse for moving the boom closer to the line than authorized minimums.

Workers in aerial baskets often fail to wear a body belt and lanyard attached to the boom. Adding to the fall hazard is the possibility of unexpected contact with an object that might strike and perhaps sweep the worker out of the basket or off the platform. Echoing the hazards avoidance principles of Chapter 3, such unusual hazards point to the importance of training for personnel who work in aerial baskets. Other unsafe procedures are failure to secure the aerial ladder before traveling; climbing or sitting on the edge of the basket; or improvising a work position other than the floor of the basket.

Elevators

Elevators are everywhere, but when can you remember one falling? The catastrophic fall of an elevator is such a horrifying thought that the public long ago set up regulations for safe elevators. Jurisdiction was placed within the states, and most states administer elevator inspections through "labor" or "labor and industry" commissions. Next time you ride in an elevator, look at the certificate of inspection posted inside the elevator car.

Elevators must be inspected both when new (or altered) and periodically thereafter. Many states even require construction permits from the authorized elevator inspection agency before elevator construction is begun. Elevator operating permits and fees are also required by some states. Not every inspector must be an agency official, but state licensing procedures for elevator inspectors may be applicable.

Manlifts are used as elevators, but unlike elevators, there are federal standards for manlifts. Manlifts are much cheaper and more efficient than elevators for many plant operations and are thus sometimes used instead of elevators. As can be seen in Figure 7.5, however, a manlift is inherently more hazardous than an elevator. It is ironic that elevators—which are safer than manlifts—are governed by strict state inspection, licensing, permits, and approvals. With manlifts, though, it is up to the safety and health manager to interpret general standards and identify hazards. It is apparent from Figure 7.5 that the biggest hazards with manlifts are getting on and getting off. Exit is essential because the belt is continuous, and to stay on the belt past the top or bottom floor would be either impossible or extremely hazardous.

Boilers

Steam boilers and pressure vessels are so safe today that most people do not even think about them. It has not always been that way. Although steam boilers are not as popular today for building heating systems, many are still in use; in addition, industrial processes use hundreds of thousands of boilers and pressure vessels. So the lack of familiarity with boiler accidents is not because the boilers themselves are rare—it is the

FIGURE 7.5

Manlift: a continuously moving belt on which workers may ride both up and down.

accidents that are rare. When an accident does occur, the energy released by the explosion is so devastating that it usually produces a catastrophe.

The extreme hazard of an unsafe boiler led to the early regulation and safeguarding of these vessels. As with elevators, the historical development of boiler codes has placed their jurisdiction within the states. State control has been very effective in keeping boiler accidents to an absolute minimum.

The safety and health manager needs to ensure that boilers and pressure vessels in the plant are being inspected and that state procedures are being followed. One question that immediately comes to mind is "What pressure vessels are covered by the regulations?" Most states exempt containers for liquefied petroleum gases (LPG), as these are covered by other regulations. The same can generally be said of vessels approved by the Department of Transportation for public highway transportation of liquids and gases under pressure. Some states also exempt vessels used in connection with the production, distribution, storage, or transmission of oil or natural gas. Note, however, that the foregoing does not exempt refineries or chemical plants that produce petroleum *products*. In the industry, the term *oil production* refers to the drilling and extraction, or "mining," of petroleum, not its refining. The only way to be sure about exemptions in a given state is to check with that state's agency of authority.

The time to stop and think about regulations for boiler and pressure vessel safety is whenever such a vessel is to be purchased, installed, modified, moved, or sold. Welding on such vessels may weaken them and is scrutinized carefully by the inspector, although welding is not absolutely prohibited. Even hot-water storage containers should be installed or reinstalled by persons properly licensed to do the work.

SANITATION

The sanitation of lunchrooms seems straightforward and obvious, but sanitation decisions can be trickier than they appear. If a decision is made to allow employees to eat in the plant, principles of hygiene must be observed. The safety and health manager should be sure that a sufficient number of waste receptacles is provided to avoid overfilling. But before going overboard, the safety and health manager should realize that too *many* waste receptacles can be provided also. If too many receptacles are provided, maintenance personnel will become lax about emptying containers that receive little use, resulting in additional sanitation problems.

The presence of toxic materials complicates the whole problem of food service, consumption, and storage. Certainly, food and beverages must not be stored in areas where they will be exposed to toxic materials. This rule may seem obvious, but the safety and health manager should consider not only the plant cafeteria or lunchroom but also the employee who brings snacks from home and stores them in areas in which they could be exposed to toxic materials.

Some toxic materials, such as lead, are particularly susceptible to exposure by ingestion during food consumption. Some toxic materials, such as vinyl chloride and arsenic, are of such concern that there are strict, specific standards for their control. Toxic substances will be discussed in more detail in Chapter 9.

SUMMARY

Safety and health managers who are willing to plan ahead can save their companies a great deal of money by heeding building and facility codes *before* commencing construction or expansion of plant space. Planning ahead is the key to compliance with standards for floors, aisles, exits, and stairways. Guardrails, ladders, and platforms may be added, but they, too, deserve some advance consideration to ensure that installations meet requirements.

Safety and health managers need to be careful not to get too enthusiastic and provide for too many aisles or exits. An extra aisle or exit that is improperly maintained or marked can easily lead to problems and indeed is not even in the interest of safety.

The subject of buildings and facilities may not be the most exciting topic for the safety and health manager's attention. However, even ordinary matters such as housekeeping and sanitation deserve careful consideration and judgment to promote safety and health at reasonable cost.

EXERCISES AND STUDY QUESTIONS

7.1 What is the height of a standard railing for walking and working surfaces?

7.2 A portable ladder is needed to climb onto a rooftop 14 feet high. How long should the ladder be?

7.3 What are the two principal federal requirements for aisles in industrial plants?

7.4 Explain why the title for a major OSHA subpart is "Walking and Working Surfaces" instead of simply "Floors." Name 10 different walking and working surfaces.

7.5 What is deceptive about the danger of an open-sided floor or platform only 4 feet high?

7.6 Suppose that in your plant, welders must stand on top of a 10-foot-high tank to complete manufacturing operations. What should be done, if anything, to protect them from falling?

7.7 What is the purpose of a toeboard?

7.8 Explain under what conditions, and how, roofing workers should be protected from fall hazards.

7.9 When must an open-sided floor or platform be equipped with a railing?

7.10 When would OSHA require guarding for a service pit for use in vehicle maintenance? How would you guard such a pit?

7.11 Explain how poor housekeeping could result in an OSHA citation that classifies the violation as "serious."

7.12 As a safety and health manager, how would you deal with the problem of housekeeping?

7.13 Explain the danger of marking too many permanent aisles.

7.14 Explain the difference between a handrail and a railing.

7.15 Describe the purposes and requirements for stairway landings.

7.16 Describe important points to consider in safety with portable ladders.

7.17 Under what conditions are ladder cages specified? When may alternative ladder safety devices be used? What are the advantages and disadvantages of these devices?

7.18 Define the term *means of egress*.

7.19 What are the biggest problems with exits?

7.20 What are vehicle-mounted work platforms? What are their chief hazards?

7.21 Explain how either too many or too few waste receptacles can lead to sanitation problems.

7.22 Describe the case history in which locked exits resulted in multiple fatalities, criminal indictments, company bankruptcy, and the permanent closing of a plant.

7.23 The decade of the 1990s saw an increase in the significance of buildings and facilities design. What landmark legislation caused this increase?

7.24 What new responsibility was placed into the hands of many safety and health managers in the decade of the 1990s, and why?

7.25 According to the National Safety Council, how do falls rank as a cause of workplace fatalities?

7.26 Discuss the problem that is associated with promulgation of federal standards for building code issues.

7.27 What is "tribology," and in what way is it important to occupational safety and health?

7.28 Compare the Life Safety Code standard with the OSHA standard for aisle width in industrial plants. Which is worded more in performance language and which is worded more in specification language?

7.29 In what way has OSHA's aisle-marking standard changed from specification to performance wording?

7.30 Explain the purpose of a "footladder board."

7.31 Explain why it is acceptable in general industry to secure the worker lifeline to the scaffold itself if the scaffold is "type T."

7.32 Is it OK for just anyone to install a hot-water tank? Why or why not? Explain.

RESEARCH EXERCISES

7.33 Check the Internet for details about the Imperial Foods disaster in 1991.

7.34 Check the Internet for details about the Triangle Shirtwaist disaster in 1911. What lessons could have been learned from this disaster that would have prevented the Imperial Foods disaster 80 years later?

STANDARDS RESEARCH QUESTIONS

7.35 This chapter referred to "one of the most frequently cited standards in all of the OSHA standards." Find this standard and search the NCM database to determine the frequency of citation. Compare the citation frequency of this standard with the frequency of citation of the General Duty Clause.

7.36 The maintenance of aisles is an important consideration in controlling the hazards of work areas. Find the OSHA General Industry standard for aisle maintenance and determine its citation frequency using the NCM database.

7.37 Search the OSHA General Industry standard for the provision prohibiting locked exits. Determine whether it is frequently cited by searching the NCM database. Are any of the citations for locked exits contested by the employer?

CHAPTER 8

Ergonomics

No other topic in this book saw more controversy and political attention in the closing decade of the 20th century than ergonomics. Just what is ergonomics, and why has it generated so much debate? This chapter will address not only the subject of ergonomics as a field of endeavor, but also will address why many employers fear that it is a threat to the continued operation of their businesses. The controversy will be scrutinized to determine whether the fear is of ergonomics itself or a fear of how federal enforcement authorities will apply it and whether or not that fear has any basis in fact. We will also explore the regulatory history of the embattled standard that was promulgated in the 1990s and then was immediately set aside by Congress. After failure of the standard, OSHA has continued to enforce ergonomics practices, especially in certain industries, such as health-care and poultry processing, relying upon the General Duty Clause as a basis for citation. OSHA's enforcement actions will be reviewed and some light shed on why OSHA focuses upon certain industries when it looks at ergonomics. Finally, an attempt will be made to map a strategy for success with ergonomics in the arena of occupational safety and health.

FACETS OF ERGONOMICS

A key to understanding the controversy surrounding ergonomics is to understand what ergonomics is and what activities can be considered to be a part of the field. Even the definition of ergonomics is somewhat controversial, so in this text, we will refer to the definition that has been forged by the national consensus committee, which has spent nearly a decade refining the draft ANSI standard for ergonomics:

> *Ergonomics* is a multidisciplinary science that studies human physical and psychological capabilities and limitations. This body of knowledge can be used to design or modify the workplace, equipment, products, or work procedures to improve human performance and reduce the likelihood of injury and illness.

> (*Source*: ref. Work-Related)

From the definition, it can be seen that the field of ergonomics covers a broad spectrum of activities involving human activity. Reducing the likelihood of injury or illness, as beneficial as that goal is, is only one of the objectives of the field. Improving human performance is another key objective, and, historically, may even be more important to the field of ergonomics.

Ergonomic Vehicle Design for Human Performance

Much thought goes into the design of automobiles and, especially, the cockpits of aircraft in order to improve the driver's or pilot's ability to assess problems and take appropriate action in a timely manner. The field of ergonomics has developed design principles based on research that has shown how people react to various stimuli. What we are talking about here is the appropriate interface between human and machine to achieve the best possible result of the machine performing the intended task for the operator controlling it. If the equipment being designed is a vehicle, there are obvious implications for safety, not only of the operator, but also of the general public.

There are certain conventions about machine interface that have grown up historically before the word *ergonomics* was ever coined. Some of these conventions tend to cause confusion for the vehicle operator, or the operator of other equipment for that matter. For instance, water faucet valves usually turn on and increase volume by twisting counterclockwise; however, electronic knobs, such as volume controls, usually turn on and increase by twisting clockwise. Other conventions are more consistent, such as the pushing forward of a lever, which usually means "engage" or "go." Other conventions or principles are concerned with the placement of instruments or dials, priorities for the attention of the operator, and rules to avoid inadvertent actuation of critical controls due to their proximity to other frequently used controls. The consideration of these principles, which are for the most part design decisions, are the objective of the practitioner in the field of ergonomics. Design engineers were once thought to be naturally equipped to consider such "commonsense" principles of safe and appropriate design. But notorious examples of the failure of designers to consider the human interface has led design and development teams to employ experts in the field of ergonomics to provide input to the design process.

Designing Safety Features into Workplace Machines

The workplace safety professional can see examples of the application of ergonomics principles in the design of machine controls, many features of which are specified by safety standards. One example is the design of punch press footswitches. A properly designed standard footswitch for activating a mechanical power press has a cover to prevent the operator or other personnel from accidentally stepping on the pedal or switch, thus causing an accidental cycle of the press ram, which could have disastrous results. Another example is the requirement that crane pendant controls be "dead-man" controls—that is, unless the operator is taking a positive action to depress the control, the spring-loaded button will return to a safe "no-action" position. The reader will recall that this principle was discussed in Chapter 3 under the topic "The Engineering Approach" to the control of hazards. Indeed, consideration of the principles of ergonomics should be included in any engineering approach to addressing equipment hazards.

Sometimes the ergonomics design principle has more to do with human behavioral characteristics than with physical actions, accidental or intentional. The footswitch design principle was simply to prevent accidental actuation. However, sometimes the operator will be motivated to *intentionally* take some dangerous action, to increase a production rate, for convenience, or perhaps for some other reason. An example is seen in the design of two-hand tripping devices for presses or other dangerous machines. Experience has shown that operators will want to take shortcuts in the repetitive elements of a machine cycle to increase production rates. Such shortcuts can be very dangerous. In the case of punch presses, to prevent amputations and serious injuries it is essential that both of the operator's hands are out of the danger zone when the ram is tripped. So, the designer specifies two-hand controls or trips, both of which must be depressed concurrently to activate the press ram and cause the machine to cycle. To counteract the very human motivation to bypass such controls, safety standards specify that two-hand controls be "anti-tie-down." This way, the operator cannot improvise a method of tying down one of the controls so that the machine can be activated with one hand while the other hand is free to facilitate machine loading and possibly be in the danger zone of the machine. An illustration of anti-tie-down design schemes can be seen in Chapter 15 in Figure 15.28. Note that the two-hand palm buttons are partially enclosed in cuplike housings that surround the periphery of the buttons preventing their tie-down by bars, pieces of lumber, etc. Another feature of the palm buttons is that they are often very rounded and smooth so as to not present a flat surface for mounting some type of improvised tie-down device. These design features of the machine can be considered applications of the principles of ergonomics.

Consideration of human behavior principles can also influence the design of machine logic control sequences so that a mistake or malfunction requires the operator to execute a certain prescribed sequence. It makes sense, for instance, that after a malfunction has occurred in an automatic sequence of steps, the machine might be programmed to revert to manual mode and require the operator to restart the automatic sequence from step one or other safe starting point. This seemingly elementary piece of logic has often been overlooked in the design of automatic sequential control of machine cycles, and the result is the development of a potentially dangerous situation. Case Study 8.1 will illustrate this concept.

CASE STUDY 8.1

AUTOMATIC MACHINE CONTROL SEQUENCE

Lead-acid storage batteries contain molded lead plates immersed in an acid solution. The manufacturing process in this case study required that the plates be formed in a hot, platen press operation in which the lead is pressed into shape in a hot, plastic state—too hot to handle, but not as hot as molten lead. Both heat and pressure are needed to complete the forming process.

Because of the necessity of handling dangerously hot workpieces, the process is sometimes automated. In such a process, the serious accident described in this case study injured the operator controlling the process. The workstation was set up to process the lead plate fabrication in several steps inside an enclosure that was not normally entered by the operator. Each step was triggered by sensors, such as mechanical limit switches, that told the logic controller when the previous step was completed and when it was time for the machine to index to the next step. A particular point in the cycle frequently presented a difficulty—when the molding platens opened and the completed, hot lead plate dropped out of the mold. Sometimes the lead plate would stick to the upper mold platen, requiring operator intervention. So frequent was this malfunction that the workstation was equipped with a wooden broom handle hanging near the control console so that the operator could stick it into the enclosure and knock the stuck plate out of the upper platen. The serious accident occurred when an operator was unable to knock the lead plate loose with the broomstick and he entered the enclosure to use his gloved hands to shake the plate loose. When he entered the enclosure area, somehow he accidentally brushed against a limit switch that triggered the automatic logic controller to conclude that the platen unloading operation had completed successfully. The programmed logic then set in motion actions to begin the next step in the process, which unfortunately was to close the mating platens. One of the operator's hands was in the danger zone between the platens when they closed with force and high temperature. Fortunately, the man was wearing a glove, but the hot platens still did enormous damage to his hand.

The irony in the story of Case Study 8.1 is that the programmed logic could easily have been devised by the designer to take the machine out of automatic mode whenever an operator was sensed to have entered the danger zone. Even without a sensor, the machine could have been programmed to time each step of the process and check those times so that whenever a plate stuck or some other malfunction occurred that was detected by the failure of a limit switch to be actuated in a reasonable time, the machine could stop and go into manual mode. Thus, the machine would not take another step in the sequence until the operator had returned to the console and either proceeded manually or pressed a "reset" switch. To accomplish this he or she would of necessity be out of the danger zone.

In the foregoing case, it can be argued that the designer cannot be expected to think of every careless thing the operator might do with utter disregard for his own personal safety. However, the problem here was with a frequently recurring malfunction, one in which every workstation was already equipped with a broomstick to deal with the problem. It would have been a simple matter to include a logic check in the automatic sequence to flag a situation in which a critical limit switch did not actuate when expected, signaling that the plate had become stuck in the upper platen. The programmed logic could then take the machine out of automatic mode and require the return of the operator to the safe console to restart the sequence on the control panel. This strategy would make sense whether the operator used the broomstick provided or entered the danger zone to free the workpiece. Either way, the machine would wait for operator intervention at the console before blindly going to the next step. The design feature needed in this situation is one that is intended to anticipate the human behavior aspect of the operation and take steps to mitigate the dangerous actions taken by the operator. Therefore, the ergonomic design process for the equipment should consider human behavioral factors as well as physical factors in the operation.

Controlling the Work Environment

Another facet of ergonomics that relates to the safety and health of workers focuses on the physical environment that surrounds the worker in the workplace. For most workplaces, the most important consideration in this regard is temperature. What limits of hot and cold temperatures are reasonable for the work environment, and what temperatures are optimal for various tasks? Ergonomics attempts to scientifically determine these temperature parameters and apply them to the workplace. Humidity is also a factor. Sometimes the demands of the job require a worker to work in a hot or a cold environment, and the consideration then becomes a matter of appropriate duration. How long should a worker be exposed to a work environment of a given temperature? If the temperature extremes are severe, a last resort is clothing to protect the worker and produce a microenvironment inside the clothing that is within acceptable limits. A principle studied in Chapter 3, the "three lines of defense," can be applied to this aspect of ergonomics. The control of the work environment through air conditioning or other means to control temperature and humidity to acceptable levels can be seen as the engineering approach, the strategy of first choice in dealing with the environment. If the engineering approach does not work, the strategy of administrative or work practice controls can be applied by rotating workers into and out of hot or cold environments so that the duration of their exposures is reasonable and within limits. The last line of defense is protective clothing, which is usually advisable in conjunction with the second line of defense, the rotation of workers to limit exposure.

For all of the attention that has been given to hot and cold work environments, OSHA has not been successful in promulgating a standard to address this hazard. As with many other facets of ergonomics, the hazards do not present a clear profile for absolute control by mandatory standards. Workplaces vary a great deal, and a universal standard for rigid limits of hot and cold have not received acceptance by the public. Hot or cold environments may violate the comfort level of some workers, but they do not violate legal limits. On a voluntary basis, employers usually control work environments in the interest of employee well-being and productivity, refuting the notion that employers will do nothing for employees' well-being unless required by legal standards.

Another aspect of the air, besides its temperature or humidity, is pollution. In this regard, OSHA and other regulatory agencies pay close attention to acceptable levels of pollution by various toxic air contaminants. This subject is so important that it is reserved for special treatment in Chapters 9 and 10.

Some miscellaneous considerations for special work environments should be mentioned at this point. Some workers must work under water, an environment that obviously requires special consideration and personal protective equipment. This work environment was at first overlooked by OSHA, but later a Diving standard was promulgated to protect these special workers. Even more exotic an environment is that of astronauts in outer space. Weightlessness, radiation, and the vacuum of outer space are all elements to be dealt with in the astronauts' environment. These unusual work environments may not be relevant to the mission of most safety and health managers, but they are mentioned here because the field of ergonomics does not overlook these unusual environments as scientists study the strange effects they may have upon the human body.

Manual Lifting

Chapter 14 of this text deals with all types of material handling operations, but the most basic material handling operation of all is manual lifting. Ironically, one cannot find a general OSHA standard for lifting, so most of Chapter 14 deals with safety standards for various types of material handling equipment, such as lift trucks, cranes, and slings. But ergonomics is more focused upon lifting, because it depends upon the human operator and applies stresses upon the human body. Manual lifting is one of the most studied subjects in ergonomics, but to date the studies are still inconclusive. It is not clear what weight limits a person can lift safely, and, despite all of the training "to lift with the legs, not with the back," back injuries continue to be prevalent, even in the industries that emphasize "proper" lifting techniques. Even NIOSH, the federal agency charged with the mission of studying hazards and recommending standards, has little respect for the benefits of training in proper lifting. This is evident in the following quote from the NIOSH "Work Practices Guide for Manual Lifting":

> Yet so long as it is a legal duty for employers to provide such training or for as long as the employer is liable to a claim of negligence for failing to train workers in safe methods of MMH (Manual Material Handling), the practice is likely to continue despite the lack of evidence to support it.

It is no wonder that OSHA has not seen fit to promulgate a general standard on lifting or on training for lifting!

The success of training in controlling manual lifting hazards is unclear. Another method of controlling the hazard is preemployment physical testing or screening to select personnel for lifting tasks. However, there are problems with this strategy also. In any preemployment test, the employer must use care not to discriminate against classes of employees. There is also the difficulty of devising a test that is really representative of the lifting operations of the actual job. If the test does not determine whether a person can do the actual job, it is ineffective and may be unfairly discriminatory. And since the passage of ADA (see Chapters 2 and 4 of this text), employers may be required to make reasonable accommodations for applicants who are unable to meet the lifting requirements of the job. This leads us to the next general facet of ergonomics, as will be seen in the section that follows. First, however, we will address the issue of personal protective "back belts."

Back Belts

Supporting belts worn around the waist are often worn by persons who do heavy lifting as a part of their job. The implication is that such belts prevent injury to the lower back. NIOSH decided to test this assumption in a preliminary study reported in 1994 and again in the late 1990s in what has been described as "the largest study of its kind ever conducted." The study examined incidence rates for workers' compensation claims for back injuries. The following comparisons were made:

(a) workers who wear back belts everyday vs. those who never wear them or, if they do wear them, do so only occasionally (once or twice per month);

(b) employers that require back belts vs. employers for whom back belt use is voluntary.

Both hypotheses showed no statistical significance in the difference between the groups in the incidence rates for workers' compensation claims. Besides the workers' compensation claims, the study also examined "self-reported back pain" and again the results showed no statistical significance in the difference between the groups. The study lasted two years and involved interviewing 9377 employees at 160 stores nationwide (ref. Back Belts).

In light of the NIOSH findings, why do workers still wear back belts? One reason may be that the worker or employer is uninformed or unconvinced by the NIOSH study. Another explanation is that workers derive some other benefit from wearing the back belts. Workers may gain comfort in the back belts, or perhaps they like to project the image that they have a strenuous job. Remember that, although the NIOSH study found that the back belts could not be shown to be of any benefit in preventing back injuries, it did not find that the back belts did any harm either. People just may like to wear them, even if they aren't the answer to preventing back injuries. Since back belts are not the answer to preventing back injuries, we now examine what else can be done for this and other special problems in which the human cannot comfortably handle the task.

Accommodating Individual Worker Characteristics

In the section on manual lifting, we saw that it is difficult to eliminate the hazards of manual lifting by training in proper lifting techniques or in the screening of personnel for the job. Also, personal protective equipment in the form of back belts does not appear to be the answer. Once again, the engineering approach to eliminate the need for manual lifting or to assist the worker by providing lifting aids is a desirable solution as it eliminates the hazard to the employee. For heavy lifting tasks, simple carts and dollies, as shown in Figure 8.1, have been necessary and desirable in general industry for decades, even before the age of ergonomics.

More recently, lift tables, as shown in Figure 8.2, have been used to facilitate the manual loading of workstations without requiring the worker to lift from the floor or pallet. The field of ergonomics has provided considerable motivation to use the types of devices shown in Figure 8.2. Sometimes the solution can be as simple as delivering material to be processed onto a platform raised to the proper height without the necessity of employing a lift table. The benefits for large, awkward workpieces is evident in Figure 8.3.

Besides the objective of changing the workplace so that workers in general can handle lifting tasks, there are many special requirements for accommodating special groups of people. The age of ADA has brought about a heightened sensitivity to the limitations and variations among the population of human workers. One very visible variation *not* to consider is gender. Granted, it is undeniably true that there are *general*

FIGURE 8.1

Simple carts and dollies

(average) physical differences between males and females, but beware of generalizing upon these physical differences and excluding a worker because of gender alone. For instance, the average man can lift a heavier load than the average woman, but there are exceptions. Some men are weaker than the average woman and cannot lift as heavy a load, and some women can lift more than the average man. Qualification for a job should be based upon measurable physical characteristics, not upon gender. The Civil Rights Act of 1964, the Americans With Disabilities Act, and other societal movements that began in the latter part of the 20th century have made the once-common practice of gender stereotyping appear ludicrous. Witness the following classic excerpt from a serious book on safety written in 1943:

> Mechanical aptitude makes boys play with trains, and lack of it makes girls play with dolls. To women, machinery is foreign to everything they have ever known or experienced, and quite bewildering... A woman should not be expected to acquire more than a few simple mechanical skills, but the task given her may be designed to require high performance standards because of her finger dexterity... This does not mean that in the training of

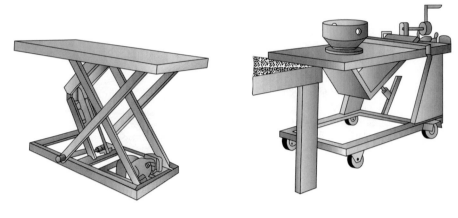

FIGURE 8.2
Lift tables

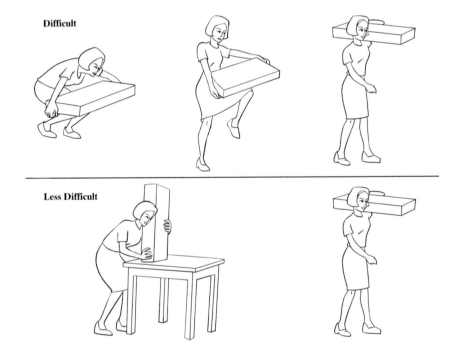

Difficult

Less Difficult

FIGURE 8.3
Store materials at the proper height.

women they will intuitively find the safe way to do their work. On the contrary, they seem always to fall naturally into unsafe and awkward habits. However, when the plant designs the safe manner of operation and so instructs them, they will respond favorably to it, but they require more constant supervision (ref. Foremanship).

My, how the world has changed since these words were written only a few decades ago! The foregoing quotation is not intended to ridicule the author, but rather to illustrate the dramatic changes in attitudes that have occurred in scarcely half a century. Authors, including the author of this book, run the risk of becoming dated, as society and future discovery make their writings obsolete.

Other human characteristics might need accommodation in the workplace. Examples are height—either short or tall—vision problems, and hearing problems. Age, like gender, is another sensitive characteristic to avoid in making generalizations about worker capability. Although many research studies have shown general differences among age groups, there are always exceptions, and specific physical capabilities of individual applicants should be measured and fit to the job, avoiding general, discriminating judgments about persons because of their age.

In summary, we have considered five facets of the broad field of ergonomics. They are ergonomic vehicle design for human performance, designing safety features into workplace machines, controlling the work environment, manual lifting, and accommodating individual worker characteristics. But there is one more facet that we have not considered, and it is the most important one for the safety and health manager. The remainder of this chapter will deal with this very important aspect of ergonomics.

WORKPLACE MUSCULOSKELETAL DISORDERS

Most of the activity and most of the controversy surrounding the enforcement of ergonomics in the workplace has surrounded the field that is currently designated "Musculoskeletal Disorders" or simply "MSDs." In Chapter 5, we dealt with the MSDS—the Material Safety Datasheet—not to be confused with MSDs, musculoskeletal disorders—the subject of this chapter. This complicated term is actually a generalization of more specific maladies that have been experienced in the workplace and have received significant attention on the part of both industrial safety and health managers and enforcement authorities. It is this part of ergonomics that has led to so much controversy and subsequent political action reaching a level as high as the U.S. Congress. A little history of the jargon surrounding these specific ailments leading up to the current emphasis on MSDs is in order.

Carpal Tunnel Syndrome

Most of the general public today knows something of the term *carpal tunnel syndrome*. This is a good starting point because much of the early history of the regulation of ergonomics resulted from worker complaints in this category. Carpal tunnel syndrome is a painful, possibly disabling dysfunction of the wrist. The condition is not clearly defined, but is generally thought to result from activities that require repetitive hand motion especially when the hands are required to be in an awkward posture. Tasks involving rapid

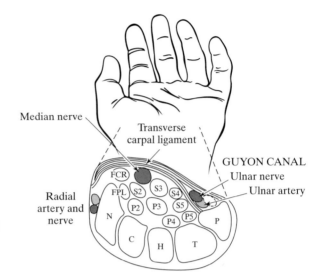

FIGURE 8.4

Schematic view of the carpal tunnel with the tendons of the superficial (S) and profound (P) finger flexor muscles, flexors of the thumb (FCR, FPL), nerves and arteries, carpal bones (P,T,H,C,N) and ligaments.

production, such as assembly or typing, are often associated with carpal tunnel syndrome. Kroemer, et al. (ref. Kroemer) report that carpal tunnel syndrome has been observed for more than 100 years. However, only recently have attempts been made to explain it and control conditions that may affect it. Kroemer credits Robbins with the first explanation of the anatomical basis for carpal tunnel syndrome in the early 1960s. Figure 8.4 illustrates a cross section of the wrist showing the crowding together of tendons, bones, and nerves within a sheath enclosed by the carpal ligament. The parts of the wrist must move within this sheath to give motion to the fingers in repetitive operations. It is understandable that awkward postures of the hand and wrist would further constrict the carpal tunnel area and give rise to discomfort from the moving parts. It also makes sense that highly repetitive motions would exacerbate the condition.

In response to increasing concern in the 1970s that repetitive type work was causing carpal tunnel syndrome, OSHA began to investigate firms in which operators were required to do such work with their hands—especially if the work was intricate or required awkward hand postures. However, OSHA had no standard to rely on for writing citations against employers whose employees were experiencing carpal tunnel syndrome. Lacking a specific standard for this situation, OSHA turned to the Section 5(a)(1) General Duty Clause, which was covered in Chapter 4. As is often the case with citations of the General Duty Clause, employers often contested the citations on the grounds that carpal tunnel syndrome is not really what Congress intended back in the 1960s when it drafted the clause to provide general protection against "hazards that are likely to cause death or serious physical harm" (to employees). However, OSHA was successful in making some of the citations stick using the General Duty Clause of the OSHA act.

Repetitive Strain Injuries

After scoring some successful citations of companies in which workers were experiencing carpal tunnel syndrome, OSHA began to enlarge its perspective with respect to the phenomenon of sore joints, muscles, and tendons. The carpal tunnel in the wrist was the classic example of the problem, but OSHA ergonomists and the body of professional practicing ergonomists reasoned that the carpal tunnel is not the only part of the body that could be irritated by repetitive motion. What about the neck, for instance? Did not workers get sore necks from jobs that required repeated motion of the head? And then there were sore elbows and sore shoulders. Therefore, the target "hazard" was shifted from "carpal tunnel syndrome" to a much broader term: *repetitive strain injuries* (RSIs)[1]. The term *carpal tunnel syndrome* went completely out of vogue among practicing professionals in the field because it limited the perspective of the practitioner as well as the enforcement powers of the regulatory officials. By the 1990s, the term *carpal tunnel* had become so out of style that it was conspicuously omitted from the definitions in the ANSI standard for ergonomics, and it is only briefly mentioned in the body of the standard as one of several different manifestations of disorders resulting from ergonomics hazards (ref. Work-Related).

Cumulative Trauma Disorders

Even the term RSI seemed too limiting in scope. Suppose that a worker experienced symptoms even when the job did not involve rapid, repeated motions? Certainly rapid, repeated motions were the most common exposures associated with sore tendons and joints, but some workers were found to experience symptoms even when their jobs did not require this type of activity. An even broader term was needed that would cover any type of trauma resulting from an accumulation of exposure over a period of time, though the worker is not injured from an occasional exposure. Thus, the term *cumulative trauma disorders* (CTDs) replaced RSIs. There is somewhat of a contradiction in the term *cumulative trauma*. The word *trauma*, used alone, means injury, usually violent injury. In Chapter 1, the term *injury*, which relates to safety, was identified as dealing with an acute exposure. Chronic exposures were identified as health hazards. However, the word *cumulative* suggests a building up of chronic exposures, not a single, violent injury. The word *cumulative* apparently carries more weight than the word *trauma*, because CTDs are generally considered a chronic exposure, not an acute one. The term *CTD* had a short life in the late 20th century, but has since been replaced with another term, *MSD*.

Musculoskeletal Disorders

If one could deal with the seeming contradiction in the term cumulative trauma disorders, it seemed it covered everything. However, there was still a problem with this term besides the contradiction. The problem was that the term itself implied a *cause* of the condition. It seemed inappropriate to assume that the worker had been injured from an accumulation of exposure to a hazard. Even worse, suppose a worker complained about pain in a joint and it could not be established that the worker had cumulative exposure of any kind? So as not to overlook any cause of the disorder, whatever the cause might be, the term *musculoskeletal disorders* (MSDs) became the new term used

---

[1]RSIs are sometimes identified as "repetitive stress injuries," but use of the word *strain* is preferred in this text, because, unless there is some sort of physical response to the stress, as in a strain, there is no injury.

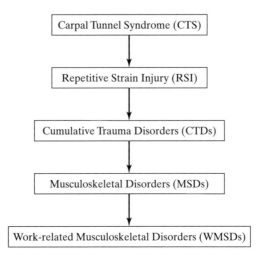

FIGURE 8.5

Historical progression of the recognition of WMSDs.

to describe all of the related worker conditions of this type, including carpal tunnel syndrome, rotator cuff syndrome, DeQuervain's disease, trigger finger, tarsal tunnel syndrome, sciatica, epicondylitis, tendinitis, Raynaud's phenomenon, carpet layers' knee, herniated spinal disc, low back pain, bursitis, and tension neck syndrome (ref. OSHA standard 1904.12(b)(1) and Work-related). The term *MSD* was modified slightly in the ANSI standard to narrow the focus to work-related exposures. Therefore, as of this writing, the term had finally morphed into *work-related musculoskeletal disorders*, WMSDs. Figure 8.5 illustrates the historical progression in recognizing WMSDs.

AFFECTED INDUSTRIES

Two landmark citations by OSHA in the area of ergonomics have been used as a model for the formulation of enforcement policy for controlling the hazard of WMSDs. The cited employers were Beverly Enterprises, a widely distributed provider of health-care services (especially nursing homes) headquartered in Ft Smith, Arkansas, and Pepperidge Farms in Downingtown, Pennsylvania, a maker of biscuits. A principal concern in the nursing home industry is heavy manual lifting on the part of nurses and health-care providers. A principal concern in the food manufacturing and processing industry is carpal tunnel syndrome and other chronic repetitive motion maladies. Meat packing, especially poultry processing, has been another industry that has had considerable Section 5(a)(1) General Duty Clause enforcement actions brought against employers by OSHA.

Certain industries may have particular problems with WMSDs, but one area of concern touches almost every industry, namely, computer terminals. It is estimated that more than half of all office workers in the United States make use of computer terminals in some way in their jobs. Occasional use is not a concern, but workers who sit at a computer terminal all day are reporting eyestrain, headaches, backaches, and neck and shoulder pain. Chapnik and Gross (ref. Chapnik) report a Wisconsin study by Sauter that shows a significantly greater incidence of these conditions among computer terminal operators than among other office workers. A consistently higher percentage of the 250 exposed subjects studied reported discomfort compared with the 84 workers who did not use computer terminals in their jobs.

The problem of eyestrain reported in studies of computer terminal operators is no surprise. What has been unexpected is the level of musculoskeletal aches and pains associated with computer terminal operation. There has been increased incidence of repetitive strain injuries, such as tenosynovitis, tendonitis, and carpal tunnel syndrome. OSHA's increased attention to ergonomic hazards, coupled with the dramatic growth in the use of computer terminals, ensures that this subject will remain of vital concern to the safety and health manager.

ERGONOMICS STANDARDS

Citation of Section 5(a)(1) of the General Duty Clause for ergonomics hazards was always a stopgap measure that was used in the absence of a relevant, specific standard for ergonomics. Therefore, throughout the decade of the 1990s, OSHA maintained a goal of developing a standard specifically focused on ergonomics.

OSHA Ergonomics Standard

A decade is a long time to develop and promulgate a standard. For ergonomics, it was never an easy process with representatives of organized labor pushing for a tougher standard and industry pointing out the costs and problems of compliance with a rigid standard. The development effort climaxed in the waning weeks of the Clinton presidential administration as OSHA officials pushed to get a standard on the books before the Bush administration took office. In this, they succeeded with the "11th-hour" final standard officially announced in the *Federal Register* November 14, 2000. The standard allowed industries 11 months to come into compliance, targeting October 15, 2001, as the effective date.

After nearly a decade of negotiation, the final standard emphasized the following main areas:

- information to employees
- quick-fix action to eliminate reported WMSDs that meet the "action trigger" defined by the standard *or* establishment of an ongoing WMSD program

The standard had a "grandfather provision" that permitted employers to continue with established ergonomics programs that met certain minimum criteria. Programs that were implemented to comply with the standard (not grandfathered) were required to have the following elements:

- Management leadership
- Employee participation
- MSD management, including access to a health-care professional (HCP), work restrictions deemed necessary by the HCP, protection of the worker's rights during the work restriction period, and follow-up evaluation of each MSD incident

- Initial and ongoing training of employees
- Program evaluation and follow-up
- Recordkeeping

One can see that documentation is important, as is always true of safety and health programs and standards.

The OSHA Ergonomics standard had a short life. As soon as the new Congress took office in 2001, the new Ergonomics standard was repealed by congressional vote, overriding OSHA's action. When Congress overrides and repeals any agency action, that agency is prohibited from resubmitting a slightly different version in a new promulgation.

Why was the standard overturned by Congress? Basically, industry was afraid of the cost speculated to be incurred in complying with the standard. Fundamental to the controversy over ergonomics is the question "What really constitutes a WMSD?" It is perfectly normal to experience some discomfort when adjusting to a new job. A key element in any ergonomics program, and one that will be noted by regulatory authorities, is whether the worker has an input. It only makes sense that the worker must provide input about which jobs entail discomfort. But when a worker complains of discomfort, who is to say that it is a genuine WMSD (that might become permanent or disabling) and not just adjusting to a new job? Another controversial issue is over the definition of injury or illness. Just what constitutes an injury or illness? Even if an injury or illness is established, there is always the question of whether it was caused by personal, off-the-job exposures or by the job itself. Finally, there is the question of remedies. How can the work environment be adequately "fixed?" And how can an injured worker be "cured?" The answer to many of these questions comes down to a judgment call—and if there is anything that an employer fears, it is the requirement to comply with a standard that is subject to interpretation or the judgment of the inspector.

OSHA Guidelines

The new OSHA administration ushered in with the Bush administration and the new Congress abandoned the strategy of promulgating a specific standard for ergonomics. In its place, OSHA unveiled a plan to issue guidelines to help control ergonomics hazards. OSHA would issue the guidelines for specific industries and encourage other general industries to construct guidelines of their own. The new strategy emphasized cooperation and the use of exemplary, successful, established ergonomics programs as models for assisting other industries. There was a provision for giving recognition for noteworthy ergonomics programs. There is no question that the new "guidelines" program emphasized the positive and relied on the judgment of the industries to take the initiative in developing programs to foster ergonomics solutions to problems and hazards. Indeed, OSHA pointed to Bureau of Labor Statistics reports that a decline in WMSDs had already been observed. The new guidelines approach de-emphasized enforcement actions, but the new administration retained an enforcement program for ergonomics using Section 5(a)(1) of the Act—the General Duty Clause. The new enforcement procedures emphasized serious hazards and were patterned after successful past prosecutions of industries in which the hazards were clearly serious.

ANSI Standard

During the entire decade of governmental action, promulgation, and OSHA enforcement developments (the 1990s), the private sector was quietly at work developing a general ergonomics standard for voluntary compliance. The form of the standard, written under the supervision and within the framework of the American National Standards Institute (ANSI), was directed at specifying a program for ergonomics in general terms and recognized the need for professional judgment to apply the program to specific work situations. The standard placed importance on the professional qualifications and training of those persons performing the required tasks. The standard specified a WMSD management program that bore a close resemblance to the program requirements specified in the former OSHA standard. The following program components were required in the draft standard:

- Management responsibilities
- Training
- Employee involvement
- Surveillance
- Evaluation and management of MSD cases
- Job analysis
- Job design and intervention

The last two components emphasized the analysis and prevention of future hazards before the occurrence of the WMSD cases. These components of the management plan in the ANSI standard went beyond the management plan requirements specified in the OSHA standard.

WMSD MANAGEMENT PROGRAMS

At this point, one may be wondering what plan to follow to have an effective Workplace Musculoskeletal Disorders program within a given company or plant. It is good strategy to have a working, documented program in place in any workplace that has exposure to hazards that can be categorized as related to WMSDs. It makes sense from the basis of

(a) worker comfort and basic well-being
(b) plant productivity
(c) reduction of workers' compensation claims
(d) compliance with the safety and health General Duty Clause

Drawing from the consensus of both the ANSI standard and the proposed OSHA standards, the general components that follow are thought to make up an effective WMSD program.

Administration and Support

The program should have both documented and real management support. Employees and supervisors should be trained in the causes and symptoms of WMSDs and be encouraged to report problems. The training also should consider the workplace and the tools and work equipment used to do the work. Proper use, maintenance, and adjustment of these tools and equipment may forestall future occurrences of WMSDs. There should be evidence of employee participation in the program. This is shown by records of employee reporting of WMSD symptoms before such symptoms become a serious problem. Employees also should be seen actually doing the work in accordance with the training they have been given in proper procedures for preventing the occurrence of WMSDs. Employee participation can also be shown in meetings about prevention of WMSDs, employee completion of surveys, and assistance with the design process for work, equipment, and procedures.

Surveillance

In addition to management involvement, training, and worker participation, a WMSD program should have a provision for what is called *surveillance*. This facet of the program insures that signs and cues will be used to signal the need for job analysis and the implementation of the principles of ergonomics. One sign of a need might arise from the review of injury and illness records in the facility. Another sign might be reports from employees, either of actual WMSD symptoms or perhaps just an employee concern about a situation that might carry a risk of WMSD exposure. The ultimate surveillance tool is an actual survey of any job suspected to be the cause of WMSDs.

Case Management

At the very least, an effective program should respond to WMSD cases as they are reported. This means that provision should be made for diagnosis, treatment, and recognition of necessary time for sufficient and timely recovery from the symptoms of WMSD exposures. To provide effective diagnosis and treatment of WMSDs, competent health-care providers (HCPs) must be identified by the employer and must be provided the data and support to do their jobs. This might entail visits to the workplace by the HCP or perhaps videotapes of the jobs in question so that the HCP can make knowledgeable recommendations to the employer. Based on his or her knowledge of the symptoms and of the workplace, the HCP may have an opinion about whether the MSDs are work related. Also, the HCP may have suggestions for modifications to the workstation to eliminate the exposures. The work of the HCP in dealing with specific cases of MSDs and in determining that they are work related culminates in the identification of the need for job analysis, which is explained in the section that follows.

Job Analysis

Earlier, the concept of job surveys was introduced as part of the overall administration of a WMSD management program. The primary objective of job surveys is fact finding. Job analysis, by comparison, is a more detailed and comprehensive study of the workstation and task and is triggered by medical evidence that the workstation or job is the cause of WMSD exposure. With job analysis comes the evaluation of "risk factors" that contribute to the problem. A possible risk factor would be an unusually cold temperature ambient to the workstation. Another might be the posture required for a particular job. Certainly, the amount of force required to be applied and the number of repetitions of a given motion have possible effects upon the incidence of WMSDs. If the workstation is subject to constant vibration, this brings another risk factor to the exposure. The data provided by the in-depth job analysis are the basis for taking corrective action to resolve the problem. Such action, which should be a part of any effective program for WMSDs, is described in the next section.

Job Design and Intervention

Once a problem has been identified, the responsibility is on the employer to rectify the situation. One may recall from Chapter 3 the "three lines of defense" for dealing with health hazards. In the case of WMSDs, only the first two seem relevant: engineering controls and administrative, or work practice, controls. Personal protective equipment, although promoted by some vendors, is not universally accepted as useful in reducing WMSDs.

Engineering controls are generally preferred because they are intended to physically eliminate the hazard instead of specifying procedures to be followed by workers or managers to deal with or mitigate the hazard. The three-lines-of-defense rule applies to health hazards in general, and certainly WMSDs are no exception. The nature of the engineering control will be dictated by the process itself and can be represented by widely varying engineering solutions to problems.

Administrative controls may be easier to achieve than engineering controls for some difficult processes. Konz and Johnson (ref. Konz) have recommended a procedure called *ramp-in* that allows new workers (or existing workers on new jobs) to gradually acclimate to highly repetitive tasks, such as assembly. The new worker is not required to do the job at the same pace as the experienced worker. Then, as the worker learns to perform the task more efficiently and becomes more physically adept at handling the particular repetitive motions required, the work standard can be increased. Johnson states that "(ramp-in) has been shown to be one of the most effective methods of reducing unnecessary discomfort for new employees who perform repetitive tasks." Another attractive feature of the ramp-in procedure is that it accommodates the setting of high standards for productivity; standards that may not be achievable for the new worker, but nevertheless are reasonable for the worker who has adapted to and become skilled at the task. Since ramp-in involves a change in procedure or the pace of the job, it should be considered an administrative, not an engineering, control. Another common administrative control is to rotate workers on and off difficult, repetitive jobs to reduce the exposure time to the hazard. This common practice is a widely recognized remedy for many different types of health hazards, including noise exposure and exposure to toxic air contaminants, as will be seen in Chapters 9 and 10.

SUMMARY

The field of ergonomics is an opportunity and also a problem for the safety and health manager. The field has much to offer in relieving the discomfort of workers doing repetitive, sometimes disabling, tasks. It also has the potential of raising productivity to new levels. However, there are problems. The causes of individual cases of musculoskeletal disorders is not always clear. Often, activities that are not work related either cause or contribute to the problem. Treatment of individual cases can be difficult and outright cures illusory. Engineering and administrative controls also may be only partially effective in eliminating the problem. Personal protective equipment may have no value at all in reducing lifting injuries or other WMSDs. Problems such as these raise questions and employer doubts about whether ergonomics programs can be cost effective in reducing hazards and raising productivity. Employer uproar over a mandatory OSHA ergonomics standard resulted in a congressional override of the standard before it could ever take effect. In place of a rigid standard, guidelines have been suggested that give industries a great deal of latitude in devising programs that fit their own operations. The challenge to the safety and health manager will be to retain the good features and potential of ergonomics in an environment of voluntary compliance, not rigid enforcement.

EXERCISES AND STUDY QUESTIONS

8.1 Describe the general categories of the field of ergonomics.

8.2 What is the primary ergonomics hazard addressed by occupational safety and health regulatory authorities?

8.3 Regulatory authorities have focused on workplace musculoskeletal disorders for approximately two decades; however, the terminology has changed over this period. What term was first used to refer to these disorders, and what other terms have been used as the focus has evolved up to the present?

8.4 In the absence of a specific standard for enforcement, what authority has OSHA used to cite ergonomics hazards?

8.5 What is the relationship between the ANSI standard and the OSHA standard for ergonomics?

8.6 What are the primary objections to an OSHA standard for ergonomics?

8.7 Are there gender differences in the physical qualification to do manual labor? If so, how should these differences be handled by employers qualifying workers for jobs?

8.8 What has research shown about the utility of back belts worn by workers for protection against back injury?

8.9 Describe the role ergonomics plays in the design of vehicle cockpits.

8.10 Describe some examples of the application of ergonomics as found in the OSHA standards.

8.11 Explain some problems with the rule "Lift with your legs, not with your back."

8.12 Is the incidence of workplace musculoskeletal disorders increasing or decreasing? Justify your answer and explain possible causes.

8.13 What two quite different meanings does the term *MSDS* represent to safety and health managers?

8.14 What role did the U.S. Congress play in the enforcement powers of the government with respect to ergonomics hazards?

8.15 Should employers require employees to wear back belts? Explain why or why not.

8.16 What is a good policy for employers with respect to employee decisions to buy back belts and wear them to work?

8.17 Describe some design features for punch presses that incorporate the principles of ergonomics.

8.18 Discuss some problems associated with training employees in proper lifting techniques.

8.19 Outline some basic elements in a program for workplace musculoskeletal disorders.

8.20 Compare management responsibilities for firms that have employees who have reported musculoskeletal disorders compared to firms who do not have employees who have reported such disorders.

8.21 Name some industry classifications that have been reported to have employee exposure to workplace musculoskeletal disorders.

8.22 What landmark cases in federal enforcement of citations for workplace musculoskeletal disorders have been used as models for current enforcement activities?

8.23 Describe two advantages of the "ramp-in" procedure for jobs susceptible to WMSD hazards.

8.24 Would you consider "ramp-in" to be an engineering control or an administrative control. Explain your reasoning.

8.25 Describe the application of the classical "three lines of defense" to workplace musculoskeletal disorders.

RESEARCH EXERCISES

8.26 Search the Internet for the latest developments of the ANSI standard for ergonomics or workplace musculoskeletal disorders.

8.27 Find several distributors of back belts and list the benefits claimed for these products.

8.28 Examine literature promoting the sale of automobiles. Can you can find any reference to ergonomics?

STANDARDS RESEARCH QUESTION

8.29 The OSHA standard for ergonomics was overturned by the U.S. Congress, so the OSHA General Industry standard contains no standard for ergonomics. Search the OSHA Web site for OSHA's current strategy or approach for dealing with ergonomics. In your own words, summarize this approach in a brief paragraph. In your summary, address whether OSHA intends to continue citing ergonomics hazards, and, if so, under what standard.

CHAPTER 9

Health and Toxic Substances

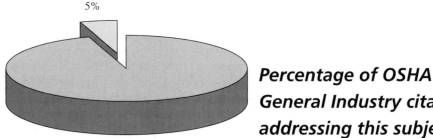

5%

Percentage of OSHA General Industry citations addressing this subject

Health hazards carry a great deal of impact because the potential harm to exposed employees is great and the cost of correction of a single health hazard can run into millions of dollars. Industrial hygienists have been saying for many years that health hazards deserve more attention, and in response to this pressure, a shift from safety toward health activities has been evident almost since OSHA began. At first, OSHA did not have a sufficient cadre of qualified health professionals to assess health hazards, so the natural focus was on safety. But the proportion of health specialists has increased a great deal since the early days of OSHA.

Chapter 1 discussed the bases of competition and even friction between safety professionals and health professionals. The two groups have common objectives, but their backgrounds are usually vastly different, which can result in conflict. The more successful professionals on both sides are busy emulating the characteristics typical of the opposite side, and thus the differences between the two professions are beginning to dissolve.

Health hazards will always tend to be more subtle to detect than safety hazards, by the very definitions of *health* and *safety*. As was noted in Chapter 1, health deals with the long-term chronic exposure effects, whereas safety deals with the more obvious acute effects that do their damage immediately.

BASELINE EXAMINATIONS

Almost everyone has taken a preemployment physical examination, but few understand the importance of the physical exam to the overall safety and health program.

Each employee's baseline health status is established by this examination. This status is important in placing the employee in the right job and in detecting any health deterioration due to job exposures. Occupational health exposures are perhaps the most important reason for preemployment physical examinations because of the chronic nature of health hazards. If an employee already has emphysema or some other lung disorder, it is essential that this fact be established at the time of employment. The same can be said of hearing impairment, as will be discussed in Chapter 10.

TOXIC SUBSTANCES

Exposure to toxic substances is the classic "health problem," and this topic will be used to model the entire subject of health and environmental control. The choice of words is important here. For instance, the term *hazardous materials* is sometimes used to refer to toxic substances, but the term *hazardous* is a much more general term that would include such safety hazards as presented by flammable and combustible liquids and explosives. This book follows popular convention, which tends to associate the term *materials* with safety hazards and the term *substances* with health hazards. Chapter 11 deals almost exclusively with safety hazards.

The safety and health manager needs to have a general knowledge of what various types of toxic substances can do to the body. Such general knowledge will be useful in convincing workers and management alike that toxic substances must be controlled for the health of the workers as well as to avoid OSHA citation. The discussion in this chapter will describe some of the various types of toxic substances by considering their effects on the body.

Irritants

Irritants inflame the surfaces of the parts of the body by their corrosive action. Some irritants affect the skin, but more of them affect the moister surfaces, especially the lungs. Even a weak irritant to the upper respiratory tract will be easily detectable by the victim, but irritants to the lower respiratory tract may go unnoticed.

When the irritant is some type of dust, the lung disease that results is called *pneumoconiosis*. This is a general term that includes reactions to simple nuisance dusts as well as fibrosis, a more serious reaction that includes the development of fibrous scar tissue that impairs the efficiency of the lungs. Examples of pneumoconioses include siderosis (from iron oxide dust), stannosis (from tin dust), byssinosis (from cotton dust), and aluminosis (from aluminum dust). The more dangerous fibroses are asbestosis (from asbestos fibers) and silicosis (from silica).

Everyone is familiar with the strong odor of ammonia. Ammonia gas and moisture on the mucous membranes of the body combine to form ammonium hydroxide, a strongly caustic agent. It is easy to understand why this would irritate and injure the delicate tissues of the nose, trachea, lungs, and other parts of the body with which it might come in contact. By similar logic, any of the gases that combine with water to produce acids would be irritants, as would airborne particles of the acids themselves.

Plating operations easily impart acid mists to the air, since the plating tanks are often splashing, hot, and acidic. Particularly offensive is chromic acid mist, which causes an ominous-sounding malady, *chrome holes*. Chromic acid also destroys the nasal septum—the tissue that separates the two nostrils.

Another well-known irritant is chlorine gas, a widely used industrial chemical. Chlorine's halogen relatives, fluorine and bromine, are also irritant—especially fluorine, the strongest of all halogens. Even the soluble salts of fluorine are poisonous. Less well known are those substances that are irritants deep within the lungs, such as the oxides of nitrogen and phosgene. Phosgene is best known as a chemical warfare gas, which is evidence of its toxicity. But phosgene can also be generated inadvertently in the workplace when chlorinated hydrocarbon solvents are exposed to welding radiation.

Chronic exposure to such irritants over a long period can cause scar tissue to develop in the lungs. Some of these substances do not produce an appreciable immediate irritant effect, but are dangerous in the long run. The most notorious of these scarring agents is asbestos fibers. Coal dust is also a scarring agent. Scarring agents are in the form of tiny solid particles, and their action on the lungs is mechanical, as contrasted with the systemic poisons discussed in the next section.

Systemic Poisons

More insidious than irritants are poisons that attack vital organs or systems of organs, sometimes by toxic mechanisms that are not understood. The chlorinated hydrocarbons common in solvents and degreasers, for example, are blamed for liver damage.

Perhaps the best known systemic poison found in occupational settings is lead. Lead is disappearing from paint pigments because of its reputation as a poison. This author formerly worked in a tetraethyl lead plant, and decades ago the workers in that plant had a high awareness of what lead can do to the body. Lead attacks the blood, the digestive system, and the central nervous system, including the brain. Autopsies have also shown damage to kidneys, liver, and reproductive systems, but these results are inconclusive. Other toxic metals are mercury, cadmium, and manganese. Magnesium, sometimes confused with manganese, is less toxic.

Another important systemic poison is carbon disulfide. Carbon disulfide is unusual in that its hazards are extreme from the standpoint of both safety (fire and explosions) and health. It is widely used in industry as a solvent, a disinfectant, and an insecticide. As a systemic poison, carbon disulfide attacks the central nervous system. Methyl alcohol (methanol), a popular solvent, also is a systemic poison to the central nervous system, but is a much milder poison than is carbon disulfide. Indeed, methanol is even acceptable in small quantities as a food additive! Methyl alcohol is also a fire and explosion hazard.

Depressants

Certain substances act as depressants or narcotics on the central nervous system and as such can actually be useful as medical anesthetics. Unlike the systemic poisons discussed earlier, the effect of the action of depressants on the central nervous system is usually temporary. However, some substances, such as methyl alcohol, are both systemic poisons and depressants. Besides affecting health, depressants can have an adverse effect on safety because they interfere with the concentration of workers who operate machinery.

The most familiar depressant is *ethyl alcohol* (the "drinking" variety of alcohol), sometimes called *ethanol* in industry. Its harmful effects as an industrial hazard are minimal compared with its effects from drinking it. In fact, ethanol's greatest on-the-job hazard is without a doubt from "voluntary ingestion" from bottles brought on the premises by employees. Ethanol is not as toxic as methanol.

Acetylene, the most widely used fuel gas for welding, is a narcotic, but its health dangers are minimal compared with its safety hazard as highly flammable and explosive. Acetylene has been used in medical anesthesia.

Benzene is a very popular industrial chemical used principally as a solvent. Benzene acts as a depressant on the central nervous system, an irritant, and a systemic poison, and has recently been recognized as causing leukemia. In addition, benzene is a dangerous fire and explosion hazard. OSHA has a movie that dramatically depicts the hospital-bed testimony of a young benzene worker who is dying from leukemia.

Asphyxiants

Asphyxiants prevent oxygen from reaching the body's cells, and in the general sense, any gas can be an asphyxiant if there is enough of it to crowd out the essential proportion of oxygen in the air. Many people have committed suicide by breathing natural gas, which is essentially methane. But methane is merely a simple asphyxiant, in that it displaces the proportion of oxygen in inhaled air. Methane can be encountered in industrial environments, as it is a product of fermentation. Other simple asphyxiants commonly encountered are the inert gases such as argon, helium, and nitrogen used in welding.

It may seem incorrect to classify nitrogen as an air contaminant and an asphyxiant when it is the principal constituent (78%) of normal air. But too much nitrogen will reduce the normal proportion of oxygen (21%) in the air. Any proportion of oxygen less than 19.5% is considered oxygen deficient. Oxygen deficiency is very dangerous, a more serious condition than most people think. Case Study 9.1 is an accident description quoted from federal records of workplace fatalities (ref. Fatal).

CASE STUDY 9.1

A contract employee was assigned to sandblast the inside of a reactor vessel during turnaround activities at a petrochemical refinery. Instead of relying on the contract company's own air compressors in accordance with the contractor's policy, the contract foreman connected the employee's supplied air respirator to a hose containing what he thought was plant air, but was actually nitrogen. Both hoses were identical except for markings at the shutoff valve. The worker entered the vessel, descended to the bottom, placed the respirator hood on his head, and was overcome.

The cause of death in this accident case study was oxygen deficiency, and the irony is that the worker was breathing nearly pure nitrogen, the primary constituent of normal air. Oxygen deficiency is a serious hazard to consider when workers must enter a tank, vessel, or other confined space. Many fatalities occur every year in just this way. OSHA focused on confined space entry in the 1990s, and this topic will be covered in detail in Chapter 12.

In another case study (Case Study 9.2), we find a hazard in which a seemingly innocuous use of nitrogen can lead to trouble.

CASE STUDY 9.2

NITROGEN AS A HAZARD

An electrician found moisture in an electrical box. As a precaution, he decided to attach a nitrogen drop to the box to eliminate the moisture and prevent an accidental ignition due to arcing in the box. The problem was that the nitrogen filled up the box and then quietly propagated through the electrical conduit system to other areas of the plant, including various confined spaces, such as operating line pulpits, mezzanines, crane cabs, and other areas in which operators were present. In one pulpit, workers began to complain of being overly tired. The oxygen level was tested and found to be much below safe minimums. The workers were being quietly asphyxiated by the relatively benign nitrogen that was entering their workspaces through the electrical conduit system. The same thing can happen with argon, a well-known inerting agent for welding operations.

The accident described in Case Study 9.2 took place in a large steel mill. The same or similar situations have developed in other types of plants. Fortunately, in the steel mill case, the plant was alert to safety and health problems. Management took worker complaints seriously and the breathing air was tested, which led to the solution of the problem before something more serious happened. By observing this case study and eliminating the practice that leads to nitrogen propagation throughout the electrical conduit system, future accidents or hazardous situations of this type can be avoided.

Carbon dioxide is one of the most important simple asphyxiants, although in normal quantities it is a harmless constituent of air. Fire is the primary source of dangerous industrial concentrations of carbon dioxide. Carbon dioxide is heavier than air, which causes it to accumulate in low, confined spaces, increasing its hazard potential. Confined spaces have a great deal of hazard potential, not only for carbon dioxide, but for all air contaminants.

The asphyxiants discussed so far are *simple asphyxiants*, essentially nontoxic substances that replace the essential oxygen content in the air. The other type of asphyxiant is the *chemical asphyxiant*, which interferes with the oxygenation of the blood in the lungs or of the body's tissues. The most notorious chemical asphyxiant is carbon monoxide, a substance for which blood hemoglobin has a stronger affinity than it does for oxygen (over 200 times the affinity). The resulting compound, carboxyhemoglobin, is a very stable substance that prohibits the life-critical exchange of oxygen and carbon dioxide by their vehicle, blood hemoglobin.

Another well-known chemical asphyxiant is hydrogen cyanide, an industrial insecticide, but better known as the gas used in prison execution chambers. The gas is formed by dropping sodium cyanide pellets into a small container of acid. Some work places have the potential for becoming execution chambers. In a workplace inspection in California, a laboratory was found in which strong acids in glass bottles were stored on shelves located directly above sodium cyanide salts!

Carcinogens

Carcinogens are substances that are known to cause or are suspected to cause cancer. A great deal of attention has been given to carcinogenesis since the advent of OSHA, but the source of the emphasis is not OSHA alone. NIOSH, the Consumer Product Safety Commission (CPSC), and other agencies have focused on carcinogens. Public awareness is now high, and many workers and consumers alike are becoming very cautious about exposure to carcinogens. One of the frightening things about carcinogens is that cancer has such a long latency period. Sometimes a lapse of 20 or even 30 years occurs between exposure and the appearance of a cancerous tumor.

New carcinogens are being labeled every year, and many of the substances indicted are very commonly used industrial materials, such as benzene and vinyl chloride. The familiar saying "Everything I do is either illegal, immoral, or fattening" might be modified to include "or causes cancer." So many substances have been indicted by laboratory tests on animals that it is possible that a sort of complacency could be developing in the minds of the public.

Vinyl chloride, mentioned earlier as a type of carcinogen, is extremely dangerous from many aspects. It is a severe explosion hazard, and when it burns, it is very difficult to extinguish. Added to this is the hazard of highly toxic phosgene being liberated during vinyl chloride fires. Acute skin exposures can result in injury from skin freezing due to rapid evaporation of vinyl chloride. Chronic inhalation is now recognized to cause a form of cancer of the liver, angiosarcoma. Vinyl chloride is also believed to be a teratogen, a type of substance that affects a fetus. (See next section.)

Before discussing teratogens, a distinction should be made between the extremely hazardous vinyl chloride and polyvinyl chloride (PVC). PVC is a type of plastic synthesized by polymerizing the unstable vinyl chloride monomer into the stable polyvinyl chloride polymer. Figure 9.1 illustrates the arrangement of the atoms in the dangerous vinyl chloride monomer and in the harmless and very stable polyvinyl chloride polymer.

Teratogens

Teratogens affect the fetus, so their toxic effect is indirect. Women should be careful about exposures to certain substances during pregnancy, especially in the first trimester (first three months). Teratogens should not be confused with mutagens, substances that attack the chromosomes and thus the species instead of the individual. Teratogens do their damage after conception, but before birth, whereas *mutagens* do their damage before conception. Mutagens can affect the chromosomes of either potential fathers or potential mothers.

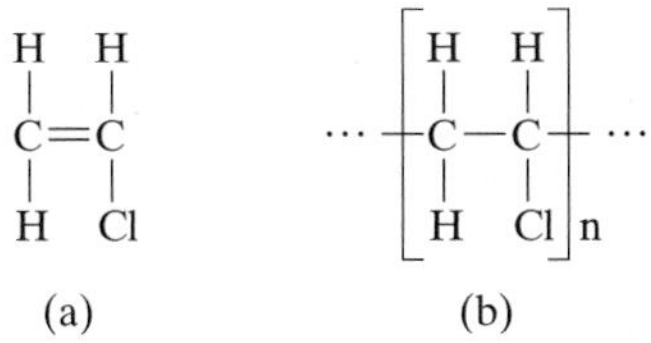

FIGURE 9.1

Comparison of the vinyl chloride molecule with the polyvinyl chloride molecule—(a) vinyl chloride monomer; (b) polyvinyl chloride polymer.

A tricky legal question is whether an industry can prohibit women of childbearing age from working in jobs in which they may be exposed to teratogens. The question is whether this constitutes sex discrimination. The affirmative contingent says that the teratogens must be controlled by the industry so that pregnant women can have a safe and equal opportunity to work in the same jobs as those offered to men. The negative contingent says that it makes more economic, safety, and social sense to employ women in other jobs during pregnancy.

Routes of Entry

The term *toxic substance* can be considered synonymous with the term *poison*, a word familiar to all of us who were taught as children not to eat or drink poison. Poison is even in our fairy tales, such as "Snow White and the Seven Dwarfs." It may be true that the greatest danger from poisons in the home is from ingestion (swallowing), but on the job, the greatest danger is from breathing. In fact, it has been said that the order of importance of routes of entry into the body for toxic substances on the job is the exact opposite as the order at home, as shown in Figure 9.2.

The various routes of entry into the body for toxic substances are more related than most workers realize. Inhalation of toxic substances results in accumulations on the mucous membranes; mucus is later coughed up, and some of it is inevitably swallowed. Skin contact with toxic materials can also result in ingestion as substances become embedded beneath fingernails and on hands that later come into contact with food. Toxic dusts in the air are also picked up in the hair and are later deposited on the pillow during sleep and thus gain entry into the body indirectly.

From knowledge of routes of entry for toxic substances, it is easy to see the importance of sanitation. Some of the principles of sanitation discussed in Chapter 7 take on more importance when toxic substances are considered. Proper storage of food and the availability of shower and washing facilities may be essential in the control of the quantities of toxic substances entering the worker's body.

One other route of entry into the body should be mentioned before closing this section: the eyes. Although the eyes are not considered a principal route of entry, it should be noted that the eyes are particularly sensitive to toxic substances. Especially with the advent of the hazard of exposure to the HIV virus and AIDS, more attention has been given to the eyes as a route of entry. Examples of hazardous occupations with respect to exposure to HIV through the eyes are dental technicians and dentists.

Air Contaminants

The greatest concern with toxic substances is with air contamination, and this is as it should be (as shown in Figure 9.2). Air contaminants take on many physical forms, and most people confuse these forms in everyday language. The safety and health manager

FIGURE 9.2

Toxic substances' routes of entry into the body. Note reverse sequence of importance at work vs. at home.

Routes of entry	Importance	
	At home	At work
Ingestion	Most important	Least important
Skin contact	Of moderate importance	Of moderate importance
Inhalation	Least important	Most important

should know the difference, for instance, between vapors and fumes. Although air is essentially a combination of gases, the *contamination* of that air can consist of any of the three states of matter: solids, liquids, or gases.

Gases easily contaminate the air because air consists of gases and gases readily mix. The most familiar toxic gas is carbon monoxide. Also dangerous in the industrial environment are hydrogen sulfide and chlorine. Even "harmless" gases, such as carbon dioxide and inert nitrogen, can become dangerous if allowed to accumulate in large quantities so that they become asphyxiants by crowding life-giving oxygen out of the air.

Vapors are also gases, but these substances are normally liquids or even perhaps solids that release small quantities of gases into the surrounding air. Some of our most useful industrial liquids, such as gasoline and solvents, have a strong tendency to release vapors.

Mists are really tiny droplets of liquids, so small that they remain suspended in the air for long periods, as in clouds. Since liquids are heavier than air, they eventually settle out or coalesce into larger droplets, which fall in the manner of rain. Long before that happens, however, mists can be inhaled by the worker. Fine mists are generated as vapors condense into clouds. Coarse mists are produced by splashing or atomizing operations, such as from machine tool cutting oils or from electroplating. Pesticide spray is also usually a mist.

Dusts are recognized as solid particles. Technically speaking, dust particles range in size from 0.1 to 25 micrometers (0.000004 to 0.001 inch) in diameter. Everyone is exposed to dust, and some dust is relatively harmless. Dangerous dusts include asbestos, lead, coal, cotton, and radioactive dusts. Silica dust from grinding operations is also becoming recognized as a hazard, although ordinary earth dust is principally silica. Asbestos dust particles are fiber shaped rather than round, which contributes to their hazard.

Fumes are also solid particles, but are generally too fine to be called dusts. Actually, the sizes for classifying fume particles and dusts overlap, as can be seen in Figure 9.3. Whereas dust particles are typically finely divided by mechanical means, fumes are formed by the resolidification of vapors from very hot processes such as welding. Chemical reactions can also produce fumes, but gases and vapors, although also produced by chemical processes, are not to be confused with fumes. Metal fumes are the most dangerous, especially those of the heavy metals.

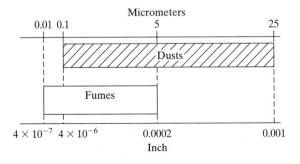

FIGURE 9.3

Comparison of dust and fume particle sizes (not to scale).

Particulates make up a general classification that includes all forms of both solid and liquid air contaminants (i.e., dusts, fumes, and mists). Particle sizes thus can vary greatly; some are visible to the naked eye, but most are not. Figure 9.4 displays some example particle sizes over the range from tiny visible particles down to submicroscopic large molecules.

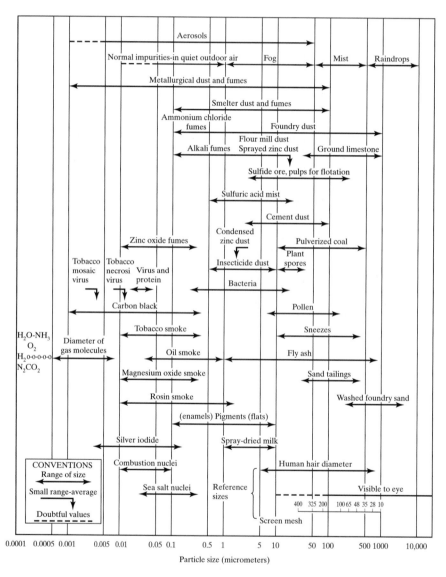

FIGURE 9.4

Sizes of airborne contaminants. (Note the logarithmic scale.)(*Source*: Courtesy of the Mine Safety Appliances Company.)

It is probably already obvious that industry and technology do not eliminate exposures to toxic substances; they merely control these exposures to keep them within acceptable levels. It is both naive and unnecessary for the safety and health manager to adopt a strategy of complete elimination of worker exposure to toxic substances. There is no known poison to which a human being cannot be exposed without significant harm, provided that the exposure is tiny enough and distributed over a long enough interval for the body to assimilate or counteract it. On the other hand, even the mildest of poisons can become lethal if the worker is constantly exposed to it in massive doses—the kind that can be found in industrial exposures more so than in any other environment.

Threshold Limit Values

Since no poison is lethal in small enough doses, and all poisons are lethal in large enough doses, no clear-cut line separates the harmful from the benign worker environment. But some line has to be drawn to serve as a basis for action to control toxic substances. Particularly with regard to airborne contaminants, it becomes necessary to identify some levels of concentration below which one need not worry about worker exposures. Thus, the term *threshold limit value* (TLV) evolved and refers to the level of concentration to which the worker could be exposed during the entire workday without significant harm. The TLV varies drastically with the toxicity of the contaminant, of course, and every toxic substance has its own TLV.

So far, this discussion of TLVs is in a generic sense, and thus every toxic substance has a TLV even if it is unknown and even if no one knows that the particular substance is toxic. But for *known* toxic substances, there is a *listed TLV*, which is a value agreed on by a committee of the American Conference of Governmental Industrial Hygienists (ACGIH) and is listed in the *TLV booklet*. When the abbreviation TLV is used instead of spelling out the words *threshold limit value*, the ACGIH listed value is usually the meaning intended.

Just because a committee decides on the TLV for a given substance, it does not mean that employers must control worker environments to comply with that level. Some TLVs are based on hard scientific data and are very robust criteria for action. Others are based on much more sketchy data, and thus professional judgment is necessary to determine the actions that should be taken to control the worker environment. The TLVs themselves may change from year to year as more information becomes available. Thus, to be strictly correct, one should refer to the date of the list when naming a given TLV—for example, "the 2003 TLV for carbon monoxide was ..." Incidentally, when TLVs change, their levels usually decrease as new information about hazards is uncovered.

Permissible Exposure Limits

The focus of this book is on what enforcement agencies require, not what ACGIH lists, but the two are definitely related. At OSHA's beginning, when it was permitted to adopt national-consensus standards without formal promulgation, the agency adopted hundreds of TLVs, most of which represented the TLV levels published in 1968 by ACGIH. Since the OSHA-published list assumed a regulatory character, the term *permissible exposure limit* (PEL) was used to distinguish between the OSHA-prescribed level and the ACGIH term *TLV*. For the most part, PELs have remained static, as OSHA has had difficulty getting the public to accept stricter and stricter levels of control. However, the TLVs remain dynamic, as ACGIH continues to revise the list whenever the collective professional judgment of the committee determines that a new TLV should be added to the list or an old one adjusted, usually downward.

Over the years, the disparity between the OSHA PELs and the ACGIH TLVs grew to such proportions that OSHA decided on a bold plan to revise all of the PELs in a single promulgation, instead of painstakingly revising each in substance-specific rulemaking. Thus, in 1989, OSHA added 164 new listings to the air contaminants list and at the same time tightened the PELs on 212 substances that were already on the list. OSHA laid its groundwork carefully because behind each of the new PELs (which they labeled *final limits*) were backup *transitional limits* at the old PEL levels. The transitional limits were to remain in effect during a specified implementation period. Then as a legal safety measure, OSHA added a footnote that retained the transitional limits in case opponents of the new PELs were able to mount a court challenge and strike down the new rules.

Because this move was so comprehensive and revolutionary, opponents to the new PELs seemed unable to cope with so many changes at once—so the strategy seemed to work. But the victory was an illusion, because nearly four years later, in 1992, the Eleventh Circuit Court of Appeals vacated the entire PEL list revision, in what Labar (ref. Labar, 1993) said was "perhaps the agency's biggest defeat in its 22-year history." A stunned OSHA was expected to appeal to the U.S. Supreme Court, but in 1993, it announced that the U.S. Solicitor General would not attempt to appeal. OSHA was compelled to prove to the public that each new PEL and each tightening of each existing PEL would be justified. OSHA had no choice but to return to its original list of 264 PELs adopted by the national consensus process in 1971.

The entire table of PELs, as revised back to its earlier state in 1993, is included in this book as Appendices A.1, A.2, and A.3, which correspond to OSHA's designations Z.1, Z.2, and Z.3, respectively. Appendix Table A.1 is the main table and contains most of the PELs, listed alphabetically by substance name, with *CAS No.* (chemical abstracts) for reference. A common mistake is to ignore the hazard when the substance is not listed in the main table, but a few of the most hazardous and frequently encountered substances in industrial exposures are those found in the second table, Appendix Table A.2. This table has its origins in a version formerly (prior to OSHA) published by ANSI for certain substances. Appendix Table A.3 is for mineral dusts, which are considered separately because solid particles are sampled and measured by different means than are toxic gases, mists, and vapors.

MEASURES OF EXPOSURE

The tables of Appendix A.1 are long and complicated, and the many columns specified for each toxic substance warrant explanation. The reason for this complication is that it is difficult to measure levels of contamination of the atmosphere in the workplace. The problem is compounded by the various physical states—solid particles, liquid droplets, mists, gaseous molecules—in which the contaminant can exist in the atmosphere. Further, the medical data incriminating a particular poison may point to danger in a single short-term exposure or may suggest deleterious effects from long-term exposures.

Time-Weighted Averages

The most popular measure of air contaminant exposures is the *time-weighted average* (TWA). The PELs are understood to be TWAs unless otherwise specified. The TWA is a computed weighted-average concentration over an 8-hour shift. Such a calculation recognizes that concentrations of air contaminants change over time and that it is sometimes permissible for a workplace concentration to exceed the threshold limit value (TLV) if at other times during the workday the exposure is sufficiently lower than the TLV, such that the average exposure for the workshift is lower than the specified level.

The following formula is used to compute the TWA:

$$E = \frac{\sum_{i=1}^{n} C_i T_i}{8} = \frac{C_i T_i + C_2 T_2 + \cdots + C_n T_n}{8}, \tag{9.1}$$

where E = equivalent 8-hour time-weighted-average concentration
C_i = observed concentration of the contaminant in time period i
T_i = length of time period i
n = number of time periods studied

The calculation will now be illustrated in Case Study 9.3.

CASE STUDY 9.3

Calculate the 8-hour, full-shift TWA for the concentrations shown.

Time period, i	Observed concentration, C_i	Length of period, T_i (hr)	$C_i \times T_i$
1	2	$1\frac{1}{2}$	3
2	4	$2\frac{1}{2}$	10
3	7	1	7
4	5	2	10
5	3	1	3
Total		8	$33 = \sum_{i=1}^{5} C_i T_i$

Solution

$$E = \frac{33}{8} = 4.125$$

The Case Study 9.3 calculation shown is quite adequate if there is only one toxic substance present in the industrial atmosphere. However, mixtures present a different problem. Suppose, for instance, that an industrial atmospheric TWA concentration of nitric acid was barely below the specified PEL of 5 milligrams per cubic meter. Suppose further that the same atmosphere showed TWA concentrations of just under the prescribed limits of 1 milligram per cubic meter for sulfuric acid and 25 milligrams per cubic meter for acetic acid *at the same time*. Taken separately, none of these three acid concentrations violates the standard, but common sense says that the three concentrations present at the same time are dangerous.

The synergistic effect of combinations of toxic substances is a complicated subject. Most research has concentrated on direct effects of substances acting alone. Some mixtures of contaminants would tend to neutralize each other and be beneficial. For instance, caustics mixed with acids may produce benign salts. In the example just described, however, the three acids working together are bound to have a combined effect. For some mixtures, the combined effect may be far worse than the sum of the individual effects.

OSHA takes a moderate approach by requiring simple combinations of toxic substances to be considered, but generally ignores the complex synergistic effects. The method is to sum the ratios of concentrations of each substance to its own PEL. The resulting sum should not exceed unity. The following formula summarizes the computation:

$$E_m = \sum_{i=1}^{n} \frac{C_i}{L_i} = \frac{C_1}{L_1} + \frac{C_2}{L_2} + \cdots + \frac{C_n}{L_n}, \tag{9.2}$$

where E_m = calculated equivalent ratio for the entire mixture

C_i = concentration of contaminant i

L_i = permissible exposure level (PEL) for contaminant i

n = number of contaminants present in the atmosphere

E_m may not exceed 1. The calculation is demonstrated in Case Study 9.4.

CASE STUDY 9.4

An industrial process produces exposure in accordance with the following table:

	Nitric acid	Sulfuric acid	Acetic acid
Contaminant, i	1	2	3
Concentration, C_i	4	0.9	22
Limit, L_i	5	1	25

> **Solution**
>
> $$E_m = \sum_{i=1}^{3} \frac{C_i}{L_i} = \frac{4}{5} + \frac{0.9}{1} + \frac{22}{25} = 2.58$$
>
> ---
>
> Since 2.58 > 1, the concentration of the mixture exceeds the PEL, even though the individual PELs are not exceeded.

Ceiling Levels and STELs

Most of the PELs listed in the main table (Appendix A.1) are to be considered TWAs, but for some substances, the concern is for short-term exposures. A "ceiling" value, sometimes abbreviated *C* or *MAC* for *maximum acceptable ceiling*, is an exposure limit that should *never* be exceeded. Another convention is to specify an *STEL, short-term exposure limit*, recognizing the danger of acute exposures, but allowing short excursions above a level that on an 8-hour-shift basis would clearly be hazardous. The STEL states a maximum concentration permitted for a specified duration, usually 15 minutes. For instance, Table A.2 lists the following PELs for toluene:

TOLUENE

TWA	200 ppm
MAC	300 ppm
STEL	500 ppm for 10 minutes

Note that the STEL for toluene is much higher than the MAC. One would think that the STEL would lie somewhere between the TWA and the MAC, but the standards consistently list STELs higher than MACs. This suggests that if the exposure duration is shorter than the duration specified by the STEL, there is no limit to the allowable concentration! This seems to be a contradiction of the definition of the MAC, but in reality it is not feasible to measure the concentration for so short a time period except for "grab samples," a very unreliable method of measurement. Indeed, even the STELs are often impractical to check with current state-of-the-art instruments, so STELs are sometimes ignored by industry and enforcement officials alike.

Units

Regardless of the type of limit against which the exposure is being measured, the analyst must be concerned with the units of the measure. For most of the substances in Appendix A.1, the table lists for each limit a pair of values, which are really two different measures of the same limit expressed in different units. Gases are usually more conveniently measured by volume, and thus the first column, labeled p/m (parts per million), is usually used for these substances. Liquids and some solids are more conveniently measured by weight, and thus the second column, labeled mg/m^3 (milligrams of particulate per cubic meter), is preferred for these substances. If the molecular weight of the substance is known, conversion can be made by using the formula

$$p/m = \frac{mg/m^3 \times 24.45}{MW}$$

where MW is the molecular weight of the substances. Parts per million is sometimes abbreviated ppm rather than p/m.

Action Levels

One other level, the action level (AL), deserves mention. If control measures are taken only after TLVs are exceeded, it may be too late to prevent serious harm and also perhaps too late to prevent citation by authorities. ALs are somewhat of a lock-the-barn-before-the-horse-runs-away measure, which anticipates the problem before TLVs or other measures are exceeded. ALs are set arbitrarily at one-half PEL. A great deal of statistical and instrumentation variation prevents precisely accurate surveys. The difference between the AL and the PEL provides a margin for error to ensure that worker exposures do not exceed the PEL, by implementing controls before PELs are reached.

STANDARDS COMPLETION PROJECT

The preceding sections have described the method of enforcement of, and compliance with, standards prescribed for hundreds of toxic substances covered by PELs listed in tabular form. This is OSHA's general approach to air contaminants and applies to a large majority of substances that can be found in the workplace environment. However, for a few substances, OSHA has taken a more comprehensive approach by issuing detailed standards, each of which is devoted to the control of one particular hazardous substance. The standards for these substances have been formulated as a series, and all the standards in the series have been issued several years after the passage of the OSHA act. This means that the standards were scrutinized by the public promulgation procedures and survived the controversy of opposing factions. Several of the substances experienced stormy promulgation, including the invoking of "emergency temporary" promulgation procedures and subsequent court challenges. Much of the background research to justify the issuance of these separate standards for individual substances was conducted by NIOSH. The effort has been called the "standards completion project" because it has been envisioned by some that every toxic substance should eventually have its own unique standard instead of simply being included in a list of TLVs. A list of standards completed, along with the associated PEL for each substance covered, appears in Table 9.1. As the table shows, some of these substances are extremely hazardous and no PEL is specified. For these substances, the standard is very specific regarding handling procedures to be followed, personal protective respirators to be used, and other precautionary measures.

OSHA standard 1910.130 has to do with bloodborne pathogens and is not included in Table 9.1. There is no question that bloodborne pathogens are toxic, but the standards completion list refers to toxic substances that were originally listed in the air contaminants list but for which full standards have since been promulgated. The standard for bloodborne pathogens does not belong in the list because it was not included

TABLE 9.1   OSHA Standards for Specific Toxic Substances ("Standards Completion Project")

Substance	Permissible exposure limit (PEL) TWA
1910.1001 Asbestos	0.1 fiber/cm^3 (longer than 5 μm) Ceiling: 1 fiber/cm^3 (longer than 5 μm)
1910.1002 Coal tar pitch volatiles	0.2 mg/m^3
1910.1003 13 Carcinogens (4-Nitrobiphenyl, etc.)	Extremely hazardous. Refer to standard.
1910.1004 α-Naphthylamine	Carcinogen see (1910.1003)
1910.1005 (Reserved)	
1910.1006 Methyl chloromethyl ether	Carcinogen see (1910.1003)
1910.1007 3,3-Dichlorobenzidine (and its salts)	Carcinogen see (1910.1003)
1910.1008 bis-Chloromethyl ether	Carcinogen see (1910.1003)
1910.1009 β-Naphthylamine	Carcinogen see (1910.1003)
1910.1010 Benzidine	Carcinogen see (1910.1003)
1910.1011 4-Aminodiphenyl	Carcinogen see (1910.1003)
1910.1012 Ethyleneimine	Carcinogen see (1910.1003)
1910.1013 β-Propiolactone	Carcinogen see (1910.1003)
1910.1014 2-Acetylaminofluorene	Carcinogen see (1910.1003)
1910.1015 4-Dimethylaminoazobenzene	Carcinogen see (1910.1003)
1910.1016 N-Nitrosodimethylamine	Carcinogen see (1910.1003)
1910.1017 Vinyl chloride	1 ppm STEL 5 ppm/15 min
1910.1018 Inorganic arsenic	10 μg/m^3
1910.1025 Lead	50 μg/m^3
1910.1027 Cadmium	5 μg/m^3
1910.1028 Benzene	1 ppm STEL: 5 ppm/15 min
1910.1029 Coke oven emissions	150 μg/m^3
1910.1043 Cotton dust	200–750 μg/m^3 (depending on process; see standard for details)
1910.1044 1.2-Dibromo-3-chloropropane	1 ppb (parts per billion)
1910.1045 Acrylonitrile	2 ppm Ceiling: 10 ppm/15 min
1910.1047 Ethylene oxide	1 ppm Excursion: 5 ppm/15 min
1910.1048 Formaldehyde	1 ppm STEL: 2 ppm/15 min
1910.1050 Methylenedianiline	10 ppb STEL: 100 ppb/15 min
1910.1051 1,3-Butadiene	1 ppm STEL: 5 ppm/15 min
1910.1052 Methylene chloride	25 ppm STEL: 125 ppm/15 min

Source: OSHA Safety and Health Standards (29 CFR 1910).

in the original air contaminant list, and indeed, the principal exposure to bloodborne pathogens is not through the air.

In fiscal year 2003, OSHA set out to improve the standards completion list by revising the standards, eliminating excessive reporting requirements, updating or eliminating obsolete medical requirements that do not conform to current medical practices, reducing the frequency of medical examinations when warranted, and eliminating excessive red tape in general (ref. OSHA 10/30).

DETECTING CONTAMINANTS

It is nice to have a list of toxic substances with permissible exposure levels for each, but more is needed to determine whether a problem exists. There are simply too many substances on the list to keep tabs on all possibilities. Safety and health managers need to have knowledge of the processes within their plants so that they know where to look or at least whom to ask. Air sampling and testing are the way to determine concentrations as accurately as possible, but before the test is done, some suspicion of possible contamination needs to be provided by other evidence.

One of the most common ways a potential problem is first detected is by the sense of smell. People seem to believe that they can smell an air contaminant, and they *usually* can smell either the toxic substance or some odorous agent that may accompany toxic substances. But sense of smell is not enough to detect some of the most dangerous contaminants. The most notorious example is carbon monoxide. But carbon dioxide, nitrogen, and methane are also essentially odorless and can be dangerous simply by displacing the oxygen in the air. Some readers will question the statement that methane is odorless because they know that methane is the principal ingredient in natural gas. But the odor in "natural" gas is from a stenching agent deliberately introduced as a safety precaution so that people can detect leaks by sense of smell. Even hydrogen sulfide, a gas that is both dangerous and has a very strong rotten odor, cannot be detected reliably by sense of smell. The odor is so foul that it quickly overcomes the olfactory system, and victims begin to block out the smell sensation so that they are not aware of the degree of their exposure.

Another approach is to examine technical literature to determine what industries might release what substances. Table 9.2 provides some information from NIOSH literature regarding possible contaminants in various industries.

Yet another approach is to analyze the processes within the plant to determine potential leaks into the industrial atmosphere. Such an analysis can become quite technical and involves not only an understanding of the machines, pumps, valves, sumps, and tanks, but also a knowledge of the materials used, quantities, intermediate phases, volatilities, and other characteristics that affect the amount of contaminant that becomes airborne. A chemical engineering qualitative flow process chart (see also Chapter 6) and perhaps a quantitative flow process chart may be necessary to determine whether air-contaminant potentials are present. The safety and health manager should not hesitate to involve a chemical engineer in the analysis of potential air contaminations. A technical analysis at this phase may preclude a lot of expensive and laborious sampling experiments later.

Measurement Strategy

Once the existence of an air-contamination risk has been determined, a procedure is needed to go about taking samples, measuring employee exposures, and instituting controls. NIOSH recommends a strategy for this purpose, which is displayed in the form of a decision chart in Figure 9.5.

Measurement Instruments

Federal regulation by both OSHA and EPA of permissible exposure levels for toxic substances in the atmosphere has stimulated the electronics and instrumentation industries to develop new and more precise instruments for determining concentrations. Interest in parts per million is shifting toward closer scrutiny to detect parts per *billion*. Such demands are pressing the physics of the instruments, which can result in large-scale inaccuracies.

TABLE 9.2   Potentially Hazardous Operations and Air Contaminants

Process type	Contaminant type	Contaminant examples
Hot operations		
Welding	Gases (g)	Chromates (p)
Chemical reactions	Particulates (p)	Zinc and compounds (p)
Soldering	(dust, fumes, mists)	Manganese and compounds (p)
Melting		Metal oxides (p)
Molding		Carbon monoxide (g)
Burning		Ozone (g)
		Cadmium oxide (p)
		Fluorides (p)
		Lead (p)
		Vinyl chloride (g)
Liquid operations		
Painting	Vapors (v)	Benzene (v)
Degreasing	Gases (g)	Trichloroethylene (v)
Dipping	Mists (m)	Methylene chloride (v)
Spraying		1,1,1-Trichloroethylene (v)
Brushing		Hydrochloric acid (m)
Coating		Sulfuric acid (m)
Etching		Hydrogen chloride (g)
Cleaning		Cyanide salts (m)
Dry cleaning		Chromic acid (m)
Pickling		Hydrogen cyanide (g)
Plating		TDI, MDI (v)
Mixing		Hydrogen sulfide (g)
Galvanizing		Sulfur dioxide (g)
Chemical reactions		Carbon tetrachloride (v)
Solid operations		
Pouring	Dusts	Cement
Mixing		Quartz (free silica)
Separations		Fibrous glass
Extraction		
Crushing		
Conveying		
Loading		
Bagging		
Pressurized spraying		
Cleaning parts	Vapors (v)	Organic solvents (v)
Applying pesticides	Dusts (d)	Chlordane (m)
Degreasing	Mists (m)	Parathion (m)
Sandblasting		Trichloroethylene (v)
Painting		1,1,1-Trichloroethane (v)
		Methylene chloride (v)
		Quartz (free silica, d)
Shaping operations		
Cutting	Dusts	Asbestos
Grinding		Beryllium
Filing		Uranium
Milling		Zinc
Molding		Lead
Sawing		
Drilling		

Source: NIOSH (ref. Occupational Diseases).

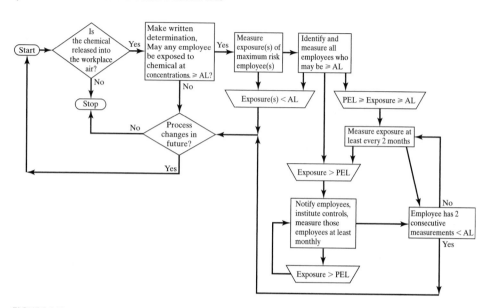

FIGURE 9.5

NIOSH exposure measurement strategy. Each individual substance's health standard should be consulted for detailed requirements: AL, action level; PEL, permissible exposure limit.

With such demands by high technology on atmospheric measurement devices, one would think that air-contamination monitoring would be a new field. But historically there have been other, more crude means for monitoring breathing air. Animals formerly were used to test for toxic gases or oxygen deficiency. A canary or a mouse in a cage was often carried into mines. If the animal died, the workers were alerted to the hazard. A flame safety lamp was used to test for oxygen deficiency; the flame would die if the oxygen proportion in the atmosphere was too low. A brighter burning lamp was supposed to be an indication of the presence of methane.

These methods were crude, but they did provide some essential indications—for *acute* exposures. With the recognition of threshold limit values and the rise in importance of chronic exposures, the canary test and the flame test became totally inadequate. By the time the canary developed cancer or scarred lung tissue, the workers would also be victimized. Furthermore, animal life spans are too short to deal with the chronic effect to which humans might be susceptible.

Today, there are basically four approaches to measuring air-contaminant exposures:

1. direct-reading instruments
2. sampling with detector tubes
3. sampling with subsequent laboratory analysis
4. dosimeters

For a few commonly encountered problems, such as oxygen deficiency and natural-gas leaks, devices have been invented that are capable of metering and registering actual

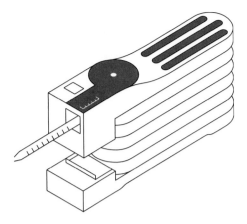

FIGURE 9.6

Air-contaminant detector tube inserted in hand-operated bellows pump(*Source:* Courtesy of Draeger Safety, Inc.).

concentrations on a display such as a digital readout. Such a direct-reading instrument is sometimes essential for confined space entry when it is necessary to obtain an on-the-spot reading to determine whether an atmosphere is safe from dangerous acute exposures. So convenient are the direct-reading instruments that instrument manufacturers are constantly trying to press the frontiers of physics to devise and patent new instruments for measuring concentrations of as many varieties of air contaminants as possible.

Less precise than direct-reading instruments, but still feasible for on-site assessment of existing concentrations, is the drawing of a sample of the atmosphere through the "detector tube" that contains a chemical that reacts to the suspected contaminant if present. Recent developments in detector tube technology have made direct measurement possible for as many as 350 airborne chemicals (ref. Color). The procedure is to use a hand bellows pump to draw an air sample of known volume through a glass tube that contains a reagent that changes color in the presence of the target contaminant. Both qualitative and quantitative determinations are possible because the length and intensity of the color band are related to the quantitative concentration of the contaminant in question. Sometimes such tubes are used for quick screening for potential problems, with subsequent sampling and laboratory analysis for exact quantitative measurement. Figure 9.6 shows a sample detector tube inserted in the hand bellows pump to collect a sample. See Case Study 9.5.

CASE STUDY 9.5
DETECTOR TUBE TEST

Figure 9.7 diagrams a detector tube used for testing atmospheric concentrations of acetaldehyde. The following specifications apply to this tube:

Standard measuring range	100 to 1000 ppm
Number of strokes	20
Standard deviation	±15–20%
Color change	Orange to brownish green

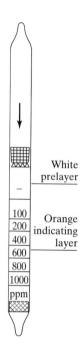

White
prelayer

Orange
indicating
layer

FIGURE 9.7

Acetaldehyde detector tube (*Source:*
Courtesy of Draeger Safety, Inc.).

The problem is to determine whether this detector tube can be used to test for the
PEL or AL for acetaldehyde.

Solution

Appendix A.1 reveals that the PEL for acetaldehyde is 200 ppm. This concentration
is within the range of sensitivities for the tube illustrated. The tube is also capable of
detecting concentrations of 100 ppm, which is the AL for acetaldehyde. The method
is to snap off the ends of the glass tube and attach it in-line to a flexible tube attached to a pump. The pump is actuated for 20 strokes. The depth of penetration of
the brownish-green band into the orange calibrated region of the tube shown in
Figure 9.7 determines the approximate concentration (in ppm). There is some error
associated with the experiment, of course, and other petroleum hydrocarbon impurities can sometimes contaminate the results. Still, the test is widely used to assess
suspected problem areas (ref. Draeger).

For more obscure contaminants and for rarer concentrations there is no choice;
sampling devices and laboratory analysis must be used. Either these devices pump a
prescribed quantity of air past a filter or a sorbent that collects the contaminant, or
they merely collect a precise volume of the air itself. The filter, the sorbent, or the air
sample is then sent to a laboratory for analysis.

Dosimeters are the most convenient device of all, especially for gathering TWA
data. A dosimeter is a small collector worn on the worker's body or clothing and that

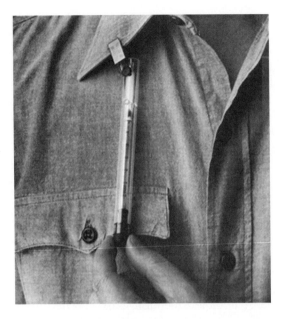

FIGURE 9.8

Dosimeter for gathering TWA data automatically on a cumulative basis (*Source*: Courtesy of the Mine Safety Appliances Company).

collects a time-weighted average exposure over a specified time period, such as a full shift. Unfortunately, accurate dosimeters are still beyond the state of the art for most toxic substances. Figure 9.8 shows one type of dosimeter used for air monitoring.

Exposure Comparisons

Before leaving the subject of exposures to toxic contaminants, it is appropriate to put the whole subject in perspective by comparing the levels of exposures due to industrial pollution and those encountered in our everyday personal lives. The focus in the closing decades of the 20th century was on the outdoor pollution of the environment by industrial sources of emission. The focus in the 21st century may well be on indoor pollution—not necessarily only that within factories, but also in our homes and everyday environment. Researchers Ott and Roberts (ref. Ott) have concentrated, not on the *emissions* of highly toxic substances, but on the *levels* to which humans are exposed to such substances in their daily activities. Some of their conclusions have been surprising:

> Most citizens were very likely to have the greatest contact with potentially toxic pollutants not outside but inside the places they usually consider to be essentially unpolluted, such as homes, offices and automobiles. The exposures arising from the sources normally targeted by environmental laws—Superfund sites, factories, local industry—was negligible by comparison.
>
> Even in the New Jersey cities of Bayonne and Elizabeth, both of which have an abundance of chemical processing plants, the levels of 11 volatile organic compounds proved much higher indoors than out. (Concentrations of the other volatile compounds tested were found to be insignificant in both settings.) The chief sources appeared to be ordinary consumer products, such as air fresheners and cleaning compounds, and various building materials.

This chapter used benzene as an example of a toxic substance associated with serious health risks such as leukemia. Benzene is of such concern that it is one of the few contaminants for which comprehensive standards have been promulgated in the

OSHA "standards completion project." But research by Wallace (ref. Wallace) has shown that it is not industrial, but personal, exposures to benzene that are responsible for the major portion of the exposure. According to his calculations, 45 percent of the total U.S. exposures to benzene are from smoking (or breathing smoke exhaled by others)! Another 36 percent are from inhaling gasoline vapors or from using various common products (such as glues), and another 16 percent from miscellaneous home sources. All three of these sources of benzene exposure taken together add up to 97 percent of total U.S. exposures. That leaves only three percent of the average person's benzene exposure from industrial pollution. In light of these statistics, is it any wonder that emphasis on air contaminants in the 21st century is shifting away from industrial pollution to everyday, personal exposures?

SUMMARY

Exposure to toxic substances is the classic health problem, but workplace exposures are different in nature from what most people think of as poisons. Off the job, poisons are thought of as deadly potions that are ingested. On the job, poisons may be deadly, but their entry into the body is usually through the lungs in minute concentrations, and the effects may take years to recognize. A notable exception is oxygen deficiency, which can have quick and fatal outcomes. The ways that poisons affect the human body can be somewhat classified into seven categories: irritants, systemic poisons, depressants, asphyxiants, carcinogens, teratogens, and mutagens. Each of the seven types of poisons can be mild or deadly, depending on concentration. Many substances fall into several of the seven categories.

An important first step in getting control of health hazards is to make effective use of baseline examinations for all new employees. Such examinations determine any preexisting health conditions that might be aggravated by workplace exposure to toxins. The examinations also permit measurement of deterioration of the worker's health, if any, during the period of employment.

Measuring the deterioration of worker's health during the period of employment provides valuable information, but even more worthwhile is the measurement of toxic exposures before they do their damage. This chapter has explored the science and instrumentation to measure tiny concentrations of toxic substances in industrial atmospheres. It has also introduced the system of standards against which these concentrations are compared. The most important standards for air contaminants are the permissible exposure levels (PELs). Most of the PELs are for 8-hour time-weighted-average (TWA) exposures. Some toxic substances are so dangerous as to require setting a ceiling (C) maximum concentration. Action levels (ALs) are derived from the PELs and indicate when air contaminants are reaching levels that need control before limits are exceeded. Special formulas are needed for considering the combined effects of multiple contaminants.

Having studied toxic substances, their effects on the human body, and the methods and standards for the measurement of their concentrations in industrial atmospheres, Chapter 10 will now consider ways that the industrial environment can be controlled to minimize the effects of these substances. In addition, Chapter 10 will introduce the hazard of industrial noise and its control.

EXERCISES AND STUDY QUESTIONS

9.1 Besides being a good basis for placement of employees in appropriate jobs for their physical characteristics, what is a secondary reason for baseline physical examinations?

9.2 How would you distinguish between the terms *toxic substances* and *hazardous materials*?

9.3 Identify the seven principal classes of toxic substances.

9.4 What is the technical name for the "drinking" variety of alcohol? To what toxic substance class does it belong?

9.5 Identify the three principal routes for toxins to enter the body.

9.6 Identify a career field in which workers are particularly vulnerable to contamination from biohazard exposure to the eyes.

9.7 What is the "standards completion project?" Name at least two common substances that are covered by such standards.

9.8 Name at least three common, odorless gases that can be dangerous.

9.9 What extremely foul-smelling toxic gas has the potential of overcoming a person's olfactory system? What does *olfactory* mean?

9.10 Explain the relationship between AL and PEL. What do these terms represent?

9.11 What is the definition of the word *fumes*?

9.12 How much carbon *mon*oxide in the air is permissible, given a normal (0.033%) concentration of carbon *di*oxide (and no other contaminants)?

9.13 Air samples show an industrial atmosphere to contain 0.001% methyl styrene during the morning half of the shift and 0.015% during the afternoon half. Calculate the TWA. Assuming that no other contaminants are present, does the exposure exceed the PEL? Does it exceed the AL?

9.14 Air samples show the following contaminant concentrations in an 8-hour shift (from 8:00 A.M. to 4:00 P.M.):

  1. Trifluorobromomethane, 0.1% from 11:00 A.M. to 2:00 P.M.

  2. Propane, 0.05% all day

  3. Phosgene, 1 part per million at 2:00 P.M., with a duration of 15 minutes

    (a) Assuming that no other contaminants are present, does the full-shift atmosphere meet OSHA standards?

    (b) To *exactly* meet OSHA standards, how much more or less time for the phosgene exposure would be permissible provided other contaminants remain as previously stated?

9.15 Two color detector tubes for testing for atmospheric concentrations of nitrogen dioxide have the following specifications:

Tube A standard measuring range:
    5 to 25 ppm (using 2 pump strokes)
    0.5 to 10 ppm (using 5 pump strokes)

Tube B standard measuring range:
    5 to 100 ppm (using 5 pump strokes)
    2 to 50 ppm (using 10 pump strokes)

Which tube provides the greater precision for the test? Which tube would be preferable for checking for concentrations close to the PEL? How many pump strokes should be used? Which tube would be preferable for checking for concentrations close to the AL? How many pump strokes should be used?

9.16 A particular gas welding process in a confined space is suspected of producing dangerous concentrations of carbon monoxide, carbon dioxide, iron oxide particulate, and manganese fumes. Atmospheric sampling produces the following exposure data:

	ppm		mg/m^3	
Period	CO	CO$_2$	Iron oxide	Manganese
8:00 A.M.–10:00 A.M.	10	1000	1	1
10:00 A.M.–12:00 noon	20	1000	4	1
12:00 noon–1:00 P.M.	25	1000	2	0
1:00 P.M.–4:00 P.M.	30	1000	3	1

Do calculations to determine whether the combined exposure represents an OSHA violation.

9.17 Name at least five pneumoconioses. Which are most dangerous?

9.18 How do fibroses differ from other pneumoconioses?

9.19 Name the two basic classes of asphyxiants and give examples of each.

9.20 Explain the following terms:

(a) *mutagen*

(b) *carcinogen*

(c) *teratogen*

9.21 In what ways are the threats of poisons at work different from those at home?

9.22 What is the difference between *fumes* and *vapors*?

9.23 Compare the particle sizes of the following substances:

(a) zinc oxide fumes

(b) tobacco smoke

(c) diameter of human hair

(d) bacteria

9.24 Explain the following terms:

(a) *TLV*

(b) *PEL*

(c) *TWA*

(d) *MAC*

(e) *STEL*

(f) *AL*

9.25 Name some traditional methods of detecting the presence of dangerous air contaminants and explain advantages and disadvantages of each.

9.26 Name three basic approaches to measuring air-contaminant exposures.

9.27 Suppose that an industrial process produces the following air-contaminant concentration for the periods shown:

Period	Methanol (ppm)	Nitric oxide (ppm)	Sulfur dioxide (ppm)
8:00 A.M.–10:00 A.M.	50	5	0
10:00 A.M.–11:00 A.M.	150	10	1
11:00 A.M.–1:00 P.M.	100	5	1
1:00 P.M.–4:00 P.M.	200	10	1

Taken collectively, would these concentrations exceed permissible exposure levels?

9.28 In Exercise 9.27, suppose that the solvent ethanol could serve to replace the methanol in this process, but at the expense of double the concentrations of the solvent in the atmosphere. Would this help or hinder matters? Explain.

9.29 Two solvents, benzene and chlorobenzene, are under consideration by process engineers for use in a plant for which you have responsibility for safety and health management. What information can you provide to the process engineers regarding the comparative hazards of these two solvents?

9.30 A man and woman spill a bottle of 150 proof rum on the kitchen floor in their small one-bedroom apartment. The total apartment area is 600 square feet and has standard 8-foot ceilings. By the time they clean up the spill, approximately 5 cubic feet of alcohol vapor has entered the air through evaporation. Noting the strong smell of alcohol in the air, they open the window and, feeling drowsy, go to bed and sleep all night (8 hours). The open window permits a gradual dilution of the alcohol in the air, and by morning, the concentration of alcohol in the air is down to 500 parts per million. Assuming a constant rate of decline in alcohol content of the air all night, was the alcohol concentration a hazard? If the exposure had been occupational, would PELs have been exceeded?

9.31 In Exercise 9.30, the couple's original intent was to bake rum cake. Had they gone ahead and made the cake, suppose that the hot oven had caused an additional 25 cubic feet of alcohol vapor to be liberated into the apartment. How would this concentration compare with the PEL?

9.32 A direct-reading gas detector tube is available for sampling concentrations of the toxic gas sulfur dioxide. Tube 5H is specified for concentrations in the range of 0.05 to 8.0% and tube 5M is for concentrations in the range 20 to 3600 ppm. Which tube is the more sensitive of the two tubes?

9.33 The following gas detector tubes are available for direct reading of concentrations of the toxic gas hydrogen sulfide:

Tube	Concentration
4HT	1–40%
4HH	0.1–4.0%
4H	10–3200 ppm
4M	12.5–500 ppm
4L	1–240 ppm
4LL	0.25–60 ppm

Which of these tubes would be satisfactory for detecting the OSHA-specified ceiling concentration for hydrogen sulfide? Of the satisfactory tubes, which one tests the narrowest range of concentrations?

9.34 A detector tube is available for testing concentrations of isopropyl acetate in the range 0.05 to 0.75%. Would this tube be capable of detecting concentrations at the OSHA PEL? Would it be a satisfactory device for testing whether the AL was exceeded?

9.35 A particle of coal is determined in the laboratory to have a diameter of 17 micrometers. What is the diameter in centimeters? Calculate the diameter in inches. Is the particle classified as dust or fume?

9.36 Consider the following concentrations of air contaminants observed together on the same day:

Contaminant	TWA (ppm)
Isopropyl ether	200
Ethyl benzene	40
Chlorobenzene	25
Chlorobromomethane	50

Taken separately, do any of the concentrations exceed the PEL for that substance? the AL? Taken as a whole, does the mixture exceed the PEL? the AL?

9.37 **Design Case Study**. A process engineer proposes a new solvent that will reduce the quantities required by the process and significantly reduce the quantities of solvent vapors released into the air inside the plant. The new solvent is perchloroethylene, and it is expected to reduce the solvent vapors absorbed into the plant air by 20% by volume as compared to the old solvent (Stoddard solvent). You are called in as a Certified Safety Professional to evaluate the proposed change to the process. Do you support the proposed process change? Explain your position.

9.38 **Case Study: Rayon Manufacturing Plant**. Measurements are taken and airborne contaminant concentrations are as shown in Table 9.3. Determine which of the substances are listed by OSHA as having PEL limits and perform computations to determine whether the given exposures, taken separately and together, exceed OSHA PELs and ALs.

9.39 **Design Case Study.** Process design engineers for the rayon manufacturing plant in Exercise 9.38 suggest introducing a new process that uses the solvent formaldehyde. Preliminary evaluations suggest that the new process will add a slight amount of formaldehyde

TABLE 9.3 Exposure Levels		
	Morning exposures (4 hour)	Afternoon exposures (4 hour)
Acetic anhydride	0.5 ppm (p/m)	1 ppm
Sodium hydroxide	0.2 mg/m^3	0.3 mg/m^3
Ammonium sulfide	3 ppm	4 ppm
Calcium bisulfide	5 ppm	8 ppm
Carbon disulfide	4 ppm	6 ppm
Sodium sulfide	0.7 mg/m^3	0.8 mg/m^3
Sodium sulfite	0.5 mg/m^3	0.5 mg/m^3

vapors to the plant atmosphere, perhaps 1 part per million by volume, in addition to current levels of the other air contaminants listed in Table 9.3. The plant engineer has had the foresight to invite the safety and health manager to join the design team for an additional perspective into the design process. Perform calculations and evaluate the potential impact of the new process proposal on the safety and health of the workers in the plant. What recommendation would you make to the design team?

RESEARCH EXERCISES

9.40 Mercury and its compounds are toxic materials addressed in this chapter. In 1995, an industrial release of mercury compounds into a river in Russia threatened the health of people in the area and environmental damage to the Arctic Ocean. Research the particulars of this accident and the extent of the damage. Why is the Arctic Ocean more vulnerable to damage from this type of accident than would be more temperate oceans?

9.41 On January 10, 1997, OSHA issued a final rule on methylene chloride. Research this standard to determine the following:

(a) the standard number

(b) the effective date of the standard

(c) the date of the end of the implementation (start-up) phase

(d) the 8-hour TWA PEL

(e) the STEL

(f) the length of time exposure records must be kept

9.42 Study the impact of the methylene chloride standard. How many lives per year are expected to be saved as a result of the promulgation of this standard? How much will cancer risks be cut for workers who use this solvent? How much will worker exposures be decreased? How many workers use methylene chloride?

9.43 In 1996, OSHA reduced the permissible exposure limit (PEL) for 1,3-butadiene. How much was the PEL reduced? What is the estimated increased annual cost to industry to comply with this more stringent requirement? How many cancer deaths are expected to be saved over a 45-year working lifetime?

STANDARDS RESEARCH QUESTIONS

9.44 This chapter listed all of the chemicals identified in the "standards completion project." Use the NCM database to determine which of these chemicals generates the most OSHA citation activity.

9.45 Study the OSHA General Industry standard for dealing with toxic air contaminants in general. Find the provision that requires air contaminants to be held within their respective PELs. Determine the frequency of citation of this provision using the NCM database. Determine the average dollar level of proposed penalties for violation of this provision.

9.46 Find the provision of the OSHA standard that deals with mixtures of various toxic air contaminants. Use this standard to verify Equation 9.2 in this chapter.

C H A P T E R   1 0

Environmental Control and Noise

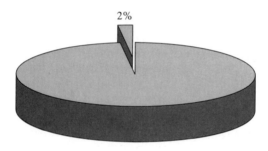

2%

Percentage of OSHA
General Industry citations
addressing this subject

In Chapter 9, we explored the important task of measuring and evaluating air contaminants to determine the degree of exposure to health hazards. Once having found that an air contaminant does exist, there are a variety of strategies available for dealing with it. In Chapter 3, it was stated that from the profession, there has emerged a definite hierarchy of strategies that have come to be known as the "three lines of defense." These strategies are engineering controls, work-practice controls, and personal protective equipment, in that order. In this chapter, we will examine methods to provide engineering solutions to the air-contaminant problem, chiefly through ventilation. Noise hazards will also be examined, along with both engineering and work-practices controls to control these hazards.

VENTILATION

Ventilation may be the most obvious engineering solution to an air-contaminant problem, but before acceding to this solution, it should be recognized that other ways to deal with the problem may be even better. In Chapter 3, a series of approaches was enumerated as "engineering design principles." In this chapter, some applications of these engineering principles will be seen.

The most desirable way to deal with an air contaminant is to change the process so that the contaminant is no longer produced. This is so obvious that it is sometimes overlooked. It may be that the process cannot be changed, but if it can, there may be tremendous gains in store, not only in health and safety, but in production cost and efficiency as well. For instance, it may be found that machined parts can be machined dry, avoiding cutting oil exposures to the machinist's skin as well as solvent contamination of the air as the parts are later cleaned. Forging, die casting, or powdered metal technology may eliminate several machining processes, changing the process in beneficial ways. Some of these ideas may generate more disadvantages than advantages in given situations, but each application should be checked for potential benefit. There is no better way for the safety and health manager to win recognition from top management than by generating a clever idea that cuts costs or increases production while it also enhances safety or health.

For more ideas, consider the hazards of toxic air contaminants from welding operations. Sometimes the principal source of the contaminant is the surface coating on the metal to be welded. Perhaps as a process change, this surface coating can be removed prior to the commencement of welding. Better yet, perhaps the material does not require welding at all. Possibly a crimping operation could produce an effective joint, eliminating the need for welding or soldering.

Chemical processes are classified as batch or continuous. The choice between the two usually involves many considerations, including investment cost, length of production run expected, volumes to be produced, and the important factor of air contamination. Continuous processes generally reduce the exposure of materials to the air because open handling is reduced, and batches of materials are not sitting idle awaiting processing. However, the mechanical handling equipment used for continuous processes may increase the contamination levels. Each situation must be studied to determine the best solution, keeping in mind safety and health aspects.

One way a process can be changed is to isolate or enclose it. If a particularly contaminating process is in the plant, perhaps it should be located in a separate building so that it does not contribute to the overall ventilation problem.

A slight variation to changing the process is to change the materials used. Carbon tetrachloride has been found to be a health hazard, so other solvents have been substituted. The chlorinated hydrocarbon solvents substituted, such as trichloroethylene and perchloroethylene, are also being found to be hazardous, but fortunately not as hazardous as carbon tetrachloride. New solvents may be found to reduce hazards even further. Labar (ref. Labar, 1993) classifies many dangerous hydrocarbon solvents as "volatile organic compounds"(VOCs) and suggests substituting water-based solvents for VOCs. An extra benefit of water-based solvents is that sometimes they are not as slippery as VOCs if spilled on the floor. They may be expensive and usually take longer to dry than VOCs. Another problem can be corrosion. Everyone knows that water causes rust, and if water-based solvents are used for wash-downs, stainless steel nozzles may be required. Finally, water-based solvents may themselves be health hazards, such as causing urinary tract infections. Substitution of materials may be a good idea, but all of the pros and cons should be identified and evaluated.

Another possible substitution is in sand blasting. Silica sand is often used for blasting to improve surface characteristics. But airborne silica causes the lung disease called silicosis. Perhaps the silica sand could be replaced by steel shot, removing the silica contamination.

A classic example of changing materials to reduce hazards is the switch from hazardous lead-based paints to substitute materials such as iron oxide pigments. Another classic was the switch from Freon to propane as a propellant for aerosol cans. In this case, the materials switch was intended to protect the environment (the ozone layer), but the solution may be more hazardous to the individual because propane is a flammable gas.

Design Principles

If the process cannot be changed or materials substituted, a well-designed ventilation system may be the best solution to the problem. OSHA has a standard that deals with this subject, but it must be emphasized that ventilation is a very technical subject, and the safety and health manager may want to turn to a professional engineer to design an adequate ventilation solution to an air-contamination problem. Exhaust ventilation is not the same as ordinary heating and air conditioning, and design errors can be made if this difference is not considered. An example is shown in Figure 10.1. Most heating and air-conditioning ducts have right-angle bends, which may be fine for gases, but greatly impair the ability of the ducts to transport particulates.

Another questionable ventilation system is an ordinary household fan used to blow away smoke from a contaminant source. It is true that a fan can dilute the concentration of a contaminant in a given place, and dilution ventilation is a recognized method of reducing concentrations to levels lower than the PEL. But the question is "Where is the fan blowing the contamination *to*?" It is adding to the overall background level of air contamination in the plant and may later have to be dealt with if other processes are also producing contaminations.

A basic objective of exhaust ventilation is to isolate and remove harmful contaminants from the air. The more these contaminants are concentrated into limited plant areas, the easier they are to separate from the air. Dilution ventilation, although expedient in some cases, is counterproductive to the goal of removing the contaminant. Dilution ventilation can be likened to "sweeping the dirt under the rug."

Focus is an important consideration for ventilation systems. Sheer volume or flow velocity is not enough. Ventilation technology is producing some very fine local exhaust systems that focus the intake right on the contaminant, much as a vacuum cleaner

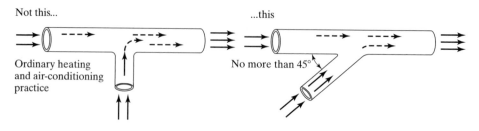

FIGURE 10.1

Avoid sharp angles at duct entry.

is designed to do. Even if sufficient flow could be achieved with a general ventilation system, the flow could be a nuisance because such a wind might blow away papers and other materials, making the job awkward and inefficient.

The earlier mention of a vacuum cleaner might give some safety and health managers the wrong idea. Particle size for fumes or other contaminants will usually be too small to be effectively trapped in the bag of an ordinary vacuum cleaner. If the process does not trap the contaminant, it is merely blowing it around, perhaps increasing exposure.

The best exhaust ventilation systems are the "pull" types, not the "push" types. Even within the exhaust duct, the fan should be placed at the end of the duct if possible, as shown in Figure 10.2. Leaks in the duct then merely draw in more air rather than pump contaminated air back into the plant environment.

Makeup Air

With an exhaust ventilation system or systems, some source of *makeup air* is essential. The traditional way to supply makeup air was simply to open windows and doors. Today, however, it has become increasingly attractive to recirculate the exhaust air after filtration and decontamination. Not only does such a solution save energy, it also reduces external atmospheric pollution, a very important point considering regulations of the Environmental Protection Agency (EPA) and the general environmental concerns of the public.

Figure 10.3 diagrams a recirculating system in which the objective is to remove dust by means of a high-efficiency filter. It is essential to recognize the importance of the condition of the filter to the overall effectiveness of the system. The filter will clog up over time if it is doing its job, and thus it must be serviced, cleaned, or changed. Note the bypass damper, which permits selective proportioning of the system from full recirculating to full exhaust. This bypass can save energy when weather conditions are mild and recirculation is unnecessary. Also note the inclusion of a manometer to detect a pressure differential across the filter as well as an alarm to sound if this differential becomes too great. Both are intended to provide indication or warning to the operator that the filter is in need of service.

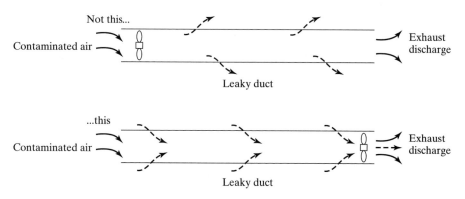

FIGURE 10.2

Fan location—keep negative pressure within the duct.

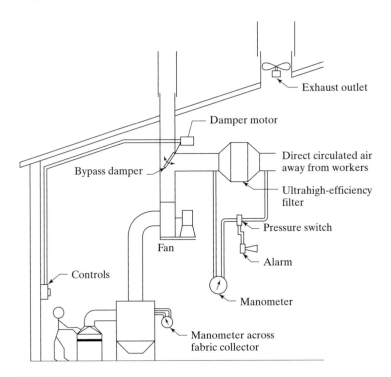

FIGURE 10.3

Example of recirculation from air-cleaning devices (dust) (*Source*: American Conference of Governmental Industrial Hygienists Committee on Industrial Ventilation).

Safety and health managers are warned to monitor recirculation systems closely. Frequently in industry a sophisticated ventilation system is installed for a process, and then it is ignored. Some filter alarms are merely red lights, which operators often ignore. Even audible alarms such as buzzers and horns are sometimes disabled by disconnecting electrical leads. Both operations and maintenance personnel have been found guilty of disabling these devices, which are designed to protect the workers' health.

Besides using recirculating systems, there are some other ways around the problem of energy losses due to the introduction of makeup air into the building. One method is to introduce the makeup air right at the point at which the contamination is taking place. With this strategy, the makeup air may need no air conditioning—no cooling in summer and no heating in winter. The exhaust system will merely suck up the unconditioned makeup air together with the contaminants, and workers will have little exposure to either.

Another solution to the problem is to use a heat exchanger to recapture the energy of the exhaust air and transfer it to the incoming makeup air. It is difficult to make this solution practical, however, because the heat differential between makeup air and exhaust air is usually too low to make the heat exchanger effective. Also, the heat exchanger approach necessitates positioning the makeup air duct close to the exhaust air duct, which introduces the possibility of cross-contamination. Finally, heat exchanger

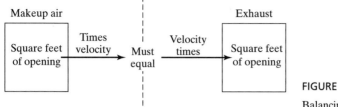

FIGURE 10.4

Balancing makeup and exhaust air.

systems can be expensive, both to install and to maintain effectively—too expensive in many cases to amortize by means of energy cost savings over the life of the system.

Besides the energy problem, another problem with supplying makeup air is the presence of contaminated air from the outside. This is an unusual problem, but it has presented itself on occasion. At one plant, the makeup air inlet was adjacent to a major freeway, which caused carbon monoxide and other automobile exhaust emissions to be drawn into the building. In another poor design, the makeup air inlet was so close to the exhaust system discharge that contaminants were being drawn back in and recirculated around the plant. That way, if workers managed to escape breathing the contaminated air the first time through, they got another chance to become exposed!

A quick check to determine whether there is a sufficient makeup air supply is to check atmospheric pressure both inside and outside the plant. The pressure inside should be only slightly lower than the pressure outside. If the pressure inside is substantially lower, then the makeup air supply is insufficient. The basic relationship between makeup and exhaust is illustrated in Figure 10.4. The cross-sectional area of makeup openings multiplied by the velocity of flow through those openings must equal the cross-sectional area of the exhaust openings multiplied by the velocity of flow through the exhaust.

The provision for adequate makeup air and a sufficient volume of general exhaust ventilation is sometimes the only practical solution to the problem of reducing air-contaminant exposures to specified levels. Case Study 10.1 will illustrate the principle of this solution to the problem.

CASE STUDY 10.1

An industrial process liberates 2 cubic feet of chlorobenzene per hour into a room that measures 20 feet by 40 feet and has a ceiling height of 12 feet. What minimum general exhaust ventilation in cubic feet per minute is necessary to prevent a general health hazard in this room?

Solution

A subtle facet of this problem is that for a continuously operating process, the dimensions of the room are really irrelevant to the solution. It is true that for a short-duration exposure, the size of the room will affect the dilution of the chlorobenzene within the confines of the room. But to deal with a continuous process, one must provide sufficient ventilation to yield an ample supply of makeup air to continuously dilute the chlorobenzene to levels within limits, *regardless of room size.*

The PEL for chlorobenzene is 75 ppm. Let X = the total ventilation necessary to dilute the chlorobenzene. Then,

$$\frac{2}{X} = \frac{75}{1,000,000}$$

$$X = \frac{2 \times 1,000,000}{75} = 26{,}667 \text{ ft}^3/\text{hr}$$

$$= \frac{26{,}667 \text{ ft}^3/\text{hr}}{60 \text{ min/hr}}$$

$$= 444 \text{ ft}^3/\text{min}$$

Purification Devices

If the exhaust air is clean enough to meet external standards, no filtration or purification may be necessary once the air gets outside the plant. But often some type of purification device is necessary on the outside as is required indoors for recirculating systems, particularly for the removal of particulates. The paragraphs that follow describe some of the basic types of particulate removal devices.

Centrifugal devices, often called *cyclones* (see Figure 10.5), take advantage of the mass of the contaminant particles, causing them to collect on the sides of the cyclone in

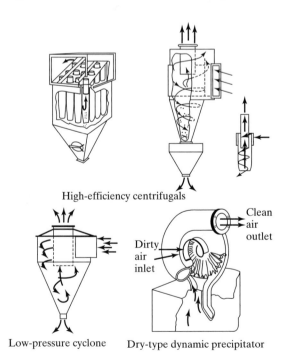

High-efficiency centrifugals

Clean air outlet

Dirty air inlet

Low-pressure cyclone   Dry-type dynamic precipitator

FIGURE 10.5

Cyclone and other dry-type centrifugal collectors for removal of particulates from exhaust air (*Source*: ACGIH Committee on Industrial Ventilation).

the swirling air and then slide to the bottom and settle in the neck of the funnel, where they can be periodically emptied. Another type of centrifugal device causes the dirty air to strike louvers, whereupon particles separate from the air. A typical application for cyclones is grain dust removal for grain elevators and mills. Cyclones are also used for woodworking sawdust, plastic, dusts, and some chemical dry particulates.

Electrostatic precipitators place a very high (e.g., 50,000-volt) electrical charge on the particles, causing them to be attracted to an electrode of opposite electrical charge. The collecting electrode can consist of plates, rods, or wires, all of which can be agitated to shake off the collected dust and cause it to settle in the bottom of the chamber. Figure 10.6 illustrates one type of high-voltage electrostatic precipitator. Electrostatic precipitators are used in the steel, cement, mining, and chemical industries. Electrostatic precipitators are also used in smokestacks to reduce fly ash.

Wet scrubbers include a wide variety of devices that employ water or chemical solution to wash the air of particulates or other contaminants. Some types pass the dirty air through standing water or solution. Another type forces dirty air to rise in a tower packed

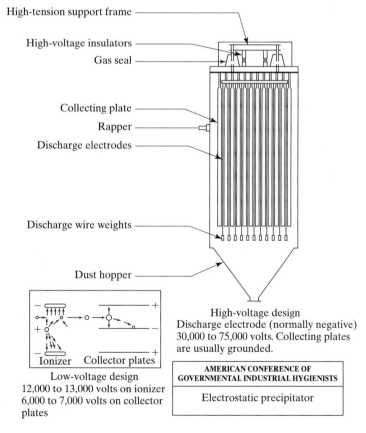

FIGURE 10.6

Electrostatic precipitators for removing particulates (*Source*: ACGIH Committee on Industrial Ventilation).

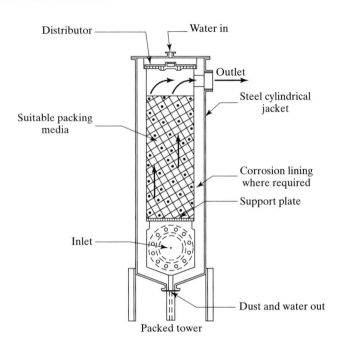

FIGURE 10.7

Packed tower type of wet scrubber
(*Source*: ACGIH Committee on Industrial
Ventilation).

AMERICAN CONFERENCE OF GOVERNMENTAL INDUSTRIAL HYGIENISTS
Wet-type dust collector

with a filler media through which water falls, as shown in Figure 10.7. Note in the figure that the cleanest water is at the top of the tower, where the exiting air is at its cleanest. The bottom of the tower is the dirtiest, where the water vehicle is exiting the packing and is contacting the dirtiest, untreated exhaust air. Wet centrifugal types, like dry centrifugal types, take advantage of the mass of the particles by causing them to impinge on blades, plates, or baffles. Wet scrubbers are seen in the chemical industries, where they are able to remove gases and vapors in addition to particulates. Other industries that use wet scrubbers are the rubber, ceramics, foundries, and metal-cutting industries.

Fabric, or bag-type, filters are essentially like a vacuum cleaner bag. Some are huge and are located in separate buildings called *baghouses*. Figure 10.8 shows three types of fabric filters. Fabric filters are used in the refining of toxic metals such as lead and in the woodworking, metal cutting, rubber, plastics, ceramics, and chemical industries.

INDUSTRIAL NOISE

Noise exposure is another classic health problem because chronic exposures are the ones that typically do the damage. A single acute exposure can do permanent damage, and in this sense, noise is a safety problem, but noise exposures of this type are extremely rare.

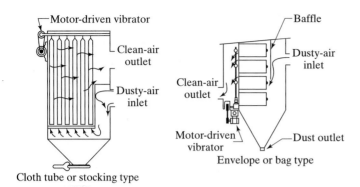

Cloth tube or stocking type

Envelope or bag type

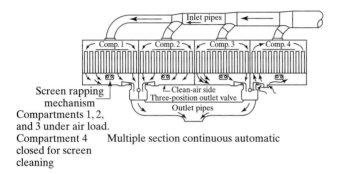

Screen rapping mechanism
Compartments 1, 2, and 3 under air load.
Compartment 4 closed for screen cleaning

Multiple section continuous automatic

| AMERICAN CONFERENCE OF |
| GOVERNMENTAL INDUSTRIAL HYGIENISTS |
| Fabric collectors |

FIGURE 10.8

Fabric filters for exhaust air (*Source*: ACGIH Committee on Industrial Ventilation).

As with other health hazards, noise has a threshold limit value, and exposures are measured in terms of time-weighted averages. To understand the units of these measures, some background is needed in the physical characteristics of sound.

Characteristics of Sound Waves

Noise can be defined as unwanted sound. In the industrial sense, noise usually means excessive sound or harmful sound. Sound is generally understood as a pressure wave in the atmosphere. In liquids, sound is also a pressure wave; in rigid solids, sound takes the form of a vibration.

Two basic characteristics of sound waves important to the subject of noise control are

1. the amplitude, or pressure peak intensity, of the wave;
2. the frequency in which the pressure peaks occur.

Our sense of hearing can detect both of these characteristics. Pressure intensity is sensed as loudness, whereas pressure frequency is sensed as pitch. Figure 10.9 illustrates the wave

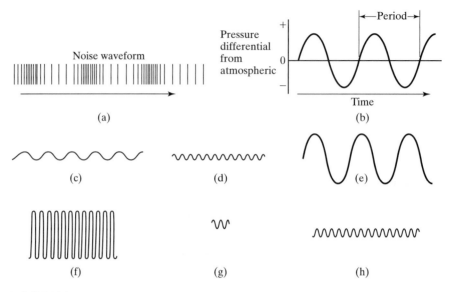

FIGURE 10.9

Characteristics of sound waves—(a) pressure wave is longitudinal (in the direction of travel of the sound); (b) relationship between pressure and time at a given point of sound exposure; (c) low pitch, soft sound; (d) high pitch, soft sound; (e) low pitch, loud sound; (f) high pitch, loud sound; (g) short-duration sound; (h) sustained sound.

form of sound and also graphs the relationship between pressure and time. Note in Figure 10.9(b) that the *period* is the length of time required for the wave to complete its cycle. In the graph, it is measured at the point at which the pressure differential becomes zero and is starting to become negative. However, it could have been measured from peak to peak, valley to valley, or at any other convenient reference point in the cycle. These periods are always short, too short to count their occurrences while listening to the sound. If we could count the occurrences of these wave cycles, the resulting count per unit time would be the *frequency*, usually measured in numbers of cycles per second (*hertz*). A typical sound is at a frequency of 1000 cycles per second—that is, 1000 hertz (Hz). Obviously, we could never count 1000 pulses of pressure in a single second, but our ears have surprising sensitivity to variations in this frequency count. The sensation is known as *pitch*, and skilled musicians have trained their ears to hear very slight variations in sound-wave frequency. Frequency is important in analyzing the sources of occupational noise exposure.

Even more important than pitch in industrial settings is the pressure intensity of the sound wave. High peaks of pressure in the waves can do permanent damage to the delicate mechanisms in the human ear, causing permanent hearing loss. The ear is delicate enough to pick up the tiny pressures of the faintest audible sounds and also is able to withstand an incredibly large range of pressures. The human ear can withstand, *without damage*, a sound pressure 10,000,000 times as great as the faintest sound it can hear! A necessary result of this incredible range of pressures is that the ear is not very sensitive to shades of differences in these pressures, especially as the pressures get into the upper part of the range. In other words, as sounds get fairly loud, the human ear cannot readily detect a large increase in the intensity, even a doubling or tripling of the intensity.

Decibels

It is difficult to talk sensibly about such a large range of audible pressures, and it is especially difficult to set standards. Imagine a noise meter reading in the millions. The situation is further complicated by the lessening of the ability of the human ear to detect pressure differences as sounds get louder. To deal with these problems, a unit of measure called the *decibel* (dB) has been devised to measure sound-pressure intensity. The decibel has a logarithmic relation to the actual pressure intensity, and thus the scale becomes compressed as the sound becomes louder, until in the upper ranges, the decibel is only a gross measure of actual pressure intensity. But this is appropriate because, as was mentioned before, the human ear only hears gross differences anyway when the sound becomes very loud. Figure 10.10 relates the decibel to familiar sound levels.

The logarithmic decibel scale is convenient, but it does give rise to some problems. If a machine in the plant is very loud, putting a second machine just like it right beside it will not make the sound twice as loud. Remember that the range of sound pressures is tremendous and that the human ear hears only a slight increase in loudness, when the

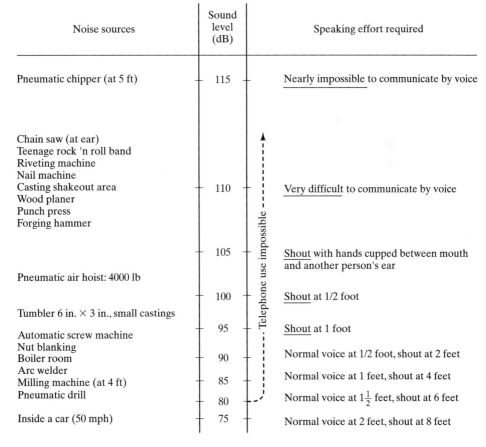

Noise sources	Sound level (dB)	Speaking effort required
Pneumatic chipper (at 5 ft)	115	Nearly impossible to communicate by voice
Chain saw (at ear) Teenage rock 'n roll band Riveting machine Nail machine Casting shakeout area Wood planer Punch press Forging hammer	110	Very difficult to communicate by voice
	105	Shout with hands cupped between mouth and another person's ear
Pneumatic air hoist: 4000 lb		
	100	Shout at 1/2 foot
Tumbler 6 in. × 3 in., small castings		
Automatic screw machine	95	Shout at 1 foot
Nut blanking Boiler room	90	Normal voice at 1/2 foot, shout at 2 feet
Arc welder Milling machine (at 4 ft)	85	Normal voice at 1 feet, shout at 4 feet
Pneumatic drill	80	Normal voice at $1\frac{1}{2}$ feet, shout at 6 feet
Inside a car (50 mph)	75	Normal voice at 2 feet, shout at 8 feet

Telephone use impossible

FIGURE 10.10

Decibel noise levels of familiar sounds (*Source:* NIOSH).

TABLE 10.1 Scale for Combining Decibels

Difference between two decibel levels to be added (dB)	Amount to be added to larger level to obtain decibel sum (dB)
0	3.0
1	2.6
2	2.1
3	1.8
4	1.4
5	1.2
6	1.0
7	0.8
8	0.6
9	0.5
10	0.4
11	0.3
12	0.2

Source: NIOSH (ref. Industrial Noise).

actual sound pressure may have doubled due to the addition of the extra machine. The decibel scale recognizes the addition of the new machine as an increase in noise level of only 3 decibels. Conversely, if the noise level in the plant exceeds allowable standards by very much, shutting off half the machines in the plant—an obviously drastic measure—may have very little effect in bringing down the total noise level on the decibel scale. Table 10.1 provides a scale for combining decibels to arrive at a total noise level from two sources. If there are three or more sources, two sources are combined and then treated as one source to be combined with a third, and so on, until all sources have been combined into a single total. An example is helpful in illustrating the method.

CASE STUDY 10.2

Suppose that noise exposure at a workstation is essentially due to four sources, as follows:

Machine A	86 dB
Machine B (identical to machine A)	86 dB
Machine C	82 dB
Machine D	78 dB

First, the two identical noise sources, Machines A and B, are combined to produce a noise level of 89 dB. Then Machine C is added as follows:

$$dB \text{ difference} = 89 \text{ dB} - 82 \text{ dB} = 7 \text{ dB}.$$

From Table 10.1, a difference of 7 dB between two sources results in the addition of 0.8 dB to the larger source. Therefore, the combined sound of machines A, B, and C is

$$\text{combined sound (A,B,C)} = 89 \text{ dB} + 0.8 \text{ dB} = 89.8 \text{ dB}.$$

Adding Machine D, we have

$$\text{dB difference} = 89.8\ \text{dB} - 78\ \text{dB} = 11.8\ \text{dB} = 12\ \text{dB}.$$

Returning to Table 10.1, a difference of 12 dB between two sources results in the addition of 0.2 dB to the larger source. Therefore, the combined sound of all machines is

$$\text{combined sound (A,B,C,D)} = 89.8\ \text{dB} + 0.2\ \text{dB} = 90.0\ \text{dB}.$$

Figure 10.11 diagrams the computation of Case Study 10.2. It should be noted that the measurement of both the total combined sound and the individual machine contributions toward that total are made from the position of the operator. Otherwise, distance factors would affect the results. The computation of combined noise levels is useful in considering the potential benefits of removing machines, enclosing machines, or changing the process.

We have been emphasizing sound intensity measured in decibels, but frequency (or pitch) also plays a role in noise control. Industrial noise is typically a combination

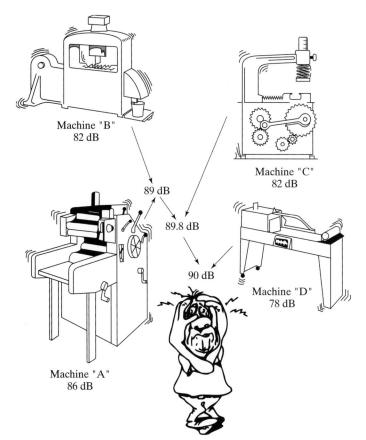

Machine "B"
82 dB

Machine "C"
82 dB

89 dB

89.8 dB

90 dB

Machine "D"
78 dB

Machine "A"
86 dB

FIGURE 10.11

Combining noise from several sources.

of sound frequencies from each of several sources. The total range of sound frequencies audible to the human ear is from about 20 to about 20,000 Hz. The ear is more sensitive to some of these frequencies than others, particularly the upper middle range from about 1000 to about 6000 Hz. Thus, sound-level meters have been devised to bias the decibel reading slightly to emphasize the frequencies from 1000 to 6000 Hz. This biased reading is called the A-weighted scale,[1] and resulting readings are abbreviated dBA instead of simply dB. OSHA recognizes the A-scale, and OSHA PELs are expressed in dBAs.

OSHA Noise Standards

Noise is unusual in that it is a hazard for which OSHA has set both a PEL and an AL. The best known is the PEL, which is set at 90 dBA for an 8-hour TWA. The action level (AL) was established in the early 1980s at 85 dBA for an 8-hour TWA, about 10 years after the 90-dBA PEL was established. It is widely recognized that workers can tolerate short periods of noise higher than the 8-hour TWA without damage, so OSHA specifies a range of decibel exposure levels for various exposure periods. The range of OSHA PELs for noise exposure is given in Table 10.2.

The range of permissible exposures in Table 10.2 makes possible a computation of a time-weighted-average exposure, relating each exposure time to the limit permitted for that sound level. The procedure is very similar to the calculation used earlier when multiple contaminants are present in the atmosphere. The formula used is

$$D = 100 \sum_{i=1}^{n} \frac{C_i}{T_i} = 100\left(\frac{C_1}{T_1} + \frac{C_2}{T_2} + \cdots + \frac{C_n}{T_n}\right) \tag{10.1}$$

where D = total shift noise exposure ("dose") as a percent of PEL
C_i = time of exposure at noise level i
T_i = maximum permissible exposure time at noise level i
        (from Table 10.2)
n = number of different noise levels observed

An interesting computation is for total shift exposure exactly at the AL of 85 dBA. By using Equation (10.1), the computation is as follows:

$$D = 100 \sum_{i=1}^{n} \frac{C_i}{T_i} = 100\left(\frac{8}{16}\right) = 50\%$$

Thus, the AL is computed to be 50% of the maximum permissible PEL. However, the reader should note from the earlier discussion of sound intensity that 85 dBA represents less than one-half of the absolute sound intensity of noise at 90 dB.

---

[1] There are also a B-scale and C-scale, but these scales are seldom used.

TABLE 10.2   OSHA's Table of PELs for Noise

A-weighted sound level	Reference duration time (hr)	A-weighted sound level	Reference duration time (hr)
80	32	106	0.87
81	27.9	107	0.76
82	24.3	108	0.66
83	21.1	109	0.57
84	18.4	110	0.50
85	16	111	0.44
86	13.9	112	0.38
87	12.1	113	0.33
88	10.6	114	0.29
89	9.2	115	0.25
90	8	116	0.22
91	7.0	117	0.19
92	6.2	118	0.16
93	5.3	119	0.14
94	4.6	120	0.125
95	4	121	0.110
96	3.5	122	0.095
97	3.0	123	0.082
98	2.6	124	0.072
99	2.3	125	0.063
100	2	126	0.054
101	1.7	127	0.047
102	1.5	128	0.041
103	1.4	129	0.036
104	1.3	130	0.031
105	1		

Source: Code of Federal Regulations 29 CFR 1910.95.

CASE STUDY 10.3

Noise-level readings show that a worker exposure to noise in a given plant is as follows:

8:00 A.M. – 10:00 A.M.	90 dBA
10:00 A.M. – 11:00 A.M.	95 dBA
11:00 A.M. – 12:30 P.M.	75 dBA
12:30 P.M. – 1:30 P.M.	85 dBA
1:30 P.M. – 2:00 P.M.	95 dBA
2:00 P.M. – 4:00 P.M.	90 dBA

Adding up the noise durations for each level, we obtain

AT noise level 90 dBA;	2 + 2 =	4	hours
AT noise level 95 dBA;	1 + 1/2 =	$1\frac{1}{2}$	hours
AT noise level 75 dBA;		$1\frac{1}{2}$	hours (ignore)
AT noise level 85 dBA;		1	hours
Total		8	hours

The reason that the $1\frac{1}{2}$-hour exposure at 75 dBA was ignored is that 75 dBA is below the range of Table 10.2. In other words, workers may be exposed to noise levels of 75 dBA for as long as desired with no adverse effects, at least as far as safety standards are concerned.

Computing the ratios at each level and summing, in accordance with Equation (10.1), yields

$$D = 100 \sum_{i=1}^{n} \frac{C_i}{T_i} = 100 \left(\frac{4}{8} + \frac{1\frac{1}{2}}{4} + \frac{1}{16} \right)$$
$$= 100(0.5 + 0.375 + 0.0625)$$
$$= 93.75\%.$$

Since 93.75% is less than 100%, the PEL is not exceeded. However, since 93.75% is greater than 50%, the AL of 85 dBA (8-hour TWA) is exceeded.

Sometimes noise is percussive or intermittent so that, technically speaking, there are tiny intervals of silence between sharp reports. Some employers have followed the scheme that these tiny intervals of silence can be counted toward the quiet time, reducing the observed duration of noise in excess of 90 dBA, but this interpretation is incorrect. Any variations in noise levels that have maxima less than 1 second apart are to be considered continuous. The slow response scale of modern noise-level meters tends to ignore such tiny interval variations, and thus "slow response" is specified in noise-level metering.

The standards do have a specification for peak impulse or impact noise at 140 dBA, but this, of course, is much higher than the PELs for continuous noise. Thus, the 140-dB OSHA specification can be considered a ceiling, or C, value. The 140-dB ceiling should be considered a limit for acute exposure and is accordingly a safety hazard. However, such exposures are so rare and difficult to measure after the fact that ceiling violations are virtually never cited. Ordinary sound-level meters are not very effective in measuring impact noise. Even for continuous exposures, measurement can be a problem that will now be addressed.

Noise Measurement

As a first check for potential noise problems, the safety and health manager should take a walk through the plant and listen. As a rule of *thumb*, if you can reach out and touch someone with your *thumb* but still cannot hear and understand that person's conversation (without his or her shouting), either your hearing is already damaged or deficient or there is excessive noise in the area. If the noise is continuous throughout the working shift but is no louder than a continuously running vacuum cleaner, there is probably no violation of standards. But if the noise is as loud as a subway passing through a station continuously throughout the full shift, a violation probably exists. If subways are unfamiliar to you, imagine a fast-moving freight train passing within 20 feet; such a noise level would easily constitute a violation if the exposure were continuous for a full 8-hour shift. Some exposures louder than the train might be permissible if they are of short duration, as was obvious in the calculation of average exposures in the preceding section. Anything between the vacuum cleaner and the train in sound level would be a gray area that ought to be measured with accurate meters.

Accurate measurement of sound levels requires instruments such as the sound-level meter (SLM) illustrated in Figure 10.12. The meter registers the sound intensity in decibels. Sound-level meters are delicate instruments and must be handled carefully. Accuracy is a problem, and the safety and health manager should not expect performance better than ±1 dB. Calibration is extremely important, and no sound-level meter is complete without a calibration device (known sound source) nearby. Variations in battery level must be compensated for, and humidity and temperature conditions can cause distortions.

Some skill is required in using the sound-level meter to obtain reliable readings. Naturally, the microphone receiver of the instrument must be held in the vicinity of the subject's ear in order to be representative of the exposure. However, the instrument should not be held too close, as the subject's body may affect the reading. The instrument's microphone receiver should be shielded from wind currents, and the instrument itself should not be subjected to direct vibrations. The safety and health manager may find the sound-level meter most useful in comparing sound levels in various locations in the plant and in comparing different machines and operating modes.

Determining the full-shift, time-weighted average exposure with a sound-level meter is tedious and requires a great number of samples. A convenient substitute is a cumulative device called a *dosimeter*, which is worn on the person of the subject under study. Dosimeters may seem to be a panacea, but they have their drawbacks. The wearer can easily bias the dosimeter by holding it close to a loud mechanism, or by simply rubbing, tapping, or blowing on the microphone. They are useful survey devices, however, and are used to monitor exposures if the firm's noise levels have exceeded the AL of 85 dBA for an 8-hour TWA.

Once it is determined that a problem exists, a more sophisticated measurement of noise level may be necessary to isolate the sources of the undesirable noise. An octave-band analyzer permits decibel readings to be taken at various frequencies over the audible range, as shown in the example in Figure 10.13. Various noise-reduction media have characteristic frequencies at which they are most effective. The octave-band

FIGURE 10.12

Example of sound-level meter.

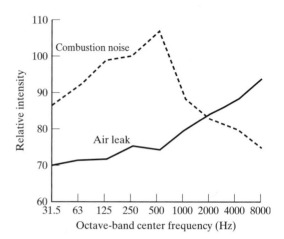

FIGURE 10.13

Example of octave-band analysis (*Source:* NIOSH, ref. Industrial Noise).

analysis will help to delineate the problem frequencies as well as provide evidence to identify the problem sources. Finally, octave-band analysis is useful in determining the frequency characteristics of a particular industry's noise sources so that damage from these sources can be distinguished from damage from noise exposures that have occurred off the job.

Engineering Controls

Once the instruments have proven that a problem exists, the safety and health manager needs physical solutions to the problems. If noise levels exceed the PEL, federal standards require that feasible engineering or administrative controls be used. If these measures fail to reduce noise exposures to within the PEL, personal protective equipment must be provided and used to reduce sound levels to within the PEL. Engineering controls should be considered a more thorough and permanent solution to the problem.

As with the control of toxic substances, the simplest solutions may be so obvious that they are overlooked. Process modification or elimination should always receive consideration. Another very simple solution, if feasible, is merely to move the operator away from the primary source of the noise. This idea has more merit than it would intuitively appear because the noise intensity from a given source goes down as the square of the distance, in the absence of reflective walls and other distorting factors. The reason for this relationship can be seen in Figure 10.14.

It must be remembered that it is absolute sound intensity, not decibels, that varies inversely with the square of the distance from the source. The logarithmic decibel scale results in a 3-dB change whenever the sound intensity is changed by a factor of 2 (doubled or halved). This leads to a rule of thumb for distance. Since sound intensity varies as the square of the distance from the source, a doubling of distance results in a *fourfold* reduction in sound intensity, which in turn reduces the decibel level by 6 dB. The effect is shown in Case Study 10.4.

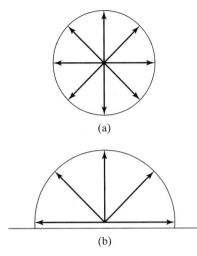

(a)

(b)

FIGURE 10.14

Distribution of sound intensity over the surface of a sphere as it radiates from a single source. (a) Sound emanates from a point source in all directions distributing over the surface of a sphere, the area of which is calculated by the formula $4\pi r^2$. Thus, the sound intensity is reduced as the square of the radial distance from the source. (b) Sound emanates from a point source located on the floor or other surface. The floor either absorbs or reflects the sound, but the resultant sound is still distributed over a hemisphere, the area of which is $2\pi r^2$. The squared-distance relationship still holds approximately.

CASE STUDY 10.4

A worker's machine is located at a distance of 2 feet from the operator and produces a noise exposure of 95 dB to the operator. How much is to be gained by moving the operator to a position 4 feet from the machine? How much reduction could be achieved by a move to 8 feet?

Solution

A move from 2 to 4 feet is a doubling of distance and results in a 6-dB reduction in sound level. The resultant level would be

$$95 \text{ dB} - 6 \text{ dB} = 89 \text{ dB}$$

which would probably be within the 8-hour PEL of 90 dB even after considering reflections and other sources, if these sources are not very significant.

A move to 8 feet would be a second doubling, resulting in a reduction of another 6 dB to a resultant 83 dB, ignoring reflections and other sound sources. This would reduce noise exposures to less than the AL of 85 dBA for an 8-hour TWA.

It is not likely to be feasible to move operators away from their own machines, and even if it is, to move away from one machine may place the operator close to a neighboring machine. Distance factors work best in separating operators from noise arising from adjacent machines or other processes in the area. A general spreading out of the plant layout can be beneficial in this regard.

If spreading out the plant layout is infeasible or too costly, the installation of sound-absorbing barriers between stations can increase their virtual separation as far as noise is concerned. The gain to be achieved by such barriers is variable and complicated

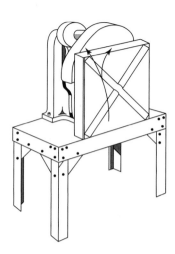

FIGURE 10.15

Application of welded stiffeners to reduce vibrations in a sheet metal component (*Source:* Arkansas Department of Labor, ref. Lovett).

to estimate in advance. An acoustics expert is recommended for advice in this area, and even the expert is likely to experiment with various temporary barriers, measuring the "before" and "after" sound levels with each. Heavy materials absorb sound vibrations, a fact that makes curtains or shields containing lead a popular choice.

Sheet metal surfaces on machines are susceptible to mechanical vibrations and may act as sound-amplifying surfaces. Metal gear contact in the drive mechanism can sometimes be eliminated by substitution of nylon gear wheels for metal gears or by the use of belt drives instead of gears. Simpler still would be a stepped-up preventive maintenance schedule to lubricate the gears more often, perhaps reducing noise levels. Useful principles for engineering controls to reduce noise are illustrated in Figures 10.15 to 10.20.

Perhaps more expensive than any of the engineering control approaches discussed so far would be the isolation of the offending machine by means of an enclosure. The effectiveness depends on the type of material used to construct the enclosure and also depends to a surprising degree on the number and extent of openings or leaks in the

FIGURE 10.16

Enlarge or reduce size of a part to eliminate vibration resonance (*Source*: Arkansas Department of Labor, ref. Lovett).

FIGURE 10.17

Rubber cushions on both sides of vibrating sheet metal surfaces where they are joined. (*Source*: Arkansas Department of Labor, ref. Lovett.)

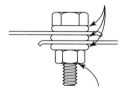

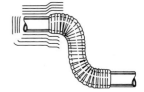

FIGURE 10.18

Flexible section in rigid pipe isolates vibrations (*Source*: Arkansas Department of Labor, ref. Lovett).

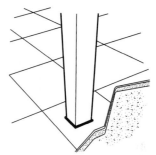

FIGURE 10.19

Resilient floor (*Source*: Arkansas Department of Labor, ref. Lovett).

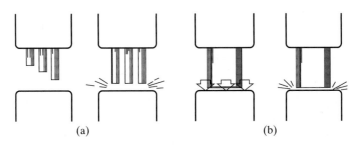

(a)   (b)

FIGURE 10.20

Power press improvements to reduce noise levels: (a) design die to result in a dull, crunching sound instead of a sharp bang; (b) replace the sharp impact action of a mechanical press with the relatively quiet pressure action of a hydraulic press.

enclosure. Figure 10.21 shows the relationship between opening size and loss of effectiveness for an example noise enclosure that has a capability of a 50-dB reduction if there are no leaks. Note that most of the effectiveness of the enclosure is lost if there is a hole in the enclosure wall of less than one-half of 1% of the total area of the enclosure.

Administrative Controls

It was stated earlier that either engineering or administrative controls are specified where feasible for excessive noise levels and that engineering controls are preferable. However, the administrative controls alternative was left unexplained. Administratively, management can schedule production runs so that noise levels are split between shifts and individual workers are not subjected to full-shift exposures. Other tricks are to interrupt production runs with preventive maintenance to give workers quiet time.

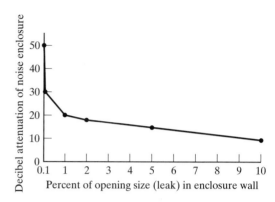

FIGURE 10.21

Loss in effectiveness of a noise enclosure due to leaks. For a sample enclosure that has an ideal (airtight) noise-reduction (attenuation) capability of 50 dB, leaks in the walls reduce the decibel attenuation by remarkable amounts, as shown.

During normal shift breaks, workers can be removed to a quiet rest area. Sometimes workers can share a loud job and a more quiet one by trading jobs at midshift. All of these practices can be used to bring down noise exposure levels to within the PEL for the given time of exposure as determined from Table 10.2. Since the term *administrative control* is somewhat vague, the term *work-practices control* has become preferable to refer to the various methods of shifting employee exposures to comply with Table 10.2.

Hearing Protection and Conservation

Personal protective equipment is required when engineering and administrative controls both fail to reduce noise to legal levels. Specifically, hearing protection is to be provided to all employees exposed to the 85-dBA TWA AL. In addition to providing protection, employers must permit employees to select protectors from a variety of suitable types and must train employees in the proper use and care of the protectors. One aspect of proper use is fit, and employers must ensure proper fit. These actions are mandatory when ALs are exceeded, but the reader may have noted that the actual wearing of the protectors by the workers is not among the mandatory steps to be taken if ALs are exceeded. However, there are conditions under which workers must actually wear the protectors. These are as follows:

1. whenever worker exposures are greater than the PELs (see Table 10.2)
2. whenever worker exposures are greater than the AL of 85 dBA (TWA) *and* the worker has experienced a *permanent significant threshold shift*

When testing shows a threshold shift in a worker's hearing, the implication is that the worker's hearing has been damaged and needs special protection. In these cases, the noise levels are required to be reduced by the protectors to 85 dB (TWA), not 90 dBA (TWA).

Personal protective equipment should not be considered a final solution, because elimination of the source of the noise provides a more satisfactory work environment. Sometimes workers are lax in the wearing of personal protective equipment, and injurious exposures result. The careful selection and fitting of various types of personal hearing protective equipment is discussed in Chapter 12.

Whenever the AL of 85 dBA (8-hour TWA) is exceeded, a "continuing, effective hearing conservation program" should be administered, including audiometric testing, noise monitoring, calibration of equipment, training, warning signs for noisy areas, and recordkeeping of audiometric tests and equipment calibration.

Audiometric testing involves a small, perhaps portable, listening station in which a subject listens to recorded sounds and an audiologist measures the subject's hearing acuity at various frequencies. Audiometric testing can be very useful in determining sources of hearing loss, or more particularly in providing data to determine whether hearing loss is due to occupational or off-the-job exposure. If plant noise levels are high, it is foolhardy to hire new employees without first testing their *baseline* hearing acuity. Without any evidence of hearing deficiency at the time of employment, any hearing deficiencies that show up after employment appear strongly to be work related.

When observing the frequency profile of a worker's hearing acuity, the audiologist often looks for a *4000-Hz shift* for evidence of occupational exposure. Experience has shown that much industrial noise occurs in the 4000-Hz frequency range, which causes audiologists to suspect occupational exposure when hearing acuity becomes deficient in this range.

RADIATION

A natural progression from the subject of noise is to the subject of radiation. Noise is, in fact, a form of radiant (wave) energy, but the term *radiation* is generally considered to mean electromagnetic radiation such as X rays and gamma rays, or highspeed particles such as alpha particles, protons, and electrons. Federal standards separate radiation into the categories *ionizing* and *nonionizing*.

Ionizing radiation is the more dangerous of the two types and is the type most associated with atomic energy. By far the most important category of ionizing radiation, from the occupational exposure standpoint, is the X ray. X rays are no longer the exclusive domain of the medical and dental professions. X rays are being widely used in manufacturing operations, especially in inspection systems.

Nonionizing radiation is somewhat of a misnomer, but applies to a more benign type of radiation in the electromagnetic spectrum. Included are radio and microwave frequencies. These phenomena are also increasingly important in industrial applications. Some workers are concerned about radiation exposure from the constant use of computer terminals. Computer terminals are of concern, but their principal hazard is musculoskeletal disorders, not radiation. Musculoskeletal disorders are considered an ergonomics hazard, and were covered in Chapter 8.

SUMMARY

To achieve a safe and healthful workplace and to comply with federal standards, the employer must devise engineering or administrative control solutions to air-contaminant and noise problems, if feasible. For air contaminants, the employer should first seek to eliminate the source of the toxic substances or find more benign substitutes for such process materials. If these attempts fail, ventilation is usually the answer. Ordinary heating and air-conditioning systems are designed for a different purpose and are generally not acceptable for eliminating air contaminants, especially particulates. A basic principle is to attempt to focus the ventilation in the form of local exhaust ventilation. The supply of makeup air also is an important consideration. A variety of filtering or particle removal mechanisms is available for purifying the air and either returning it to the plant atmosphere or exhausting it into the outside environment.

Industrial noise is a phenomenon that requires an understanding of the physics of sound-wave energy, the way in which noise is heard, and how noise affects human hearing. Human hearing is capable of sensing an incredible range of amplitudes (loudness) of wave energy while employing a very fine degree of discrimination among frequencies (pitch). So great is the range of amplitudes that a logarithmic scale is used to measure the absolute sound pressures and describe the levels of sound that humans actually hear. Computations of noise levels can be done by logarithmic manipulations or by using formulas and tables provided by OSHA standards. The basic OSHA noise standard (PEL) is 90 dB as an 8-hour time-weighted average (TWA). The action level is 85 dB, a level that is actually less than half the PEL in absolute sound intensity, due to the logarithmic nature of the decibel scale. A rule of thumb is that a doubling of absolute sound intensity results in an increase of 3 decibels on the decibel scale.

Radiation is another physical phenomenon that has similarities to noise. Two categories of radiation are recognized: ionizing radiation and nonionizing radiation. Ionizing radiation is the more dangerous of the two. The most widely encountered form of ionizing radiation in industry is the X ray.

Chapters 9 and 10 have shown that the subjects of occupational health and environmental control can be quite technical. The safety and health manager probably will find it beneficial to employ experts to take time-weighted averages with appropriate meters and sampling instruments. Subsequent control measures, such as ventilation systems for toxic substances and acoustic panels for noise control, may require the design capabilities of experts in their respective fields. Dealing with these experts demands a degree of understanding of their methods and terminology, but does not require the safety and health manager to duplicate the experts' capability in each field. Such an understanding of methods, terminology, and basic principles of occupational health and environmental control are what these chapters have attempted to provide.

EXERCISES AND STUDY QUESTIONS

10.1 Identify the comparative hazards of batch vs. continuous processes for chemicals.

10.2 Explain the benefits and disadvantages of using an ordinary household fan for ventilation in the presence of toxic substances.

10.3 What are the comparative advantages and disadvantages of using a vacuum cleaner to exhaust toxic air contaminants by placing the hose intake close to the source of the contaminant.

10.4 Why should toxic air contaminant ventilation systems "pull" instead of "push?"

10.5 What is the purpose of a "manometer" in exhaust ventilation systems?

10.6 Explain three solutions to the problem of energy losses in the supply of makeup air for exhaust ventilation systems.

10.7 Identify four different types of devices for the purification of exhaust air before releasing to the outside air.

10.8 What two basic characteristics of sound are easily detected by human hearing? Which of these two is more dangerous?

10.9 Identify the two principal types of hazardous radiation. To which class do X rays belong?

10.10 What is the principal, recognized hazard from working at computer terminals?

10.11 A plant has two identical standby generator units for emergency use. In the area of the generators, the normal noise level registers 81 dBA on the sound-level meter with the generators turned off. When one generator switches on, the SLM needle jumps to 83.6 dBA.

(a) Perform calculations to determine what the dBA reading will be when the second generator also turns on (so that both generators are on).

(b) If both generators are on for a full 8-hour shift, will OSHA's PEL be exceeded? Will the AL be exceeded? Explain.

(c) If one generator is on for half the shift and both are on for the other half, will the PEL be exceeded? Will the AL be exceeded? Explain.

(d) In the absence of any plant background noise, what would be the sound level contributed by a single generator? by both generators? Show calculations to explain.

10.12 Four machines contribute the following noise levels in dB to a worker's exposure:

Machine 1	80 dBA
Machine 2	86 dBA
Machine 3	93 dBA
Machine 4	70 dBA

(a) Calculate the combined noise-level exposure for this worker.

(b) The offending machine is obviously Machine 3. Suppose that Machine 3 was at a distance of 5 feet away from the worker when the 93-dBA noise level was measured. Determine how far away would Machine 3 have to be moved to bring the worker's continuous 8-hour combined exposure from *all* machines down to the OSHA PEL?

10.13 What are some desirable alternatives to industrial ventilation to remove air contaminants?

10.14 What is makeup air?

10.15 Ten machines all contribute equally to the noise exposure of one worker, whose exposure level is 99 dB for a full 8-hour shift. When all the machines are turned off, the noise level is 65 dB. How many of the 10 machines must be turned off to achieve a full-shift noise-exposure level that would meet standards if the worker wears no personal protective equipment?

10.16 A worker stands on a factory floor and a sound-level meter shows a reading of 55 dB at that point. A machine 3 feet away is turned on, and the meter jumps to 90 dB. What will the SLM read if the machine is moved to a point 12 feet away?

10.17 A paper mill uses liquid chlorine, delivered in 90-ton railroad tank cars, as a pulp bleaching agent. One volume of liquid chlorine produces approximately 450 volumes of vapor under normal atmospheric temperature and pressure. The density of liquid chlorine is 103 pounds per cubic foot. In the event of rupture and vapor release of 20% of the tank car contents, how much vapor by volume would be released? If the release were in a closed building with a 30-foot ceiling height without ventilation, how large would the building have to be (in square *miles* of floor space) to contain the thoroughly mixed vapor–air ratio within the OSHA PEL? The logical conclusion to this exercise is that, with or without ventilation, it is more practical to unload chlorine tank cars outdoors.

10.18 A worker exposure to noise in a given plant is measured, resulting in the following readings for various time periods during the 8-hour shift:

8:00 A.M. – 9:00 A.M.	86 dBA
9:00 A.M. – 11:00 A.M.	84 dBA
11:00 A.M. – 12 noon	81 dBA
12 noon – 1:00 P.M.	101 dBA
1:00 P.M. – 4:00 P.M.	75 dBA

(a) Perform computations to determine whether maximum PELs have been exceeded.

(b) Have ALs been exceeded?

(c) Given the noise exposure just described, would the employer be required to furnish hearing protectors?

(d) Would employees be required to use the hearing protectors?

(e) Suppose that an engineering control could be devised that would cut the noise level (sound pressure level) in half either in the morning or in the afternoon, but not both. Which would you select? Why?

10.19 From the perspective used by the federal enforcement agency, rank the following solutions to a worker noise-exposure problem (from most effective, "1," to least effective, "4"):

- *Solution A*   Enclose the noise source with a barrier that reduces the noise level by 3 dBA.
- *Solution B*   Position the operator at a distance twice as far from the source of the noise.
- *Solution C*   Rotate personnel so that each worker is exposed to the noise source for only one-half shift.
- *Solution D*   Provide ear protection that cuts the absolute sound pressure in half. Justify your choices with calculations, analysis, and in light of established priorities.

10.20 A certain drying process produces 5 cubic feet of ethanol vapors per hour. If general exhaust ventilation is used, calculate the flow in cubic feet per hour needed to keep the ethanol vapors within OSHA limits. What is another name for ethanol?

10.21 For each of the following materials, suggest substitutes that are feasible for some operations and that prevent certain hazards:

(a) Silica (for blasting)

(b) Lead-based paint

(c) Freon (as a propellant)

(d) Acetylene (for welding)

10.22 What problem frequently develops with ventilation filter alarms that indicate pressure differential across the filter?

10.23 What is the purpose of using a heat exchanger for makeup air? What is the disadvantage of this approach?

10.24 If the air outside is at substantially higher pressure than the air inside the plant, what ventilation problem probably exists?

10.25 What form of ionizing radiation is most commonly encountered in industrial exposures?

10.26 **Design Case Study**. A glue-making process releases ethylene glycol that becomes generally diluted and interspersed throughout the plant atmosphere. The rate of release is 2.4 cubic feet per hour vapor volume at standard plant temperature and pressure. The plant ventilation system is of the general dilution type, with makeup air being supplied through windows and doors throughout the plant area. The plant area is 12,000 square feet and the average ceiling height is 16 feet. The problem is to specify the capacity of the general ventilation system required to maintain a steady-state condition throughout this process area that protects against both health and safety hazards due to ethylene glycol. For your information in performing calculations, the following data are provided:

ETHYLENE GLYCOL (CH_2OHCH_2OH)
    Molecular weight 62.1
    Boiling point: 197.5° Celsius
    LEL: 3.2%
    Firepoint: −13° Celsius
    Flashpoint: 232° Fahrenheit
    Autoignition temperature: 752° Fahrenheit
    Vapor pressure: 0.05 mm at 20° Celsius
    PEL: 50 ppm (ceiling)

(a) How much exhaust ventilation (in cubic feet per hour, general dilution type) is required to maintain safety hazards below explosive levels?

(b) How much exhaust ventilation (in cubic feet per hour, general dilution type) is required to maintain health hazards below OSHA-specified action levels?

(c) How many plant area room changes per hour would the level of ventilation calculated in part (b) represent?

10.27 **Design Case Study.** A particularly noisy process is operated by a single operator working at a control console. The 8-hour TWA exposure level for this operator is 96 dBA. The company has initiated an engineering project to alleviate the problem and has two plans:

- *Plan A*. Move the operator's control console from its current position 5 feet from the source of the noise to a point 10 feet away.
- *Plan B*. Enclose the noise source in an enclosure that would be effective in reducing the *absolute* sound pressure by 75%.

Evaluate the effectiveness of each of these plans in reducing the noise-exposure level. Suppose that both plans were executed; calculate the combined effect of both plans on the noise-exposure level.

RESEARCH EXERCISES

10.28 An employer is having difficulty meeting the OSHA asbestos standard using engineering controls and is considering using administrative (work-practice) controls. Study current OSHA standards for asbestos and prepare a professional recommendation to this employer, citing appropriate sections of the OSHA standard to justify your position.

10.29 While visiting an industrial site where asbestos is being removed, you observe that air hoses are being used to blow dust off work clothing. Comment on this procedure, citing specific portions of the OSHA standards to justify your position.

10.30 Sometimes industries voluntarily initiate actions to control exposures to levels even lower than those specified by OSHA standards. Examine a noteworthy example of this initiated by the Lead Industries Association, Inc. (LIA). What other industry association joined with LIA in this initiative? Specifically what target improvement became the 5-year goal of this initiative?

STANDARDS RESEARCH QUESTIONS

10.31 This chapter has dealt with noise. Search the General Industry OSHA standard and use the NCM database to determine how seriously OSHA takes the general hazard of noise exposure. Specifically, find the frequency of citation, the percentage of the citations designated as "serious," and the dollar level of proposed penalties for noise citations.

10.32 The effectiveness of exhaust ventilation systems depends on the design of the hood for enclosing the operation to be ventilated. Search the OSHA standards for provisions pertaining to exhaust hoods. Use the NCM database tool to determine whether OSHA ever writes citations for exhaust hoods. If any citations are found, what percentage of the citations are designated as "serious?"

10.33 Examine the OSHA standards for "audiometric testing" and "hearing conservation programs." What is the action level trigger for these programs? Are the standards for these programs frequently cited? How does their citation frequency compare with the frequency of citation for ventilation hazards? How do these standards compare for "seriousness" (percentage of citations that are designated as "serious")?

CHAPTER 11

Flammable and Explosive Materials

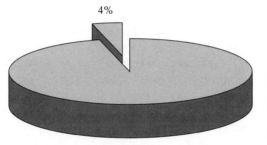

4%

Percentage of OSHA General Industry citations addressing this subject

In Chapter 5, materials hazardous to the environment were considered. The health hazards of toxic materials have also been studied in Chapter 9. In this chapter, we examine a more traditional type of hazardous material, flammable and explosive substances, and their use in factory processes, such as spray finishing areas and dip tanks. We now know that most of these substances are dangerous from a health hazard perspective, but their safety hazards are the traditional concerns with flammable and explosive materials, and safety hazards take center stage in this chapter.

FLAMMABLE LIQUIDS

Flammable liquids such as gasoline are familiar to everyone, and the new safety and health manager might expect applicable standards for flammable liquids to be the easiest to learn and to apply. But, unfortunately, the standards are quite complicated as a natural result of the fact that flammable liquids occur in industry so frequently and in such widely varying quantities and applications. To illustrate this point, procedures for handling gasoline in a petroleum refinery in which gasoline is *manufactured* are vastly different from procedures for storing and handling flammable liquids in an ordinary factory or office. Therefore, no simple set of rules for flammable liquids is appropriate.

As familiar as flammable liquids are, most people do not really understand many of the commonly used terms, such as *flashpoint, Class I liquid, flammable, combustible,* and *volatile.* Much confusion also surrounds the sources of ignition for flammable liquids and the circumstances under which such substances will burn, explode, or not burn at all. This chapter will therefore focus first on definitions and principles of flammable liquids ignition and then will turn to discussion of some of the problems in complying with appropriate standards.

Perhaps the most basic term should be defined first, and this is the term *liquid.* Almost everyone knows what a liquid is, but on the other hand, virtually every flammable substance exists as both a liquid and a gas, depending on temperature or pressure. A good practical rule to follow is that if the substance is normally a liquid, it is defined as a liquid. Where one can get into trouble is in classifying propane and butane, which are gases and are not intended to be considered flammable liquids, although they can be liquefied. The National Fire Protection Association (NFPA) standard definition of flammable liquid excludes propane and butane by excluding all "liquids" having a vapor pressure in excess of 40 pounds.

The term *flashpoint* is very important to the safety and health manager because it is the principal basis for classification of flammable and combustible liquids. Therefore, it is primarily flashpoint that determines the various amounts of the liquid that are permitted to be stored in various types of containers. Flashpoint is the point to which a flammable liquid must be heated so that it will give off sufficient vapor to create a flash at the surface of the liquid when there has been a spark or flame applied to it. Flashpoint is not the same as *firepoint;* firepoint is a higher temperature and is the temperature at which a fire on top of the liquid is sustained.

Three principal testing methods are in use for determining flashpoint. The Cleveland open-cup test is simple, but is not often used because it is intended for heavy oils. The most frequently used method is the Tag closed-tester method. The term *Tag* is simply an abbreviation for a French name *Tagliabue.* The third method is the Pensky–Martens closed-tester method. This method uses a small stirring rod and is used for viscous liquids and liquids that form a film on the surface. This method is also rarer than the Tag test. The open-cup method best simulates the in-plant situation in which open vats are being utilized. The closed-cup situation best simulates the situation for flammable liquids in storage.

Classification of flammable liquids also depends on *boiling point,* but even this can be confusing because liquid does not normally boil at only a single temperature point. This range is acknowledged in the standards by designating the 10% *point* as the key. The 10% point is the temperature at which 10% of the liquid has become gas. *IBP* refers to initial boiling point and is the temperature at which the first drop of liquid falls from the end of the distillation tube in a standard ASTM[1] test distillation.

Volatility refers to how readily a liquid will evaporate; it is closely related to boiling point. *Light* and *heavy* refer to high volatility and low volatility, respectively.

Flammable liquid is a term that means the same thing as *Class I liquid.* However, such liquids are further divided into IA, IB, and IC. *Combustible liquid* is the general

---

[1]American Society for Testing and Materials.

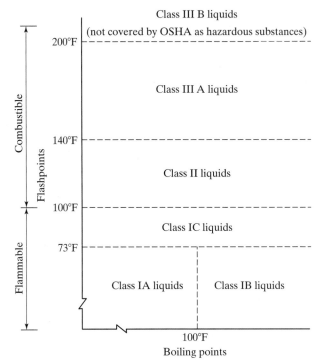

FIGURE 11.1

Classification of flammable and combustible liquids.

term for both Class II and Class III liquids. The entire classification scheme, which is based on flashpoint and boiling point, is explained in Figure 11.1.

Gasoline is the most widely used and plentiful flammable liquid. Because of its widespread use and because of horrible explosions and fires in its history, gasoline is blamed for a great deal of the fire hazard in the workplace and elsewhere. Some of the fears of gasoline hazards arise from ignorance rather than knowledge-based prudence, and this is detrimental to the cause of safety. In Chapter 3, it was explained that overzealous safety rules contribute to worker apathy for the rules and in turn work against the cause of safety rather than helping it. Operational and safety rules for working around gasoline and other flammable liquids are sometimes in this unfortunate category. Admittedly, gasoline can be extremely dangerous, but there is no substitute for knowledge of the mechanism of its hazards so that the worker can take sensible precautions.

Knowledge begins with exposing the falsehood of the many myths surrounding the subject. Gasoline and other flammable liquids have their share of such myths, and this book will attempt to dispel some of them here.

Perhaps the wildest myth about gasoline is the one discussed next.

Flammable Liquid Myth 1

A lighted cigarette when brought into contact with the surface of a container of gasoline is sure to ignite it.

To the contrary, it is almost impossible to ignite a tank of gasoline at the surface with a lighted cigarette. As with any ordinary fire, three ingredients are essential to support combustion:

1. Fuel
2. Oxygen (usually from the air)
3. Sufficient heat

There is plenty of fuel at the surface of a container of gasoline, but both of the other two ingredients are generally insufficient to support combustion. A concentration of gasoline vapors in excess of 7.6% is too rich and will not burn. At the surface of gasoline standing in still air, the concentration is much higher than 7.6%. Also, a glowing cigarette is not hot enough in most cases to permit ignition.[2] In fact, dramatic demonstrations have been performed in which a lighted cigarette is extinguished by drowning it in a cup of gasoline! Incidentally, there are hazards involved in such demonstrations, and experimentation is not recommended. Things can go wrong, such as tiny flame on the cigarette paper being hot enough for ignition. Also, there is the problem of getting the cigarette *through* the region in which the vapors are not too rich and causing ignition before the rich area near the surface can be reached. Also, small quantities of gasoline in the surrounding area can cause vapor–air mixtures just right for combustion. These are the reasons for "no smoking" rules around gasoline.

Gasoline has a burnability range of 1.4 to 7.6% gasoline vapors in dry air. Some other flammable liquids have wider burnability ranges and are more easily ignited than gasoline. Burnabilities for some commonly used highly flammable liquids are shown in Figure 11.2. Note that though gasoline is easier to ignite at "lean" concentrations than

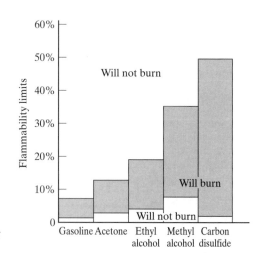

FIGURE 11.2

Burnability ranges of some popular flammable liquids.

_____

[2]The ignition temperature for gasoline (536 to 853° F) is higher than that for wood (approximately 400° F)!

alcohol is, alcohol will ignite at much richer concentrations. Also note the extremely broad and dangerous burnability range of carbon disulfide. The upper limit, above which concentrations of flammable vapors are too rich to ignite, is designated as the *upper explosive limit* (UEL). The corresponding lower limit, below which concentrations of flammable vapors are too lean to ignite, is the *lower explosive limit* (LEL).

Another myth regarding gasoline has to do with service station fires and other fires around underground tanks.

Flammable Liquid Myth 2

> Fires in underground gasoline tanks burn or explode with such intensity as to destroy much life and property around service stations.

Actually, fires do not burn in underground tanks of gasoline, even when a serious fire occurs above the ground. John A. Ainlay[3] states that in a 45-year long study of petroleum-fire reporting by the National Fire Protection Association (NFPA) and the American Petroleum Institute (API), there has never been a fire reported in an underground tank of gasoline currently in use. The vapor mixture in the tank is too rich to support combustion.

An abandoned empty tank is, curiously, more dangerous than a full or nearly full tank in use. The abandoned empty tank has had a chance to dry out, and vapors have gradually dispersed until the mixture may have become lean enough to result in an explosive mixture. The same is true of any *empty* gasoline drum. But a drum that is full or partially filled with gasoline will normally have a vapor–air ratio that is too rich to support combustion. Figure 11.2 shows that fire has a much greater hazard potential inside drums of alcohol or carbon disulfide.

With aboveground tanks, the hazard takes on a different dimension. The aboveground tank is exposed to intense heat and possible rupture during a service station fire. When a tank ruptures or explodes, tremendous quantities of fuel are suddenly added to the fire in the presence of abundant supplies of oxygen and heat.

The preceding paragraph describes ample reason for the standards that prohibit aboveground tanks for service stations, except when special conditions are met. There is, however, still another hazard for aboveground tanks. The vapor density[4] of gasoline is higher than 3 to 1. This means that unlike natural gas or other lighter-than-air materials, gasoline vapors will settle in low places. An aboveground tank is an invitation for gasoline vapors to settle in low areas of the service station, such as in service pits. The vapor density of gasoline is the reason that basements in service stations are now illegal.

Despite these hazards, a large number of violations of the standard prohibiting aboveground tanks have been found. Small independent service stations and private, in-plant service stations are the most frequent violators. Many of these installations were made prior to the writing of the standard, and some feel that such a standard was

---

[3] John A. Ainlay, of Evanston, Illinois, is a nationally recognized authority on the chemistry of petroleum fires.
[4] Vapor density is the ratio of the weight of the vapor to the weight of the same volume of air.

intended to act as a "building code" and apply to all future installations—not requiring all existing noncomplying stations to remodel.

Misconceptions about *octane rating* are at the root of the third myth.

Flammable Liquid Myth 3

> High-octane "aviation gasoline" or "premium gasoline" is much more hazardous than regular gasoline.

Octane rating has to do with the preignition characteristics of gasoline within internal combustion engines and has nothing to do with fire safety. High-octane gasoline should be given the same fire precautions as regular gasoline—no more, no less.

SOURCES OF IGNITION

Dispelling myths about flammable liquids should make personnel more cautious. The chemistry of a petroleum fire explains some of the seemingly peculiar instances of nonoccurrence of such fires. But at the same time, this understanding can highlight the extreme hazard involved when conditions are right for a fire. A broken light bulb seems innocent, but can result in a disastrous fire. In the instant before a hot light bulb filament burns out after the glass breaks, that filament is hot enough to ignite gasoline vapors. This is why it is important to protect light bulbs in the presence of flammable vapors as required in both the *National Electrical Code®* and further references in federal standards.

Welding sparks are another important ignition hazard for flammable vapors. The usual temptation is to speed up a repair operation, and welding is often commenced before sources of flammable vapors are removed from the area and existing vapors purged. Welding around flammable vapors has cost the lives of many inexperienced personnel who were not aware of the hazards involved.

Associated with welding is the grinding of the finished weld. Sparks generated are not to be trusted. It is true that many grinding sparks do not reach the required ignition temperature for gasoline vapors, but some sparks do. Therefore, spark-producing grinding should not be performed in the presence of flammable vapors.

The hazards of static electrical discharge around flammable vapors are well publicized. After all, it is electrical discharge that ignites gasoline vapors with precise reliability in most internal combustion engines. Electrical arcing or static electrical discharge is easily a source of ignition.

To prevent ignition hazards, electrical interconnection is required between nozzle and container when dispensing Class I liquids. But this raises questions regarding dispensing Class I liquids into containers made of plastic or other nonconductive material. It makes no sense to bond a plastic container to the nozzle because such a bond would be ineffective in equalizing the static charge. (See Figure 11.3.) The NFPA recognized this fact when it exempted nonconducting containers from electrical bonding requirements in NFPA standards.

Some readers may wonder why there is so much fuss about static electricity during dispensing operations. But filling tanks or containers *generates* static electricity due to the flow of the liquid—a little known phenomenon regarding fluid flow. The rapid flow of carbon dioxide from a fire extinguisher in use can cause static electrical shocks

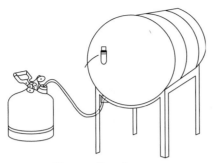

Nonmetallic safety can
needs no bonding wire
during filling operation

FIGURE 11.3

Bonding not necessary for nonconducting
container.

that make the extinguisher very uncomfortable to hold. During the loading of fuel oil into tank trucks at night, an astonishing phenomenon has been observed in which "lightning displays" are seen flashing around inside the tank!

To prevent buildup of static electricity during loading operations, the flow should be kept as quiet and smooth as possible. Filters are big static generators and are best placed as far back in the line as possible, remote from the fill spout. Another measure that can be taken to reduce static electricity is to retard the flow. Also, *splash loading* should be avoided—that is, the fill spout should extend down to a point close to the bottom of the compartment to avoid excessive splashing, which generates static electricity. When loading multicompartment trucks, the front and rear compartments are the most likely to present problems from splash loading because of the arrangement of some types of loading apparatus. The reason for this can be seen in Figure 11.4. Another way to eliminate static is to place a rest area in the delivery line, as shown in Figure 11.5. The rest area is an expansion area in the pipe that allows static charge to bleed off the liquid before rapid flow continues.

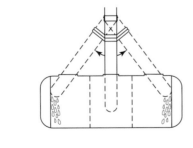

FIGURE 11.4

Splash loading occurs more often
while loading front and rear
compartments than while loading a
middle compartment.

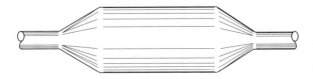

FIGURE 11.5

Static eliminator rest area.

STANDARDS COMPLIANCE

An understanding of the principles underlying the hazards with flammable liquids is useful when applying appropriate standards. Having covered some of the basic definitions and principles, we now turn to a direct analysis of the standards and helpful procedures to be followed by safety and health managers to bring their facilities into compliance.

Federal codes for tank storage are quite complicated and are mostly the concern of layout designers of petroleum tank farms, bulk plants, dikes and drainage schemes, refineries, and service stations. The codes also cover such elements of design as tank construction and proper venting. Most safety managers do not need to concern themselves with mastering the details of tank construction. It is sufficient to know where to find the requirements and to alert designers and other planners that strict codes must be followed in specifying tanks for flammable liquids.

One criterion for distance requirements between tanks is whether the tank roof is fixed or floating. The general public does not usually realize that the roofs of many petroleum tanks rise and fall with the level of the liquid inside. (See Figure 11.6.) A tank with a fixed roof will not fill unless it is vented, and such venting causes a costly loss of vapors. But the safety and health manager should understand that the floating roof also protects against the fire hazards of releasing vapors to the atmosphere. The vapor–air space inside an empty or nearly empty tank with a fixed roof is also more hazardous than for the floating-roof tank, which has little or no vapor–air space. The floating-roof tank is a dramatic example of an industrial improvement that saves production cost while promoting a safer workplace.[5]

One curious provision of appropriate safety standards requires accurate *inventory records* for Class I liquid storage tanks. Inventory records are usually for accounting and cost control, so what business has a *safety* standard in dealing with accurate inventory records? The answer is that accurate inventory records can be used to detect dangerous leaks. Unfortunately, inventory discrepancies are often attributed to "clerical error" or "unexplained loss" and are simply ignored. After a serious fire has occurred, an investigating team sometimes probes the past inventory records only to discover that the evidence of a dangerous leak had been in the records for many days. It may seem incredible

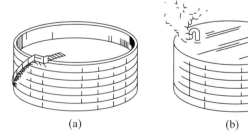

FIGURE 11.6

Two types of storage tanks: (a) floating-roof type; (b) conventional vented type. The floating roof rises and falls with the level of the liquid inside the tank, eliminating the necessity of costly and dangerous venting.

(a)　　　　　(b)

---

[5]The disastrous effects of improper design of flammable liquid tanks are illustrated in Exercise 11.22 at the end of this chapter.

that a leaking underground tank of gasoline would go undetected, but it happens, as will be seen in Case Study 11.1

CASE STUDY 11.1

SERVICE STATION UNDERGROUND LEAK

In the 1980s, a resident of Fayetteville, Arkansas, complained of the odor of gasoline vapors in the basement of his home. The local fire department investigated and found that indeed flammable gasoline vapors did exist there. Other homes in the same block were also found to have traces of gasoline vapors. Further investigation revealed that a service station in the neighborhood (but not adjacent to the house) had a leaking underground gasoline tank. It was necessary that the residents of the unsafe homes be evacuated until necessary repairs could be made. The repair required an extensive excavation to resolve the problem.

With respect to the hazards of leaking tanks, the safety and health manager must follow the regulations of two federal agencies: OSHA and EPA. The EPA cracked down on underground storage tanks in late 1988 (ref. Cross) and required systems for monitoring tank and pipe leaks, automatic shutoffs for pressurized systems, specified tank and pipe construction, spill protection, and systems to prevent overfill. In addition, there are reporting requirements to notify local or state authorities whenever a new tank is installed, an old one is closed permanently, or a leak is discovered. A notable exemption to EPA's tank rule is *above*ground tanks, provided that less than 10% of the product is stored in subsystem piping.

In summary, the safety and health manager should approach the flammable liquids problem with a knowledge of the principles of ignition. Any new facilities should be designed and constructed with recognition of the hazards of flammable liquids. After construction and installation, commonsense rules for eliminating sources of ignition and preventing dangerous leaks should be established and enforced by the safety and health manager. Training of personnel to understand the underlying principles of the hazards will help a great deal with the problem of safety around flammable liquids.

COMBUSTIBLE LIQUIDS

Thusfar, the emphasis of this chapter has been on flammable liquids, not combustible liquids. There is a distinction between the two, as was shown in Figure 11.1. Since the flashpoints of combustible liquids are higher than normally encountered temperatures in most plants, the hazard of ignition is much lower than for flammable liquids. However, a false sense of security is often engendered by commonplace association with combustible liquids at room temperatures. In the event that temperatures are elevated by some exceptional circumstance or even by normal process operations, the gravity of the hazard can shift dramatically. Ordinary kerosene, a combustible liquid, can become even more dangerous and ignitable at elevated temperatures than is gasoline at room temperature.

Another way in which combustible liquids can present an unexpected hazard is in *switch loading*, when transport trucks are used for hauling gasoline, a flammable liquid, and fuel oil, a combustible liquid, interchangeably. Switch loading is a danger from the standpoint of ignition by static electricity. With gasoline, static electricity is not as serious a problem in loading because the vapor concentration is generally far too rich to permit ignition. Even when loading gasoline into a tank that previously contained fuel oil, the vapor concentration becomes too rich as soon as loading commences. The real danger is when fuel oil is loaded into a compartment that previously contained gasoline. This is switch loading and is very hazardous. The vapor concentration in such an operation is just right for ignition and a static discharge or any other source of ignition can result in an explosion that will rip the truck apart. Remedies for the problem when switch loading is necessary are to (1) fill the tank with carbon dioxide (the cardox method), (2) use vacuum cleaners to purge the tank of gasoline, or (3) cut the loading speed down to about 30% until the tank is about one-third full.

When a choice is to be made between a flammable and a combustible liquid for a given application, the difference in costs of electrical equipment installations can be dramatic. When the normal operation of a process produces ignitable concentrations of flammable liquids in the atmosphere, explosion-proof electrical equipment approved for Class I, Division 1 hazardous locations is specified for the area in which the vapors are present. Providing explosion-proof electrical equipment is a costly undertaking and is discussed in detail in Chapter 17. Case Study 11.2 will now be used to illustrate the valuable impact a knowledgeable safety and health manager can have on a company when a decision is to be made regarding flammable and combustible liquids.

CASE STUDY 11.2

FLAMMABLE VERSUS COMBUSTIBLE LIQUIDS

A process engineer has a new idea to cut costs in an operation that strips organic coatings from metal parts before they are plated. The process currently uses Enthone Stripper S-300, but the engineer has discovered in the laboratory that Enthone Stripper S-15 is much more effective in removing the organic coatings than S-300 and can save the company both production time and money in the savings in volume of stripper required to be purchased from the vendor. How would this new idea affect fire safety?

Solution

The safety and health manager would wisely take a keen interest in a comparison of the flammability characteristics of the two strippers under consideration. Solvents and strippers are typically either flammable or combustible, and which classification a liquid turns out to be is an important consideration in the design of a process. Checking the MFPA's *Flashpoint Index of Trade Name Liquids* (ref. Flashpoint) the safety and health manager notes that Stripper S-300 has a flashpoint of 155°F, whereas S-15 has a flashpoint of 34°F. From Figure 11.1, it can be seen that these data mark S-300 as a moderately safe Class IIIA combustible liquid, whereas S-15 is identified as a very hazardous Class I flammable liquid. Choosing Stripper S-15 would definitely affect fire safety and perhaps increase insurance rates. If electrical equipment, such as conveyors and switches, is required around the stripping operation, the requirement to provide Class I, Division 1 explosion-proof equipment could make the S-15 alternative prohibitively expensive, even considering the cost savings that the process engineer is claiming.

In Case Study 11.2, the idea generator was a process engineer. The case study should serve as a model, however, for ways in which the safety and health manager, armed with knowledge of the contrasting properties of flammable and combustible liquids, can be the idea generator to make processes safer and perhaps much cheaper. This is the kind of impact on the company's bottom line that corporate management has not customarily expected from the safety and health manager, but that will certainly get top management's attention. We now turn our attention to one of the most important industrial applications of flammable and combustible liquids: spray finishing.

SPRAY FINISHING

A direct concern of safety and health managers, especially in manufacturing plants, is the installation of facilities and proper procedures for spray painting areas or booths. The subject is important not only from a compliance standpoint, but also greatly affects insurance rates and basic insurability.

   The construction and operation of a spray painting area that meets applicable code is a quite costly undertaking, and an unethical safety and health manager sometimes attempts to circumvent the rules to appease top management. The most common such circumvention of the rules is to call the paint facility a "small portable spraying apparatus not used repeatedly in the same location" and thus achieve exemption from spray painting standards. But if the small "temporary" setup becomes a more or less permanent arrangement, a continuing and serious fire hazard will result. Furthermore, neither the insurance company inspector nor an experienced government inspector will be fooled by a so-called temporary arrangement that has eventually become permanent, because spray paint residues will accumulate throughout the area in large quantities.

   There are both health and safety considerations in spray finishing operations, but standards for spray finishing principally concern safety aspects, particularly fire. The most frequent violations of spray finishing standards are in the following categories:

- Improper wiring type for hazardous location
- Exhaust air filter deficiencies
- Cleaning and residue disposal
- Quantities of materials in storage
- Grounding of containers
- "No Smoking" signs

Also frequently violated but of somewhat less emphasis are the physical construction requirements for spray booths and their mechanical ventilation requirements.

Applying the rule of attacking the easiest problems first, the safety and health manager should take steps to immediately install "No Smoking" signs in spraying areas and in paint storage rooms. This advice may sound superficial, but thousands of firms have received OSHA citations simply for failure to post such signs. The cost of complying with this rule is almost negligible.

After ensuring that "No Smoking" signs are installed, the safety and health manager should investigate the wiring in the spray area to determine whether it conforms with *National Electrical Code®* specifications for hazardous areas. A competent electrician, knowledgeable in the provisions of the *National Electrical Code*, is useful for this phase of the problem.

Appropriate wiring and electrical equipment classification in and around spray areas may be simplified somewhat by using a decision diagram like the one shown in Figure 11.7. A great deal of controversy has arisen regarding the enforcement of electrical requirements around spray areas, particularly regarding the interpretation of the

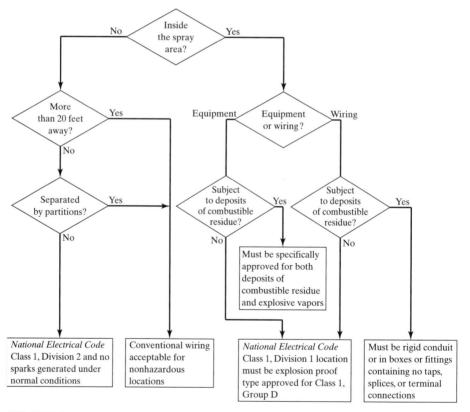

FIGURE 11.7

Decision diagram—How to follow OSHA Standard 1910.107(c)(5) and (c)(6) specifications for spray area wiring and electrical equipment.

legendary "20-foot distance" from the spray area. It is interesting to note that the *National Electrical Code*, 1975 edition, reduced the required distance from 20 feet to 5 feet from the spray area.

Besides distance of travel, another issue is the *direction* of travel of the flammable vapors after they leave the paint spray booth. Since flammable vapors can travel in any direction, it is recommended that the safety and health manager take the cautious rule of using Class I, Division 2 wiring in all directions from the open face of the spray area or booth, including vertically and around the corners alongside the booth. Additional guidance on the various classifications for electrical wiring is contained in Chapter 17.

Some people construe the standards to require automatic sprinkler systems for all paint spray areas, but the standards do not really specify this. If an automatic sprinkler system is used, however, it must meet NFPA requirements. The vendor who installs the sprinkler system should be required to ensure that the system conforms to all applicable codes for the installation to which it will be applied. If the system is installed inside the ducts, sprinklers are needed on both sides of the filter system.

Combustible residues contribute to the largest proportion of spray booth fires. Reactions between different materials can add to this hazard, especially when peroxides are used. The control of overspray residues involves both engineering and administrative controls and is an item that deserves the attention of the safety and health manager. Residue accumulation is easy to recognize and is an embarrassing testament to poor maintenance and lack of hazard control. Once removed, the residue and debris must be properly disposed of to prevent spontaneous combustion and other fire hazards from oily rags and residue debris.

It was stated earlier that automatic sprinkler systems are not required for *all* areas. However, for fixed electrostatic systems, automatic sprinklers are required "where this protection is available." Automatic sprinkler systems are preferred, and if such a system is already available close by (within approximately 50 feet), it should be extended to the electrostatic spraying area. In the absence of "available" automatic sprinkler systems, the standard requires "other approved automatic extinguishing equipment" for electrostatic spray areas. Such alternate systems would include fixed carbon dioxide or dry chemical systems, and these systems will be discussed further in Chapter 13.

The use of heating devices for drying in the spray area increases the hazard by raising the temperature of overspray residues and also by increasing the vapor level in the air. In addition, the standards are not very clear concerning the use of the spray area for a drying area. The use of the spray area as a drying area is prohibited unless the arrangement does not "cause a material increase in the surface temperature of the spray booth, room, or enclosure."

DIP TANKS

Dip tanks often contain hazardous materials and are treated separately in federal standards. Care must be exercised when consulting the standard, however, because it applies only to those dip tanks containing flammable and combustible liquids. Plating dip tanks containing hazardous acids are *not* covered by the dip tank standard unless the acid is flammable or combustible.

The following are the principal problems with dip tanks:

- Automatic extinguishing facilities
- "No Smoking" signs
- Dip tank covers

The lack of dip tank covers is the most frequent violation of the three. One problem with covers is that they must be "kept closed when tanks are not in use." It is unreasonable to expect dip tank covers to be closed during short intervals of nonuse, such as coffee breaks and other short interruptions. However, an idle period of as long as one-half shift would be considered a "not in use" period. Also, if the dip tank is discovered to be idle for any period in which the work crew and supervisor have left the area, the dip tank should be considered "not in use." Automatic closure devices for actuation in event of fire are desirable, but are not specifically required. Such closure devices "shall be actuated by approved automatic devices and shall also be arranged for manual operation." Automatically closing dip tank covers are considered among the most appropriate means of automatic extinguishing facilities specified under conditions described in Figure 11.8.

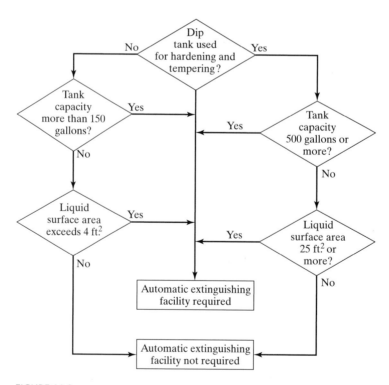

FIGURE 11.8

Decision chart—Automatic extinguishing facilities for dip tanks.

EXPLOSIVES

Everyone knows that explosives are dangerous, and the general public completely avoids contact with them. Only the well-trained professional knows which procedures are safe and what to do in each situation. The body of existing codes governing explosives pertains almost entirely to storage or to the construction of the magazines in which the explosives are stored.

As in flammable liquids, explosives are classified according to degree of hazard. Class A is the most hazardous, and most of the materials that the general public considers "explosive," such as nitroglycerin, black powder, and dynamite, fall into the Class A category. Class B explosives include propellants, photographic flash powders, and some special fireworks. Class C explosives are manufactured articles that contain explosives in restricted quantities. The safety and health manager should take advantage of manufacturers' labeling to determine explosive classes when a questionable item is concerned.

The storage magazines for explosives are divided into two groups, also called *classes*, but in the case of magazines, the class designation is a Roman numeral instead of a letter of the alphabet. The magazine class depends chiefly on quantity (weight) of explosives stored, not the explosives class. Class I magazines are for quantities more than 50 pounds, and Class II magazines are for 50 pounds or less.

Most facilities do not use explosives, and most safety and health managers can ignore the standard if they do not have explosives. But those who do have explosives within their facilities should take the precaution of assuring that handling and especially storage procedures are in compliance with applicable code.

LIQUEFIED PETROLEUM GAS

Liquefied petroleum gas (LPG) is a commonly used fuel gas, especially in areas remote from piped natural gas utilities. All petroleum gases can be liquefied if the temperature is lowered sufficiently, but natural gas, which is chiefly methane, is very difficult to liquefy, although it is otherwise cheaper than LPG. LPG is principally a mixture of propane and butane, both of which can be liquefied more easily than methane and therefore can be transported more compactly. The expansion ratio is approximately 1:270—that is, 1 gallon of liquid converts to 270 gallons of gas at normal temperature and pressure.

The choice between propane and butane is a matter of climate and economics. Butane has a higher boiling point and is unsuitable for cold climates because it will not convert to a gas at low temperatures. Butane has historically been cheaper, however, so it has been used in climates such as those in the southern United States. Recently, propane has been used almost exclusively because butane prices have increased due to butane's application in the manufacture of artificial fabrics.

Propane is a product of the refinery cracking[6] process and is odorless in the natural state. For safety's sake, the odorant ethyl mercaptan is added as a stenching agent before delivery to the customer. This stenching agent facilitates leak detection. However, as was emphasized for hydrogen sulfide in Chapter 9, a heavy exposure to

---

[6]Large molecules are "cracked" into simpler molecules of more economically useful products.

a strong odor can overcome the olfactory system, and consequently the victim simply can no longer smell it.

One of the safety hazards of propane is that it is heavier than air (approximately $1\frac{1}{2}$ times the density of air). This contrasts with natural gas (methane), which is lighter than air. This difference in properties can lead to problems when one attempts to convert a piece of equipment from natural-gas powered to propane powered.

Another hazard with propane is that the extreme cold of its liquid state can burn the flesh. In fact, the treatment is the same as for third-degree burns. One way such a burn can occur is by opening the valve too quickly and placing the hand over the valve to feel the flow. Injury from liquid propane while opening the valve does not indicate a defective valve.

Unlike gasoline and other atmospheric-vented tanks, LPG tanks are closed, and there is no opportunity for water vapor to accumulate inside the tank. Therefore, no "bleed-off"' valve for moisture is used. If water should happen to be introduced into the system, it would probably freeze the valve during expansion of the gas. The closed tank of LPG then contains no air and is a mixture of liquid and gaseous propane. The vapor pressure inside the tank depends on temperature. At 0°F, the pressure is 28 pounds per square inch (psi), and at 100°F, the pressure is almost 200 psi. The pressure-relief valve for tank trucks is set at 250 psi and for cylinders at 375 psi.

Although flesh burn from the extreme cold is a hazard to be considered, the principal hazard with LPG is fire, and when an LPG fire occurs, it is usually a disaster. The fire expands quickly, and portable fire extinguishers are generally useless, although they may be useful in extinguishing other burning materials threatening LPG facilities. Once a tank itself ignites, however, it is strictly a matter for professional firefighters and special techniques using large volumes of high-pressure water spray to protect firefighters who must approach the tank to close valves or otherwise control the fire. Large tanks, such as railroad cars, have resulted in spectacular fires, including the phenomenon called *BLEVE* (rhymes with "heavy"), which means "boiling liquid expanding vapor explosion."

The federal standard requiring the use of laboratory-approved equipment ("listed" by an approved testing laboratory) seems like so much red tape. But without this requirement, people would try all kinds of makeshift arrangements. One individual decided to use an old hot-water tank for storage of LPG. The tank exploded, killing one person and injuring another. Another temptation is to use ordinary water hose instead of approved piping. The insidious nature of this hazard is that the water hose will often withstand the LPG pressures and will seemingly work, but the LPG will attack the rubber in the hose and will eventually result in a rupture. Similar problems result from using ordinary plumbing fittings and valves that have rubber seals.

Another frequent misuse of equipment is to interchange tanks for storage of anhydrous ammonia and LPG. Anhydrous ammonia attacks brass and copper fittings in LPG tanks, making them unsafe. Particularly dangerous is damage to the tank relief valve.

Fire, welding operations, and other sources of intense heat can weaken LPG cylinders and make them no longer capable of meeting laboratory test standards. The safety and health manager should be on the alert for this danger and have LPG equipment recertified after a plant fire or other exposure to heat. As for welding, no welding should be permitted directly to the tank shell; it is permissible, however, to weld to existing brackets, plates, and lugs, which in turn were already welded to the tank in its original manufacture in the configuration that has laboratory approval.

Fire control for LPG tanks is quite different from fire control for flammable liquid tanks. Dikes are built around flammable liquid tanks to contain the burning liquid in event of tank rupture. But such dikes are hazardous to LPG tanks because they can cause burning near or beneath the tank, which might result in explosive rupture.

In any high-pressure cylinder with a valve on the end, there is danger should the valve be accidentally ruptured or even broken off. Like a torpedo in size and shape, the cylinder becomes a dangerous missile. The most dangerous are the very high-pressure oxygen cylinders used in welding. (See Chapter 16.) However, LPG cylinders at 200 psi can also be quite dangerous. This danger is in addition to the fact that the gas liberated in the rupture can ignite explosively. Therefore, valves must be protected, and there are two acceptable methods for this: by recessing the valve into the tank or by a ventilated cap or collar.

Safety and health managers often misinterpret the provision of the federal standard for LPG, which reads as follows:

> Engines on vehicles shall be shut down while fueling if the fueling operation involves venting to the atmosphere.

In-plant refueling of forklift trucks that operate on LPG normally does not involve venting to the atmosphere. It is not required that the engines be shut down in such refueling operations.

CONCLUSION

A final suggestion for safety and health managers is to take advantage of community resources, public and private, for advice and assistance in dealing with hazardous materials. Local fire departments and state fire marshals can be of assistance, particularly with regard to flammable liquids, spray finishing, and codes. Some fire or police departments have explosives experts on hand. Compressed gases, LPG, and anhydrous ammonia problems can sometimes be alleviated by consulting the local distributor for these materials. Some of these distributors are backstopped by extensive training resource centers at their company headquarters and can supply booklets, guides, warning labels, films, and slide-tape programs for in-house training to combat hazards.

EXERCISES AND STUDY QUESTIONS

11.1 What is a floating roof? What are its benefits?

11.2 Would it be proper to use a Class II magazine for storage of Class A explosives? Explain.

11.3 What is a BLEVE?

11.4 What is the difference between "light" and "heavy" oils? Which evaporate more readily?

11.5 Compare the flammability hazards of gasoline with those of ethyl alcohol.

11.6 A manufacturing process uses the powerful solvent acetone. One phase of the process is to dry out the acetone in a *drying area*. The drying area evaporates 2 gallons of acetone per hour. Every gallon of liquid acetone produces 41 cubic feet of vapor. Calculate the amount of ventilation (ft³/hr) needed to maintain vapor concentration below the burnable level. How many times per hour would the ventilation system exhaust a room 9 feet by 12 feet with a ceiling height of 10 feet?

11.7 In a one-time inspection of a facility, how will an insurance representative be able to determine that a temporary spraying area or a so-called "touch-up area" is really a more permanent arrangement?

11.8 Under what circumstances are automatic sprinkler systems required for paint spraying areas?

11.9 When are dip tank covers required to be closed? Must they close automatically in event of fire?

11.10 Compare liquefied petroleum gas (LPG) with natural gas in terms of safety hazards.

11.11 Are fire extinguishers appropriate for LPG facilities? Why or why not?

11.12 What is the hazard of using a paint spray area as a drying area? Under what conditions is it acceptable to use a paint spray area as a drying area?

11.13 Carbon disulfide has the following physical properties:

Flashpoint	$-22°F$
Boiling point	$46.5°$ C
Density	1.261
Vapor density	2.64
Lower flammable limit	1.3%
Upper flammable limit	50%
8-hour TWA PEL	20 ppm

An industrial process liberates 3 cubic feet of carbon disulfide per hour into a room that measures 10 feet by 20 feet and has a ceiling height of 8 feet.

(a) Calculate the minimum general exhaust ventilation (ft³/hr) necessary to prevent a general safety hazard for the process.

(b) Calculate the minimum general exhaust ventilation (ft³/hr) necessary to prevent a general health hazard for the process.

(c) Carbon disulfide is which of the following:

  1. Class IA flammable liquid
  2. Class IB flammable liquid
  3. Class IC flammable liquid
  4. Class II combustible liquid
  5. Class III combustible liquid

11.14 What is the difference between Class I liquids and flammable liquids?

11.15 Explain why an empty gasoline drum can be more dangerous than a full one. Why is a drum containing carbon disulfide more likely to ignite than a drum containing gasoline?

11.16 Explain the hazard mechanism that has led to a ban on basements in service stations.

11.17 Under what circumstances can kerosene become even more dangerous and ignitable than gasoline?

11.18 Why is ethyl mercapton added to propane?

11.19 Sometimes forklift trucks powered by LPG are fueled with the engines still running. Is this a violation of safety standards? Why or why not?

RESEARCH EXERCISES

11.20 Use the Internet to find special firefighting materials for use in petroleum fires.

11.21 Examine facts about the train accident near Shepardsville, Kentucky, in 1991. Describe the concern about a potential BLEVE in this accident.

11.22 **Design Case Study.** Study the design of the tanks shown in Figure 11.9 and attempt to determine the cause of an accident, described as follows. The tank arrangement shown was intended to store tetrahydrofuran liquid on the second-story level of a building in Chicago. The first day the tanks were loaded, the following events took place: A tanker delivery truck hooked up at the receiving valve at ground level and proceeded to deliver 500 gallons. About halfway through the filling operation, a company employee inside the building yelled out the window that the tank was overflowing. Soon thereafter a tremendous explosion and fire killed both the company employee and the tank truck driver. What faults do you see in the system, and how would you redesign the system to prevent this type of accident? Look up characteristics of tetrahydrofuran, including the flashpoint. To what flammability classification does tetrahydrofuran belong?

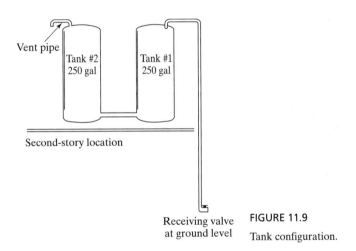

FIGURE 11.9

Tank configuration.

STANDARDS RESEARCH QUESTIONS

11.23 Do a General Industry standards search to identify the numerical designation of the spray finishing standard. Use the NCM database to determine the frequency of citation of this standard. Check average penalty levels.

11.24 Identify the provisions of the General Industry OSHA standards that deal with dip tanks using flammable and combustible materials. Use the NCM database to determine whether these provisions are cited by OSHA.

11.25 Find the OSHA General Industry standard for liquefied petroleum gas. Use the NCM database to examine the citation statistics for the provisions of this standard.

CHAPTER 12

Personal Protection and First Aid

13%

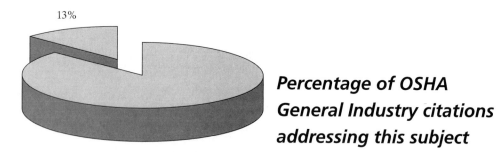

Percentage of OSHA General Industry citations addressing this subject

In a way, it is unfortunate that this chapter is even necessary. The need for personal protection implies that hazards have not been eliminated or controlled. And the need for first aid implies something even worse! When feasible, engineering control of the hazard is preferred over the use of personal protective equipment. As discussed in Chapter 1, we realize that some risks will always remain—our goal is to eliminate *unreasonable* risks, not all risks. The job of improving safety and health in the workplace will never be completely finished, so we must concern ourselves with the need to provide personal protection against hazards that have not been completely eliminated and to provide first aid when an accident does occur.

The problem of providing personal protective equipment seems straightforward and easy enough to understand. But the simplicity of the problem is an illusion, and many industrial safety and health managers fall into its trap. For instance, it would seem that if the noise level in the production area is too high, the solution to the problem would be to provide ear plugs for the workers. But anyone who has actually confronted this problem knows that the solution is not that simple. For a variety of reasons, many people do not want to wear ear protection. They may be shy about the appearance of ear protection equipment, they may feel discomfort or perhaps even pain, they may feel it interferes with their necessary hearing or efficiency, or they may feel that use of personal protective equipment is their own business, not their employer's. The bases for these complaints and what to do about them will be taken up later in this chapter, but first we will discuss the hazards for which personal protection may be needed and the types of equipment available to meet these needs.

The matter of personal protective equipment becomes very delicate when employees bring their own equipment to work. If the equipment is not properly maintained, who is responsible—employer or employee? OSHA's position is that the employer is responsible. As safety and health manager, consider the following logic: If employees bring their own personal protective equipment to work, is it not possible that the equipment itself could represent a hazard? Personal protective equipment must be properly selected to match the hazard, and employees bringing their own equipment might falsely think they are safe when their equipment could actually be malfunctioning or inappropriate.

Even if the employee works in a job that does not require personal protective equipment, the use of faulty or inadequate equipment might tempt the worker to take chances. For instance, suppose that a maintenance worker who has no need to even approach the edge of a rooftop while maintaining air-conditioning equipment wears his or her own rope attached to a leather belt as added protection, not required by company policy. Then suppose that some unusual situation tempts the employee to approach the edge of the building and the false sense of security from the improvised "safety belt and lanyard" causes the worker to proceed without caution. An accidental fall could cost the worker's life when the leather belt breaks under a 2000-pound shock load due to the fall. Alternatively, even if the leather belt withstands the shock load, the worker could be killed by the shock itself if the lanyard is too long and is nonelastic. These are facts workers might not know, and when they improvise or bring their own personal protective equipment to work, the employer should ensure that the equipment is adequate for the situation and is properly maintained.

PROTECTION NEED ASSESSMENT

The first federal standards for personal protective equipment emphasized the design and performance of the equipment itself, not when such equipment should be used. The reason was that there were really no detailed consensus standards that addressed when personal protective equipment was required. A notable exception was hearing protection, which was required when neither engineering nor administrative controls were capable of reducing noise exposure to acceptable levels.

In the mid-1990s, OSHA addressed this problem by promulgating a more specific standard for personal protective equipment that mandated a more comprehensive program of use of personal protective equipment in general. A key element in this new standard was the requirement for the employer to conduct a hazard assessment to determine whether personal protective equipment is needed for various jobs found within the plant. It is recognized that such an assessment is a judgment call, so the safety and health manager should not expect to find a standard that specifies exactly when various personal protective equipment must be used. However, there are guidelines that can be used to assist in making a decision. The nonmandatory appendices to the standards are often helpful in providing such guidance. The sections that follow will reflect the suggestions made by standard appendices and should be considered a resource for the personal protective equipment needs assessment and selection of appropriate equipment. Besides the type of equipment selected, the fit of such equipment to the particular employee who will wear it is also important. For respirators, fit is *critical* to the protection that the respirator will provide. This topic will be covered in more detail later in this chapter.

PERSONAL PROTECTIVE EQUIPMENT (PPE) TRAINING

Another important provision of the general personal protective equipment standard is the requirement for employee training in the proper use of the equipment. The basic premise of enforcement authorities is that if personal protective equipment (PPE) is needed, then the employees must be trained to use it properly. Employees need to know when PPE is necessary, what kinds of PPE are required, and how to wear it effectively. To avoid developing a false sense of security, the employee needs to know the limitations of the PPE he or she is using, including its useful life under proper care and maintenance. Even if the employee has received PPE training, workplaces often change, rendering the previous PPE obsolete for the task. If either the workplace changes or the PPE changes, the employer must take a responsible position and retrain the employee if necessary. Both the initial training and the retraining must be documented with a certificate that identifies the names of employees trained, the dates, and the subject for which the employee is certified.

Thus far, this chapter has covered general requirements for personal protective equipment, the assessment of the need for such equipment, and the required employee training for the use of such equipment. We now turn to the specific types of personal protective equipment for various hazards, starting with hearing protection.

HEARING PROTECTION

As might be expected, the greatest emphasis in providing personal protective equipment coincides with the principal problem in environmental control, covered in Chapter 10—the problem of noise. If engineering or administrative control measures are unsuccessful in eliminating the hazard of noise in the workplace, management must turn to personal protective equipment to shield the worker from exposure.

The most important factor in selecting a type of noise protection is probably effectiveness in reducing decibel level of noise exposure. However, this is by no means the only important factor, and selection can be somewhat complicated. Economics is always a factor, and if limited effectiveness is all that is necessary in a given situation, cheaper devices can be selected. Employee comfort is probably at least as important a factor as economics. The worker comfort factor goes beyond the simple goal of promoting worker satisfaction; it affects the amount of protection the worker will receive. If workers find a type of ear protection uncomfortable or awkward to wear, they will use every excuse not to wear it, which results in loss of protection. The next paragraphs consider the merits of various types of ear protection.

Cotton Balls

Ordinary cotton balls, without addition of a sealing material, are virtually worthless as a means of personal protection from noise.

Swedish Wool

Similar in feel to cotton, Swedish wool is a mineral fiber that has much better attenuation values than cotton. Swedish wool is somewhat effective as is, but is much more effective when impregnated with wax to make a better seal. One problem with Swedish wool is that it can tear when it is pulled out. To alleviate this problem, Swedish wool sometimes comes in a small plastic wrapper that is inserted with the wool. Swedish wool can be considered only fairly reusable; reuse will depend on personal hygiene, quantity of earwax, and worker preference.

Earplugs

The most popular type of personal protection for hearing is the inexpensive rubber, plastic, or foam earplug. Earplugs are practical from the standpoint of being easily cleaned and reusable. Workers often prefer earplugs because they are not as visible as muffs or other devices worn external to the ear. But within this advantage lies a pitfall: Workers may be more complacent about using the earplugs when it is not immediately obvious to the supervisor whether the earplugs are being worn. The noise attenuation for properly fitted earplugs is fairly good, falling somewhere between that of Swedish wool and the more effective earmuffs.

Molded Ear Caps

Some ear protectors form the seal on the external portion of the ear by means of a mold to conform to the external ear and a small plug. Since human ear shapes vary so widely, fit is a problem. Molded ear caps are more visible than earplugs, which has both advantages and disadvantages, as discussed earlier. Molded ear caps may be more comfortable to the wearer, but are more expensive than earplugs.

Earmuffs

Earmuffs are larger, generally more expensive, and more conspicuous than Swedish wool, plugs, or caps, but they can have considerably better attenuation properties. The attenuation capability depends on design, and more variety in design is possible with earmuffs. Although some workers object to wearing conspicuous earmuffs some workers prefer them, stating that they are more comfortable than earplugs.

Helmets

The most severe noise-exposure problems may force the safety and health manager to consider helmets for personal protection against noise. Helmets are capable not only of sealing the ear from noise, but also of shielding the skull bone structure from sound vibrations that can be transmitted to the ear as noise. Helmets are the most expensive form of hearing protection, but have the potential of offering protection from a combination of hazards. Properly designed, the helmet can act as a hard hat and a hearing protector at the same time.

It must be remembered that fit is very important for all types of hearing protectors. As in noise enclosures or sound barriers, the material itself might have excellent sound-attenuation properties, but if there is a leak or crack, most of the effectiveness of the device is lost.

EYE AND FACE PROTECTION

The use of safety glasses has become so widespread and so many different styles are now available that many safety and health managers establish a rule that they must be worn throughout the plant. A general custom in industry is to require visitors to wear safety glasses during plant tours.

There is a difference between *street* safety glasses and *industrial* safety glasses. Visitors or employees who claim that their prescription glasses are "safety glasses" probably mean that they have street safety lenses. Industrial safety lenses must pass much more severe tests to meet ANSI standards. This is not to say that street safety lenses are not adequate for some industrial environments. Eye protection standards are not specific about which jobs do and which do not require safety lenses—street or industrial. It is good to have exacting standards to be sure that safety glasses will meet uniform performance standards. But responsibility for deciding when eye protection equipment is necessary usually falls on the safety and health manager—not the optics industry, not the standards, and not enforcement agencies.

A word of caution is in order for the safety and health manager who is setting forth a policy on the wearing of safety glasses. It can be as bad a mistake to require safety glasses in those areas of the plant where there is no eye hazard as it is to not require safety glasses in areas where they are needed. The peril is that workers will not respect the safety glasses policy, and use of the glasses will not be uniform. Eye injuries can be the result—in addition to violations of code. It is easy for inspectors to establish violations when they observe workers not wearing eye protection in the presence of an established company rule requiring eye protection. But some safety and health managers say that a simple plantwide rule is easiest to enforce. The costs on both sides of the question are great, and the decision to require eye protection must be made with care.

There are some jobs for which industry and enforcement agencies alike have seemed to come to a consensus on the need for eye protection. Machining operations that produce chips or sparks are almost universally conceded to necessitate the use of eye protection. Notable among these operations are those of grinding machines, drill presses, and lathes. Both metal and wood materials can produce dangerous eye hazards when machined. Corrosive liquids or other dangerous chemicals also represent eye hazards when poured, brushed, or otherwise handled in the open. When working with such materials, face protection may be needed in addition to eye protection. Guidelines for the specification of various types of eye protection for typical hazards are offered in Table 12.1

More important than when to *require* eye protection is how to *educate* workers to be alert to eye hazards and the long-term consequences of eye injury. The National Safety Council has some films that bring this point home. One effective film shows emergency eye surgery in progress. Another is a poignant personal testimony of a man who was blinded on the job and the hardships he and his family have suffered since.

A final note about eye protection is a behavioral factor. Historically, people have avoided wearing glasses because of their appearance. Males and females alike have sought contact lenses to correct their vision, often for no other reason than personal appearance. The feelings about wearing glasses are complex. For men, the notion has been that glasses detract from the ruggedness of their appearance. For women, historically, glasses have somehow been thought to detract from sex appeal. Toward the end of the 20th century, those notions began to change and the wearing of glasses began to become fashionable.

TABLE 12.1   Appropriate PPE for various hazards (ref. Nonmandatory)

Source	Assessment of Hazard	Protection	
IMPACT—Chipping, grinding machining, masonry work, woodworking, sawing, drilling, chiseling, powered fastening, riveting, and sanding.	Flying fragments, objects, large chips, particles, sand, dirt, etc	Safety glasses with side shields or face shields, depending upon the hazard and its severity	
HEAT—Furnace operations, pouring, casting, hot dipping, and welding.	Hot sparks	Goggles, safety glasses with side shields or face shields for severe exposures	
	Splash from molten metals	Face shields worn over goggles	
	High temperature exposure	Screen face shields or reflective face shields	
CHEMICALS—Acid and chemicals handling, degreasing plating.	Splash	Goggles, eyecup and cover types face shields for severe exposure	
	Irritating mists	Special-purpose goggles	
DUST—Woodworking, buffing, general dusty conditions.	Nuisance dust	Goggles, eyecup and cover types	
LIGHT or RADIATION		See Chapter 16 for welding, cutting, and brazing operations.	
Glare	Poor vision	Spectacles with shaded or special-purpose lenses.	

Television series in which beautiful models wore glasses served to usher in this change in attitudes toward glasses. Glasses manufacturers were quick to pick up on this trend, and especially safety glasses manufacturers began to feature new styles or looks in advertising and promotions. Figure 12.1 is an advertisement for safety glasses that illustrates the appeal of physical appearance in the sale, adoption, and use of safety glasses for eye protection.

FIGURE 12.1

Models feature safety eyewear that promotes a desirable image for workers who wear safety glasses (Courtesy Aearo Company, used with permission).

RESPIRATORY PROTECTION

Of even more vital importance (in the literal sense of the word *vital*) than eye and hearing protection is the need for respiratory protection from airborne contaminants. Chapter 9 discussed the different types of problems with industrial atmospheres, and determination of the type of atmospheric problem is essential in selecting the correct respiratory equipment. A well-designed and expensive gas mask is useless and might be more properly designated a "death mask" if the atmospheric problem to be tackled is oxygen deficiency, for example.

Particularly hazardous atmospheres may be referred to as IDL or IDLH, which stand for "immediately dangerous to life" and "immediately dangerous to life or health," respectively. Recently, the acronym IDLH has become more widely used. If a single acute exposure is expected to result in death, the atmosphere is said to be IDL. If a single acute exposure is expected to result in *irreversible* damage to health, the atmosphere is said to be IDLH. Some materials—hydrogen fluoride gas and cadmium vapor, for example—may produce immediate transient effects that, even if severe, may pass without medical attention but are followed by sudden, possibly fatal collapse 12 to 72 hours after exposure. The victim "feels normal" after recovery from transient effects until collapse. Such materials in hazardous quantities are considered to be "immediately" dangerous to life or health (ref. Industrial).

It should be apparent to the reader at this point that there is more to respiratory protection than simply handing out respirators to workers who might be exposed to hazards. Effective protection demands that a well-planned program be implemented, including proper selection of the respirators, fit testing, regular maintenance, and employee training.

Some firms supply respirators to employees without bothering with a comprehensive respirator program, relying on the excuse that the respirators are not really required anyway because contaminants in the plant atmosphere do not exceed maximum permissible exposure limits (PELs). The safety and health manager is really asking for trouble, though, by toying with a partial program. The atmospheres are no doubt marginal, or the question of a partial program would never have arisen. The marginal atmospheres might later deteriorate without warning. Employees would be lulled into a false sense of security by the superficial respirator program. Bad habits, such as negligent maintenance, inadequate fit testing, or improper equipment usage, could be developed. A whole feeling of complacency toward the use of respirators can be engendered by the use of such equipment when it is not really needed.

As with general personal protective equipment discussed earlier, the safety and health manager is often placed on the spot when employees bring their own respiratory protection equipment onto the job site. In this situation, the employer should take a responsible position and ensure that these employees use their respiratory equipment properly. If the employee resents what he or she perceives as the employer's interference, he or she should be reminded of the employer's continuing responsibility to eliminate hazards in the workplace, including misuse of personal protective equipment. If the equipment or its improper use can be hazardous, the employer can forbid the employee from bringing the equipment onto the premises. The safety and health manager may feel some hesitation in exercising such authority over the employee's own personal property, but instances have already occurred in which employers have exercised such authority when necessary to prevent hazards.

Before going further into the subject of respiratory protection, a classification of the various devices is in order. The two major classifications are *air-purifying* devices versus *atmosphere-supplying* devices. The air-purifying devices are generally cheaper, less cumbersome to operate, and the best alternative if they are capable of handling the particular agent to which the user will be exposed. But some materials simply cannot be reduced to safe levels by air-purifying devices, and a supplied-air system is required. Another important consideration is oxygen deficiency. No amount of filtering or purifying is going to make an oxygen-deficient atmosphere safe. The only way to go in this situation is with atmosphere-supplying respirators. A summary classification of respiratory protection devices is as follows:

1. Air-purifying devices
   (a) Dust mask
   (b) Quarter mask
   (c) Half mask
   (d) Full-face mask
   (e) Gas mask
   (f) Mouthpiece respirator

2. Atmosphere-supplying respirators
   (a) Air line respirator
   (b) Hose mask
   (c) Self-contained breathing apparatus

Further description of each type of device follows.

Dust Mask

The most popular respirator of all is also the most misused. Approved only for particulates (suspended solids), the dust mask (Figure 12.2) is not approved for most painting and welding hazards, although it is frequently used improperly in this way. Some dust masks are approved for mild[1] systemic poisons, but generally these masks are limited to *irritant dusts*—those that produce pneumoconiosis or fibrosis. (See Chapter 9.) One of the principal limitations of the dust mask is fit. Even the best-fitting models have approximately 20% leakage. A rule of thumb is that approval is valid for particulates no more toxic than lead.

Despite its disadvantages, the dust mask is popular because it is inexpensive, sanitary, and can be thrown away after each use. Its low cost and general availability make the dust mask attractive for purchase at the local drugstore and for personal use. Therefore, this is the type of respirator that the safety and health manager is most likely to find employees bringing from home for use on the job. Care should be taken to educate employees about the limitations of the dust mask.

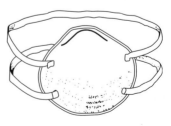

FIGURE 12.2

Disposable dust mask.

---

[1]Dusts that have a threshold limit value of not less than 0.05 mg/m^3.

FIGURE 12.3
Quarter mask.

Quarter Mask

The quarter mask, sometimes called the type B half mask, is shown in Figure 12.3. The quarter mask looks very much like a half mask except that the chin does not go inside the mask. The quarter mask is better than the dust mask, but it, too, is generally approved for toxic dusts no more toxic than lead.

Half Mask

The half mask, shown in Figure 12.4, fits underneath the chin and extends to the bridge of the nose. This mask must have four suspension points, two on each side of the mask, connected to rubber or elastic about the head.

Full-Face Mask

Actually, the gas mask is also *full face*, but the name *full-face mask* generally refers to a mask in which the filtering chamber attaches directly to the chin area of the mask. The filters may be either dual *cartridges* or single *canisters*. Both types are shown in Figure 12.5. Canisters contain granular *sorbents* that filter the air by adsorption, absorption, or chemical reaction.

Gas Mask

The gas mask is designed for filtering canisters that are too large or too heavy to hang directly from the chin. In the gas mask, the canister is suspended by its own harness and is typically connected to the face mask by a corrugated flexible breathing tube. The gas mask is shown in Figure 12.6.

FIGURE 12.4
Half mask.

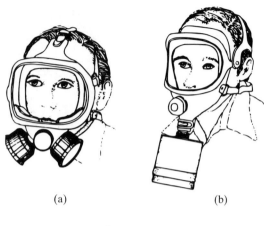

FIGURE 12.5

Full-face masks: (a) dual-cartridge full-face mask;
(b) canister-type full-face mask.

(a)   (b)

FIGURE 12.6

Gas mask.

Mouthpiece Respirator

Perhaps the mouthpiece respirator should be omitted from this discussion because this type of device is not intended for normal use. But emergencies will occur from time to time, and enabling the user to be prepared to escape during such an emergency is the purpose of the mouthpiece respirator. Breathing is accomplished through the mouth by means of a stem held inside the teeth. A nose clip must be used to prevent inhaling through the nose. It is possible to form a good seal with the mouth and lips, but the effectiveness of the mouthpiece respirator is greatly dependent on the knowledge and skill of the user.

Air Line Respirator

The air line respirator is an atmosphere-supplying respirator and derives its name from the way in which air is supplied to the respirator mask. The air is supplied to the mask by a small-diameter hose (not over 300 feet long), which is approved together with the mask. Ordinary garden hose is not acceptable. The air is supplied by either cylinders or compressors. The method of delivery of air to the user results in three different modes of air line respirator: continuous flow, demand flow, and pressure demand.

In the *continuous-flow* mode, the air line respirator receives fresh air without any action on the part of the user—that is, the flow is forced by the apparatus. The air flow must be at least 6 cubic feet per minute to qualify a hood for use with this mode. The flow should not be much greater than this because it can induce a high noise level inside the hood. One of the advantages of the continuous-flow mode, however, is that it permits use of a somewhat leaky, loose-fitting hood. The positive pressure differential between the inside and outside of the hood keeps the flow outward, preventing toxic agent entry. The continuous-flow mode needs an unlimited supply of air, so a compressor is used instead of tanks.

In the *demand-flow* mode, air does not flow until a valve opens, caused by a negative pressure created when the user inhales. Exhalation, in turn, closes the valve. This mode has the advantage of using less air, so it is feasible with cylinders. But the disadvantage is the need for a tight-fitting facepiece. Since inhalation causes a negative pressure differential to develop, a leaky facepiece will readily draw in the toxic ambient atmosphere. In fact, if the facepiece is too leaky, the inhalation valve will fail to open, making use of the facepiece even more hazardous than not using personal protective equipment at all. For this reason, the demand-flow mode is becoming obsolete and is being replaced by the third mode: pressure demand.

The *pressure-demand* mode has features of both the continuous-flow and pressure-demand modes. As in the continuous flow, a positive pressure differential is maintained by a preset exhalation valve. Despite its advantages, the demand-flow mode still requires a well-fitting mask; use by a person with a beard is not acceptable.

Hose Mask

A hose mask is a somewhat crude form of an air line respirator. The diameter of the hose is larger than in the air line respirator, permitting air to be inhaled by ordinary lung power. A blower is sometimes used as an assist. The hose mask is declining in popularity.

Self-Contained Breathing Apparatus

In this type of respiratory protection, the user carries all of the apparatus with him, usually on his back. This has the advantage of increasing the distance the user can roam because there is no umbilical cord to drag along and perhaps sever or crush. But a disadvantage is that the large pack on the back may restrict the passage of the user through a manhole or other close passage. Engineers should consider this problem when designing manholes for vessel entry when toxic atmospheres may be a problem. Many fatalities have occurred when rescue breathing packs were rendered useless because the rescuer was unable to enter the vessel with a breathing pack on his back as Case Study 12.1 will illustrate.

CASE STUDY 12.1

CONFINED SPACE ENTRY

An employee of a zinc refinery was working in a zinc dust condenser when he collapsed. Another employee donned a self-contained breathing apparatus (SCBA) and attempted to enter the condenser to rescue the downed employee. He was not able to fit through the portal wearing the SCBA, so he removed it, handed it to another employee, and then entered the condenser. He planned to have the other employee hand the SCBA to him through the portal, redon it, and then continue with the rescue. He collapsed and fell into the condenser before he could redon the SCBA. The first employee was declared dead at the scene; the would-be rescuer died two days later. The toxic air contaminant was later determined to be carbon monoxide (OSHA fatality case history).

FIGURE 12.7

Open-circuit self-contained breathing apparatus.

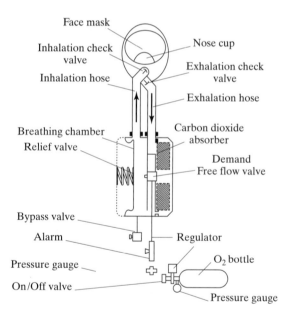

FIGURE 12.8

Closed-circuit self-contained breathing apparatus (oxygen bottle type).

The scenario described in Case Study 12.1 is not unusual. The hazards of confined space entry have prompted OSHA to promulgate a standard on this subject, to be covered in a later section.

Most self-contained breathing apparatus units at the present time are *open circuit*—that is, the exhaled breath is discharged to the atmosphere. (See Figure 12.7.) *Closed-circuit* units recycle the exhaled breath, restoring oxygen levels. The advantage of the closed-circuit types is that the pack can be much smaller and lighter per minute of maximum permissible use. Some closed-circuit types (see Figure 12.8) have a small high-pressure oxygen tank for restoring oxygen levels after carbon dioxide is removed. Other closed-circuit types use a chemical reaction to restore oxygen levels, making possible a very compact unit. The source of the oxygen is a superoxide of potassium in which oxygen is liberated by mere contact with water. The water is supplied by the moisture in the user's exhaled breath. It takes a little while for the chemical process to function and become balanced, so the user should prime the system while still in clean air before entering the hazardous zone. A hazard with the chemical oxygen-generating unit is that the potassium superoxide must be kept sealed from moisture, except for the small amount of moisture in exhaled breath. A water flooding of the potassium superoxide unit's interior is almost sure to cause an explosion. Another hazard with the chemical oxygen-generating unit is that it supplies an oxygen-rich atmosphere, and this can be a fire hazard.

Respirator Plan

The long list of available types of respiratory protection may bewilder the safety and health manager who must choose which type of protection is optimum for each application. Adequate planning for respirators and their appropriate use, maintenance, and the training of employees who will wear them is an important OSHA requirement. Expert consultants can help in this situation, but one simplifying factor is that the hazard usually dictates the choice of device or at least greatly narrows the field of choice. Devices are approved for particular concentrations of particular substances, and unapproved devices must not be used.[2] Consultants can help to select between approved respirators, considering efficiency, cost, convenience, and other factors.

Sometimes the excuse is offered that no approved device exists for a particular toxic substance. But this excuse is a poor one because the worker must somehow be protected from dangerous atmospheres. There are many different kinds of respiratory equipment, just as there are many different types of respiratory hazards. It may be true that there is no *approved* device of a particular type for a given hazard. For instance, the employer may want to use some type of respirator for a hazard that has no approved respirator as such, but another type of device *could* be used. An example would be the hazard of mercury vapor. No air-purifying device is acceptable for protection against the hazards of mercury vapor. Mercury vapor offers no warning odor or other means to detect when the canister is no longer effective. So with mercury vapor, the employer must turn to another type of protection, such as air line breathing equipment. Thus, for every hazardous atmosphere in which an employee must work, there must be an approved device of *some type* to protect the worker. If there is no approved device for the given hazardous contaminant, workers must be prohibited from the area.

Determining the approval status of a device for a particular respiratory hazard is no easy task. The approval authority is complicated and has frequently changed from agency to agency. Furthermore, the equipment approval lists are continually being updated. A practical solution to the problem is to seek the advice of the manufacturer of the equipment, for the obvious reason that the equipment manufacturers are likely to be most aware of the approval status of their own equipment. Not as worthy of trust is the local

---

[2] For approval of respirator devices, see *NIOSH Certified Equipment, Cumulative Supplement* (NIOSH 77–195), June 1977, and subsequent issues available periodically from NIOSH Laboratories in Morgantown, West Virginia.

equipment distributor. With no reflection on the character of the local equipment distributor, such persons often carry so many lines of safety equipment that it is impossible for them to be as intimately aware of the various toxic substance approval limitations of each respiratory protection device as will be the manufacturer of the given device.

Complicating the situation is the misleading nomenclature. For instance, *organic vapor respirators* are not necessarily acceptable for commonly encountered organic vapors. Methanol, for instance, is not readily absorbed by any cartridge, and only air line equipment is effective for protection against hazardous concentrations of methanol. It is unfortunate that federal regulations require a respirator to be identified as an organic vapor respirator because it has passed a certain prescribed test, even though the respirator may be useless for certain organic vapors. Workers are not even protected by warning labels that their organic vapor respirators may not be as effective for the organic vapors with which they are working. The same is true for gas masks.

Why does such a confusing and misleading system exist? Part of the problem is the complexity of organic chemistry and the myriad hydrocarbon compounds that exist. If the manufacturer attempted to label each cartridge with all the organic compounds for which it was effective, there would soon be no space on the cartridge for anything else. Any kind of shorthand coding system might confuse users in the field. Furthermore, an attempt to itemize would imply that the list was exhaustive, and the user might check no further to see whether a particular cartridge might be effective against a given substance. So again the best course of action is to check with the manufacturer for more complete and detailed lists of substances for which the organic vapor respirator cartridge is effective.

A basic principle of respirator selection is never to select a gas-absorbing respirator for use with a gas that has no distinguishing warning properties. A moment's reflection will reveal the logic behind this principle. All respirator cartridges eventually become saturated or loaded to the extent that they are no longer effective. The users will know automatically when that point is reached through their own senses of smell, taste, or perhaps an irritation or other sensory warning, depending on the properties of the dangerous gas. But if the toxic gas has no sensory warning property, the users will never know and may be subjected to dangerous exposures while wearing an approved respirator.

A good example of such misuse is use of half-mask respirators by a typical roofing company when applying expandable foam. Unfortunately, the organic substances present in expandable foam do not have adequate warning properties, and there is no air-purifying device approved for such use. Therefore, air line respirators should be used.

There is one exception to the rule that air-purifying devices cannot be used with gases that do not have warning properties capable of being sensed by the user: when canisters can be equipped with an effective end-of-service-life indicator. Currently, such canisters are available for carbon monoxide (type N canisters).

Paint, lacquer, and enamels cartridges offer protection from both the particulates and the organic solvents in the paint. Actually, the organic vapors represent the real hazard with paint, lacquer, and enamels respirators. Except for lead-based paint, the particulates in paint are not really a problem.

Welding operations present special problems because toxic gases or fumes, or both, are often encountered, while at the same time, the welder must be protected from harmful rays. Snap-on lenses are available for full-face respirators to protect against harmful rays.

Another alternative is to wear the respirator beneath the hood. This requires a special welding hood, and only a half-mask respirator can be worn under this hood.

Respirator cartridges are not like vacuum cleaner bags; they cannot be changed from manufacturer to manufacturer without approval. The cartridge and the respirator must be approved together as a unit, although it would be possible for a respirator of one make and a cartridge of another make to be approved together if the specific combination was subjected to approval testing. Public controversy over cartridge refill approvals has charged that manufacturers have monopolistically promoted safety regulations to protect sales of their own products. On the other side of the issue is the argument that cartridge fit and proper seal are critical to the effective operation of the respirator.

Personnel Screening

For jobs requiring the use of respirators, a determination should be made to be sure that persons are "physically able to perform the work and use the equipment. The local physician shall determine what health and physical conditions are pertinent." This sounds costly and time consuming, but there is a shortcut that can expedite the personnel screening process, eliminating a large majority of the problem cases. The shortcut is to use a questionnaire that can identify obvious problems before employing a person in a job that requires respirators. Any questionnaire that is used to place employees properly should be used carefully so as not to be construed as a mechanism for discrimination against the handicapped. In Chapter 4, we examined the provisions of the Americans with Disabilities Act (ADA) (ref. Public Law 101–336) that prohibit discrimination in preemployment medical examinations that work to discriminate against the handicapped. However, if the employer properly documents the job-specific requirements in occupations that require respirators and ensures that all candidates are examined for these physical characteristics, problems with ADA will be avoided.

A variety of physical conditions makes a person unfit for respirator use, whereas that person might be quite well suited to a job that does not require such use. The U.S. Nuclear Regulatory Commission lists some of the physical conditions that should be included in a questionnaire for screening workers from jobs requiring the use of respirators. Asthma or emphysema are pulmonary problems that may result in problems with respirators. If the environment requires self-contained breathing apparatus, the potential employee may not be fit for carrying the heavy equipment due to a back injury history or heart disease. Another physical problem is ruptured eardrums. A person with a ruptured eardrum actually breathes through his or her ears. Thus, a face mask respirator that does not cover the ears may offer little in the way of protection. Epilepsy is another potential problem. Epileptics are often on medications that themselves can present some hazards. Also, the possibility of a seizure presents the threat of the respirator being removed at the wrong time. Diabetics may also be on medications, and the use of respirators may interfere with such medications. These individuals, for instance, would probably be a poor choice for a rescue team. Even psychological factors such as claustrophobia may be important to consider. The employee questionnaire can be a source of protection from future liability in case physical problems are present. The questionnaire should be developed with the assistance of a physician. If physical problems or variations are revealed by the preliminary questionnaire, the employee should be examined by the physician for a final determination.

Fit Testing

An important element of an effective respirator program is fit testing or leak testing. Facial characteristics vary significantly, and fit tests are necessary to determine which particular model fits each worker best without appreciable leaking. Even the best respirators on the market will fit only a percentage of workers. Other respirators must be selected for other facial shapes.

One facial shape that is impossible to fit is the bearded one. A beard should be prohibited for any worker who must wear a fitted respirator for his safety and health. Two firemen in Los Angeles filed a discrimination suit when they were laid off because they wore beards. Their job required the possible use of self-contained breathing apparatus. The firemen lost their case. The argument against facial hair on jobs requiring respirators is well documented.

Respirator fit testing seems to imply expensive environmental chambers, but such chambers are not required. Some consultants recommend a simple plastic clothes bag suspended over the wearer's head. The subject's respirator should be equipped with an organic vapor cartridge, and the air inside the bag is contaminated with isoamyl acetate (available from the local pharmacy). If the subject smells the familiar "banana oil" odor, his or her respirator is leaking. Once fit-tested for a particular make and model of respirator, every respirator of the same make and model is acceptable for that user.

One ironic development in the field of respiratory protection is NIOSH's discovery of evidence suggesting that the substance di-2-ethylhexyl phthalate (DEHP) is a potential carcinogen. The irony is that DEHP had been used as a test agent for determining the fit of respirators!

Systems and Maintenance

For air supplied by compressors, care should be taken to select a *breathing-air* type of compressor. What to avoid is the use of the ordinary mechanical tool air compressor for this purpose. Pneumatic tools are sometimes lubricated through a system of injected lubricants into the supplied air. This, of course, would not be satisfactory for breathing purposes.

Some confusion exists over "alarms" required for compressor failure or overheating. The reason alarms for overheating are needed is that a hot compressor can deliver deadly carbon monoxide to the air line respirator user. Some breathing-air compressors have automatic shutdowns instead of alarms, and this is entirely satisfactory for some types of industrial environments. For instance, abrasive blasting atmospheres are not IDLH (immediately dangerous to life or health), and if the compressor overheats and shuts down, workers can merely take off their respirators and immediately leave the area. But in an IDLH atmosphere, a compressor shutdown could result in a fatality, and an alarm is needed.

Respirators are subject to deterioration from improper maintenance, and, in addition, they can simply become unsanitary. A routine inspection before and after each use can be performed by the users themselves. These routine inspections could include a check for cleanliness, deterioration, and obvious function.

Self-contained breathing apparatus units should have an effective monthly inspection. The regulator should be pressurized to determine whether the low-pressure warning device operates. Cost for loss of pressurized air should not be an important factor because it should take less than 50 pounds of air to perform a complete check. Tanks should be pressurized to at least 1800 psi. Both the cylinder and the regulator have a gauge. Instead of trusting the cylinder gauge, the regulator should be pressurized and the gauge verified.

The self-contained breathing apparatus unit is usually reserved for emergency use, not ordinary on-the-job use. The whole idea of emergency respirators is serious business because in an emergency, there is a high probability of IDLH atmospheres. Emergency rescue equipment in IDLH atmospheres demands high-quality equipment, carefully inspected and maintained and effectively used by trained rescue teams.

CONFINED SPACE ENTRY

One of the most dangerous jobs in industry is cleaning, repair, or maintenance that requires entry inside a tank or other confined space. The hazard is heightened by the fact that the inside of a tank is not designed for continuous occupancy; therefore, the environment is usually suspect. Since the operation is temporary, the temptation is to take chances and hope for the best. Even if the atmosphere is known to be marginally dangerous, workers are tempted to "get in and get out in a hurry" to get the job done without the time and expense of extensive and comprehensive personal protective equipment. With this frame of mind, let us examine Case Study 12.2.

CASE STUDY 12.2

HYDROGEN SULFIDE POISONING

A maintenance worker entered a sewer manhole to repair a pipe and collapsed at the bottom. A coworker, who had been observing the initial entrant, entered the manhole, lost consciousness, and fell to the bottom. A supervisor looked into the manhole, saw the would-be rescuer, and entered to attempt rescue. The supervisor became dizzy, climbed from the manhole, and passed out. When he regained consciousness, the supervisor summoned rescue and emergency services. Both the initial entrant and the first would-be rescuer died of hydrogen sulfide poisoning (ref. Federal).

By using hindsight, it seems that the second and third workers should have known better than to enter the confined space after the first worker was overcome. However, Case Study 12.2 is not an isolated incident. Double- and even triple-fatality incidents of this type are fairly common. In the heat of the emergency, there is a strong tendency to try to save the victim, and somehow our reasoning process tells us that what happened to the first worker would not happen to us. Apparently, we think that we will be more alert to the symptoms than was the first victim, and we will get out quickly as soon as we become aware that we are falling into the same fate.

OSHA has taken a keen interest in this hazard, and for many years, it collected data, opinions of industry and labor representatives, and suggested wordings for a standard specifically addressing confined space hazards. Meanwhile, it continued to investigate such fatalities and cite employers, usually reverting to the General Duty Clause of the OSHA law. In 1993, OSHA promulgated a confined space standard that crystallized industry thinking regarding what must be done to prepare for and avoid the hazards of confined spaces.

Identifying the Hazard

Confined spaces have more hazards than most people think. The principal hazard is the atmosphere that the worker breathes inside the space, but this is by no means the only problem to consider. Some confined spaces represent a mechanical hazard, such as the nightmare of a descending and ever-tightening constricting space that may trap the worker, making any movement to escape only a further aggravation of the problem until the worker is hopelessly ensnared in a space of suffocating dimension. The hazard of *entrapment* in such places as silos, tapered storage bins, hopper–feeders, and cyclone collectors (see Figure 10.5) is real and even commonplace in agricultural and material processing industries.

Another hazard that has nothing to do with the atmospheric quality in a confined space is *engulfment*. Sand, grain, and other granular dry solid material can have fluidlike properties, and a person can become effectively captured or surrounded by it while sinking deeper with every movement. Death can come in at least two ways: The worker can breathe the dust or other material particles and the breathing passages can become blocked, or the worker may even be crushed by the weight of the material closing in around him or her. See Case Study 12.3.

CASE STUDY 12.3

SAND ENGULFMENT

Two Ohio foundry employees entered a sand bin to clear a jam. While they were working, sand that had adhered to the sides of the bin began to break loose and fall on them. One employee quickly became buried up to his chest, just below his armpits. The other employee left the bin to obtain a rope, intending to use it to pull his coworker out of the sand. He returned to the bin, tied the rope around the partially buried employee and tried to pull him free. He was unsuccessful. During his attempted rescue, additional sand fell, completely covering and suffocating the employee who had been only partially buried (ref. Preamble).

Although engulfment is a serious hazard to be considered, simple oxygen deficiency (less than 19.5% oxygen in breathing air) is likely the biggest killer in confined spaces. Oxygen deficiency often arises from chemical processes that react with oxygen in the air, such as fermentation, combustion, and even corrosion. Sewers and sewer processing facilities are often found to be oxygen deficient. It is ironic that *inerting*, a procedure intended for enhancing fire safety, causes a different hazard—oxygen deficiency. Nitrogen or other inert gas is used to displace oxygen in an atmosphere that might have dangerous concentrations of flammable gases or vapors. Too little oxygen is a hazard, but so is too much. Oxygen has a density slightly heavier than normal breathing air, so oxygen enrichment (oxygen content greater than 23.5% in breathing air) can present problems in such confined spaces as missile silos. In Chapter 16, we shall see how this hazard led to one of the worst industrial accidents in the history of the United States. Wherever welding is done, and it is often done in confined spaces, the possibility of fires being generated in oxygen-enriched spaces must be considered.

Earlier in this chapter, we considered the usual definition of the term *IDLH*; now, with respect to confined spaces, we must add another facet to this definition—the problem of escape. Thus, even if an atmosphere has no immediate effects on life or health, if it temporarily paralyzes the worker or impairs his or her ability to escape, it becomes IDLH for the confined space.

Isolation of the Space

It is obvious that a space that normally contains a liquid, especially a liquid that issues toxic vapors, must be removed from the tank or other confined space before entry. In addition, it is elementary that piping valves that normally deliver the dangerous liquid to the space must be closed. However, the hazard is more insidious when a confined space is involved. Even though a pipe valve may be closed, if there is high pressure behind the valve, there may be a small amount of leakage or *bleeding* into the space to be entered. This has led to a recognized safety procedure known as *double block and bleed*, described further in Figure 12.9.

Another practice that accomplishes the same objective as double block and bleed is to sever or separate the line and misalign it to physically break continuity between the

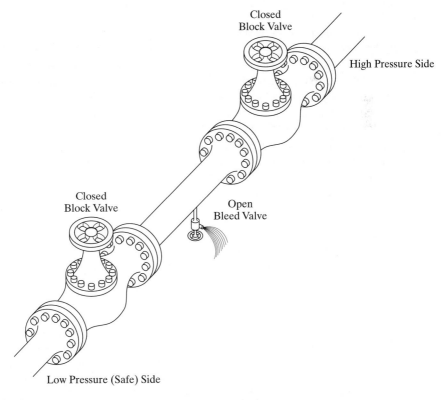

Closed
Block Valve

High Pressure Side

Closed
Block Valve

Open
Bleed Valve

Low Pressure (Safe) Side

FIGURE 12.9

Double block and bleed.

confined space and the hazardous material. Another solution is *blanking* or *blinding*, which means the absolute closure of a pipe, line, or duct by the fastening of a solid plate that completely covers the inside cross section of the pipe and is capable of withstanding the maximum pressure of its contents without leakage.

One of the most hazardous operations in confined spaces is welding. Often, tanks must be entered for welding repairs; and the potential for igniting hazardous atmospheres when welding in confined spaces adds an additional hazard beyond all the other hazards associated with confined spaces. Because welding has special hazards and is an important subject from both a health and a safety perspective, Chapter 16 is devoted entirely to welding processes and associated hazards including confined spaces.

HEAD PROTECTION

A primary symbol for OSHA, corporate safety departments, and just about anything related to occupational safety and health is the familiar profile of the "hard hat." So important is the symbol that many zealous safety managers have established sweeping hard-hat rules throughout large general work areas. There is nothing wrong with such rules if a genuine hazard exists. But when workers sense that no hazard exists and that the hard-hat rule exists as a promotional device or for window dressing, they often show their defiance by refusing to wear the hard hat.

Hard-hat rules should be carefully formulated with ample consideration for the consequences both ways. Once it has been decided that a hard-hat rule is necessary, the safety and health manager should take steps to ensure its implementation. The evidence that was used to prove the need for the hard-hat rule should be compiled into organized training packages to convince workers. After training and launching of the implementation phase, follow-up checks should be used to ensure that the rule is being followed. Corrective steps should be taken to overcome individual violations of the rule, including disciplinary actions if necessary.

Hard hats seem to have won wider acceptance than hearing protection. Besides being a symbol for occupational safety and health, the hard hat has become a symbol for rugged, physical jobs. This image has appealed to males for centuries and is becoming an increasingly appealing image for the female worker also. Management personnel have also coveted the image conveyed by the hard hat. A hard hat worn by a manager seems to imply that the manager is well founded in the operations of the enterprise and is action oriented, doing more than talking on the telephone, attending meetings, and sitting behind a desk.

MISCELLANEOUS PERSONAL PROTECTIVE EQUIPMENT

Safety Shoes

Safety shoes are a more expensive undertaking than hard hats because safety shoes are worn out faster and are more expensive per item. Employees may buy their own shoes at attractive discounts in some arrangements, and this encourages actual use. Safety shoes come in a wide variety of appealing styles, and employee resistance to wearing safety shoes is largely a thing of the past.

The safety and health manager is usually saddled with the decision as to which jobs require safety shoes and which do not. Although applicable national standards are explicit about the design and construction of safety shoes, as with almost all personal protective equipment, the decision of where such shoes must be worn is left up to either the user or management.

One place safety shoes are clearly needed is on shipping and receiving docks. This should be obvious, but there has been some legal controversy over this issue. The courts have settled the question, and safety and health managers should ensure that shipping and receiving dock personnel wear safety shoes.

Protective Clothing and Skin Hazards

Occupational skin disease, especially contact dermatitis from irritants to the skin, represents a significant number of all occupational diseases reported. The safety and health manager should be alert to several skin hazard sources, such as welding, special chemicals, open-surface tanks, cutting oils, and solvents.

Most welders know the value of heavy-duty protective aprons and flameproof gauntlet gloves. Leather or woolen clothing is more protective than cotton from a burnability standpoint. Nomex™ is a treated flame-retardant fabric.

Another concern for protective clothing is chemical exposures from open-surface tanks. Gloves must be impervious to and unaffected by the liquid being handled and long enough to prevent the liquid from getting inside. If the gloves are not long enough, they can be more hazard than benefit. Many workers' hands have become more irritated than their unprotected arms simply because the gloves they were wearing permitted liquids to get inside, turning the gloves themselves into dip tanks for the hands!

Discussion of personal protective equipment for open-surface tanks is not complete without mention of *chrome holes*, an ominous-sounding affliction that comes by its name honestly. Vapors and mists from chromium plating tanks can cause open ulceration, particularly in moist, tender parts of the skin. The inner septum of the nose—that is, that part that divides the nostrils—is particularly susceptible to ulceration from chromic acid, sodium chromate, and potassium dichromate. In plating plants, it is not unusual to find workers whose nasal septums have been completely destroyed by these chromium compounds. Proper ventilation and engineering controls are the most appropriate means of prevention. Worker education and regular examination can also be helpful, including periodic examination of the nostrils and other parts of the body for workers exposed to chromic acids. As with other hazards, the last line of defense would be personal protective equipment in the form of respirators with cartridges to remove the dangerous chromic acid mists.

Easily the most common hazards among skin irritants are the cutting oils used in metal machining operations. Cutting oils are useful and sometimes essential to lubricating the tool, reducing cutting temperature, removing chips, permitting a higher-quality cut, and lengthening tool life. But cutting oil is not always required and is even not desirable in some situations. The manufacturing engineer usually makes the cutting oil decision for the various manufacturing processes, but there is no reason that the safety and health manager should not have some input in this decision, especially since it affects safety.

Cutting oils are basically of two types: natural and synthetic. The natural oils are petroleum based and are the chief culprits in a very common industrial skin disease: oil folliculitis. Oil folliculitis is basically a clogging of the hair follicles in the skin, which results in acnelike lesions. The synthetic oils are easily recognized by their familiar milky white appearance and have gained widespread use. Although the synthetic oils are not likely to cause folliculitis, they do have a nasty reputation for becoming contaminated with bacteria, presenting the hazard of skin infections. If antibacterial agents are added, these, too, can be skin irritants. Technology is working to improve cutting oils, but the job is not complete. It appears that personal protective equipment is still in order.

Protective barrier creams for the skin are an alternative to gloves or protective clothing, but these creams are by no means a panacea. The creams must be removed and reapplied at least every break, every lunch hour, and every shift. Applying and reapplying creams cost labor time as well as the cost of the creams themselves. Also, creams are not considered as effective as gloves, even when properly applied. Creams can be used, however, when the requirements of the job make gloves infeasible.

A chief concern of the safety and health manager with regard to protection of the skin is the use of various solvents within the plant. Solvents are essential for removal of grease and cutting oils, and herein lies another reason for doing without cutting oils if practical. A familiar solvent is trichloroethylene, and it is bad practice for workers to wash piece parts in trichloroethylene with their bare hands. Alternatives would include the use of wire baskets for handling the parts in the solvent, or perhaps simply substituting soap and water for trichloroethylene in some situations. For the most part, soap and water will not be effective against the oils and greases encountered, although in some situations, soap and water washing is effective. The safety and health manager is not doing the job unless he or she seeks out these situations and calls them to the attention of management and engineering. When alternative solvents, wire baskets, and other engineering controls are infeasible, personal protective equipment, such as gloves, is in order.

It seems appropriate to conclude a discussion of skin hazards, gloves, and protective clothing by mentioning one of the simplest personal protective measures of all—personal cleanliness and hygiene. Workers who are fastidious about washing their hands and bodies have frequently been found to enjoy a lower incidence of skin disease. It is easy to understand why this is so. With skin irritants, the amount of injury is generally directly related to the duration of exposure, other factors being equal. It is easy to forget how effective soap and water are in removing injurious elements of all kinds.

What can the safety and health manager do to motivate workers to adopt good habits of washing and personal hygiene? The obvious answer is training and motivational reminders in the form of signs and posters around the plant. But this obvious answer is not the only answer and perhaps is not even the best answer. The safety and health manager can attempt to influence the selection and layout of convenient, well-maintained, pleasant restrooms that will boost employee morale while encouraging workers to wash regularly. There is nothing more discouraging to employees than an encounter with a washroom without warm water, soap, or towels.

FIRST AID

The safety and health manager will frequently be responsible for the first-aid station and may supervise a plant nurse. The first-aid station may satisfy several additional functions besides providing immediate care for the injured. The first-aid station is often used for medical tests, screening examinations, and monitoring of acute and chronic effects of health hazards. Also, the plant nurse or other first-aid personnel may be responsible for performing some of the recordkeeping and reporting functions discussed in Chapter 2.

One adequately trained first-aid person is required in the absence of an infirmary, clinic, or hospital "in near proximity" to the workplace. No one seems to be able to determine authoritatively what constitutes *near proximity* in this context. Various interpretations around the country have been compared on this point, and the opinion has for the most part varied from 5 to 15 minutes of driving time. The interpretation has sometimes depended on whether the route to the hospital crosses a railroad track. If the workplace is not itself a hospital or clinic or is not directly adjacent to one, the safety and health manager is advised to be sure that at least one, preferably more than one, employee is adequately trained in first aid.

A first-aid kit or first-aid supplies should be on hand, and the safety and health manager should seek a physician's advice regarding the selection of these materials. Unfortunately, medical doctors are hesitant to give such advice, probably because they fear subsequent involvement in litigation should an accident occur for which adequate materials are not available. Safety and health managers should do their best to obtain such advice and then document what was done to obtain information.

Another first-aid consideration is the provision of emergency showers and emergency eyewash stations on job sites where injurious corrosive material exposure is a possibility. Almost everyone has seen the deluge-type shower, which is activated by grabbing and pulling a large ring attached to a chain that activates the valve. Eyewash facilities are similar to a drinking water fountain in which two jets are provided, one for each eye.

Noting the cost and permanence of a conventional eyewash facility, some enterprising and innovative individuals have marketed a simple plastic water bottle and plastic tube hanging on a frame that identifies its purpose as an emergency eyewash. Such a solution is probably not what the drafters of the standard had in mind, but the little water bottle is not without merit. For one thing, water bottles can more easily be deployed to the best spots to permit instant use in case of an accident. Most workplace layouts change rather frequently, making the permanent eyewash installation cumbersome, expensive, and often in the wrong place. Furthermore, the water-bottle approach permits convenient introduction of antidotes or neutralizing agents for specific corrosive materials. However, woe be to the injured who uses an antidote for acid when the exposure was caustic, or vice versa. Also, the volume of water in the bottle is usually limited to 1 or perhaps 2 quarts. This may not be enough water if a first-aid manual specifies flushing the eyes "with copious amounts of water for 15 minutes."

CONCLUSION

Reflection on this chapter leaves the impression that the proper provision and maintenance of personal protective equipment and first aid is not an easy task. Even after the proper equipment has been selected and procedures set up for maintenance, employees must be trained and disciplined to use the equipment properly. Inspectors can easily recognize simple violations, such as failure of individual employees to wear prescribed personal protective equipment. One pitfall for the safety and health manager is the blanket specification of personal protective equipment when its necessity is marginal. Such blanket specification is a trap that will lead to employee apathy and subsequent violation of the rule.

Returning to the principle that personal protective equipment is a last resort method of protecting the workplace, engineering the hazard out of the workplace is definitely preferred. Personal protective equipment seems to be the easy and less costly way out, but an examination of the principles and pitfalls of personal protective equipment is a strong motivation toward engineering controls.

EXERCISES AND STUDY QUESTIONS

12.1 Who is responsible for determining whether personal protective equipment (PPE) is needed in a particular workplace?

12.2 Besides needs assessment, what other general requirements for the employee use of personal protective equipment are required?

12.3 What conditions might precipitate the need for retraining in the use of personal protective equipment?

12.4 Explain possible hazards associated with attaching a safety line to a worker's belt to protect the worker from an accidental fall.

12.5 Identify resources to assist the employer in determining when and which types of personal protective equipment would be needed for a particular work site.

12.6 Under what circumstances is the employer required to train workers in the proper use of personal protective equipment?

12.7 One limitation of personal protective equipment is adequacy for the hazard for which it is intended. What is another limitation of such equipment for which the user needs training?

12.8 How might the employer be called upon to prove that employees have been properly trained in the use of personal protective equipment?

12.9 How effective is the use of cotton balls in the ears to protect against the hazard of workplace noise?

12.10 Identify the most effective form of PPE for hearing protection in the most severe noise environments. What additional benefit, besides hearing protection, is offered by this type of PPE?

12.11 Roofing workers often apply expandable foam materials using half-mask, canister respirators. Why is this basically an incorrect and unsafe practice?

12.12 What is a chemical oxygen-generating unit? Under what conditions will it explode?

12.13 As a safety engineer, under what circumstances would you recommend a closed-circuit respirator?

12.14 In an IDLH situation, would you favor demand-flow mode or pressure-demand mode if both are feasible choices? Why?

12.15 In a not-so-unusual accident, two workers were killed when an employee was cleaning out a tank accessed by a manhole. A second employee saw the first employee collapse. While trying to rescue the first employee, the second employee was also overcome, and both died. What preventive measures can you suggest to prevent accidents of this type?

12.16 What are the two basic types of "safety lenses?" Which is more durable?

12.17 What is the argument against requiring safety glasses and hard hats throughout all areas of the plant, including areas in which they are not needed?

12.18 Name some jobs for which eye protection should be provided.

12.19 Why are "organic vapor respirators" not to be trusted to protect against all organic vapors?

12.20 In an industrial plant, a sign reads "PREVENT ACCIDENTS: WEAR HARDHAT." What principles of personal protective equipment does this sign fail to recognize?

12.21 Why is the provision of personal protective equipment not a very satisfactory solution to the problem of protecting workers?

12.22 What are the hazards of employee-owned personal protective equipment being brought onto the premises at work?

12.23 In what way have undersized manholes for vessel entry resulted in a large number of multiple fatalities in the United States?

12.24 Why must a safety belt for fall protection be much stronger than is required to support the wearer's body weight?

12.25 Name several alternatives to washing piece parts in trichloroethylene by workers placing their bare hands in the solvent.

12.26 Explain the multiple-fatalities hazard associated with entry into confined spaces.

12.27 Describe the principal hazards of confined spaces.

12.28 What is the phenomenon of engulfment? How does death come to the victim?

12.29 Describe the typical hazard mechanism for confined spaces that contain mechanical hazards.

12.30 The safety procedure of inerting is intended to alleviate what hazard? What other hazard does it sometimes create?

12.31 What is the principal hazard associated with oxygen enrichment?

12.32 How do confined spaces impact the definition of *IDLH*?

12.33 Describe some procedures for isolation of a confined space.

12.34 Name one type of commonly encountered dangerous atmosphere for which a gas mask will not help.

12.35 Explain the peculiar aspects of the effects of IDLH exposures to materials such as hydrogen fluoride and cadmium vapors.

12.36 Explain the problems that can develop with a partial program of respiratory protection when respiratory protection is not really required.

12.37 Explain the term *double block and bleed*—that is, what is *blocked* and what is *bled* and why?

RESEARCH EXERCISES

12.38 Examine the safety of service pits for display waterfalls, such as those seen in shopping malls. Determine whether any accidents or incidents have been reported and what hazards have been found to exist.

12.39 Some workplace environments call for the use of air-supplied suits supplied by either self-contained breathing apparatus units or by supplied air lines. An emergency situation occurs when an "air-off" condition occurs, especially when the worker is still in the dangerous atmosphere. Examine this hazard and determine whether any guidance has been published to deal with it.

12.40 Case Study 12.3 discusses the hazards of entrapment encountered on entering sand bins. Research similar hazards encountered in grain bins. How quickly does a victim, standing in the bin of grain, become trapped when an unloading auger starts at the bottom of the bin?

How quickly does the victim become completely immersed in the sinking pile of grain? How much does a 1-foot-deep pile of corn lying on a typical man, 6 feet tall and lying down, weigh (ref. Loewer)?

12.41 Study the various hazards of methane in confined spaces. How does concentration enter the picture? Compare health and safety aspects.

12.42 NIOSH studied the hazards of confined spaces in the late 1980s, prior to OSHA's promulgation of the confined space standard. How did NIOSH define *confined space*? Identify the three "classes" of confined spaces as seen by NIOSH. What three employer "problems" with respect to confined spaces were identified by NIOSH? What percentage of confined space fatalities consists of "would-be rescuers?"

12.43 What industries did OSHA exempt from coverage under the "permit-required" confined space entry standard? What was the rationale behind these exemptions?

12.44 Find a published estimate of the number of telecommunications manholes in the United States. Are these manholes "confined spaces"? Why or why not? GTE is a major company in this industry. Approximately how many GTE employees enter manholes each year? How many times per year on average do these employees enter manholes? What OSHA standard regulates safe entry into telecommunications manholes?

STANDARDS RESEARCH QUESTIONS

12.45 Study the OSHA General Industry standards to compare enforcement activities surrounding respiratory protection as compared to general personal protective equipment.

12.46 Examine the OSHA General Industry standards for "medical services and first aid." From enforcement statistics for these standards, comment on how seriously OSHA takes this subject. Justify your answer with statistics gleaned from the NCM database.

CHAPTER 13

Fire Protection

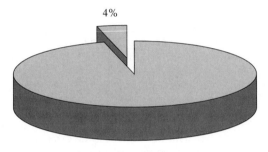

4%

Percentage of OSHA
General Industry citations
addressing this subject

This chapter deals with perhaps the oldest topic in occupational safety and health. But modern developments in the field of fire safety place it in a very dynamic phase. More so than for other categories of safety and health, fire safety presents the safety and health manager with a wide variety of alternatives for dealing with hazards.

Industrial fire protection standards, when dealing with OSHA, formerly consisted of little more than dealing with fire extinguishers, their selection, placement, marking, inspection, and maintenance. True, there were a few obscure standards about such topics as "standpipe and hose systems," but almost all of the activity centered around fire extinguishers. Today, the field of industrial fire protection is much more sophisticated, recognizing such alternatives as emergency action plans, fire prevention, fire brigades, fire alarm signaling systems, fixed extinguishing systems, and automatic sprinkler systems. Rather than blindly following old specific standards for fire extinguishers, safety and health managers now have the opportunity to explore alternative strategies or combinations of strategies to accomplish the most cost-effective method of fire protection for their own circumstances. Even the federal standards have been changed to delegate such decision-making authority to industrial managers.

It is easy to oversimplify fire protection as meaning only fire extinguishment, but it really encompasses three fields: fire prevention, fire suppression, and personal protection (escape). Much of the criticism of old fire safety standards has been that they emphasized fire suppression and fire extinguishers, which may not be the safest alternative in some fire situations. Some firms wanted no extinguishers at all and desired to instruct employees to escape without attempting to extinguish fires. Their argument was that fire extinguishers are mainly for property protection, whereas the safest thing for workers to do is escape. The current standards recognize this rationale together with several shades of combination strategies. Before discussing these strategies in detail, some facts about fires should put the fire protection problem in perspective.

INDUSTRIAL FIRES

According to the National Safety Council (ref. Injury), fire is a leading cause of accidental death. One might think that the high-technology fire detection, protection, and suppression systems used in the United States would make possible one of the lowest fire death rates in the world. Exactly the opposite is true! The United States has one of the worst fire death rates in the industrial world, about 16 deaths per million people.

The U.S. fire death rate is a bad one, but anyone who attempts to pin the responsibility on industry is in for a disappointment. According to NFPA statistics (ref. Fire), 80% of U.S. fire deaths are residential, not industrial. Workplace fatality statistics show that approximately 3% of all workplace fatalities are attributed to fire. Workplace violence accounts for many times the number of fatalities as does fire. These facts strengthen the conclusion that industry, more than most other elements of our society, has done a great deal to control fire hazards. Considering the incredible exposure to flammable liquids in refineries and chemical plants and the billions of labor hours spent in industrial plants every year, it is amazing that the number of fire deaths in all industrial plants is no more than the number who die in fires in taverns and prisons. In fact, more people died in a single fire in a supper club in Kentucky in 1977[1] than in all the multiple-death industrial fires in the nation in that year and the two succeeding years combined.

The most dangerous industries from a fire hazard standpoint are mines, grain elevators, grain mills, refineries, and chemical plants. The fire fatalities from these four industries dwarf the total for all remaining industries combined. For general manufacturing industries, the number of fire fatalities is extremely low. A major tragedy occurred in 1991 in Hamlet, North Carolina, where 25 people lost their lives in 35 minutes when a fire swept through the Imperial Foods poultry processing plant. An even worse tragedy occurred in New York in 1912, in the Triangle Shirtwaist Company fire, which claimed 145 lives. The Triangle Shirtwaist fire received so much publicity that it had a profound impact on regulations to control industrial fires in the United States, with the result that an excellent fire record was achieved in the decades following this terrible tragedy. The Imperial Foods tragedy has again reminded the public of the consequences of locked exits and complacency about fire hazards.

FIRE PREVENTION

The best way to deal with fires is to prevent their occurrence, just as engineering controls were found in Chapter 12 to be preferable to personal protective equipment. Effective fire prevention requires anticipation of fire sources. Each facility is different

---

[1]Beverly Hills Supper Club fire, Southgate, Kentucky, 165 lives lost.

and requires an individual analysis of potential fire sources. Once the hazards are identified, decisions must be made as to who has responsibility for controlling the hazards. These decisions should be documented in a fire prevention plan.

A principal cause of industrial fires is overheated bearings or hot machinery and processes. Another cause is clogged or dirty ventilation filters or ducts, especially when the clogging material is a flammable or combustible air contaminant. Some of these causes can be averted by adopting an effective preventive maintenance program. Such a program, while decreasing the likelihood of fire, may also extend the life of equipment. The safety and health manager may see an opportunity in this strategy to save production costs while furthering the cause of fire safety.

Another component in a fire prevention plan is a strategy for housekeeping. Accumulation of combustible dusts in grain elevators and paint residues in spray painting operations are good examples of how poor housekeeping can contribute to fire hazards. Even ordinary combustible paper and material waste can be a fire hazard.

EMERGENCY EVACUATION

Using the *escape strategy* for dealing with fires or other emergencies, the employer must prepare a written *emergency action plan*. The emergency action plan concept has been around for many years for hospitals, schools, and institutions, and more recently, has been extended to industries in general.

Alarm Systems

Crucial to an emergency action plan is an employee alarm system. But alarm systems are not as simple as they may seem. There are searching questions that must be asked: Will persons recognize the signal as a fire alarm? and What about deaf or blind employees? Audible, visual, and tactile systems must be considered, or perhaps combinations of these systems. In small workplaces, even direct voice communication may be the best fire alarm medium. Public address systems may be used in larger facilities, but the system should provide that emergency messages take priority.

System reliability is important to fire alarms because a failure within the system may not be immediately obvious. Some sophisticated systems have built-in monitor circuits to supervise reliability. Such systems do not need testing as often as do simple alarm systems, which have no such monitoring circuits. When repairs are being made, some type of backup system is needed to provide continuous protection. The backup system might even employ "runners" or telephones or other informal systems, but the safety and health manager should document what backup system is in place. The importance of both alarm system testing and a backup system when the primary system malfunctions is illustrated in Case Study 13.1.

Fire Detection Systems

Smoke alarms and other detection devices may be used to trigger the alarm system. However, it should be noted that automatic smoke alarms are not mandatory in American industries in general. Even manual or visual systems can be considered alarm systems.

If automatic detection systems are employed, care must be taken to maintain and protect the equipment. Most detection systems are delicate instruments and will not withstand the rigors of the industrial environment. Conditions to be considered are dust, corrosive atmospheres, weather exposure, heat from processes, and mechanical damage.

CASE STUDY 13.1

HIGH-RISE OFFICE FIRE EVACUATION

A ten-story office building was equipped with a cafeteria in which a fire started in an exhaust flue over the cooking surface. The cafeteria worker who discovered the fire properly actuated the fire alarm switch to signal evacuation of the building, but the alarm system failed to operate. A disaster was averted by telephoning each of the ten floors, and a successful evacuation of the building was effected. In the aftermath of the emergency, a new system was installed consisting of both audible and visual alarms, and, in addition, an emergency telephone was installed on each floor.

People are sometimes reluctant to sound fire alarms, with tragic consequences. Hotel managers are particularly unwilling to alarm occupants. The typical response from a hotel front desk when an occupant calls in a fire alarm is to send a bellman to investigate. It is rational to consider the hazards of panic when a fire alarm is sounded. However, this rationale is sometimes permitted to be an excuse for failing to take action.

FIRE BRIGADES

Some firms may adopt a strategy in which employees are organized into brigades to fight fires themselves. Such strategies should be carefully scrutinized because in the scramble to protect property, these fire brigades can be a danger to employees.

Employee Fitness

Volunteering to join the fire brigade is not sufficient to qualify the worker to fight fires. Conditions that may be hazardous include heart disease, epilepsy, or emphysema. Other conditions, such as ruptured eardrums or the wearing of a beard, may make the use of respiratory equipment ill advised. The safety and health manager should be sure that the fire brigade volunteers are screened, and a physician's certificate may be necessary for questionable cases. Volunteers unfit for interior structural firefighting may be used in other tasks.

Firefighter Training

Many states have fire training academies, and safety and health managers should find out what schools and academies are available for their fire brigade members. Interior structural firefighting is more demanding, and fire brigade members assigned to such tasks should be trained at least quarterly. Other fire brigade members should be trained at least annually. Also, firefighting equipment to be used by the fire brigade should be inspected annually, and fire extinguishers should be inspected monthly.

Protective Clothing and Apparatus

If the firm elects to have fire brigade members fight interior structural fires, protective clothing and respirators must be provided. This includes protective boots or shoes, fire-resistant coats, gloves, and head, eye, and face protection.

One concept stressed for self-contained breathing apparatus units is the mode for air flow into the mask. At this point, the reader may want to review the three modes of respirator air flow discussed in Chapter 12. The preferred mode for firefighting is one of the positive-pressure types: pressure demand or continuous flow. The only valid argument for using simple demand flow is that this mode permits exposures of longer duration for a given charge. If the employer believes that the demand-flow mode is essential, quantitative fit testing is necessary for each firefighter.

FIRE EXTINGUISHERS

Fire extinguishers are still the most effective method of immediately controlling a very local fire before disastrous consequences ensue. The safety and health manager needs to understand the various fire classes and the type of extinguishers appropriate for each class.

Fire Classes

The fire protection field classifies fires into four categories. Application of the wrong extinguishment medium to a fire can do more harm than good.

Table 13.1 describes the four classes of fires, sample appropriate extinguisher media, and the maximum travel distance specified for extinguishers for each type of fire. Liquefied petroleum gas (LPG) fires, although technically Class B, are really not adequately addressed by any of the four classifications. Such fires are extremely dangerous,

TABLE 13.1   Four Classes of Fires and Appropriate Extinguishing Media

Fire class	Description	Example extinguishing media	Maximum OSHA-authorized travel distance to nearest extinguisher
A	Paper, wood, cloth, and some rubber and plastic materials	Foam, loaded stream, dry chemical, water	75 feet
B	Flammable or combustible liquids, flammable gases, greases, and similar materials, and some rubber and plastic materials	Bromotrifluoromethane, carbon dioxide, dry chemical, foam, loaded stream	50 feet
C	Energized electrical equipment	Bromotrifluoromethane, carbon dioxide, dry chemical	Nonspecific maximum; distribute "on the basis of the appropriate pattern for existing Class A or B hazards"
D	Combustible metals such as magnesium, titanium, zirconium, sodium, lithium, and potassium	Special powders, sand	75 feet

and fire extinguishers are not appropriate for their control. LPG fires should be extinguished by professional firefighters using powerful water-spray systems.

The key to determining whether an extinguisher is appropriate for a given class of fire hazard is to check the approval marking on the extinguisher itself. Some types of extinguishers have been found to be hazardous and are forbidden regardless of prior approval markings. These types are listed in Table 13.2. Some extinguishers may be approved for more than one classification of fire. These multipurpose fire extinguishers typically employ a dry chemical medium. Although dry chemical extinguishers are growing in popularity, they are not a panacea. Expensive equipment such as computers can be fouled or even ruined by the application of a dry chemical extinguisher when a CO_2 extinguisher would have been satisfactory. Foam or water extinguishers may also be cheaper for the more ubiquitous Class A fires.

Inspection, Testing, and Mounting

OSHA has abandoned the rule requiring fire extinguisher tags indicating their inspection status, but still requires the employer to maintain records of annual maintenance for each fire extinguisher for up to one year after the last entry or the life of the shell, whichever occurs earlier. In addition, a visual inspection is required monthly. Many employers have found it expedient to retain the tag system even though it is no longer required. Thus, when anyone requests to see the inspection record for a given fire extinguisher, it is immediately available on the tag attached to the extinguisher.

In addition to the monthly and annual inspections, fire extinguishers must receive a hydrostatic test according to a prescribed test schedule. Fire extinguisher shells deteriorate from mechanical damage or corrosion and may be unsafe for containing pressures inside. The hydrostatic test places the extinguisher under a test pressure to determine whether it can safely contain the pressures to which it will be subjected in use. The test has technical specifications and must be done by a trained person using suitable equipment and facilities. The safety and health manager invariably leaves the hydrostatic tests for the professional fire extinguisher service to perform.

Another big OSHA enforcement issue used to be the mounting height and identification of mounting of fire extinguishers. OSHA citations for mounting, identification, and inspection tagging of fire extinguishers once amounted to more than half of all OSHA violations named in fire protection citations. These requirements have all been eliminated or drastically changed by rewording them in performance language. The current standard permits the employer the latitude of mounting extinguishers high on the wall, out of the way of forklift truck traffic, accessing them by rope and pulley.

TABLE 13.2   Forbidden Fire Extinguishers

---

1. Carbon tetrachloride
2. Chlorobromomethane
3. Soldered or riveted shell self-generating soda acid or self-generating foam or gas cartridge water-type portable fire extinguishers that are operated by inverting the extinguisher to rupture the cartridge or to initiate an uncontrollable pressure-generating chemical reaction to expel the agent

---

Source: Code of Federal Regulations, 29CFR 1910.157.

The employer may select any convenient mounting scheme provided that the extinguishers are readily accessible without subjecting employees to possible injury.

Training and Education

A facility looks well equipped and protected when fire extinguishers are placed about the workplace, readily accessible for use in an emergency. But the appalling reality is that few employees know how to use a fire extinguisher effectively, and some would even be afraid to use the extinguishers if they knew how. This is especially true of fire extinguishers or hose systems located behind glass doors. For most people there is a great reluctance to break glass even in an emergency. A study of the nursing staff conducted in an Ohio hospital revealed that most knew nothing about, and were afraid to use, fire extinguishers. Accordingly, the safety manager in charge set out to train the nurses to use fire extinguishers in an emergency. A hospital bed was taken outdoors and set afire to provide a simulated fire emergency for the nurses to practice on. The field of industrial safety has now recognized the need for fire extinguisher training, and such training has been accepted as standard for general industry. Training is required on initial employment and at least annually thereafter.

STANDPIPE AND HOSE SYSTEMS

Some employers choose to install standpipe and hose systems for firefighting, and these systems can be employed in lieu of general distribution of fire extinguishers in most cases. Standpipe and hose systems come in various ratings or classes. Large diameter hose ($2\frac{1}{2}$ inch) is difficult or even dangerous to handle and is intended for professional firefighters. Large diameter hose is designated as Class I and is exempted from coverage by the OSHA General Industry standard for standpipe and hose systems. Smaller diameter standpipe and hose systems are for employee use, and federal enforcement authorities take an interest in the adequacy of the equipment and in its maintenance and use.

Equipment

Some standpipe and hose systems can be quite ancient. These old systems can usually be retained as long as they are serviceable and meet annual test requirements. But when replacements are made, new equipment should conform to all current standards. Examples of modern system changes are as follows:

1. *Shutoff*-type nozzles
2. *Lining* for hose
3. Dynamic pressure *minimums at the nozzle*
4. Hydrostatic *testing* upon installation

The water supply for standpipe and hose systems can be provided either by elevated water-supply tanks or by pressure tanks. The supply must be sufficient to provide 100 gallons per minute flow for at least 30 minutes. This calculates to 3000 gallons per use period, but remember that it takes considerably more than 3000 gallons to provide the head pressure to maintain the 100-gallon-per-minute flow throughout the 30-minute period. One seemingly clever idea is to forget about standpipes or pressure tanks and connect hose systems to city water supplies. But the flaw in that theory is the adequacy (pressure and flow) of the supply, which usually cannot meet the 100-gallon-per-minute flow requirement.

Maintenance

The ancient systems described earlier can create some unpleasant surprises when the hose is deployed for use. Hemp and linen systems are especially subject to deterioration. After hanging on the rack for many years without use, the hose may break apart or disintegrate when taken off the rack for use in an emergency. Hose systems should be checked annually and after each use. In the case of hemp or linen hose systems, the hose must be reracked using a different fold pattern.

AUTOMATIC SPRINKLER SYSTEMS

Automatic sprinkler systems present a paradox because they affect employee safety, but are usually installed principally to protect property and to lower insurance rates. If such a system is *voluntarily* installed by the employer to protect property, should it be required to meet personal safety standards? And if an existing system does not meet current standards, should it be dismantled and withdrawn from use? This would hardly be in the interest of safety.

Sometimes a good sprinkler system is installed only to be made ineffective by incorrect usage of the space protected. One error is to allow the sprinkler heads to become fouled by materials such as paint spray residues. If a spray area is protected by an automatic sprinkler system, a good way to protect the sprinkler heads is to cover them with paper bags. If a fire does occur, the paper bags will either burn away or be washed away by the water spray so that they do not interfere with the fire suppression action of the sprinklers.

Another error with automatic sprinklers is to stack material too close to the ceiling. This interferes with the distribution of spray from the sprinkler head. At least an 18-inch clearance must be used, as shown in Figure 13.1, to permit the sprinkler spray to be well distributed.

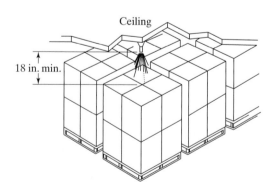

FIGURE 13.1

Minimum vertical clearance between stacked material and sprinkler head is 18 inches to permit distribution of spray.

FIXED EXTINGUISHING SYSTEMS

Technically speaking, automatic sprinkler systems are fixed extinguishing systems, but what is usually meant by a fixed extinguishing system is a more local system for controlling special fire hazards such as kitchen grills or tempering tanks. Again, the principal objective may be property protection and insurance-rate reduction, but steps must be taken to prevent the system, which may discharge dangerous gases or other agents, from becoming a hazard to employees. Thus, if the discharge of the system is not obvious, it is necessary to warn employees, perhaps with a discharge alarm, that dangerous agents are being expelled into the atmosphere. If a strategy of "total flooding" is used with a dangerous agent, an emergency action plan is necessary to assure that personnel escape. Some agents are so dangerous as to be prohibited altogether as extinguishing media; examples are chlorobromomethane and carbon tetrachloride.

Fixed extinguishing systems are somewhat like giant, fixed, automatic fire extinguishers. Many of the maintenance procedures appropriate for portable fire extinguishers are also appropriate for fixed extinguishing systems. Like portable extinguishers, fixed systems are required to be inspected annually. Corresponding to the monthly visual inspections of portable extinguishers is a somewhat more comprehensive semiannual inspection for fixed systems to determine whether containers are charged and ready for operation. If the containers are factory charged and have no gauges or indicators, they must be weighed to determine charge. A weight decrease of 5% or a pressure decrease of 10% is considered within tolerance. Discussion of specific systems follows with requirements for each.

Dry Chemical Systems

The word *chemical* in dry chemical systems should be remembered, and the question to be asked is whether the extinguishing chemical will produce any undesirable reaction with process reagents or perhaps foams and wetting agents also employed. There is more than one type of dry chemical available, and generally these chemicals are not to be mixed when filling the cylinders or containers. Mixing is permitted if the chemical to be added is "compatible" with the chemical stated on the approval nameplate of the system.

Dry chemicals used for extinguishing agents are usually not dangerous to the health or safety of personnel. However, the actual distribution of the chemical powder during an emergency may obscure vision, preventing escape. A possibility such as this calls for a predischarge employee alarm system as described earlier.

The biggest problem with dry chemical systems is the caking or lumping of the agent. Humid climates or moisture-producing processes subject the system to a greater risk of caking. Caking can render a dry chemical useless, so the chemical should be sampled annually to be certain moisture is not causing caking.

Other Fixed Systems Agents

Many fixed extinguishing systems employ carbon dioxide, Halon 1211, or Halon 1301 gases. These systems have the advantage of not requiring as much cleanup after the emergency as other systems, but all three of these gases can be dangerous to unsuspecting employees, especially if a total flooding strategy is employed. Standards must be followed in planning predischarge warning systems, employee egress routes, and maximum concentrations of gases to be released.

Water spray and foam agents are less dangerous to employees, but necessary volumes required to be effective may introduce egress hazards. Drainage must be directed away from work areas and must not obstruct egress paths.

SUMMARY

The many strategies for dealing with industrial fire hazards can be grouped under the general categories of prevention, suppression, and escape, or combinations of these categories. Current applicable industrial standards encompass all of these strategies.

It is necessary to keep the hazards of industrial fires in perspective. Industrial fires cause very little loss of life and injury now, compared with fire deaths elsewhere and occupational fatalities and injuries from other causes. In light of this perspective, it may seem that emphasis on equipment specifications, regular inspection of extinguishers, personnel training, and written plans is somewhat misplaced. However, the excellent record of industry in controlling fire hazards on the job should not be allowed to induce complacency now that success has been achieved. There is no doubt that the adherence to strict fire codes is what has helped industry achieve such a superior control of fire hazards compared to residential and other fire exposures. An example of the tragic consequences of not adhering to fire codes and life safety codes was the fire at the Imperial Foods poultry processing plant in 1991.

The beginning of this chapter identified fire protection as the oldest of topics in occupational safety and health. There is perhaps a correlation between the facts that industrial fire protection is an old endeavor and that it is also a very successful endeavor. Perhaps with time, even the newer fields of occupational safety and health covered in other chapters of this book will achieve the same high degree of safety and health as has already been achieved for fire protection.

EXERCISES AND STUDY QUESTIONS

13.1 Should fire extinguishers be required in all industrial plants? Why or why not?

13.2 What are the three major fields of fire safety?

13.3 What argument does industry offer against providing fire extinguishers?

13.4 Which accounts for more workplace fatalities, fire or workplace violence?

13.5 How does the United States compare with other nations of the world in numbers of fire deaths per million people?

13.6 What is the occupancy category for most fire deaths in the United States?

13.7 What percentage of the total number of industrial fatalities is attributed to fire?

13.8 Name some items to be included in fire prevention plans.

13.9 How is preventive maintenance related to fire hazards?

13.10 How can automatic audible alarm systems fail to warn employees of a fire emergency?

13.11 Are automatic smoke alarms required in industrial plants?

13.12 Why does an employee sometimes notice a fire in the plant, but fail to sound the alarm?

13.13 Are industrial plants required to have fire brigades?

13.14 How often must fire brigade members be trained?

13.15 Name some conditions that would make an employee unfit to serve in a fire brigade.

13.16 Identify the four classes of fires. Give example extinguishing media for each.

13.17 What extinguishment medium is used for LPG fires?

13.18 Identify the typical fire extinguisher medium (fill substance) for multipurpose use (e.g., for use in a Class A, B, or C fire).

13.19 What are the advantages and disadvantages of dry chemical as a fire extinguishment medium?

13.20 How often must fire extinguishers be inspected?

13.21 Are fire extinguishers required to be mounted at a certain distance from the floor?

13.22 How often are employees required to be trained to operate fire extinguishers?

13.23 Name some modern requirements for standpipe and hose systems.

13.24 Why are city water supplies usually unacceptable for directly supplying hose systems for fire protection?

13.25 Are industries required to have automatic sprinkler systems? Explain.

13.26 Why are paper bags placed over sprinkler spray heads?

13.27 How close to the automatic sprinkler head in a warehouse is it permissible to stack material?

13.28 If a fixed extinguishing system has no gauges or indicators, how can its charge condition be determined?

13.29 Name some gases employed in fixed extinguishing systems.

13.30 Explain why ruptured eardrums would represent a hazard to firefighters.

13.31 Name three acceptable media for use in fire alarm systems.

13.32 What are the two principal ingredients of a fire prevention plan?

13.33 Describe the 1991 industrial fire tragedy that resulted from failure to follow fire and life safety codes.

13.34 In terms of lives lost, what was the worst fire of the 20th century in a U.S. manufacturing plant?

13.35 A company has a standpipe system that has a capacity of 3000 gallons of water. Pressure is maintained by gravity head. Is this system sufficient to meet standards for standpipe systems? Why or why not?

13.36 Explain the difference between dynamic pressure and static pressure.

13.37 Are fire extinguishers required to have attached tags that document their inspection status? Explain.

13.38 How long are fire extinguisher inspection records required to be maintained?

13.39 What is the specific purpose of hydrostatic tests for fire extinguishers (i.e., what specific hazards are the tests intended to guard against)?

13.40 Why does the safety and health manager usually choose not to have hydrostatic tests for fire extinguishers performed in-house?

13.41 Name two reasons why a fire extinguisher might fail a hydrostatic test.

13.42 Of the three modes of respirator flow, which two are preferred for firefighters? What is required if the third mode is used?

13.43 **Design Case Study**. You are called on as a consultant to specify the type of fire extinguisher to be used for extinguishing LPG fires. What is LPG and to what fire class does it technically belong? What type of fire extinguisher, if any, would you specify for LPG? What other recommendations might be appropriate?

RESEARCH EXERCISES

13.44 Examine current fatality statistics to determine how each of the following causes of death rank among workplace fatalities:

(a) Falls

(b) Electrocutions

(c) Oxygen deficiency

(d) Exposure to caustic, noxious, or allergenic substances

(e) Motor vehicle accidents

(f) Homicide or other workplace violence

(g) Fire

What percentage of the total is attributed to each?

STANDARDS RESEARCH QUESTIONS

13.45 Search the OSHA General Industry standard for OSHA's current strategy for dealing with fire extinguishers. Explain OSHA's general position regarding fire extinguishers.

13.46 What are OSHA's general concerns about fire brigades? Justify your answer with reference to OSHA fire protection standards. Include a discussion of OSHA's enforcement activity pertaining to fire brigades. Cite inspection statistics.

13.47 Study OSHA General Industry standard for fire alarm systems. Examine enforcement activity for such systems by using the NCM database.

C H A P T E R   1 4

Materials Handling
and Storage

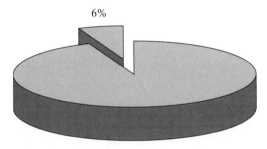

6%

Percentage of OSHA General Industry citations addressing this subject

The usual concept of a factory is a place where things are made or materials are processed, but often the major activity in a factory consists of moving things and materials around. Lifting, a most basic material-handling activity, accounts for most back injuries, one of the largest workplace injury categories of all. At Liberty Mutual Insurance Company, low back pain alone accounts for 33% of the firm's total workers' compensation claims (ref. Lorenzi). Industrial trucks, tractors, cranes, and conveyors all have the simple mission of moving materials, and they all cause injuries and fatalities every year.

The National Safety Council (NSC) charges materials handling with 20 to 25% of all occupational injuries. The size of the problem is emphasized in NSC's *Accident Prevention Manual for Industrial Operations* as follows:

> As an average, industry moves about 50 tons of material for each ton of product produced. Some industries move 180 tons for each ton of product.

In materials handling, masses are usually measured in tons or pallet loads instead of ounces, pounds, or kilograms. The human body is light and frail by comparison, so bulk materials' masses can easily pinch, fracture, sever, or crush its parts. Contributing to the hazards of large masses is the reality that materials handling includes *motion* for these masses.

To illustrate the general *mass/motion* hazards of material-handling equipment, consider the following comparison of processing versus material-handling equipment. To be struck by a moving part of a processing machine may or may not cause injury, depending on the size of the machine, the motion of the moving part, and the shape or surface characteristics of the part. But being struck by an industrial truck or conveyor is almost certain to cause injury. More indirectly, the mass/motion hazards of materials handling can affect safety by impacting facilities such as gas lines or electric lines or by overloading structural components of buildings.

Another general hazard of materials handling is its automatic or remote-control nature. Materials pumps and conveyors are often started automatically on demand or from a manual switch that is located far away. Conveyor accidents are often caused by this remoteness characteristic. In another example, railroad cars often move about a yard, far from the eye of the engineer, or worse yet, they may roll almost silently, coasting independently into position from the momentum of a locomotive's momentary push with no local control at all.

An indirect general hazard from materials handling is fire. This hazard emphasizes the storage aspect of materials handling. Warehouse fires are costly in terms of property loss, but they can also be dangerous to workers.

MATERIALS STORAGE

Materials-handling standards say that bags, containers, or bundles stored in tiers shall be "stacked, blocked, interlocked, and limited in height so that they are stable and secure against sliding or collapse." For general materials, the standard does not explain what constitutes "blocking" or "interlocking" and does not specify height limits for stacks. Some industry practices, however, seem to defy the standard. Since the standard for materials storage is not specific in its requirements but does address results (e.g., "so that they are stable and secure against sliding or collapse"), the standard should be recognized as a performance standard.

Housekeeping is another consideration for materials in storage. Sloppy warehouse practices can lead to trip hazards or fire. Pest harborage can result from certain accumulations of materials and can constitute a hazard. Outside storage can become grown over in weeds and grass, which can be a fire hazard in dry weather.

The safety and health manager should monitor the ups and downs of the company's sales and production fortunes. If there is a recent buildup in production in anticipation or in response to booming sales, there is likely to be a warehousing problem. Additional warehouse space is likely to be added later, after the upturn appears more permanent. Even after an expansion decision is made, a certain amount of lead time is necessary to plan and build add-on warehouse space. Meanwhile, the existing warehouse space becomes crowded and unsafe. Judgmental limits for material stacking are cast aside as creative ideas are explored to squeeze more material into a small space. Some of these new ideas will be good ones; it is not basically unsafe to try to conserve space. But during a period of warehouse crowding, special attention should be given to new hazards that may result from new procedures in the warehouse. Aisle and exit blockage are likely to occur during such periods of warehouse stress, as is the incidence of storage stack collapse. Any accidents that occur during such periods will point to some of the new hazards being generated, and these accidents should be analyzed to eliminate cause.

INDUSTRIAL TRUCKS

This category of material-handling equipment is exemplified by the forklift truck. There is no question that OSHA considers safety with forklift trucks important. OSHA's own estimate of the number of forklift trucks nationwide is over one million (ref. OSHA commits). When you consider that each forklift truck may have more than one employee within the plant authorized to drive it, it is easy to see that forklift truck safety is very important to worker safety. Typically, forklifts either are powered by electric motor or have internal combustion engines. Besides forklifts, there are tractors, platform lift trucks, motorized hand trucks, and other specialized industrial trucks. Not included are trucks powered by means other than electric motors or internal combustion engines. Also not included are farm tractors or vehicles primarily for earth moving or over-the-road hauling.

Truck Selection

"Ignorance is bliss" for most industrial managers who set out to buy a forklift truck. Little known is the fact that there are 11 different design classifications by type of power and by degree of hazard for which approved. These 11 design classifications are distributed over at least 26 different classifications of hazardous locations to which the forklift truck may be exposed.

One can easily wonder why it must be so complicated to select a forklift truck. The basis for this complication is that engines and motors can be dangerous sources of ignition for flammable vapors, dusts, and fibers. To design these engines and motors to effectively prevent ignition hazards is a costly matter, and the marketplace will not support the purchase of an explosion-safe forklift truck for use in an ordinary factory location. Thus, a complicated array of classifications and regulations is set up to permit the specification of the right industrial truck for the right job—no more, no less.

Industry standards for industrial truck classifications and their corresponding hazardous location codes are a maze of abbreviations and definitions. To understand these abbreviations, keep in mind that the objective is safety from *fires and explosions*. Whether the industrial truck is diesel, gasoline, electric, or LP-gas powered, the more fire-safe models are designed to prevent ignition of accidental fires, and thus they are more expensive. A simple summary is contained in Table 14.1. More details are contained in the standards, but most safety and health managers will need only the general idea. Another thing to keep in mind is that it is legal to use a safer, higher classified industrial truck than the minimum required, but of course it usually will not be economical to do so. If a firm already has an EE-approved electric forklift, however, it may be expedient to go ahead and use it instead of buying a new ES- or E-approved unit that would suffice for the given application.

Summarizing these principles and several pages of applicable regulations, Table 14.2 gives the safety and health manager a perspective into the approval classes of various industrial truck designs. The *classes*, *groups*, and *divisions* represent varieties of hazardous locations in which the definitions correspond *roughly* to those of the *National Electrical Code* and are covered in more detail in Chapter 17. *Class* and *group* refer to the type of hazardous material present, and *division* refers to the extent or degree to which the hazardous material is likely to be present in dangerous quantities.

TABLE 14.1   Summary of Industrial Truck Design Classifications

Diesel	Electric	Gasoline	LP gas
D	E	G	LP
Standard model	Standard model	Standard model	Standard model
Cheapest	Cheapest	Cheapest	Cheapest
DS	ES	GS	LPS
Safer model	Safer model	Safer model	Safer model
More expensive	More expensive	More expensive	More expensive
Exhaust, fuel, electrical systems safeguards	Spark prevention Surface-temperature limitation	Exhaust, fuel, and electrical systems safeguards	Exhaust, fuel, and electrical systems safeguards
DY	EE		
Safest diesel	Safer still		
Most expensive diesel	More expensive		
No electrical equipment	All motors and electrical enclosed		
	EX		
Temperature-limitation feature	Safest of all		
	Most expensive electrical truck		
	Even electrical *fittings* designed for hazardous atmospheres		

Source: Summarized from Code of Federal Regulations 29 CFR 1910.178.

There are so many categories of "approvals" for industrial trucks that it is easy to lose sight of the overall objective in the approval process: to prevent fires and explosions from improper use of the wrong truck in a hazardous atmosphere. The authority for approval of industrial trucks is delegated to nationally recognized testing laboratories, such as Underwriters' Laboratories, Inc., and Factory Mutual Engineering Corporation. The prudent safety and health manager will leave the application-for-approval

TABLE 14.2   Allowable Categories for Industrial Trucks for Various Hazardous Locations

Class	Group	Division 1	Division 2
I	A	No industrial truck permitted	DY, EE, EX
	B	No industrial truck permitted	DY, EE, EX
	C	No industrial truck permitted	DY, EE, EX
	D	EX	DS, DY, ES, EE, EX, GS, LPS
II	E	EX	EX
	F	EX	EX
	G	EX	DY, EE, EX, DS, ES, GS, LPX
III		DY, EE, EX	DS, DY, E[a], ES, EE, EX, GS, LPS

[a] Permitted to continue if previously in use.

Source: Summarized from Code of Federal Regulations 29 CFR 1910.178.

process to the manufacturer of the equipment and simply look for the UL or FM approval classification, such as DY, EX, GS, and so on. For nonhazardous locations, even an *unapproved* truck may be used if the truck *conforms* to the requirements of type D, E, G, or LP.

In these days of high energy costs and the search for alternatives, some company managements may desire to convert an industrial truck from one energy source to another. But tampering with the design or altering the truck may invalidate the approval. Industrial truck conversions can be made, but the process is a bit tricky. The conversion equipment *itself* must be approved, and there is a right way and many wrong ways to carry out the conversion.

The hazard of fires and explosions may be the most complicated factor in the selection of forklift trucks, but it certainly is not the only factor. Far more important than the fire hazard rating of the design of the truck are the operations, fueling, guarding, training of drivers, and maintenance—subjects discussed in the next section.

Operations

One of the first items of interest to the safety and health manager should be the refueling or recharging area of forklifts. Smoking is prohibited in these areas, and this fault is found more than any other. Also a problem is the charging of forklifts in a *nondesignated area*. Hazards in the area include spilled battery acid, fires, lifting of heavy batteries, damage to the equipment by the forklifts, and battery gases or fumes. All of these hazards need to be addressed by the safety and health manager in some way.

Federal standards forbid pouring water into acid when charging batteries. Perhaps a sign in the area would achieve compliance with the rule. A better way to promote safety, though, would be to include in an employee training program an explanation of the violent and exothermic reaction that occurs when water is poured into a concentrated strong acid.

Everyone knows that arcs and sparks frequently fly when battery connections are made. What most do not know is that gases liberated during charging processes can reach ignitable concentrations. Fire is little or no hazard when a battery is simply "jumped" to another. But a forklift charging area is a different matter: Large volumes of gases are liberated; therefore, adequate ventilation is essential. In addition to ventilation, personal protective equipment and emergency eyewash and shower should be provided in the battery charging area due to potential exposure to acid.

With hazards of gases and acids in battery-charging areas, the alternative of the internal combustion engine seems attractive. But each of the internal combustion engine choices—diesel, gasoline, and LP gas—emit another dangerous gas: carbon monoxide. Since forklifts usually operate indoors, carbon monoxide gas levels can be a problem. The 8-hour time-weighted-average exposure limit for carbon monoxide is 50 ppm.

If the safety and health manager determines that a carbon monoxide problem exists in the plant and that the forklift trucks are the culprits, several alternatives are possible. The obvious one is to switch to electric forklifts. Another solution might be to alter the building or to install adequate ventilation systems. Perhaps the cheapest solution of all would be to review procedures and operations to determine whether sources of emissions can be reduced or perhaps eliminated entirely. The following are key questions:

1. Are operators leaving engines running unnecessarily?
2. Can the layout of warehouses or plant facilities be revised to reduce concentrations?
3. Are faulty or worn-out lift trucks creating more emissions than necessary?

Although there are no universal general minimum lighting requirements for industrial plants, where industrial trucks are operated, safety demands that the trucks themselves have directional lights if the plant area is too dark. Truck lights are required if the general lighting is less than 2 lumens per square foot. This is really a quite low level of light, since an ordinary 100-watt incandescent bulb can produce 1700 lumens. Even in an all-black room with nonreflective walls, a 100-watt bulb could produce more than 2 lumens per square foot in an 8 by 12 by 16 foot room. Wall or other surface reflections help the overall situation, so the requirement for 2 lumens per square foot is not difficult to meet. A lighting consultant can help in this determination.

Forklift Driver Training

The original National Consensus Standard adopted by OSHA emphasized the proper selection and hazard rating of forklift trucks, as well as the procedures surrounding their operation and maintenance. But far more important to the safety of operators and other workers around forklift trucks is how they are driven. Ironically, only one small general paragraph in the standards required operator training for forklift trucks. Employers were able to show compliance to this very general requirement by merely exposing new forklift drivers to a videotape. The result was a very haphazard approach to forklift driver training and a continued high incidence of fatalities.

Recognizing the gravity of the problem, OSHA promulgated a major change to the Powered Industrial Truck standard in 1999. To safety and health managers in general industry, the most important part of this standard was the dramatic change it brought to the operator training provision of the standard. Responsibility was placed on the employer to ensure that forklift truck drivers are competent. Content of the training must include formal instruction, which can consist of classroom training or videotape training, but this is only part of the requirement. The training must also include practical training, including driving demonstrations, and, perhaps most important of all, evaluation of the operator's performance in the workplace.

The specific content of the formal instruction is outlined in the standard and relates to both the equipment itself and the special hazards of the given workplace in which the equipment will be used. Therefore, it is not enough just to use a standard training program supplied by the equipment manufacturer. The instructor must also be qualified and capable of training the operators to deal with the specific hazards to which they will actually be exposed—for example, ramps that could cause tipping accidents, pedestrian traffic areas within the plant, or narrow aisles and places in which turning and driving will be restricted. These are items that would not be addressed in a general equipment operations manual.

The standard provides for evaluation of the operator's actual driving experience on the job, and if accidents or dangerous near misses occur, the driver must receive refresher training. Additional training may also be required if driving conditions change within the plant, such as those brought about by a remodeling of the work or traffic areas.

The final, clinching requirement of the forklift truck operator training requirement is certification. The employer or employer representative must certify and document the training, the evaluation, the dates of both the training and evaluation, and the identities of both the drivers and the trainers. The certification requirement traces the responsibility for the training and its effectiveness.

There is no doubt that OSHA has stiffened the requirements for forklift truck driver training and has made these requirements more specific. There is also little doubt that OSHA will continue to give this subject a great deal of attention. The problem is a serious one, and resolution will take years to achieve. As recently as the autumn of 2002, more than three years after the new standard on operator training, OSHA was reacting to new reports of large numbers of fatalities from forklift trucks. In one four-state region, 86 fatalities were reported within a 4-year period from forklift truck accidents alone (ref. OSHA Commits).

Just why are forklift trucks so dangerous? Some insight to this hazard may be revealed by considering the special hazards associated with the stability of forklift trucks and the lack of knowledge of this characteristic on the part of forklift drivers.

Many workers feel that because they know how to drive an automobile, they also basically know how to operate a lift truck. Unfortunately, many employers are inclined to take their word for it. But the operation of a lift truck takes a great deal more skill than the operation of an automobile. Compared to an automobile, a lift truck has a much shorter wheelbase, and when the load is lifted the center of gravity is very high. This creates stability problems to which the operator may be unaccustomed. Compounding the stability problem are the small-diameter wheels found on lift trucks, making chuckholes and obstructions more hazardous. When loaded, the center of gravity of the lift truck and load together can be shifted dangerously forward. Picking up and depositing loads require skill in proper manipulation and safe positioning. An off-center load presents a special hazard in which the load might tip in transit even though the truck itself is in a stable position.

Figure 14.1 diagrams the center of gravity of a forklift truck. The center of gravity shifts forward when the forks are loaded. If the load is too great, the center of gravity will shift forward of line BC and the forklift will tip dangerously forward, possibly dumping the load and overturning. Counterweights are added to the back of the forklift to counteract the forward-tipping forces, but if too much counter-weight is attached, the forklift becomes unstable when it is not loaded. Because of the three-point suspension scheme, the forklift may become laterally unstable when traveling while unloaded and tip sideways, especially when traveling on uneven terrain. Recognition of these physical factors facilitates understanding of the special hazards of driving a forklift truck and the need for the training specified by OSHA.

Besides stability problems, visibility can be a problem with lift trucks. The load itself can block the view and necessitate driving with the load trailing. Driving in worker aisles presents problems of pedestrian traffic, especially at corners where visibility is limited. Although lift trucks are not silent, they may seem so in a noisy factory environment. This increases the hazard to pedestrians and the need for greater visibility for lift truck operators.

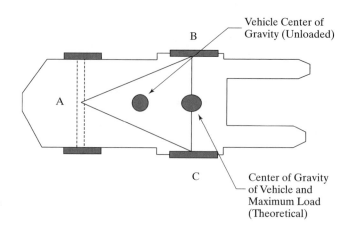

FIGURE 14.1

Top view of forklift truck illustrating the locations of the center of gravity when the forklift is unloaded and when it is loaded to capacity (*Source*: OSHA standard 1910.178, Appendix A).

One of the greatest hazards with forklifts and other industrial trucks is the transition between dock and cargo vehicle. Figure 14.2 shows precautions to be taken. Although a highway truck is shown in the figure, the hazard also exists for loading railroad cars.

Passengers

Passengers on a lift truck can be a hazard in more than one way. For one thing, the truck is often equipped to seat only the driver, and there may be no safe place for a passenger to ride. Hitchhikers also can distract the driver, whose attention is even more important on a lift truck than in an automobile. One practice particularly frowned on is riding on the lift *forks*. The temptation is great to use a forklift truck as a

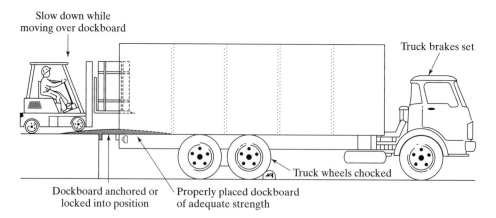

FIGURE 14.2

Preventing forklift hazards: transition from dock to cargo vehicle.

personnel elevator. Actually, this practice can be safe *if* all of the following conditions are obtained:

1. A safety platform is firmly secured to the lifting fork
2. The person on the platform can shut off power to the truck
3. Protection from falling objects is provided if needed

An ordinary wood pallet is not generally considered to be a safety platform, although it is often used for lifting personnel.

Some readers may consider the rule that the person on the platform be able to shut off power to the truck to be unreasonable. This rule also stands in the way of use of an ordinary pallet as a lifting platform. Workers, too, may resist this rule, and the safety and health manager needs to be in a position to counter this resistance with training programs that effectively explain the reasons for the rule. As a rationale for the rule, ask workers to consider the hazard caused by unexpected obstructions. A small obstruction could do any or all of the following:

1. Damage the platform
2. Tip the platform, causing the worker to lose balance
3. Injure the worker on the platform
4. Knock the worker off balance.

The forklift driver is really in a poor position, due to distance or angle, to see all obstructions and to judge their distance from the lifted platform. One might argue that the lift may be entirely clear of any obstructions, but such lifts would be unusual. There is usually no reason for employees to be lifted unless the lift is adjacent to equipment, stacks of material, or building structure. Any of these items can present obstruction hazards.

A variation of the *forklift rider* is the *carpet pole rider.* In carpet warehouses, the lift truck is equipped with a single pole, which is guided into the spool of a roll of carpet to lift the roll and transport it around the plant. Workers have been known to ride these poles rodeo style to gain access to the top of a stack of carpet rolls. Normally, there should be no reason for a worker to ride the pole because the pole can be raised and guided into position by the driver without assistance

When manipulating loads, the forklift is often operating close to observers, supervisors, or assistants giving directions to the operator. A dangerous place to stand is beneath the elevated fork, whether it is loaded or not. Another dangerous position is between an approaching lift truck and a fixed object or bench.

Parking and Maintenance

One thing to check is parked, unattended forklift trucks. The first thing to do is determine whether the lift truck is really unattended. If the operator cannot see the truck, it should be considered unattended. Even if the operator can see the truck, if the truck is more than 25 feet away, the truck is unattended. If the truck is unattended, the motor should be turned off. Even if the operator *is* close by, if the operator is *dismounted*, the fork must be fully lowered and the controls neutralized. The next thing to check is whether the brake is set, and if the truck is on an incline, whether the wheels are blocked. Finally, the horn should be tested.

Federal standards take very seriously the question of maintenance, inspection, and service for industrial trucks. No latitude is permitted concerning defective industrial trucks, to prevent them from being operated until the next regular overhaul. Any condition in a forklift, such as an inoperative horn, a defective brake, or a broken headlight, is cause to remove it from use until it is repaired.

Most people are amazed to learn that federal standards require industrial trucks in use to be inspected *daily* for safety. Compare this rule with procedures for automobile safety inspections, which most states require *annually*. If the industrial truck is used on a round-the-clock basis, safety inspections are required *after every shift*. It would be wise for the safety and health manager to institute some type of procedure or record to ensure that this job gets done and that proof of performance will be kept on file.

A final note on the subject of industrial trucks is the provision of an overhead guard to protect the operator from objects falling from the elevated load. More and more forklifts are being equipped with such guards for protection against falling objects such as small packages, boxes, and bagged material. They are not designed for protection against the impact of a full capacity load. These overhead guards are not to be confused with the much sturdier rollover protective structures (ROPS) described in Chapter 18, although the same structure can be used for both purposes if it is constructed properly. Some loads are unitized, and objects within the load are secured from falling back on the operator. For such applications, the hazard to the operator from falling objects disappears, and the overhead guard becomes unnecessary.

CRANES

An industrial truck is convenient for general material handling of palletized loads. But some material-handling jobs cannot be handled by an industrial truck. Larger, heavier, more awkward loads require the versatility of a crane, especially if the path of travel is complicated.

A crane is a construction industry device that is typically seen lifting heavy steel beams to lofty places. Although this image is accurate, it is incomplete. Cranes are also used extensively in general industry, although they usually take a different form. Cranes in industrial plants are generally limited in travel by a track or overhead runway structure, typified by the *overhead traveling crane* shown in Figure 14.3. Such cranes are popularly called *overhead bridge cranes* or simply *bridge cranes* by the workers who use them. Some models, such as those shown in Figure 14.3, are operated from a cab mounted on the crane itself. Others are operated from the floor by means of a hanging cord control called a *pendant*, or from a fixed remote station called a *pulpit*. *Gantry* cranes have legs that support the bridge above the railway. *Cantilever* gantry cranes have extensions on one or both ends of the bridge; these extensions extend the reach of the crane outside the area between the rails on which the crane travels. One characteristic common to all overhead and gantry cranes is that the trolley, which carries the hoisting mechanism, rides *on top of* the rail on which it travels. Overhead cranes whose trolleys are not so mounted are called *underhung cranes* or *monorails*,

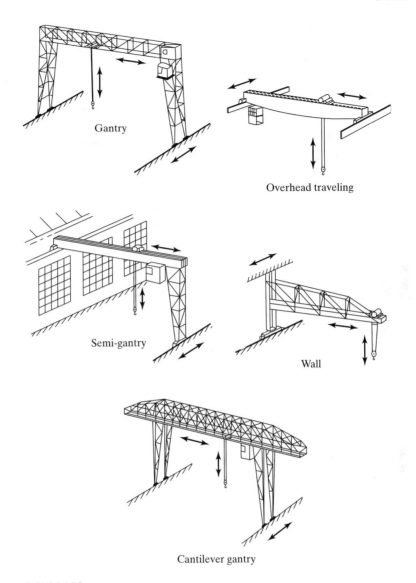

FIGURE 14.3

Various configurations for overhead cranes (*Source*: ANSI Standard B30.2.0-1976, reprinted with permission of ANSI and ASME).

depending on type or application. Safety standards for overhead and gantry cranes are different from standards for monorails.

The safety and health manager's chief concern with regard to overhead cranes should be that workers will overload the crane. The rated load should be plainly marked on each side of the crane, and if the crane has more than one hoisting unit, each hoist must have its rated load clearly marked. So easily recognized are crane load markings that their absence is conspicuous.

Even if the rated load marking *is* on the crane, workers will sometimes be tempted to exceed the rated capacity. Everyone knows that engineers provide for a safety factor in their designs, so virtually every crane will support more than its rated load, even without damage to the crane. But uncertainties in the actual weight of the load, dynamic loads during transport, shock loads during lifting, variations in crane components, and unavoidable design variabilities can combine to result in a very dangerous situation—even when rated load capacities are exceeded "slightly" or only "occasionally." As with running a stop sign, an alert driver can almost always avoid an accident, but once in a great while, another driver will be approaching the intersection at a high rate of speed, the visibility will be obstructed, or the stop sign violator will be complacent or distracted, and a serious accident will occur. It becomes very difficult for the safety and health manager or anyone else in the plant to police the proper use of cranes within their rated load limits at all times. This is where training becomes important so that workers will understand the risks and consequences of their actions. Recalling the principles set forth in Chapters 2 and 3, another way to control this problem is to take the time to make a big example out of any accident or near accident that occurs as a result of overloading a crane, even if no injuries occur.

Many overhead cranes operate outdoors, where wind can be a hazard. Wind alone is usually not dangerous, but a wind load in combination with a working load can result in dangerous structural damage to the crane. Automatic rail clamps are required for outdoor storage bridge cranes. The purpose of these clamps is to lock the bridge to the rail if the wind exceeds a certain velocity. This idea sounds good because it protects against the types of wind hazards just described and also protects the bridge from unintentional and uncontrolled rolling in a high wind or a brake failure. But like some other devices, the safety device carries with it a hazard of its own. Think about it for a moment: In what mode of crane operation would the tripping device most likely engage the automatic rail clamp? The answer is when the bridge is traveling *at top speed* against the wind. The sudden engagement of the rail clamp when the bridge is traveling at top speed is likely to injure the operator in the cab, dangerously swing the load, damage the crane, or all three. Therefore, the crane needs either a visible or an audible alarm, or both, to warn the bridge operator *before* the rail clamp engages.

If the crane operator rides in a cab mounted on the bridge or the trolley, the operator must somehow have a way to gain access to his or her station. One idea that comes to mind is to use a portable ladder, but this is not such a good idea. Since the bridge travels, the operator, cab, trolley, and bridge may roam far from the point at which the operator mounted. The idle ladder standing far away would be an invitation for someone to remove it for some other use or perhaps just to get it out of the way. A ladder used for crane cab or bridge access should be of the fixed type. Stairs or a platform, or both, can also be used, but must require no step over a gap exceeding 12 inches.

Overhead cranes designed or constructed in-house sometimes do not meet accepted principles for crane design and overlook some of the necessary safety features specified by applicable standards. Besides cab access, there are specifications for footwalks for safe maintenance of the trolley and bridge. These footwalks need standard toeboards and handrails, as described in Chapter 7.

Ideally, footwalks will have at least 78 inches of headroom. Sometimes, however, this is not practical because the crane may be near the ceiling of the building. The standards recognize this difficulty and permit less than 78 inches of headroom. However, a footwalk becomes rather ridiculous if the headroom is less than 48 inches and would more properly be called a "hands-and-knees" walk. In these situations, footwalks should be omitted, and a stationary platform or landing stage for crane maintenance workers should be installed.

A critical hazard to the wire rope of a crane or hoist results from drawing the load hook or hook block up too far—to the point at which the load block makes contact with the boom point of the crane or other mechanical assembly for reeving the wire rope. This event is known as *two-blocking*, the term being derived from the physical contact of two blocks in the reeving system. On the incidence of two-blocking, continued travel of the load block causes a severe tensile stress to be immediately imparted to the wire rope, and this stress usually stretches or breaks the wire rope. The blocks may also be damaged.

Two-blocking is a very serious hazard that has resulted in many fatalities. A sudden break in a load-bearing wire rope constitutes an obvious hazard, especially if personnel are under the load or are themselves being hoisted aloft by the crane. As a matter of ergonomics, it can be shown that it is very difficult for a crane operator to be sufficiently vigilant to prevent an occasional dangerous brush with two-blocking, especially for construction cranes. So many fatalities have resulted from two-blocking that crane construction standards now address this hazard, and devices are on the market for the purpose of preventing two-blocking or the damage that can result from it. These mechanisms, usually electromechanical, are commonly called *anti two-block devices*.

An obviously serious hazard associated with the operation of an overhead bridge crane is overtravel. Devices to control overtravel would include trolley stops, trolley bumpers, and bridge bumpers. Bumpers and stops are somewhat different in that bumpers absorb energy and reduce impact, whereas a stop simply stops travel. The stop is simpler and can consist merely of a rigid device that engages the wheel tread. For safety, such a stop needs to be at least as high as the wheel axle center line. Bumpers soften the shock by absorbing energy, but are not as positive as a true stop. For example, bridge bumpers need only be capable of stopping the bridge if it is traveling at 40% or less of rated load speed. If the crane travels only at slow speeds, shock loads are not significant, and bridge bumpers and trolley bumpers are not necessary. There may be other circumstances of operation, such as restriction of crane travel, that eliminate the need for bridge bumpers or trolley bumpers.

Related to bumpers and stops are rail sweeps. If a tool or some item of equipment happens to obstruct one of the rails on which the bridge travels, a catastrophic accident could occur. Therefore, bridge trucks are equipped with rail sweeps, projecting in front of the wheels, to eliminate this hazard. There is nothing magical about the term *sweep*; even part of the bridge truck frame itself may serve as a sweep.

One obstruction that can be encountered by an overhead crane is another overhead crane that runs adjacent to it with parallel side rails. Proper clearance should be provided between the two adjacent bridge structures. Unfortunately, such clearance may make it impossible for one crane to transfer its load to the domain of the adjacent crane. Some factories resolve this problem by using retractable extension arms on the crane. The trouble with these retractable arms is that the operator may inadvertently leave them extended, resulting in a collision between the two adjacent cranes.

Electric shock is a concern with overhead cranes in two principal areas:

- Shock from exposure to current-carrying portions of the crane's power delivery system
- Shock from a shorted connection in a hanging control pendant box. (See Figure 14.4.)

A third area of concern for electrical shock is the accidental contact of live, high-voltage overhead transmission lines. This is a hazard for mobile cranes with booms (discussed in Chapter 18) because of their wide use in the construction industry. Exposed live parts in the crane's power delivery system are usually protected by their remoteness or "guarded by location." Some older models might present hazards and need modification.

A greater hazard is the possibility of shock from a pendant control. The electrical conductors can be strained if they are the sole means of support for the pendant. The control station must be supported in some satisfactory manner to protect against such strain. If there is a failure, it is likely to occur at a connection within the box. This raises the possibility of a dangerous short to ground through the operator's body. Pendant controls take much abuse and need to be of durable construction.

Not related to electrical shock but on the subject of pendant control boxes is the requirement that boxes be clearly marked for identification of functions. Some older model or homemade cranes might have pendant control boxes without function markings. Without a great deal of trouble or expense, the safety and health manager can check the cranes in-house for compliance and get the control functions marked on the pendant controls.

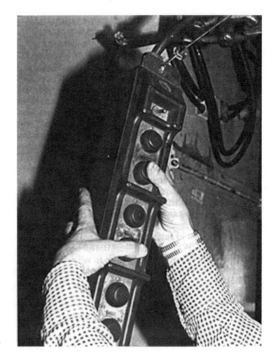

FIGURE 14.4

Hand-held pendant control for overhead crane.

One hazard to think about with overhead cranes is what would happen if a temporary power failure occurred. Suppose that the crane were in the process of lifting a heavy load by means of its hoist mechanism. Obviously, no one would want the crane to drop its load to the floor upon power failure, but the hazard does not end here. Suppose that the crane bridge is traveling in a horizontal direction, whether loaded or unloaded. Upon power failure, the bridge would stop, which might not be dangerous. But when power is *restored*, dangerous lurching action might occur. In fact, upon a power failure, the crane operator might even leave the crane cab!

Several design alternatives can protect against these hazards. One solution is to equip the control console with spring-return controllers. Pendant boxes can be constructed with spring push buttons instead of toggle switches. A disconnect device can neutralize all motors and not permit a reconnect until some sort of positive "reset" action is taken. Even if the power remains on, it could be a hazard to inadvertently switch on a lever at the wrong time. Notches, latches, or detents in the "off" position can prevent such inadvertent actions.

Brakes are of obvious importance to the safe operation of a crane, yet a large number of crane operators do not use brakes, but instead rely on a practice the industry calls "plugging." The operator merely reverses the control and applies power in the opposite direction, thereby stopping the load. Although no OSHA standard prohibits the practice of *plugging*, it should be pointed out that plugging is not as effective as applying a brake under extreme conditions, such as stopping a large, fast-moving load. Under no circumstances should a crane operator depend entirely on plugging due to the brake being inoperative. In fact, plugging would be completely useless in the event of crane motor failure.

A crane has many moving parts, many of which are located far from the operator's console in the cab or from a floor operator holding a pendant. Moving machine parts are hazardous, and the remoteness characteristic amplifies the hazard. Moving parts are hazardous not only to personnel; they can be hazardous to the crane itself, which in turn can be indirectly hazardous to personnel. Hoisting ropes, for instance, can possibly run too close to other parts in some crane configurations and in some positions of the bridge and trolley. The result can be chafing or fouling of the hoist rope. If the configuration of the equipment will permit this situation to develop, guards must be installed to prevent chafing or fouling. Such moving parts as gears, setscrews, projecting keys, chains, chain sprockets, and reciprocating components need to be checked to determine whether they present a hazard; if so, they must be guarded.

As with conveyors and other material-handling equipment, overhead cranes are often large and widely distributed items of equipment. Electrical power is distributed over long distances, sometimes by means of open runway conductors. Portions may be so far away as to be obscured from view from the location of the power supply switch. Picture the insecurity of the maintenance worker who is compelled to repair a crane and is in direct contact with an exposed 600-volt (but deenergized) runway conductor, but the power supply switch is so far away that it is out of sight! Therefore, such switches should be so arranged as to be *locked* in the open or "off" position. This is an example of a *lockout/tagout* requirement that was in place before the general lockout/tagout standard was promulgated by OSHA in 1989. The general lockout/tagout standard will be covered in more detail in Chapter 15.

Ropes and Sheaves

Safety standards for wire rope strength state that "rated load divided by the number of parts of rope shall not exceed 20% of the nominal breaking strength of the rope." The term *rated* implies that a safety factor has been applied, which is numerically equivalent to 5, as can be derived from the standard as follows:

$$\frac{\text{Rated load (including load block)}}{\text{Number of parts of rope}} \leq 20\% \times \text{(nominal breaking strength)} \quad (14.1)$$

It will later be explained that the "number of parts of rope" is a multiplying factor that enables a multiple-sheaved block-and-tackle assembly to be loaded much higher than the wire rope load. Thus,

$$\text{Wire rope load} = \frac{\text{rated load (including load block)}}{\text{number of parts of rope}} \quad (14.2)$$

From equations (14.1) and (14.2)

$$\text{Wire rope load} \leq 20\% \times \text{(nominal breaking strength)} \quad (14.3)$$

Multiplying each side of the inequality by 5, we have

$$5 \times \text{(wire rope load)} \leq 100\% \times \text{(nominal breaking strength)} \quad (14.4)$$

Rearranging to create a ratio yields

$$\frac{\text{Nominal breaking strength}}{\text{Wire rope load}} \geq 5 \quad (14.5)$$

Since the ratio of the strength to the load is at least 5, the *safety factor is 5*.

The term *parts of rope* refers to the mechanical advantage provided by the block-and-tackle assembly. *Parts of rope* is computed by counting the number of lines supporting the load block. Of course, all the lines make up one continuous line that is reeved through several sheaves to achieve mechanical advantage. The concept is best explained by a picture; Figure 14.5 shows five different reeving combinations. Note that the mechanical advantage is numerically equivalent to the number of parts of line.

One additional caution is in order when determining the appropriate maximum load to be applied to a given reeving setup. The weight of the load-carrying sheave must be added to the weight of the load to be picked up to arrive at the total load of the line. The weight of the load block cannot be ignored, as is emphasized by the massive load block displayed in Figure 14.6. Case Study 14.1 will now be used to illustrate calculations used to determine the safety of a wire rope reeving application.

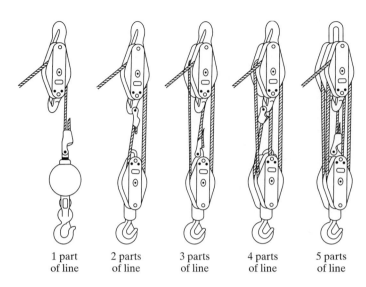

1 part
of line

2 parts
of line

3 parts
of line

4 parts
of line

5 parts
of line

FIGURE 14.5

Five different reeving combinations.
The mechanical advantage is equal to
the number of "parts of line"
supporting the load block.

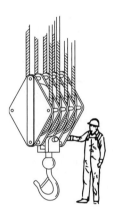

FIGURE 14.6

Massive load block. The load block becomes
part of the total load on the line.

CASE STUDY 14.1

BLOCK-AND-TACKLE SAFETY FACTOR

The lower block of a block-and-tackle assembly has three sheaves and is thus supported by six parts of line as the wire rope is wrapped around the sheaves, plus a seventh part as the wire rope is tied off on the lower block. The wire rope has a nominal breaking strength of 4000 pounds. The lower block (the load block) itself weighs 80 pounds. Calculate what maximum payload this block-and-tackle assembly can pick up safely.

> *Solution:*
>
> $$\frac{\text{Rated load (including load block)}}{\text{Number of parts of rope}} < 20\% \times (\text{nominal breaking strength})$$
>
> $$\frac{\text{Rated load (including load block)}}{7} < 20\% \times 4000 \text{ lb}$$
>
> $$\text{Rated load (including load block)} < 7 \times 20\% \times 4000 \text{ lb}$$
>
> $$< 5600 \text{ lb}$$
>
> $$\text{Maximum payload} = \text{rated load} - \text{weight of the load block}$$
>
> $$= 5600 \text{ lb} - 80 \text{ lb} = 5520 \text{ lb}$$

One way to prevent overloading the wire rope and the crane itself is to provide a hoist motor that can develop insufficient torque to overload the wire rope. The combination of such a hoist motor and correct reeving for the crane design will result in no overloading. Under this arrangement, the crane simply will be unable to lift any load that would damage the crane or exceed its safety factor. Most overhead cranes today are designed this way. How fortunate it would be if the human back had this design feature.

Any time a rope is wound on a drum, the rope-end anchor clamp bears very little of the load when several wraps are on the drum. The friction of the rope on the drum holds the load. But if the drum is unwound to less than two wraps, dangerous loading of the anchor clamp may result in failure—the wire rope will break free from the drum. Usually, the overhead crane is set up so that, even if the load block is lying on the floor, several wraps remain on the hoist drum. The floor may not be the extreme low position for the crane, however. The safety and health manager should look around for pits or floor openings into which the overhead crane might operate, resulting in dangerous unwinding of the hoisting drum.

"Don't saddle a dead horse" is a familiar safety slogan that refers to the improper mounting of wire rope clips employing U bolts. Such a clip assembly bears a resemblance to a saddle, and the U bolt represents the cinch strap. The U bolt places more stress on the wire rope and has less holding power than the clip. Therefore, it should not be placed on the live portion of the rope when a loop is formed. The "dead" end of the rope gets the U bolt, and the "live horse" gets the clip. Right and wrong methods are illustrated in Figure 14.7. Unfortunately, some workers in the field, unsure of the correct method, place the clips *both ways* in alternating fashion, thinking they are "playing it safe." Such an arrangement may be even more unsafe than "saddling dead horses" with every clip.

Before leaving the subject of wire rope, the hazard of rope whip action should be emphasized. Wire rope "cable" seems so heavy and inflexible that it seems unnatural

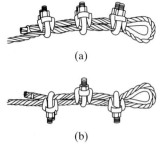

(a)

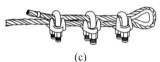

(b)

(c)

FIGURE 14.7

"Don't saddle a dead horse." Right and wrong ways to secure wire rope loops using U-bolt clips. (a) Incorrect—"saddle" is on dead end of rope; (b) incorrect—clips are staggered both ways; (c) correct—all clips are placed with the saddle assembly on the live portion of the rope and the U bolt on the dead end.

that it could crack like a whip or any fiber rope. It is difficult for anyone to visualize the tremendous tension forces on a wire rope during use in material-handling operation, until that wire rope breaks. Most workers have never seen what happens when a wire rope breaks; perhaps this explains why so many workers stand too close while the rope is drawn taut by the load. The hazard is very serious, and an injury accident is very likely to be a fatality.

Crane Inspections

Almost everyone is aware of the long, perhaps tedious checklists for inspection of an aircraft every time it flies. An aircraft must not fail in any catastrophic way during operation, and this makes frequent inspections worthwhile, even if they are repetitious and rarely uncover any defects. In some ways, a crane is like an aircraft; it, too, must not fail.

With regard to crane inspections, the standards use the terms *frequent* or *periodic* to specify when various items on the crane should be checked. Such usage is an attempt to avoid being overly specific in telling the employer *what* to do and *how often* to do it. Some broad guidelines describing the meaning of these terms are illustrated in Figure 14.8. Note that there is some overlap, as monthly inspections can be considered either frequent or periodic.

The crane manufacturer is a good source for detailed guidance in what to look for in the frequent inspections. This type of inspection is performed by the crane operator, just as a pilot inspects his or her aircraft before a flight. This analogy between aircraft and cranes might be a starting point for a safety theme in a training program for crane operators.

The frequent-inspection routine should include a daily visual inspection of hoist chains, plus a monthly inspection with a signed report. In the field, the term *hoist chains* has been widely misinterpreted to include chain *slings* for handling the load. A separate standard exists for slings. Figure 14.9 identifies which chain is hoist and which is sling.

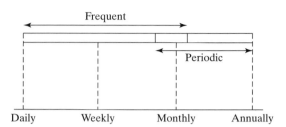

FIGURE 14.8

Inspection intervals for overhead cranes.

Crane hooks take a great deal of abuse and are critical to the safe operation of the crane. Although they usually contain considerable overdesign, damage or wear can reduce the margin of safety. Telltale signs of an abused and dangerous hook are illustrated in Figure 14.10.

A more thorough inspection of crane components is needed for "periodic" intervals. Whereas daily inspection of crane hooks is merely visual, the periodic inspection calls for a more scientific approach, such as using magnetic-particle techniques for detecting cracks. More extensive checks for wear are appropriate also, such as the use of gauges on wire rope sheaves and chain sprockets.

For most kinds of plant equipment, safety testing is customarily done by an independent testing laboratory such as Underwriters' or Factory Mutual. But the safety of an overhead crane is in part a function of the installation method and proper adjustments at the site. Therefore, an actual rated-load test is needed prior to initial use to

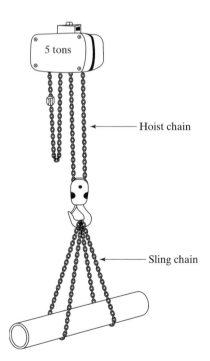

FIGURE 14.9

Hoist with sling. Hoist chain is not to be confused with the sling chain.

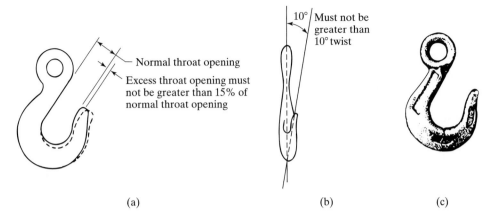

FIGURE 14.10

Defective crane hooks: (a) bent hook; (b) twisted hook; (c) cracked hook.

confirm the load rating of the crane. This is a bit tricky, because if the crane is subjected to too heavy a load, the crane may fail; but if it is not tested with a heavy enough load, why conduct the test? Standards specify that the maximum load during the test should go 25% higher than the crane's load rating. This will provide some assurance that the crane will withstand its rated load. During use, however, the crane should not be loaded with more than its rated load. Any extensive repair or alteration of the crane may subject it to a retest.

Wire Rope Wear

The two principal moving parts of an overhead crane are the wire rope and the drum and sheaves on which they travel. In addition, the wheels for the bridge and trolley move, as do a few other items on the crane. Whenever parts move and contact other parts, they are subjected to wear. Wear on the bridge and trolley wheels may eventually lead to problems, but is not likely to cause crane failure. Wear on the drum and sheaves is more dangerous because of the harm this can bring to the wire rope. But worn sheaves and drum *alone* are generally not the *direct* cause of crane failure. The wire rope remains as the most critical moving part. With continued use, every wire rope will eventually wear out and fail, a hazard that generally is not tolerable. Some way has to be devised to predict the failure of wire rope and withdraw it from use before a catastrophe.

It is not a simple matter to determine when a wire rope needs replacing. A wire rope has many individual wires, and it is easy to cut or break away any one of these small wires. Almost everyone has seen broken wires in an old wire rope (or "cable," as laypersons often say), which raises the question, "Is such a wire rope dangerous?" Safety and health managers do not survive long in their companies if they go around ordering wire rope replaced every time they find a broken wire. And if they did survive, their companies would not survive. Before delving into this issue, some basics about wire rope need to be examined.

A wire rope should really be considered a machine because the individual wires move on each other while the rope flexes, resulting in friction and wear. Furthermore, unless the individual strands are able to move properly during flexure, tremendous tensile stresses may be placed on some of the wires, causing them to break. Rust, kinks, and other types of abuse can interfere with the movement of the wires and lead to the stresses that cause wire breakage. As some wires break, the tensile stresses on other wires increase, leading to their breakage as well. Eventually, the force of the crane load will be sufficient to overcome the tensile strength of all remaining wires combined, and the wire rope will fail.

Even a well-maintained wire rope will be subject to wear on the individual wires, especially the outside wires. As the wear causes the diameter of the wire to become smaller, the tensile force increases the tensile stress on that single wire due to its reduced cross-sectional area. Thus, the worn wire, too, may break due to stress concentration, even if there are no kinks or rust, and even if the wires move properly to distribute the load among all wires.

For obvious purposes, a wire rope is overdesigned and will withstand more than its rated load. Equally obvious is that all wire rope will eventually develop broken wires with continued use. Broken wires are permissible to an extent, but beyond a certain point, the rope becomes dangerous. A means of evaluating the *degree* of wire rope deterioration is awkward and difficult, but nevertheless *necessary*, to prevent catastrophic failure.

The ANSI standard recommends[1] a procedure for counting the broken wires. Figure 14.11 diagrams the components of wire rope, defining the terms *strand* and *lay*. If there are more than 12 randomly distributed broken wires in a single strand in a single lay, there should be cause for questioning the continued use of the rope. A good place to look for broken wires is around end connections. Sometimes an expert can evaluate the remaining strength in a deteriorated rope after inspection. This possibility is acknowledged by the ANSI standard, which advocates the use of good judgment.

Another measure of rope condition is the amount of reduction of rope diameter below nominal. Figure 14.12 shows that when gauging a wire rope, the caliper can be placed on a small diameter or a larger diameter. The convention for wire rope terminology is to use the larger diameter for designation as the nominal diameter of the rope. Most people pay little attention to wire rope, but wire rope is no small matter considering that a quarter million tons of wire rope is sold annually.

Operations

The actual handling and moving of the load by the crane is a function of the skill, knowledge, and performance of the crane operator and the workers who attach and secure the sling or lifting device. As is the case with motor vehicles, the *operator* of the crane is probably the most important factor in preventing accidents.

---

[1]ANSI B30.2–2.4.2.

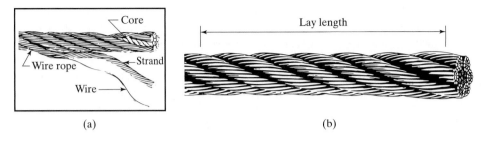

FIGURE 14.11

Wire rope components. (a) This wire rope has six strands plus an inner core. Do not confuse *strands* with *wires*. (b) Wire rope lay. *Lay* is one complete wrap of a single strand around the rope. Since this rope has six strands, lay is the length from the first hump to the seventh hump, as shown in the diagram (*Source:* courtesy of the Construction Safety Association of Ontario, ref. Rigging).

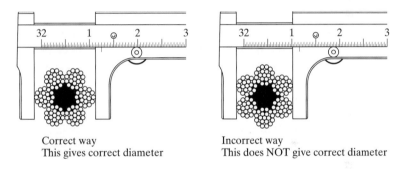

FIGURE 14.12

Gauging wire rope (rotate rope to select largest diameter) (*Source:* Courtesy of the Construction Safety Association of Ontario).

A good deal of skill is required to attach the load safely, especially if a sling is used. Slings will be discussed in the next section. The hoist rope is not intended for wrapping around the load, as this kind of abuse may damage the rope and at the same time be an inadequate support for the load. Misplacing the attachment off the line of the center of gravity can cause dangerous swings when the load is lifted. After the load is lowered, the tendency is to consider the hazards removed. But disengaging the load attachment can also result in dangerous shifts in material that can injure the inexperienced or unwary worker on the floor.

SLINGS

Slings are used to attach the load to the crane, helicopter, or other lifting device. Slings come in a great many varieties and are very important in the safety of material-handling operations. Components of the sling assembly are often subjected to much larger forces during lifting than is the hoist rope or other material-handling equipment. Because the skill of the user is so important in the proper application of the sling, slings are often misused, resulting in much more abuse and damage to slings than to the components of the crane.

The most important point to remember for the safe use of all slings is that the stress on a sling is greatly dependent on the way it is attached to the load. Figure 14.13 shows two different ways of applying a sling to pick up identical loads. If the angle of the supporting legs of a sling is sharp, as in Figure 14.13b, the advantage of multiple legs can be lost. Using too short a sling is the most common cause for this condition. The "rated capacity" of a sling is the working load limit under *ideal* conditions; if the sling is applied at leg angles other than those specified in the rated capacity table, the capacity can be drastically reduced due to the physics of the applied forces. Therefore, *rated capacity* is an incomplete term without the accompanying leg angle.

Note the following progression in capacities of 1/2-inch alloy steel chain as the number of legs increases:

Single (at vertical)	11,250 pounds
Double (at 60° from horizontal)	19,500 pounds
Triple (at 60° from horizontal)	29,000 pounds
Quadruple (at 60° from horizontal)	29,000 pounds

It might be assumed that the capacity of the sling increases as the number of legs or supporting members is increased. But note that no increase in capacity is shown in going from three to four legs. The reason for this is that, as in a four-legged chair, three legs will actually carry the load. At times, the weight distribution might be equal among legs of the sling, but usually the balance is not so perfect. As the load shifts around, it is

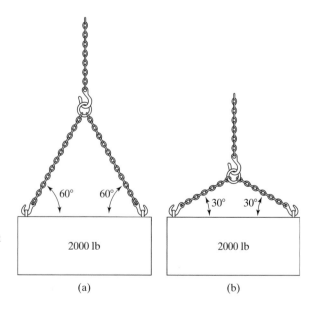

FIGURE 14.13

Comparison of sling tensile forces for two different methods of lifting identical loads: (a) tensile force on the sling is approximately 1150 pounds; (b) tensile force on this sling is approximately 2000 pounds.

entirely possible for one of the four legs to become slack and for the other three to bear the entire load.

Alloy steel chain, besides being very strong, is very durable and capable of withstanding the physical abuse that industrial slings routinely receive. The ordinary carbon steel chain commonly found in hardware stores should not be used for slings. Wire rope slings can be just as strong as steel chain slings, but wire rope is more subject to wear. Individual wires in the wire rope are more easily broken, with eventual incapacitation of the sling.

Wire rope slings also have a specification for the permitted number of broken wires; the rule is no more than ten randomly distributed broken wires in one rope lay, or five broken wires in one strand in one rope lay. Note from our earlier discussion that this is a bit more strict than the requirement for wire rope for overhead cranes.

Selection of the proper sling for a given application can consider several factors besides rated load. The nature of the item to be lifted, its surface finish, temperature, sling cost, and environmental factors must be considered. The safety and health manager is usually not the person who makes this decision, but there is increasing reason for the safety and health manager to have a voice in what is done in this area. Many factory superintendents, supervisors, and material-handling workers are not aware of the many federal standards confronting the criteria for selection of slings. In fact, many of these personnel do not even have a clear understanding of the inherent hazard mechanisms surrounding industrial slings. Therefore, it is recommended that the safety and health manager provide consultation and advice in the selection and use of industrial slings in the interest of worker safety.

For some criteria, such as load markings, repair procedures, proof testing, and operating temperatures, the requirements for various types of slings are not identical, and even vary in curious ways. Some of this is due to the physical differences in sling types, and some is due to the various origins and rationales for the requirements for various slings. Table 14.3 is intended to summarize some of the more curious differences between requirements for various types of slings, but it is by no means exhaustive. For instance, nylon web slings are not permitted in the presence of acid or phenolic vapors. In the case of caustic vapors, polyester and polypropylene web slings and web slings with aluminum fittings are not permitted. The safety and health manager can use Table 14.3 as a first check in an in-house inspection or purchase decision; then further details should be checked out in the standards.

It must be reemphasized, however, that the skill and training of the worker who uses the sling to attach the load is more important than all of the detailed specifications and sling standards combined. This is a good place for the reader to reflect on the fatality described in Case Study 14.2.

CASE STUDY 14.2

Two inexperienced workers had the assignment of hoisting a 40-foot-bundle of channel steel. The question was where to attach the hoist hooks to the load. The solution they selected was based on their experience in lifting loads with which

TABLE 14.3 Comparison of Certain Requirements for Slings

Type of sling	Rated capacity markings required?	Repairs permitted?	Employer required to maintain records of periodic inspection?	Employer required to maintain a certificate of proof test?	Safe operating temperature (°F)		Proof test of sling required?	Repair records required?[a]
					Maximum	Minimum		
Alloy steel chain	Yes	Yes	Yes	Yes	1000[b]	None specified	Yes	No, except for welding or heat treating
Wire rope								
With fiber core	No	Yes	No	Yes, for welded end attachments	200	None specified	Yes, for welded end attachments	No
With nonfiber core	No	Yes	No	No	400[c]	−60[c]	Yes	No
Metal mesh	Yes	Yes	No	No	550[d]	20[d]	No	Yes
Natural fiber rope	No	No	No	No	180	20	No	N/A
Nylon fiber rope	No	No	No	No	180	20	No	N/A
Polyester fiber rope	No	No	No	No	180	20	No	N/A
Polypropylene fiber rope	No	No	No	No	180	20	No	N/A
Nylon web	Yes	Yes	No	Yes, for repaired slings	180	None specified	Yes, for repaired slings	Yes
Polyester web	Yes	Yes	No	Yes, for repaired slings	180	None specified	Yes, for repaired slings	Yes
Polypropylene web	Yes	Yes	No	Yes, for repaired slings	200	None specified	Yes, for repaired slings	Yes

[a]N/A, not applicable.
[b]If alloy steel chains are working in temperatures exceeding 600°F, load limits are reduced.
[c]Seek manufacturer's recommendations for use outside the temperature range.
[d]Impregnated metal mesh slings have more restricted temperature requirements.

they were familiar—hand-held loads. To these workers, the heavy, steel straps used to secure the bundle seemed to be a natural attachment point. But these straps were not intended to be used as a substitute for a sling. Their strength was insufficient, and the angle of attachment was severe. The angle of attachment will always be severe when bundle straps are used in this way, because to do their job, the straps must be tight. When the load was lifted one of the straps gave way and one of the workers was killed.

CONVEYORS

Hazards with conveyors can be quite serious, and workers seem to sense these hazards more than with some other types of machines. It seems that all of us have implanted somewhere in our imaginations the vision of being tied to the conveyor in a sawmill about to be sawed in half. The truth is that sometimes workers *are* caught in industrial conveyors and not only killed, but dismembered or even pulverized beyond recognition. The horror of this truth instills a healthy respect on the part of most workers around industrial conveyors.

By contrast, some of the worst conveyor hazards are very innocent in appearance. In-running nip points that themselves might not even seriously injure a hand or arm can start an irreversible process once the employee is caught, resulting in an employee's entire body being drawn into a machine. Loose clothing in particular can get caught, and the employee, even before being injured in the slightest way, can become doomed.

Belt Conveyors

On a belt conveyor, in-running nip points are seemingly everywhere. Pulleys are required to drive the belt, change direction of the belt, support the belt, and tighten the belt. One side of every pulley is always an in-running nip point. Defense against this hazard generally consists of one of three means: isolating the nip points, installing guards, and installing emergency tripping devices.

The best method of protection is to isolate the in-running nip point so that an employee would not or could not come into the danger area. If isolation is impractical, a guard can sometimes be installed to keep out the worker's body or extremities. The design of the guard must vary with the application, and sometimes it is difficult to make the guard practical because it may interfere with the operation of the conveyor. Because of body geometry, distance from the danger zone is a design factor in building the guard, a principle that is covered in more detail in Chapter 15.

If both isolation and guarding are impossible or infeasible for the application, the workers can be protected by some type of emergency tripping mechanism. A wire or rope can be run along the length of the conveyor so that a worker falling into the conveyor can grab the trip wire and stop the machine. Unfortunately, this method of protection requires the overt action of an alert worker or coworker.

Overhead Conveyors

Large appliance parts or vehicle assemblies are often handled by overhead conveyors. Hooks attached to a moving chain support each item as it is moved. This type of conveyor is particularly suited for products that have delicate or finished surfaces because so little of the conveyor actually contacts the product. For the same reason, overhead conveyors are very useful for paint spray or finishing operations.

Overhead conveyors avoid many of the risks of belt conveyors by eliminating many of the in-running nip points and by removing the moving parts from worker access. But overhead conveyors have hazards of their own, such as dropping conveyored materials to the plant floor or onto workstations. Screens or guards can protect against this hazard, but not entirely, because the moving parts must be accessible to workstations for processing. A good rule is to place screens or shields under the conveyor whenever it passes over an aisle or other area where personnel are likely to gather. Another good place for screens is where the conveyor chain moves up or down an incline. Such movements cause the loads to shift on their hangers and increase the possibility of a falling load.

Figure 14.14 shows three different orientations for the hangers or hooks that support the work held by an overhead conveyor. Note how much safer is the orientation in which the work is held in front of the hook. If the work encounters an obstruction, it is more likely to catch and stop the conveyor if the work is in front of the hook. If the work trails the hook, an obstruction may lift and knock off the load.

Screw Conveyors

Screw conveyors can be very dangerous. The very principle of their operation is an ingoing nip point at the intake. Complicating the hazard is the fact that in order to operate at full capacity, the intake must be completely submerged in the material to be transported. *Submerged* usually also means hidden, so an unseen serious hazard exists at the intake. Finally, there may be a need for the worker to be fairly close to the screw conveyor for many applications in order to shovel or distribute material into the intake.

A simple and often effective way to protect workers from the hazards of the screw intake is to box the intake area in a small screen enclosure that allows passage of the

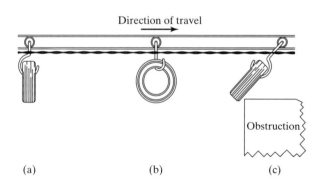

FIGURE 14.14

Three different orientations for conveyor hooks:
(a) hook trails the load; (b) side orientation;
(c) load trails the hook (dangerous).

material, but keeps out fingers, hands, and feet. If even a coarse screen mesh is too fine to permit passage of the material, an enclosure with larger openings may be necessary, perhaps with openings large enough to admit a finger or hand. This type of enclosure can be made safe, too, by making the box large enough that the worker's *reach* will not permit entry of hands or fingers into the danger zone. This follows principles of machine guarding, which are discussed in more detail in Chapter 15.

LIFTING

Before closing this chapter on materials handling, we return to the subject of lifting. It was stated earlier that back injury, mostly from lifting, is one of the biggest compensable injury categories of all. Lifting injuries are very complex and very difficult to control. Naturally, the amount of weight lifted is important, but many other factors determine whether an injury occurs. Even a lightweight lift of 5 to 10 pounds can cause serious back injuries if conditions are just right (rather, wrong). The physical condition of the person doing the lifting is also important.

Much emphasis has been placed on technique, with the most frequently heard saying being "Lift with your legs, not with your back." Unfortunately, the rule is rather difficult to follow because almost everyone is able to lift a heavier weight with the back than with the legs. Lifting with the legs requires squatting down and then lifting both the load and the lifter's body. This requires a great deal of leg strength for heavy lifts and is especially difficult when the worker is unaccustomed to lifting with the legs. Training and exercise with light loads can help in developing the technique, but there are other disadvantages of lifting with the legs. Chaffin and Park (ref. Chaffin) have shown that if the shape of the load is such that it must be brought out in front of the knees, lifting with the legs *increases* the compression force on the lower back! Also overlooked by the often-quoted rule is the fact that lifting with the legs takes as much as 50% more energy as lifting with the back, especially when the load is light and the frequency of lifting higher.

The capacity to lift varies greatly with the horizontal position of the load, which is determined largely by the shape of the object being lifted. Various independent studies of this relationship have been analyzed by NIOSH, resulting in a proposed specification for maximum weight lifted versus the horizontal distance of the load from the center of gravity of the body. This specification is summarized in Figure 14.15, but it must be remembered that the graph represents merely a NIOSH recommendation, not an established standard.

SUMMARY

This chapter has brought recognition to the fact that the handling of material in a manufacturing plant can be as hazardous as the industrial process itself. The basic nature of material-handling hazards were examined; then hazards of specific machines and equipment were discussed.

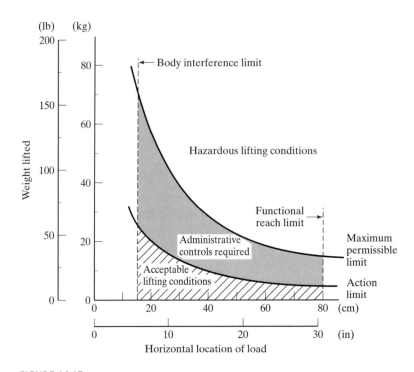

FIGURE 14.15

NIOSH-recommended specification for maximum lift weights at various horizontal distances for infrequent lifts from floor to knuckle height (*Source:* NIOSH).

The safety and health manager needs to be aware not only of the proper safety features to seek in new equipment, but also of the inspection, service, and maintenance of equipment in-house. However, for industrial trucks, cranes, slings, and perhaps *all* material-handling equipment, the operator's skill, attitude, and awareness of hazards are probably more important to the safety of the worker than are the safety features of the equipment itself.

EXERCISES AND STUDY QUESTIONS

14.1 Why are four-legged slings rated no higher than three-legged slings?

14.2 An often-heard safety slogan is "Don't saddle a dead horse!" What does this slogan mean?

14.3 What *safety* design feature do most modern overhead cranes have that is unfortunately not characteristic of the human back?

14.4 Name a part of the plant to which the safety and health manager should give special attention when production and sales are booming.

14.5 In the diagrams, which orientation of the 4000-pound load will place less stress on the sling that handles it? Explain.

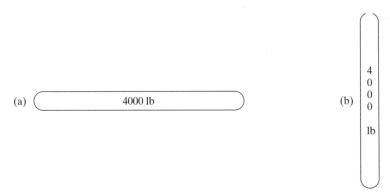

(a) 4000 lb

(b) 4000 lb

14.6 Suppose that a company was able to save some money by trading in its type LPS forklift truck for a type DY. Would this introduce some safety problems? What if the trade were from DY to LPS?

14.7 Why might spring push buttons be a better crane control than toggle switches?

14.8 In Figure 14.16, what is the mechanical advantage? If the nominal breaking strength of the rope is 5000 pounds and the load block weighs 200 pounds, calculate the maximum rated payload of the hoist (not including the load block).

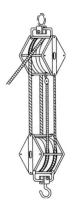

FIGURE 14.16

Blocks and pulleys for Exercise 14.8.

14.9 What is a rail sweep, and why is it needed?

14.10 Name in order of preference three methods of protection for in-running nip-point hazards on conveyor belts.

14.11 Why are overhead crane access ladders specified to be of the fixed type?

14.12 Because of rising gasoline costs, a company desires to convert existing gasoline-powered forklifts to LPG powered. What implications would such a decision have?

14.13 Identify a performance standard in the standards for materials handling. Explain why it is a performance-type standard.

14.14 Explain the following terms as applied to industrial cranes: *bridge, trolley, pendant, pulpit, gantry*, and *cantilever gantry*.

14.15 Name at least four characteristics of forklift trucks that require more s●ll for safe operation than is required for operation of an automobile.

14.16 At least four general characteristics of materials handling contribute to its intrinsic hazard
potential. Name and explain four such characteristics.

14.17 A 2000-pound payload is supported by the crane block-and-tackle assembly shown in
Figure 14.17. In addition to the payload, the load-carrying sheave weighs 100 pounds. Calculate the approximate load on the wire rope. How many parts of rope are used in the
reeving, as shown in Figure 14.17?

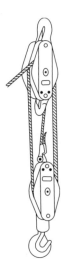

FIGURE 14.17

Blocks and pulleys for Exercise 14.17

14.18 What minimum nominal breaking strength is specified by safety standards to be suitable
for the application described in Exercise 14.17?

14.19 Suppose the wire rope in use in the reeving shown in Figure 14.16 is rated at 2000 pounds
and the load-carrying block weighs 150 pounds. The objective is to lift a load weighing
3000 pounds. Does the assembly as described meet safety standards?

14.20 What maximum payload rating would you assign to the block-and-tackle assembly described in Exercise 14.19? What is the nominal breaking strength of the wire rope?

14.21 A sling with three legs is used to pick up a 1000-pound load. The load is distributed equally among the three legs. When the load is picked up, each leg forms an angle of 30° with the
horizontal. Calculate the tensile force on each leg.

14.22 A three-legged sling distributes its load equally among its three legs. When the load is
picked up, each leg forms an angle of 60° with the horizontal. The rated load of the chain
used in the sling is 6 tons. Calculate the maximum total load this sling is rated to pick up?

14.23 From a safety standards aspect, what is the significance of whether a crane trolley rides on
top of the rail or hangs from the lower flange?

14.24 Explain the term *plugging* as it applies to the operation of an overhead bridge crane. Do
the OSHA standards prohibit plugging?

14.25 Explain the difference between a sling and a hoist.

14.26 Explain the relationship of leg angle to the stress that is placed on a sling.

14.27 What is the principal hazard of using a sling that is too short?

14.28 Explain why it is often hazardous to pick up a load by its binding straps.

14.29 Explain two complicating factors that heighten the hazard of the in-running nip point at a screw conveyor intake.

14.30 Explain how to design out the risk created by the hazards of the in-running nip point at a screw conveyor intake.

14.31 Explain why there is a concern for pits in the floor of a factory served by an overhead crane.

14.32 Give probable reasons for the fact that there is no OSHA standard for lifting.

14.33 Describe the general relationship between maximum lifting capability and horizontal distance between the load and the person performing the lift.

14.34 Explain why lifting with the legs requires more energy than does lifting with the back.

14.35 **Design Case Study.** The objective is to design a trolley hoist for an overhead bridge crane that will meet OSHA standards. The hoist must be rated at 10 tons. The hoist will be reeved with wire rope that has a nominal breaking strength of 30,000 pounds. Specify the reeving arrangement, including the number of sheaves on the load block and on the upper block. Include a reasonable estimate of the weight of the load block. Sketch the reeving arrangement showing the relationship between the blocks and the number of parts of rope.

14.36 **Design Case Study.** Specify the lighting in the aisles of a warehouse with a ceiling height of 20 feet. The objective is to specify the minimum lighting required for forklift trucks to operate without the requirement for headlamps. Specify the lumen output per overhead lamp and the spacing of the lamps.

14.37 **Design Case Study.** You are on the design team for building an overhead bridge crane for use in-house within the company. The design team is considering a proposal to place a switch box on the wall with toggle on/off switches for control of the bridge, trolley, and hoist, respectively. What would be your design input to this committee? Explain the rationale behind your recommendations.

RESEARCH QUESTION

14.38 Search for currently available simulation tools to enable a designer to use virtual reality to test various workplace designs to determine whether the human operator at a proposed workstation will exceed NIOSH recommended lifting limits.

CHAPTER 15

Machine Guarding

17%

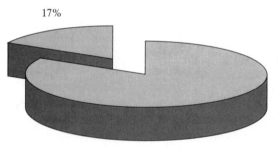

*Percentage of OSHA
General Industry
citations addressing
this subject*

Most people think of a machine guard when industrial safety is mentioned, and for good reason. More efforts and resources have been expended to guard machines than for any other industrial safety and health endeavor. To modify or guard a single machine is generally not a major project when compared with installing a ventilation system or a noise-abatement system. But although each machine guarding modification is usually small, the aggregate becomes a major undertaking involving plant maintenance, operations, purchasing, scheduling, and, of course, the safety and health manager. The safety and health manager should take a leadership role in the implementation of machine guards—enumerating problem areas, setting priorities, selecting safeguarding alternatives, and ensuring compliance with standards.

GENERAL MACHINE GUARDING

If safety and health managers are to be able to "enumerate problem areas" and "set priorities" as just suggested, they need to be knowledgeable about what makes a machine dangerous. Despite wide differences between machines, some mechanical hazards seem to be shared by machines in general, and these mechanical hazards will be discussed first.

Mechanical Hazards

The following are general machine mechanical hazards listed in approximate order of importance:

1. Point of operation
2. Power transmission
3. In-running nip points
4. Rotating or reciprocating machine parts
5. Flying chips, sparks, or parts

In addition to the mechanical hazards listed are others, such as electrical, noise, and burn hazards. These hazards are usually controlled by other methods, however, and are covered elsewhere in this book. It is predominantly the mechanical hazards that are controlled by machine guards, the subject of this chapter.

Although the sequence of priority in the foregoing list is only approximate, there is no question about which hazard should be at the top of the list. By far the largest number of injuries from machines occur at the point of operation, where the tool engages the work. This machine hazard is so important that it is discussed later in a separate section detailing various strategies and guarding devices in controlling the hazard.

The power transmission apparatus of the machine, typically belts and pulleys, is the second most important general machine hazard. Belts and pulleys are usually easier to guard than is the point of operation. Access to the belts and pulleys is usually necessary only for machine maintenance, whereas the point of operation must be accessible, at least to the workpiece, every time the machine is used. Although belts and pulleys are easier to guard, they also are easily overlooked by the safety and health manager. A later section is devoted to belts and pulleys, due to their overall importance to safety.

Machines that feed themselves from continuous stock generate a hazard where the moving material passes adjacent to or in contact with machine parts. This hazard is called an *in-running nip point* or an *ingoing nip point*. Even on machines not equipped with automatic feed, in-running nip points occur where belts contact pulleys and gears mesh. Figure 15.1 shows examples of in-running nip points. In-running nip points are not only direct hazards, but can cause injury indirectly by catching loose clothing and drawing the worker into the machine.

Rotating or reciprocating moving parts can present hazards similar to those of belts and pulleys and in-running nip points—in truth, there is overlap between these categories. But rotating or reciprocating moving parts bring to mind other parts of the machine that might need guarding. Particularly dangerous is a part of the machine that moves *intermittently*. During the motionless part of the cycle workers might forget that the machine will later move. Material-handling apparatus, clamps, and positioners are in this category, as are robots and computer-controlled machinery. The most intermittent motion of all is *accidental* motion. It pays to consider what would happen in the event of a hydraulic failure, broken cotter pin, loosened nut, or some other accidental occurrence. Would the guard on the machine protect workers in these circumstances? Is the risk of occurrence of enough significance to warrant installing a guard?

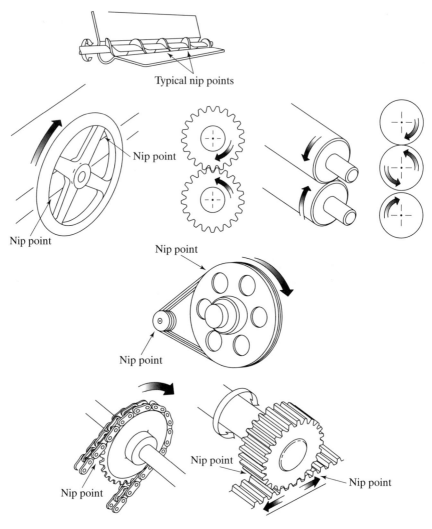

FIGURE 15.1

In-running nip points.

The fifth item listed earlier, flying chips, sparks, or parts, is not necessarily the least significant; it merely stands alone as a somewhat different category. Many machines obviously throw off chips or sparks from the area of the point of operation. Flying objects should also be included because sometimes the product being manufactured actually breaks, and pieces may be hurled at the operator. It is also possible for parts of the machine to break and fall on or be thrown at the operator. One means of protecting workers from flying chips and sparks is with personal protective equipment. But this is not nearly as effective as using machine guards to protect the operator and other workers in the vicinity. Often, the term *shield* or *shield guard* is used to describe guards that protect the operator from flying chips, sparks, or parts.

Guarding by Location or Distance

The easiest and cleverest way to guard a machine is not to use any physical guard at all, but rather to design the machine or operation so as to position the dangerous parts where no one will be exposed to the danger. This is usually in the realm of machine design, and more and more attention is being given to safety in modern machine designs. But even without altering a machine, the machine can be turned and backed into a corner so that its belts, pulleys, and drive motor are impossible to reach during normal operation. A good example would be a portable concrete mixer. Admittedly, this strategy makes the motor and drive difficult to reach for maintenance, but on the other hand, so do ordinary bolt-on guards.

Guarding "by distance" refers to the protection of the operator from the danger zone by setting up the operation sequence such that the operator does not need to be close to the danger. For some difficult-to-guard machines, such as press brakes, the distance method of guarding the point of operation (see Figure 15.2) is expressly permissible. Press brakes are used to bend sheet metal, and their long beds make it difficult to guard their points of operation. When the workpiece is a large piece of sheet metal, the operator stands well back from the point of operation out of necessity and is thus safeguarded by "distance." Although guarding by distance is recognized as an acceptable method of guarding certain difficult-to-guard machines, the safety and health manager is cautioned against generalizing the concept to other types of machines. Guarding by distance is not a positive control to keep the operator or other personnel out of the danger zone at all times.

Tagouts and Lockouts

A surprising number of industrial machine accidents occur not when the machine is in operation, but when it is down for repair or cleaning. A worker simply turns a machine back on, not realizing that it is down for repair and that a maintenance worker is still close to or inside the machine!

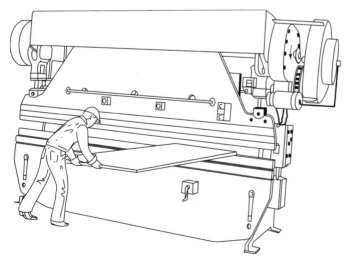

FIGURE 15.2

Press brake guarding "by distance."

Such accidents seem like freak occurrences, but that is because most of us are more accustomed to small machines around the home, where only a few persons, usually family members, may be in the area. But factory machines may be large, and their repair status may not be obvious. Large numbers of personnel may have access to the machine, and a miscommunication between operating supervisors and maintenance crews can easily occur. Case Studies 15.1 and 15.2 will illustrate what can happen when multiple workers are working independently on the same piece of equipment. Human beings have been literally mangled and digested by large industrial machines.

CASE STUDY 15.1

FLOUR BATCH MIXER

A flour batch mixer is sometimes so large as to occupy several floors of a building. In this case, a trainee employee was on the third floor, reaching in to clean a flour mixer when it suddenly turned on, pulling him into the rapidly rotating blades. The start switch for the mixer was situated on the fourth floor of the building, and it was positioned next to a similar-looking switch that controlled the flow of flour from a storage bin to a scale. At the same time that the victim was cleaning the mixer on the third floor, another employee was on the fourth floor weighing out batches of flour. When the fourth-floor employee reached for the switch to weigh a batch of flour, he accidentally pushed the mixer start switch instead, fatally injuring the employee on the third floor. There was an unwritten company procedure for locking out during all maintenance. The procedure was not followed (ref. Preamble).

Two simple safety procedures for preventing accidents of this type are the tagout system and the lockout system. In the tagout system, the maintenance worker places a tag on the on/off switch or control box, so that anyone who might have occasion to turn the machine back on will be warned by the tag to leave it alone. The lockout system (see Figure 15.3) provides positive protection to the maintenance worker because he or she is the only person who has the key to the lockout. Note in Figure 15.3 that the lockout has several positions for separate padlocks for each maintenance worker exposed. Each maintenance worker therefore has an element of personal responsibility and control for his or her safety while working on the machine.

The tagout system is simpler, but the lockout system is required when feasible. It would seem that a lock would not be necessary, because, after all, who would turn on a machine when a maintenance worker's tag warned not to do so? But factories are operated by humans, and various mistakes can lead to a tagout accident. For instance, the maintenance worker can forget to remove the tag after the repair work is complete. Operating personnel may think that the maintenance personnel have forgotten to remove the tag and ignore it. The reader can no doubt think of other scenarios that would lead to an accident with tagouts, but that would have been prevented with the lockout. When the maintenance worker has the only key to the lock, there is no way for an operator to turn the machine back on—that is, with the assumption that the maintenance worker was careful enough to actually use the lock.

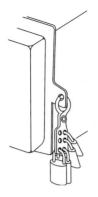

FIGURE 15.3

Lockout system for protection of maintenance workers while the machine is being repaired.

CASE STUDY 15.2

SHEAR HAZARD

An employee was cleaning scrap from beneath a large shear when a fellow employee hit the control button activating the blade. The blade came down and decapitated the employee cleaning scrap (ref. Preamble).

Some types of switches are good for normal starting and stopping of a machine, but they are inadequate for assuring that the machine cannot be turned on accidentally. Push-button switches and selector switches, for instance, do not qualify as *energy isolation devices*, since they cannot safely be locked off. To qualify for lockout, a disconnect switch or circuit breaker should be locked off to nullify the effect of a normal push-button or selector start switch.

Federal standards for lockout/tagout are among the most frequently cited standards in the nation. Not adopted in the beginning as a National Consensus Standard, the OSHA lockout/tagout standard was promulgated in 1989 and quickly rose to near the top of the frequently cited standards list in the early 1990s. Today it remains a high priority for enforcement. Perhaps even more significant is the seriousness of the citations. Nearly 80% of the alleged violations in a single year have been classified as "serious." Proposed OSHA penalties for violation have totaled millions of dollars per year. A majority of the OSHA citations for lockout/tagout have been for lack of proper training and documentation. Thus with the lockout/tagout standard, it is seen once again that the primary problem in complying with many OSHA standards is not the physical control of the hazards, but the failure to provide employee training and documentation of compliance.

Zero Mechanical State

One of the more insidious hazards of machines is that they can quietly hold dangerous energy even when they are turned off. Various forms of energy can be stored, such as pneumatic or hydraulic pressure, electrically charged capacitors, spring tension or compression, or kinetic energy from flywheel rotation. Flywheels are massive wheels that rotate continuously to deliver a uniform source of energy to a machine while running. Flywheels continue to rotate by their own momentum after the power is turned off until the energy of rotation gradually dissipates due to friction. This momentum is sometimes available to operate the machine partially, even after the power is turned off! The great mass of the flywheel often makes it impractical to brake the wheel to bring it to an abrupt stop. However, the stored energy in the rotating flywheel still represents a hazard to maintenance workers. Case Study 15.3 demonstrates what the tremendous energy of an unleashed rotating flywheel can do.

CASE STUDY 15.3

FLYWHEEL ACCIDENT

Two employees were repairing a press brake. The power had been shut off for *10 minutes!* They positioned a metal bar in a notch on the outer flywheel casing so that the flywheel could be turned manually. The flywheel *had not stopped completely.* The men lost control of the bar, which flew across the workplace and struck and killed another employee who was observing the operation from a ladder (ref. Preamble).

The hazards of stored energy in machines, even when they have been turned off, have led to a safety concept known as *zero mechanical state.* To reduce a machine to zero mechanical state, the residual sources of energy still in the machine after it is turned off must be relieved or restrained in such a way as to render them harmless. Pressure must be relieved, springs released, counterweights lowered or locked, and fly-wheels stopped so that they cannot continue to power the moving parts of the machine. Zero mechanical state thus goes beyond lockout or tagout of the power switch.

At this point, the reader should recognize that the concept of zero mechanical state relates to the general fail-safe principle studied in Chapter 3. The fact that some machines can retain dangerous energy in various forms after being turned off or after accidentally losing power is a hazard to be considered in the design of the machine.

Interlocks

Contrasted with the lockout is a safety device called an *interlock.* Modern clothes dryers stop rotation as soon as the door is opened, and thus comply with industry safety standards for revolving drums, barrels, and containers. Even if the drum itself is closed, its rotation can present a hazard unless it is guarded by an enclosure. An interlock between the enclosure and the drive mechanism is specified to prevent rotation whenever the guard enclosure is not in place.

A tumbling machine is a popular industrial machine that uses a revolving drum to rotate metal parts in the presence of an abrasive tumbling medium to improve their surface characteristics. Many tumbling machines found in industry do not have interlocked enclosure guards, and safety and health managers should check for such deficiencies.

Trip Bars

Large machinery layouts are often difficult to guard, but can be provided with trip bars that stop the machine if the operator falls into or trespasses into the danger zone. The

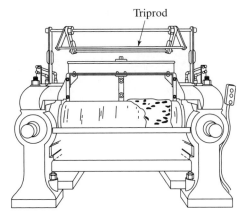

FIGURE 15.4

Pressure-sensitive body bar on a rubber mill.

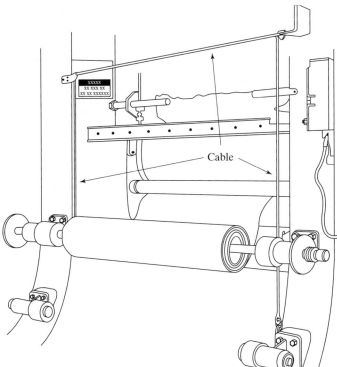

FIGURE 15.5

Safety triprods and tripwires.

operator's hand or body deflects the bar that trips a switch. Figure 15.4 illustrates an emergency trip bar on a rubber mill, a very dangerous machine.

It is sometimes impractical to place the trip bar such that any time the worker falls into the danger zone the bar will trip automatically. An alternative in these situations is to provide a triprod or tripwire for the worker to grab to switch off the machine. Examples are shown in Figure 15.5. Such devices deserve some careful study and experimentation to be sure that workers are able to reach the triprod or tripwire if they get into trouble.

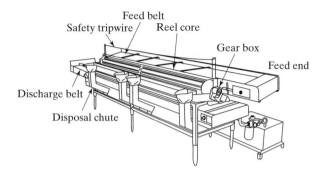

FIGURE 15.5   (*Continued*)

Fan Blade Guards

A very prominent safety standard is the one that requires fan blades to have guards whose openings do not exceed 1/2 inch. Literally millions of ventilation fans are distributed throughout industries all over the country, and many of them have guards with openings larger than 1/2 inch. New fan designs are generally built according to the latest standard, but a principal problem is the retrofit of old fans to meet the 1/2-inch-opening-size standard.

An enterprising safety equipment manufacturer got the clever idea of marketing a nylon mesh that could be wrapped around the existing noncomplying guard and then be drawn up tightly with a drawstring in the rear. (See Figure 15.6.) The nylon mesh was 1/2-inch gauge, and if it was drawn tight, it kept workers free from danger. Before rushing out to purchase these inexpensive nylon mesh guards, the safety and health manager should realize that they are not a panacea. First of all, the nylon mesh, or any guard for that matter, cuts the efficiency of the fan. Furthermore, any fan will gradually accumulate oil and lint on its blade surfaces and on the guard. The nylon mesh seems even more susceptible to oil and lint buildup than does the metal guard, generating a nuisance to maintenance in keeping the guard clean or else resulting in a gradual deterioration in the effectiveness of the fan.

To deal with the fan blade guard dilemma, many innovative guard and fan blade designs are beginning to appear on the market. Some of the new guards are removable for easy washing, and some small plastic-bladed fans have no guards at all. A very

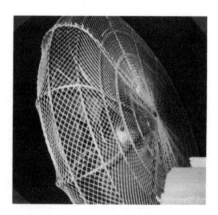

FIGURE 15.6

Nylon mesh guard for ventilation fan blades.

light-weight fan with plastic blades, powered by a small motor, presents no hazard to personnel. With no guard at all, the increased efficiency can go a long way toward making up for the low-power motor on such an unguarded fan.

Anchoring Machines

Another troublesome machine guarding standard is the rule for anchoring fixed machinery to the floor to keep it from "walking," or moving. Such anchoring is required for all machines *designed for a fixed location*. Machines that have reciprocating motions, such as presses, have a tendency to "walk" unless securely anchored. Drill presses and grinding machines can also be hazardous unless anchored.

One interpretation of the phrase "designed for a fixed location" is any machine that has mounting holes in the feet or bases of the legs. It is true that these holes are for the purpose of anchoring machines, but the mere presence of the holes is not proof that the machine must be anchored. The mounting holes might simply be a convenience feature for ease in shipping or to permit the machine to be mounted at the discretion of the user for whatever reason, such as security purposes instead of safety.

SAFEGUARDING THE POINT OF OPERATION

Injury statistics bear witness to the fact that the point of operation is the most dangerous part of machines in general. On some machines, the point of operation is so dangerous that some type of safeguarding is required for every setup; mechanical power presses are one example. Taking guidance from the specific rules for mechanical power presses, the safety and health manager can extend the principles to other machines because most of the safeguarding methods specified will work also for other machines.

A general classification of methods of safeguarding the point of operation is into guards and devices as follows:

1. Guards

    (a) Die enclosures
    (b) Fixed barriers
    (c) Interlocked barriers
    (d) Adjustable barriers

2. Devices

    (a) Gates
    (b) Presence-sensing devices
    (c) Pull-outs
    (d) Sweeps (no longer accepted for mechanical power presses)
    (e) Hold-outs
    (f) Two-hand controls
    (g) Two-hand trips

The reader will note that nowhere in this list do hand-feeding tools such as tongs appear. Hand-feeding tools (see Figure 15.7) are helpful in eliminating the *need* for

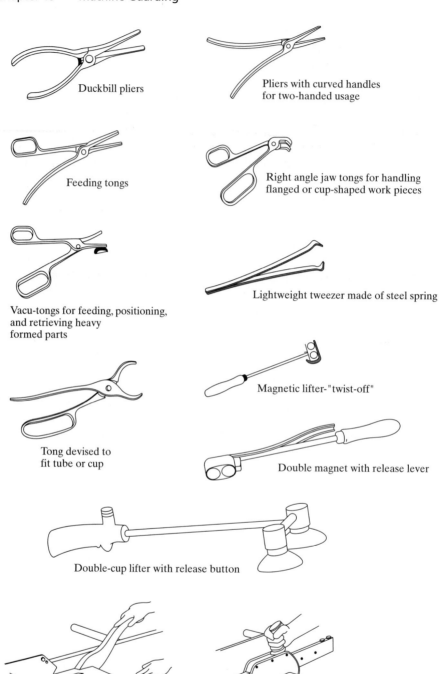

FIGURE 15.7
Hand-feeding tools help with, but do not replace, the function of a point-of-operation guard.

operators to place their hands in the danger zone, but it must be emphasized that these tools do *not* qualify as guards or devices for safeguarding the point of operation.

Guards

The function of a guard is obviously to keep the worker out of the danger area, but many guards do not succeed in this function. Some guards merely screen part of the danger area around the point of operation, but this can be hazardous. Too many workers will defeat the purpose of the guard by reaching through, over, under, or around it, exposing themselves to perhaps a greater hazard than if the guard were not present. Of course, every guard has to be removable between jobs for maintenance or setup purposes, but even this must be kept inconvenient, or operators will take it upon themselves to unfasten the guard to get it out of their way. Wing nuts or quick-release devices should not be used to secure guards. Nuts and bolts are better, but even better than ordinary nuts and bolts are recessed head fasteners, such as Allen screws.

Most guards are metal, and popular design uses expanded metal, sheet metal, perforated metal, or wire mesh as filler material. A secure frame is needed to maintain the structural integrity of the guard. When a guard panel becomes larger than 12 square feet, the rigidity of the guard is in jeopardy, and additional frame components are needed. Many types of ordinary wire mesh are unsuitable because the wires are not secure at the cross points. Ordinary window screen would be in this category. Galvanized screen is better as are some types that are welded or soldered at the cross points.

One principle of machine guarding borrowed from the power press standard is the maximum permissible opening size. A contemplation of human anatomy results in the conclusion that the farther away from the danger zone, the larger can be the openings in the guard without creating a hazard. If the guard is at arm's length from the danger zone, an opening of several inches still might not be dangerous. But if the guard is immediately adjacent to the danger zone, no opening should be large enough to permit a finger to reach through. Standard guard opening sizes are specified in Table 15.1. The

TABLE 15.1 OSHA's Specification for Maximum Permissible Guard Opening Size versus Distance from the Point of Operation

Distance of opening from point of operation hazard (in.)	Maximum width of opening (in.)
$\frac{1}{2}-1\frac{1}{2}$	$\frac{1}{4}$
$1\frac{1}{2}-2\frac{1}{2}$	$\frac{3}{8}$
$2\frac{1}{2}-3\frac{1}{2}$	$\frac{1}{2}$
$3\frac{1}{2}-5\frac{1}{2}$	$\frac{5}{8}$
$5\frac{1}{2}-6\frac{1}{2}$	$\frac{3}{4}$
$6\frac{1}{2}-7\frac{1}{2}$	$\frac{7}{8}$
$7\frac{1}{2}-12\frac{1}{2}$	$1\frac{1}{4}$
$12\frac{1}{2}-15\frac{1}{2}$	$1\frac{1}{2}$
$15\frac{1}{2}-17\frac{1}{2}$	$1\frac{7}{8}$
$17\frac{1}{2}-31\frac{1}{2}$	$2\frac{1}{8}$

Source: OSHA Standard 1910.217, Table O–10.

FIGURE 15.8

Maximum permissible guard opening should depend on distance to the danger zone.

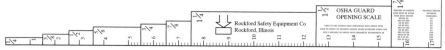

FIGURE 15.9

Guard opening size gauge.

principle behind the standard openings is illustrated in Figure 15.8. Some companies have made available a simple go/no-go guard gauge (see Figure 15.9): Insert the point of the gauge through the guard. If the gauge reaches the danger zone, the guard opening is too large.

Visibility through the guard is a problem with some guards. An old practice was to paint all guards orange. But the bright orange color makes it difficult to see through the guard to the point of operation of the machine; black is by far the superior color for point-of-operation guards. Even better than the color black would be a transparent material.

Die Enclosures

Punch presses and similar machines have mating dies that close on each other to act as the point of operation. The space between the upper and lower dies is the danger area, and the die enclosure guard is designed to enclose only this small area. The advantage over other guard types is that the die enclosure guard is small, but still it is not the most popular guard. Since dies vary widely in size and shape, the die enclosure guard must be essentially custom-made to fit the die or at least the die "shoe" that acts as a base to hold the die. Another disadvantage is that the die enclosure guard is essentially right at the point of operation, which permits no latitude for the guard mesh size or grillwork spacing. The maximum permissible opening size at this point is 1/4 inch, and this may limit visibility. Figure 15.10 illustrates a die enclosure guard.

Fixed Barriers

Fixed barrier guard is a general term for a wide variety of guards that can be attached to the frame of the machine. Figure 15.11 shows one example of a fixed-barrier guard, but remember that there is no set style or shape for such guards. Even the mesh or spacing of the bars is variable, depending on the distance of the guard to the point of operation. (Refer back to Table 15.1.) Large fixed-barrier guards can permit large distances between the guard and the point of operation and a coarser mesh for the guard material.

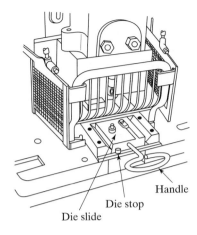

Handle
Die stop
Die slide

FIGURE 15.10

Die enclosure guard used with sliding die for feeding.

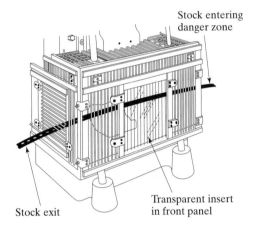

Stock entering
danger zone

Transparent insert
in front panel

Stock exit

FIGURE 15.11

Fixed-barrier guard.

Interlocked Barriers

More sophisticated is the interlocked-barrier guard shown in Figure 15.12. An interlock, typically electrical, disables the actuating mechanism whenever the guard is opened. But the interlock is not required to stop the machine if it has already been tripped, and thus it usually affords inadequate protection to an operator who is attempting to hand

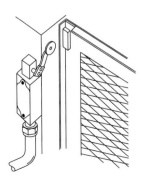

FIGURE 15.12

Interlocked-barrier guard.

feed the machine. If the barrier is so easy to open and close that the operator can open it and reach in while the machine is still moving, the interlocked barrier is not doing its job. Instead of a guard, such an arrangement would be more properly labeled a gate, a device that will be discussed later.

Adjustable Barriers

Guard manufacturers have devised some clever ways for guards to be adjusted to individual applications during the setup. Unlike the fixed-barrier guard, the adjustment is temporary, and the same guard can be reshaped later for a different setup. The trick with adjustable barriers is to make them easy enough to adjust to be practical, but not so easy as to tempt an unauthorized person to tamper with them or gain access to the danger zone. Figure 15.13 shows one type of adjustable-barrier guard.

Awareness Barriers

Some people confuse the term *adjustable-barrier guard* with the term *awareness barrier*. An awareness barrier (see Figure 15.14) is not recognized as a guard and does not meet the guarding criteria of keeping the operator's hands or fingers out of the danger zone. Although the awareness barrier is not a guard, it does provide a reminder that the hands are in danger. In the style pictured in the figure, metal rings or cylinders lie on the table and are lifted by the operator's fingers when the fingers are too close to danger. The operator at that point could go on to reach further into the machine, which could result in injury, but training and good judgment should inhibit him or her from taking this action. Contact with the awareness barrier should be a learned cue to immediately withdraw the hands. The effectiveness of awareness barriers remains in doubt, as some feel that a mere deterrent is not enough to protect the operator. An added complication is that the awareness barrier may obscure the real danger from view. Many operators believe that it is not only a matter of convenience, but also one of safety to be able to see the actual point of operation.

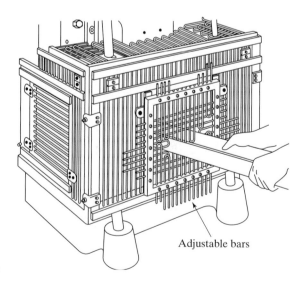

FIGURE 15.13

Adjustable-barrier guard.

Adjustable bars

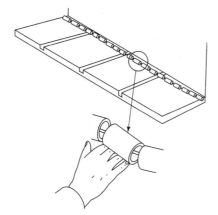

FIGURE 15.14

Awareness barrier installed on a shear.

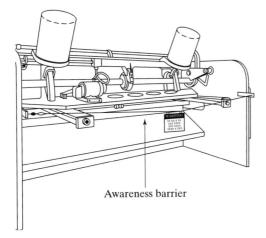

Awareness barrier

FIGURE 15.15

Rear view of power-squaring shear.

Sometimes the term *awareness barrier* is also used to describe a simple rope or chain suspended in front of a danger area with perhaps a sign hanging on it to warn personnel to keep out. An example is the back side of a metal shear, as shown in Figure 15.15. The chain will not ensure that personnel will stay out of the point of operation or danger zone, but it will warn employees of the danger.

Jig Guards

The design of a jig guard is integral to the engineering of the manufacturing operation. The guard has the function of both protecting the operator and facilitating the operation to increase productivity. There is nothing standard about a jig guard because it is designed to fit the individual workpiece and hold it in place while the machine performs the cut or other operation. Jig guards usually move with the work while the operation is performed. The jig guard depicted in Figure 15.16 is being used to notch cross members in the manufacture of four-way pallets. This clever guard keeps the circular table saw blade covered at all times, either by the jig guard between cuts or by the workpiece itself during the cut.

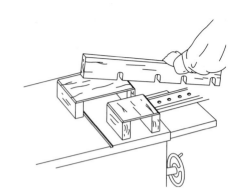

FIGURE 15.16

Jig guard for notching cross members of a four-way shipping pallet.
(Idea credited to the Occupational Center of Central Kansas, Inc.)

Point-of-operation guards are excellent when the machine can be fed effectively by automatic means or through a guard window without the operator's having to reach into the danger zone. But the only feasible way to feed some machines is manually, and some of these machines are very dangerous. This class of machine is typified by the mechanical power press, to be discussed next. The only way to assure protection of the operator while hand feeding these very dangerous machines is to use some kind of *device* that keeps the operator's hands out of the danger zone while the machine cycles and does its work. Some ingenious devices have been developed for this purpose, as will be seen in the next section.

POWER PRESSES

Punch presses are at the same time one of the most inherently dangerous and most useful production machines in industry. The epitome of mass-production machines, the press excels when huge volumes of identical products are required. Mass production depends on interchangeable manufacture, which in turn requires machines that successively produce parts that are essentially identical. The power press is superbly qualified for this task.

Figure 15.17 illustrates two popular model power presses. The term *power press* is really a general one that encompasses hydraulic-powered models and forging presses in addition to the popular mechanical punch press. The outstanding feature of a power press is the set of mating dies that close on each other to cut, shape, or assemble material, or to do combinations of these operations in one or more strokes. The more subtle features of presses, such as methods of power transmission and control of the stroke, are important determinants of permissible safeguarding. It is at the mechanical models, powered by flywheels, that the most restrictive press standards are aimed.

Press Hazards

There are, of course, reasons that so much importance is attached to safeguarding the point of operation of mechanical power presses. The injury record for power presses is not a good one. An estimate by Ryan in 1987 indicates that there are about 2000 press-operator work-related amputations each year in the United States (ref. Ryan). When feeding the press by hand, the operator is close to danger every time the dies close, and this happens thousands upon thousands of times in a press operator's career. One slip

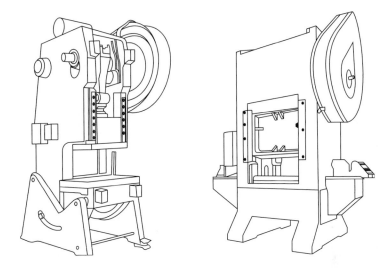

FIGURE 15.17

Typical power presses: (a) open-back inclinable (OBI) model; (b) straight-sided model.

and in a fraction of a second a finger or a hand is amputated. Such accidents were commonplace in the first half of the 20th century. Shortly before World War II, there developed an awareness that even the careful operator could become a victim of the power press, and efforts were initiated to eliminate the hazard.

To understand the nature and significance of the power press hazard, it is necessary to study the interaction between human being and machine. In a hand-feed setup, the press and the operator are alternating actions in a rhythm that cycles every few seconds, and some bench press operations may cycle in a fraction of a second. Figure 15.18 depicts the sequence of actions in a typical press cycle employing hand feeding without safeguarding. Production incentives motivate the operator to higher and higher speeds as skill develops. The operator learns to develop rhythm with the press motion. The sound of the press-tripping mechanism, the closing dies, and other press motions can become cues to the operator to make a hand or foot motion. The process involves eye–hand–foot coordination every cycle. It is easy to visualize the hazards involved in hundreds of thousands of repetitive cycles.

One of the biggest causes of accidents with power presses is an attempt on the part of the operator to readjust a misaligned workpiece in the die. The motivation is a very powerful one to reach back in to correct the error even after the ram has been tripped. If the operator lets the error go, the misaligned workpiece can wreak havoc when the dies close. At the very least, the workpiece will be ruined. More likely is that the expensive dies will be broken or ruined, and it is very possible that the press frame itself will be damaged or broken. One misaligned workpiece can result in many thousands of dollars damage to the dies and the press itself. But even worse, the misaligned workpiece or the dies can be fragmented when the dies close, subjecting the operator to injuries from flying metal pieces. It is no wonder, then, that the operator has a powerful urge to reach back in to correct a misaligned workpiece. The operator's eye will

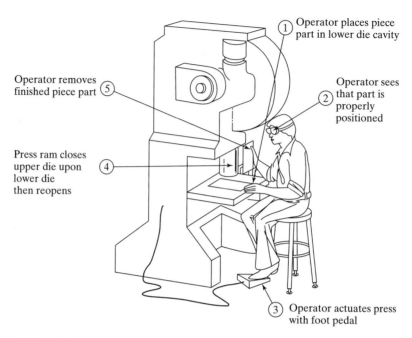

Operator places piece
part in lower die cavity

Operator removes
finished piece part

Operator sees
that part is
properly
positioned

Press ram closes
upper die upon
lower die
then reopens

Operator actuates press
with foot pedal

FIGURE 15.18

Press-cycle operation sequence (no safeguarding).

see the error, and the hand will reach in—even though the foot may have already de-pressed the pedal to actuate the press.

The somewhat insidious nature of power press hazards motivated the standards writers to initiate a policy of "no hands in the dies." The theory was that tongs or other tools and devices could be used to feed workpieces, eliminating the need for the work-er to put his or her hands into the danger zone. In addition to the tongs or other feed-ing tools, press guards or safety devices would prevent operators from placing their hands or fingers into the danger zone, even if they tried, while the dies were closing.

The theory worked for most jobs, but some applications presented awkward feeding problems that defied solution with the no-hands-in-the-dies approach. It be-came clear that the metal-stamping industry would not be able to comply with a rigid no-hands-in-the-dies rule in all situations, and support for the theory began to crumble. In its place, new rules were established for assuring the reliability of safeguarding de-vices for protecting operators at the point of operation.

Press Designs

To understand the somewhat complicated rules for safeguarding a power press, it is necessary to examine the basics of how presses work. Most presses are mechanically powered, although there are a large number of hydraulically powered models, general-ly recognizable by the presence of a large hydraulic cylinder above the ram. The cylin-der usually resembles the large cylinder in an automobile service station lift. Hydraulic power presses are explicitly excluded from coverage by standards for mechanical power presses, as are pneumatic presses, fastener applicators, and presses working with hot metal, *even if they are mechanically powered*. Also excluded is the press brake, a power press with a long bed (refer back to Figure 15.2) that is used to bend sheet metal. Shears, because they employ blades instead of dies, are not considered within the definition of mechanical power presses.

The easiest way to distinguish a mechanical power press from other types is by the presence of a large heavy flywheel that, by its rotation, carries energy that is imparted to the ram when the press is tripped. The flywheel is *usually* mounted on one side of the press near the top, as was shown in the typical presses in Figure 15.17.

One of the most important features is whether the press is a *full-revolution* or a *part-revolution* type. This refers to the method of engaging and disengaging the flywheel to deliver power to the ram. Full-revolution types make a positive engagement that cannot be broken until the crankshaft and flywheel make one complete revolution together. During this revolution, the press ram goes down, the dies close and then reopen, and the ram returns to its full-up position to await another cycle. At the end of the revolution, the flywheel is disengaged and rotates freely under the power of the motor.

The part-revolution machine typically has a friction clutch that can be disengaged at any time during the press cycle. Compressed air is used to engage or disengage the clutch instantly at the discretion of the operator. Upon disengagement of the clutch, a brake is applied that immediately or almost immediately stops the ram. It is easy to see the advantage of being able to interrupt a press stroke at any point in the cycle, but the part-revolution advantage does not stop there. The instantaneous engagement is also valuable in causing the press to cycle quickly once engaged, giving less time for the operator to get into trouble by making an afterthought reach into the point of operation.

The safety and health manager must know which presses are full revolution and which are part revolution in order to know how to equip the press with the proper safety equipment. Thousands of dollars have been foolishly spent by purchasing the wrong safety equipment for a power press. One way to get a rough idea whether the press is full revolution is by the age of the press. Most presses, and invariably the older presses, are full revolution, unless they have gone through a renovation process. Not incidentally, the average age of mechanical power presses in the United States is steadily increasing. Statistics suggest that about half of all presses in the United States are more than 20 years old!

Since part-revolution presses employ a friction clutch, the housing on the flywheel often has a bulge to accommodate the clutch, as shown in Figure 15.19. The clutch is actuated pneumatically, and this generally means that an extra line can be seen on the outside of the flywheel cover running to the center of the clutch, as shown in the figure. Full-revolution clutch machines may also have a small line running outside the flywheel housing, but this is an oil line for lubricating purposes. None of these criteria for distinguishing machines is 100% reliable, so they should be used only as a means of preliminary screening for possible troublespots. Press engineers or the equipment manufacturer's representative can be consulted for an authoritative determination.

Point-of-Operation Safeguarding

Once having determined whether a press is a full- or a part-revolution unit, the safety and health manager or engineer can proceed to a determination of the most effective

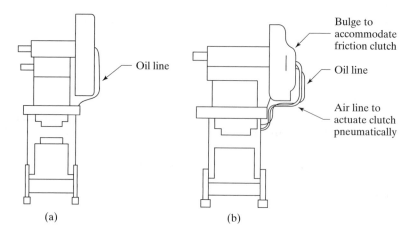

FIGURE 15.19

Full-revolution and part-revolution machines: (a) full revolution; (b) part revolution.

means of safeguarding its most dangerous zone, the point of operation. There are at least ten recognized methods of power press safeguarding, but acceptability of each depends on the press configuration and method of feeding. The various safeguarding methods can be divided into the following four categories, ranked according to degree of security:

1. Methods that prohibit the operator from reaching into the danger zone altogether
2. Methods that prohibit the operator from reaching into the danger zone any time the ram is in motion
3. Methods that prohibit the operator from reaching into the danger zone only while the dies are closing
4. Methods that do not prohibit the operator from reaching into the danger zone, but that stop the ram before the operator can reach in

Only categories 1 and 2 are to be trusted for a full-revolution press. Some category 3 methods are permissible for full-revolution presses, but with category 3, there is exposure to the hazard that the press will "repeat." So far, no antirepeat mechanism has been found to ensure absolutely that a full-revolution press will not repeat with an extra unwanted stroke. These repeat strokes are a terrifying possibility, but fortunately that possibility has been significantly reduced in recent years. Recalling that it is likely that about half of all presses are more than 20 years old, the threat of repeats is still something to consider.

Category 4 methods of safeguarding are permitted by OSHA only for part-revolution presses, and this is where many safety and health managers go wrong. It should be obvious that any method that depends on stopping the ram to protect the operator must be installed only on part-revolution presses. Full-revolution presses, by definition, cannot be stopped. It is surprising, though, how many category 4 devices are seen in industry installed on full-revolution presses. It is true that these devices can afford some protection on full-revolution presses by locking out the tripping mechanism while the operator is in the danger zone. Once the press is tripped, however, these devices are powerless to stop the ram.

Press Guards

The earlier section on guarding described types of guards for use on machines in general. Four of these types—die enclosure, fixed barriers, interlocked barriers, and adjustable barriers—are acceptable on mechanical power presses. Indeed, the die enclosure guard is used almost exclusively on mechanical power presses, although it is not as popular as some of the other types. The fixed-barrier guard is a very popular method of guarding power presses that employ automatic feeding of coils of strip stock and automatic ejection of the finished parts. The interlocked press barrier guard is not permitted for hand feeding, but a gate *device* (not a guard) can be used. Indeed, *none* of the four types of guards is permitted for hand feeding the press (putting the hands in the dies) because by definition of a press guard, "it shall prevent the entry of hands or fingers into the point of operation by reaching through, over, under, or around the guard."

With only one exception, a guard or some type of point-of-operation safeguarding device must be installed on *every* mechanical power press. That one exception is where the full-open position of the ram results in a gap between the mating dies of less than 1/4 inch, too small to permit entry of the fingers (see Table 15.1), and thus too small to be a hazard. But for all other mechanical power presses, even the automatic-feed models or the robot-fed setups, point-of-operation safeguarding is required. An accident can occur even with automatic feeding if a worker tending the automatic setup attempts to adjust a workpiece during operation.

We have already established that guards cannot be used where the operator feeds the press by putting his or her hand in the die. We have also stated that virtually every mechanical power press is required to have point-of-operation safeguarding. Does this make hand feeding illegal? The answer is no, and the key is the difference between the terms *guarding* and *safeguarding. Safeguarding* is a more general term and encompasses a variety of mechanical or electromechanical devices that protect the operator even when hand feeding is used. The standards are specific about which devices can be used with which types of machines and appropriate setups. Each of these configurations for mechanical power press devices will now be considered.

Gates

Gates look somewhat like a guard (see Figure 15.20), but are different in that they open and close with every machine cycle. In contrast to interlocked barrier guards, gates *can* be used for manual feeding. Gates are used almost exclusively with mechanical power presses.

There are two types of gates: type A and type B. The *type A gate* is the safer of the two because it closes before the press stroke is initiated, and it *stays closed* until all motion of the ram has ceased. *Type B gates* are the same except that they remain closed only long enough to prevent the operator from reaching in during the more dangerous downward stroke. Although the upward stroke is less dangerous, there is still the hazard of repeats when an operator reaches in on the upward stroke. Type B gates are not

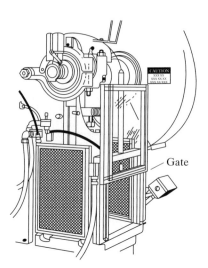

FIGURE 15.20

Gate devices.

forbidden for full-revolution clutch machines, but the occasional tendency of these machines to repeat is a sobering thought, and type B gates are not recommended for presses with full-revolution clutches.

In defense of the type B gate, however, is its efficiency over the type A gate. A substantial percentage of the press cycle time can be saved if the operator can start reaching in as soon as the ram starts back upward. The savings may be only a fraction of a second per cycle, but over hundreds of thousands of cycles, the difference can be significant. The labor and overhead cost associated with the operator's time is saved, and also conserved are productive capacity of the press and the plant floorspace in which it is housed. It is possible for these combined costs to exceed $50 an hour. This represents an incentive to modernize the power press equipment so that it qualifies for the most efficient safeguarding systems.

Presence-Sensing Devices

Modern electronic detection devices have made their way into the machine-guarding industry, and there are several devices of this type available for protection of the point of operation. One type uses a bank of photoelectric cells to set up a light screen, the penetration of which will immediately stop the ram. Figure 15.21 illustrates this type of device.

It becomes a game for workers to defeat these devices intended for their protection. It is obvious that if a worker can reach over or around the light screen, the machine will not stop. Alternate points of entry not covered by the sensing device should be guarded so that the operator cannot reach into the point of operation without tripping the device.

Another way to defeat a photoelectric device is to use ambient light to maintain the sensors in an energized condition at all times even if the worker's hand has broken the field. To compensate for the ambient-light problem, most presence-sensing devices function in the infrared frequencies rather than in the visible-light spectrum. This makes the "light screen" invisible, and this feature may have some advantages, too.

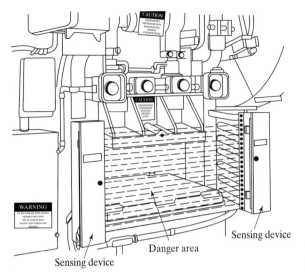

Sensing device

Danger area

Sensing device

FIGURE 15.21

Photoelectric presence-sensing screen.

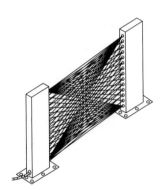

FIGURE 15.22

Programmed crisscross scan for sensing plane.

Another way to trick a light screen is to somehow squeeze between the beams. If the sources and sensors are tightly spaced, squeezing between the beams becomes impossible for the parts of the human body. The concept can be demonstrated, however, by carefully slipping a paper or cardboard edgewise between adjacent beams. Many commercially available screens have a sophisticated programmed scan that crisscrosses the field in such a complicated fashion as to prevent defeating the device. Such crisscrossing may reduce the number of sensors required to protect a given area. The concept is illustrated in Figure 15.22.

Another type of presence-sensing device uses a conductor to set up an electromagnetic field in its vicinity. Many variables affect the tripping threshold of these devices, sometimes called *radio-frequency sensors*, and this has damaged their reputation. For instance, one person's body, due to its mass or conductivity characteristics, might trip the mechanism from a distance of 2 feet from the point of operation. Another person might not trip the device until he or she has actually entered the danger zone. If this type of device is used, it should be "tuned" to have the proper sensitivity for a given operator and setup. Figure 15.23 illustrates the electromagnetic field type of

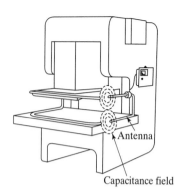

Antenna

Capacitance field

FIGURE 15.23

Electromagnetic field type of presence-sensing device.

presence-sensing device. Late-model versions of this device have been found to be very effective.

Presence-sensing devices are quite practical for hand feeding in conjunction with a foot switch. Thus, if the operator's rhythm is off and the foot switch is depressed too soon (while the operator's hand is still in the sensing field), the machine will not operate. Even more important, if the operator sees a misaligned workpiece and attempts to reach in after the ram has started its downward motion, the sensing field will detect this action and stop the ram. Thus, the device not only prevents operator injury, but it also arrests a costly smash-up of the dies and possible damage to the press itself.

From what the reader has already learned, it should be obvious that this advantage of the presence-sensing device is feasible only for part-revolution presses. Indeed, the use of presence-sensing devices for safeguarding the point of operation on full-revolution mechanical power presses is prohibited.

Although the presence-sensing device is designed to stop the ram before the operator can reach the danger zone, the worker should not tempt the machine. The author knows of a case in which a new machine[1] equipped with a presence-sensing device was being demonstrated by a proud worker to his family during an open house. The worker repeatedly thrust his hand into the machine to prove that the photoelectric cell was quicker than his hand. Finally, he succeeded in beating the machine and lost the ends of his fingers. This "accident" actually happened.

Presence-sensing devices are like gates in that they are subject to the issue of when to return access to the point of operation to the operator. Since presence-sensing devices are permitted only for part-revolution presses, it seems sensible to afford them the same production efficiencies permitted the type B gate. Therefore, it is permissible to deactivate the sensing field on the upstroke. This process of bypassing the protective system is called *muting*. As soon as the ram reaches its home position, the system is reset to its protective mode. Muting permits the same production efficiencies that the type B gate holds over the type A gate.

While we are discussing efficiency, why not go one step further and eliminate the foot switch? It would be feasible for the press to be restrained by the control system for as long as the operator's hand or arm breaks the sensing field during hand feeding.

---

[1] A printing press, not a punch press, in this instance.

Then, as soon as the operator withdraws his or her hand from the danger zone, the press would *automatically* cycle without a signal from the operator! Feasible? Yes. Legal? No. The presence-sensing device is prohibited as a press-tripping mechanism, although many European factories do employ this highly productive mode of operation for power presses.

There is a way that a presence-sensing device *can* be used to trip the press—when the presence-sensing device is used in conjunction with another safeguarding device, such as a gate. Thus, the gate represents the safety device, and the presence-sensing field acts as a tripping device. This would be a complicated and expensive system and should be considered rare.

Presence-sensing devices must be designed to adhere to the general fail-safe principle stated in Chapter 3. Thus, if the presence-sensing device itself fails, the system must remain in a protective mode. A failure in the device must prevent the press from operating additional cycles until the failure is corrected. But such a failure must not deactivate the clutch and brake mechanisms, which are essential in stopping the press. If a failure in the device causes an interruption of the main power supply to the machine, the clutch must automatically disengage. Of course, the clutch and brake system should have this characteristic, regardless of the choice of safeguarding devices; the clutch and brake are simply additional examples of systems that should be designed in accordance with the general fail-safe principle.

One final note about failures in the presence-sensing system should be mentioned. It is not enough that the failure inhibits the operation of the press; the system must also indicate that a failure has occurred. This is usually done with a trouble light on the panel.

Pullbacks

A very popular method of safeguarding a power press is by means of cables mechanically linked to the travel of the ram. These cables are attached to wristlets which pull the operator's hands out of the danger area as the ram makes its downward stroke. Figure 15.24 shows one example setup of *pullbacks*, or *pull-outs* as they are sometimes called.

One reason for the popularity of pullbacks is their versatility. They can be used with virtually any power press, regardless of power source or type of clutch. However, pullbacks do have their disadvantages.

Proper adjustment is very important to the effectiveness of pullbacks, especially with respect to close work done on bench model machines. Even the method of attachment to the wrists is important because ordinary wristlets permit too much variation in reach. Figure 15.25 is a close-up of a wristlet assembly that minimizes variations in an operator's restricted reach. Even with properly designed wristlets, proper adjustment is critical. Differences in operators' hand sizes can be a factor, but much more important are variations in die setups. A large die set will of course have a danger zone that extends closer to the operator, requiring an adjustment in the pullback limit of reach.

Recognizing the hazard of improper adjustment of pullbacks, safety standards require an inspection for proper adjustment at the start of each operator shift, following a new die setup, and when operators are changed. This inspection requirement, and especially the *frequency* of the inspections, decidedly cools the interest of employers in the use of pullback devices. Were it not for this problem and the fact that many workers dislike wearing pullbacks, these devices would be extremely popular. These drawbacks

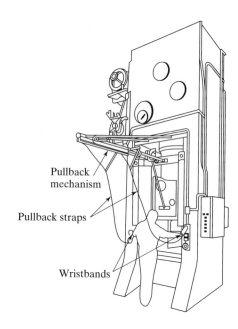

FIGURE 15.24

Pullback devices for safeguarding the point of operation.

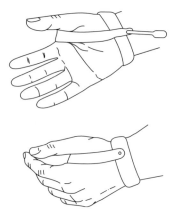

FIGURE 15.25

Close-ups of wristlet assembly for pullback device.

notwithstanding, the pullback device remains one of the most popular press-safeguarding devices.

It is usually the diesetter or operating supervisor who makes the actual check of the pullbacks for proper adjustment. The concern of the safety and health manager should be that the job is done effectively and that a record is made of the inspections. Simplicity is the key to the inspection record system for pullback devices. Simplicity will help to ensure that the job gets done and will also minimize the impact on production efficiency. One convenient method is to use a tag attached to the pullback device itself, with blank lines for indicating "date inspected" and "initials" by the responsible party. In multiple-shift operations, in new die setups, or in operator changes, more than one line would be initialed in one day on the tag, but the tag still could be easily designed to accommodate multiple entries on the same day.

Sweeps

Very popular in the past, and still seen on some power presses, are devices that sweep away the operator's hands or arms as the dies close. These devices, illustrated in Figure 15.26, have fallen into disfavor as a means of protecting the operator. Operators even fear injury from the sweep device itself as it comes swinging down in front of the machine. The design dilemma of these devices is that they must be powerful enough to inflict injury themselves in order to be effective in positively removing the operator's hands from the point of operation. But the overriding reason for their disfavor is that the design and construction of sweeps are inherently inadequate as a press-safeguarding device. Sweeps are no longer recognized as adequate point-of-operation safeguarding devices on mechanical power presses.

Hold-Outs

A simplification of pullbacks is the *hold-out* (sometimes called *restraint*) device, which is feasible only for setups in which it is unnecessary for the operator to reach into the danger area. Figure 15.27 shows that hold-outs look almost exactly like pullbacks, the

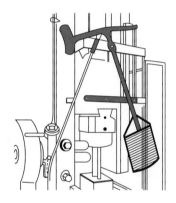

FIGURE 15.26

Sweep device. This device no longer qualifies as an acceptable safeguard for mechanical power presses.

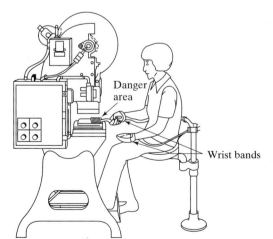

FIGURE 15.27

Hold-outs or restraints for restraining the operator's hands from reaching into the danger zone at *all* times (compare with pullbacks).

difference being that the hold-out reach is fixed and does not permit the operator to reach in at all, even between strokes of the machine. If tongs, suction cups, or other gripping devices can be used to feed a machine manually, it is feasible to use hold-outs instead of pull-outs to protect the operator. Even without these gripping devices, long workpieces can be hand fed into the machine without actually placing the hands in danger. For these applications, hold-outs are appropriate. If the operator's hands must enter the zone between the dies, however, hold-outs are infeasible as safeguarding devices.

It would seem that protection for the operator would be unnecessary for applications in which tongs or other feeding devices are used instead of the hand to feed a machine. However, this idea does not recognize the strong tendency for the worker to reach in when something goes wrong. Therefore, although hand-feeding tools are somewhat useful in promoting safety, they are not recognized as point-of-operation safeguarding devices. Another means must be used to *ensure* safety, and hold-outs are a good method when hand tools are used.

Two-Hand Controls

Since people have only two hands, neither hand will be injured by the point of operation if the machine can require both of them to operate the controls, or so the theory goes. The theory is a good one, but some sophistication is necessary to ensure that the device achieves its goal. Workers take pride in "beating the system" or in tricking the machine to make it operate without their using the controls. One trick is to use a board or rope to tie down one control so that the operator can operate the machine with one hand and feed it with the other. Another trick is for workers to use their heads, noses, or even toes to depress one of the controls. Workers have been known to try almost anything to defeat the safety features of a machine in order to achieve a production breakthrough and receive a higher production incentive payment. This points to the power of money over personal safety in motivating workers. It may also reveal some inefficiencies in safety devices as currently designed—inefficiencies that unduly compromise productivity in the name of safety. As was pointed out in Chapter 3, industry workers and managers will tolerate some slowdown of an operation for the cause of safety, but not much.

Figure 15.28 illustrates a two-hand control device and some of the features that attempt to prevent the worker from defeating the device. Note the smooth, rounded surface of the palm button, which makes it convenient for the palm, but not for tying down. Also note the cups around the button, which are intended to foul attempts at tying down the button. Control circuitry can also be used to detect foul play and stop the machine if the buttons are not depressed concurrently and released between cycles.

Controls versus Trips

The term *control* implies a more sophisticated device than a mere *trip* to actuate the machine. Within the context of safeguarding the point of operation, a two-hand control means a device that not only requires both hands concurrently to actuate the machine, but also stops the machine by interrupting its cycle if the controls are released prematurely. The nature of some machines does not permit such a degree of control over the machine cycle. Two-hand controls are infeasible for these machines.

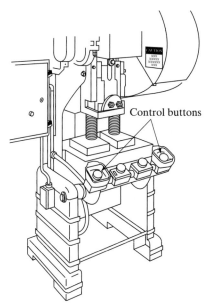

Control buttons

FIGURE 15.28

Two-hand control.

For full-revolution presses and other machines that cannot be stopped once their cycle has begun, two-hand trips are used instead of two-hand controls. Two-hand trips require both hands to start the machine cycle, but once started there is no protection for the operator. If the machine cycle is fast, and if the operator is far enough away to stay out of danger, two-hand trips effectively protect the operator. But if the machine is slow or the trip station is close enough, the operator can reach into the machine *after* it has been tripped. This gets into the subject of safety distances, covered in the next section. After studying the section on safety distances, it will be easy to see that controls are superior to trips.

Safety Distances

In reviewing the safeguarding devices for protection of operators from the points of operation of machines, we see that most of them protect the operator by making it impossible to reach into the danger zone after the machine cycle has begun. However, two of these devices—the presence-sensing device and the two-hand control—rely on a capability of interrupting a machine in *midcycle*. Every mechanical machine has inertia, so some time must elapse between the signal to stop and the complete cessation of motion of the machine in the point-of-operation area. If the inertia is great, an operator might be able to reach quickly into the danger zone before the protective device is able to completely stop the machine. Therefore, the operator station must be moved back from the point of operation a sufficient distance so that a reach into the danger zone before the machine stops is impossible.

In addition to the presence-sensing device and two-hand control, the two-hand trip device must also be located at a sufficient distance, as was mentioned earlier. Although the two-hand trip is not capable of stopping the machine, protection is possible if the distance to the danger zone is great enough to prevent the operator from reaching in after releasing the palm buttons.

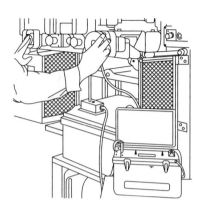

FIGURE 15.29

Brake-stop-time measurement device.

To compute a *safe distance* for a presence-sensing device or a two-hand control, it is necessary first to compute the stopping time of the machine. Figure 15.29 shows one model of a stop-time measurement system that connects one sensor to a palm button and the other to the machine motion. Upon a signal that the palm button has been depressed, the system starts counting in fractions of a second until motion of the machine ceases. Keep in mind that we are speaking here of the motion of the machine's ram, die, cutter, or other part *at the point of operation* that could cause injury. Flywheel motion or motor rotation may continue and indeed due to their inertia would usually be impossible to stop in time to be of any benefit. The resulting stopping time is generally displayed by the portable instrument. The reader should take care not to confuse the brake stop-time measurement device in Figure 15.29 with a *brake monitor* to be discussed later.

Once the stopping time has been determined, this time must be multiplied by the maximum speed the hand can move toward the point of operation, as in the following formula:

$$\text{Safety distance} = (\text{stopping time}) \times (\text{hand speed constant}) \qquad (15.1)$$

OSHA relies on a maximum hand movement speed of 63 inches per second, sometimes referred to as the hand speed constant; this constant is credited to L. Lobl of Sweden. Consider the next example.

Example 15.1

A power press is protected by an infrared-beam-type presence-sensing device. A stop-time measurement system is used to compute the time between breaking of the infrared beam and stoppage of the press ram. This stop time is found to be 0.294 second. The safety distance then is computed to be

$$\text{Safety distance} = 0.294 \text{ second} \times 63 \text{ inches/second}$$

$$= 18.5 \text{ inches}$$

Thus, all points in the plane of the infrared sensing field must be at least $18\frac{1}{2}$ inches away from the point of operation for this power press.

The calculation would have been identical if the device in the example had been a two-hand control (but not a two-hand trip!). The palm buttons of the two-hand control would have to be placed $18\frac{1}{2}$ inches away from the point of operation.

The safety distance calculation for the two-hand *trip* uses an entirely different logic from that used for the two-hand *control*. True, the same hand-speed constant of 63 inches per second is used. But with two-hand trips, the idea is for the machine to complete the dangerous part of its cycle before the operator can reach in after releasing the palm buttons. Thus, the *slower* the machine, the more dangerous it is when using trips. It is paradoxical that the *faster* the machine, the harder it is to stop and the greater must be the safety distance for two-hand *controls*, whereas for *trips*, it is just the opposite: the *slower* the machine, the greater must be the safety distance.

With many machines, hydraulic presses, for example, the commencement of the cycle is virtually instantaneous with the depressing of the palm buttons. But with machines that commence the cycle with the mechanical engagement of a flywheel, there is an inevitable delay as the engagement mechanism waits for its mating engagement point on the flywheel. The delay can be considerable for a slow-rpm machine, especially if it has only one place on the flywheel suitable for making the mechanical engagement. The process is best explained by a diagram, as in Figure 15.30. Part (a) of the figure shows a very unlucky happenstance in the position of the flywheel with its engaging point just past the tripping mechanism at the moment of actuation. This follows the principle that the *worst state* of the machine should be used to determine how to make the operation safe. Since the position of the flywheel at the moment the machine is tripped is purely a matter of chance, it is equally likely that the flywheel will be in the lucky position indicated in Figure 15.30(b). But since this position cannot be counted on, the safety distance is computed assuming the flywheel is in the position indicated in Figure 15.30(a).

One complete rotation of the flywheel completes the machine cycle for most machines. One-half of this rotation of the flywheel is the dangerous portion of the stroke, the closure motion of the machine. Thus, adding a complete revolution for engagement plus one-half revolution for closure motion, the resultant danger period is one and one-half revolutions of the flywheel for a machine whose flywheel has only one engagement point. For machines that have several engagement points evenly spaced around the flywheel, the danger period is shorter, depending on how many engagement points exist. To see this, consider an example of a machine with four engagement points. In the worst case, the farthest the engagement point could be from the tripping mechanism at the moment of tripping would be 90° or one-quarter revolution. Add this

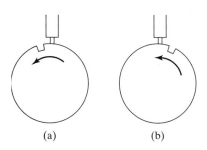

(a)   (b)

FIGURE 15.30

Two possible positions of the flywheel when a machine is tripped: (a) unlucky position—the engagement point has just passed the tripping mechanism; (b) lucky position—the engagement point is approaching and is very near the tripping mechanism.

to the one-half revolution during machine closure, and the total danger period becomes three-quarters revolution.

Summarizing the calculation of safety distances for two-hand *trip* devices is the following formula:

$$\text{Safety distance} = \frac{60}{\text{rpm}}\left(\frac{1}{2} + \frac{1}{N}\right) \times 63, \tag{15.2}$$

where rpm represents the flywheel speed in revolutions per minute while engaged, and N is the number of engaging points on the flywheel.

In Equation (15.2), the factor 60/rpm is used to determine the time in seconds required for the flywheel to complete one revolution. The computation in parentheses is the worst-case number of flywheel rotations until closure, and the factor 63 is the hand-speed constant in inches per second. The computed safety distance is in inches.

It is important to remember that Equation (15.2) is to be used for two-hand *trips*, not two-hand controls. It is easy to determine that two-hand controls are superior to two-hand trips. From a production efficiency standpoint, two-hand controls can be normally placed much closer to the machine, which facilitates hand feeding. Two-hand trips can be more dangerous unless moved to a greater safety distance, which in turn impairs productivity.

Figure 15.28 was captioned "two-hand control," but the figure could also represent a two-hand trip. The device in Figure 15.28 could have been mounted on a pedestal so that it could be moved closer or farther from the point of operation, dependent on calculation of the formula for the given setup. Two-hand devices can also be mounted directly on the machine if the safety distance is short enough to permit this.

Neither Equation (15.1) nor (15.2) is constant for a given machine; both are dependent on a given setup because of variations in die dimensions and die weight, which in turn affect flywheel speed, stopping time, and dimension of the danger zone on the machine. The pedestal is movable, but only a supervisor or safety engineer should be capable of moving it. This can be accomplished by locking the pedestal in position with a key or by bolting it to the floor and forbidding operators to unbolt it, or some other means of fixing the position. Failure to fix the position results in a temptation to the operator to move the pedestal closer to the machine to speed up production.

Because of the superiority of two-hand controls over two-hand trips, both in terms of safety and productivity, many industrial plants are converting their older machines equipped with trips to the more modern two-hand controls. Such a conversion is not a mere change in the two-hand device, but also generally represents a major modification to the machine itself and its power transmission apparatus, so that the machine can be classified as part revolution instead of full revolution. Brake monitoring and control reliability features are also required on part-revolution presses and will be explained next.

Brake Monitoring

It can be seen from the preceding discussion that the stopping time is very important in computation of permissible safety distance for part-revolution machines. But stopping time is dependent on the brake, and unfortunately brakes are subject to wear. Stopping time is also dependent on die setup, which can change from production lot to lot. Therefore, it is naive to apply the principle of safety distances and then to walk away and trust the press to always respond in the same way as it did the day it was tested for safety distances. So for every part-revolution press whose safeguarding device depends on the brake, a brake-monitoring system is needed to monitor the brake *every stroke*. Note how different this is from the brake stop-time measurement system of Figure 15.29, which would be set up only occasionally to check or set safety distances. In contrast, the brake monitor is a permanent installation on the press, monitoring the overtravel of the ram past the top stop every stroke.

Since the system is mechanical, there will be some overtravel, and a tolerance must be set that permits this overtravel. Employers can set this tolerance as high as they desire, but it will not be to their advantage to set the tolerance very high because a larger overtravel means a longer stopping time, which means a greater safety distance, which means reduced efficiency. There is no safety distance for type B gates, but the employer is still allowed to reasonably establish a "normal limit" for ram overtravel.

The brake monitor can be engineered to measure either stopping time or overtravel distance. The most popular style is electromechanical, with a pair of limit switches triggered by a cam linked to the press crankshaft. This type is commonly known as a *top-stop* monitor. (See Figure 15.31.) The first switch signals the application of the brake, and the second signals overrun. The cam must not trigger the overrun switch until a new cycle is initiated. Sooner or later, the brake will deteriorate, and the overrun switch will be tripped, which means that brake-stopping-time tolerance has been exceeded. At that point, the monitor system must provide an indication to that effect.

In addition to the brake monitor, a control system is required to ensure that the press will cease to operate after a failure in the point of operation safety system occurs, *but* the brake system will *not* be shut down due to the system failure. This requirement is a direct application of the general fail-safe principle stated in Chapter 3.

The reader should be reminded that brake system monitoring and control reliability are not required on all power presses. The logic is that on many press setups, the brake monitor and control system would have marginal benefit, whereas on others, they would be of critical importance due to the selection of safeguarding methods, the mode of operation, and the construction of the press itself. Table 15.2 summarizes the

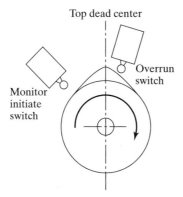

FIGURE 15.31

Top-stop brake monitor.

TABLE 15.2 Power Press Safeguarding Summary[a]

Guard type or device	Full revolution		Part revolution	
	Hands in	Hands out	Hands in	Hands out
Guards Barrier guards (fixed, adjustable or die enclosure)	Illegal	Inspect weekly	Illegal	Inspect weekly or Brake monitor and control system
Interlocked barrier guards	Illegal	Inspect weekly	Illegal	Inspect weekly or Brake monitor and control system
Type A gate	Inspect weekly	Inspect weekly	Inspect weekly or Brake monitor and control system	Inspect weekly or Brake monitor and control system
Type B gate	Inspect weekly	Inspect weekly	Brake monitor and control system must detect top-stop overrun beyond limits	Inspect weekly or Brake monitor and control system
Presence sensing devices	Illegal	Illegal	Safety distance and Brake monitor and control system	Safety distance and Inspect weekly or Brake monitor and control system
Pullbacks	Inspect: each shift, each die setup, each operator	Inspect: each shift, each die setup, each operator	Inspect: each shift, each die setup, each operator	Inspect: each shift, each die setup, each operator
Sweeps	Does not qualify as safeguard	Does not qualify as safeguard	Does not qualify as safeguard	Does not qualify as safeguard
Hold-outs (restraint)	Illegal	Inspect weekly	Illegal	Inspect weekly or Brake monitor and control system
Two-hand controls	*See* Two-hand trips	*See* Two-hand trips	Safety distance and Fixed control position and Brake monitor and control system	Safety distance and Fixed control position
Two-hand trips	Safety distance and Fixed trip position and Inspect weekly	Safety distance and Fixed trip position and Inspect weekly	Inspect weekly or Brake monitor and control system	Inspect weekly or Brake monitor and control system

[a] This summary compares inspections requirements, brake monitoring and control system requirements, safety distances, and legal and illegal arrangements. The summary does not include detailed specifications, requirements for multiple operators, and other details too numerous to include in this summary. For details, see OSHA Standard 1910.217.

various options for guarding or safeguarding presses, the need for brake monitors and control systems, and alternate permissible configurations.

GRINDING MACHINES

Grinding machines are in almost every manufacturing plant—on the production line, or in a toolroom, or maintenance shop.

There are two or three items that create the most trouble as follows (shown in Figure 15.32).

- Failure to keep the workrest in close adjustment (within $\frac{1}{8}$ inch) to the wheel on off-hand grinding machines
- Failure to keep the tongue guard adjusted to within $\frac{1}{4}$ inch
- Failure to guard the wheel sufficiently

These rules may seem "nitpicking," but there is a grave hazard with grinding machines that most people do not know about—the breakup of the wheel while rotating at high speed. It does not happen very often, but when it does, the injuries to the operator can be fatal. It is against this hazard that all three of the items just listed are aimed, even the workrest adjustment requirement.

Severe stress can be placed on the grinding wheel if the workpiece becomes wedged between the workrest and the wheel. A large gap invites the workpiece to become pinched and then drawn down by the wheel, resulting in a severe wedging action, as illustrated in Figure 15.33. The forces of this wedging action threaten the integrity of the bonded abrasive wheel, possibly causing it to break up and hurl stone fragments at the operator at near-tangential velocities. The only protection to the operator from these flying stone fragments are a good wheel guard, a properly adjusted tongue guard, and any personal protective clothing the operator might be wearing.

It is a simple matter for anyone to check the gap adjustment on the grinding machine workrest—easy to check, but not so easy to *keep* in adjustment. Since the wear of

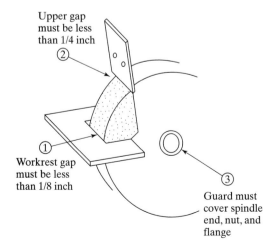

Upper gap must be less than 1/4 inch ②

Workrest gap must be less than 1/8 inch ①

Guard must cover spindle end, nut, and flange ③

FIGURE 15.32

The three biggest troublespots on ordinary grinding machines.

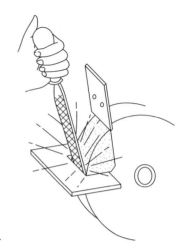

FIGURE 15.33

Severe wedging action due to a large gap between workrest and wheel.

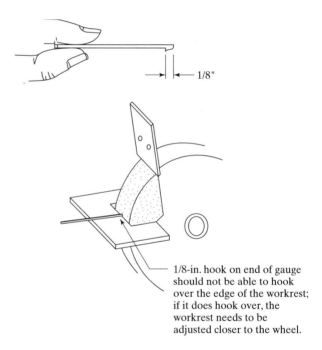

— 1/8"

1/8-in. hook on end of gauge
should not be able to hook
over the edge of the workrest;
if it does hook over, the
workrest needs to be
adjusted closer to the wheel.

FIGURE 15.34

Grinding machine workrest gauge.

the grinding wheel causes the gap to gradually widen, constant surveillance is needed to assure that the gap is maintained less than $\frac{1}{8}$ inch. There is little tolerance for over-compensating for the wear because $\frac{1}{8}$ inch leaves little margin for setting the workrest closer than required. Since there is no way to avoid frequent adjustment, some convenient means should be initiated to make adjustments quickly and efficiently. To this end, a workrest gauge is recommended, as illustrated in Figure 15.34. The go/no-go character of this gauge aids in a quick and sure decision as to whether to adjust the workrest every time it is checked.

Another easy check is the maximum spindle speed; it should not exceed the maximum speed marked directly on the wheel. Operation of an abrasive wheel above its design speed subjects the wheel to dangerous centrifugal forces that also could lead to wheel breakup.

Sometimes an abrasive wheel has manufacturing imperfections or transportation damage that makes it dangerous. Before the wheel is mounted, it should be visually inspected for such damage or imperfections. Invisible imperfections can sometimes be detected by tapping the wheel gently with a nonmetallic implement such as the plastic handle of a screwdriver or a wooden mallet. A good wheel typically produces a ringing sound, whereas a cracked wheel will sound dead. The reason nonmetallic objects are used to make the *ring test* is that metallic objects might themselves ring, giving the impression that a defective wheel is good.

SAWS

Saws have some obvious hazards and some not-so-obvious ones. Almost everyone respects the danger of a power saw, but serious injuries continue to occur, and acceptable means of guarding both obvious and subtle hazards from saws need to be considered.

Radial Saws

If the safety and health manager learns about no other saw, he or she should become familiar with this one. Radial saws or radial arm saws can be quite dangerous, and in addition they are difficult to guard. Figure 15.35 depicts a typical radial arm saw, but although the blade can be seen to be partially guarded on the top half, the lower portion of the blade is exposed. Figure 15.36 illustrates one type of lower blade guard for radial arm saws. The guards are very unpopular and are often removed by employees.

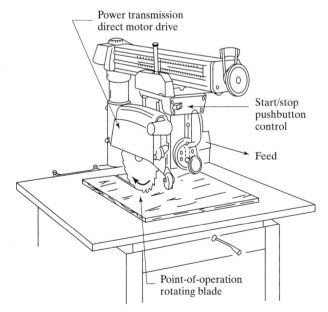

Power transmission direct motor drive

Start/stop pushbutton control

Feed

Point-of-operation rotating blade

FIGURE 15.35

Typical radial saw, not properly guarded and in violation of OSHA standard (*Source:* NIOSH, ref. Machine).

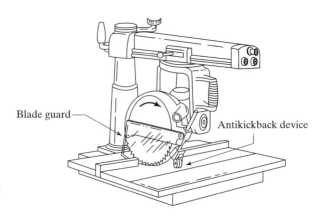

FIGURE 15.36

Radial arm saw equipped with lower blade guard.

Another problem with radial saws is the return-to-home position. The saw should be mounted such that "the front end of the unit will be slightly higher than the rear, so as to cause the cutting head to return gently to the starting position when released by the operator." A radial saw that creeps out while running is an obvious hazard, but one that is adjusted too much at an angle will "bounce" on the home position stop—another hazard. A radial saw also will have a stop that prevents the operator from pulling the saw off the radial arm or beyond the edge of the saw table. This stop should be adjustable for limiting travel of the saw head only as far as necessary.

A radial saw is usually used as a cutoff saw and is mounted as shown in Figure 15.35. It is possible, though, to reorient the cutting head 90° so that the blade is parallel to the table and thus to convert the saw to a ripsaw for long pieces of stock. In this mode of operation, the saw head is locked into position, and the material is pushed into the saw. There is a hazard, though, if the material is fed into the saw in the wrong direction. To be safe, the material must be fed *against* the rotation of the blade. If the material is fed *with* the rotation of the blade, especially if the feed rate is fast, the saw teeth are likely to grab the workpiece and pull it into the machine at high speeds, possibly pulling the operator's hands into the blade with the material being sawed. It is not uncommon that a radial saw fed incorrectly in this manner will hurl the workpiece through the machine and toss it across the room.

Table Saws

Table saw is an everyday term for a hand-fed saw with a circular blade mounted in a table. Unlike the radial saw, the table saw head always remains stationary during a cut while the work is fed into it. With table saws, the three biggest problem areas are hood guards, spreaders, and nonkickback fingers (see Figure 15.37). Antikickback protection is more important for ripsaws than for crosscut saws.

The hood guards present the most problems because the obstructed view makes the saw operator's job more difficult and awkward. Although most of the hood guards in the field are metal, most new machines come with transparent plastic guards. But the rapidly rotating saw blade can cause a static charge to develop on the nonconducting plastic guard, causing it to become covered with sawdust so that the blade cannot be seen. Also, the plastic guard is easily scratched, further reducing visibility through the guard.

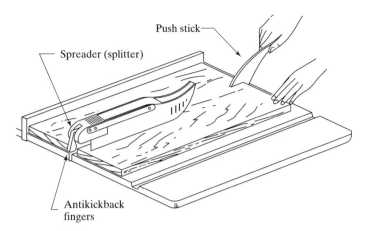

Push stick

Spreader (splitter)

Antikickback
fingers

FIGURE 15.37

Hood guard, spreader, and
antikickback fingers on a table saw.

It is true that hood guards will not absolutely prohibit contact of the operator's hands with the saw blade. The hood guard, with its spring action, acts more as an *awareness barrier*, as was discussed earlier in the section on safeguarding the point of operation on general machines. But there is another reason for using a hood guard: to protect the operator from flying objects. The saw blade rotates at 3000 rpm, and this produces large centrifugal forces and high tangential velocities. Consider the following calculation for a popular make of a table saw:

Blade speed	3450 rpm
Blade diameter	10 inches

$$\text{Tangential velocity} = (\text{rpm}) \times (\text{blade perimeter})$$

$$= 3450 \text{ rpm} \times 10 \text{ in.} \times \pi$$

$$= 108{,}385 \text{ in./min}$$

$$= 102.63 \text{ miles per hour}$$

This means that if a tooth breaks off the saw or if a small chip or block of wood breaks off and is carried on the blade for nearly a full revolution, as in Figure 15.38, it will be propelled at the face of the operator at a velocity of over 100 miles per hour. It is little wonder that eye protection is considered necessary for operating a table saw.

Kickback

Kickback is the word used to describe the situation in which the entire workpiece is picked up and thrown back at the saw operator. The energy for the kickback comes from the sawblade itself. The rotation of the blade is toward the operator. At the front of the blade where the saw first meets the work, the direction of blade motion is toward the operator and *down*. But at the back of the blade, the direction of motion is toward the operator and *up*. Because the saw teeth are slightly wider than the blade thickness, a correctly aligned workpiece will contact the blade only at the point at

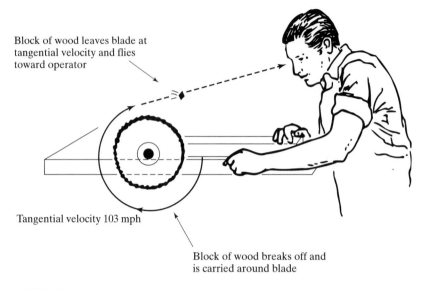

Block of wood leaves blade at tangential velocity and flies toward operator

Tangential velocity 103 mph

Block of wood breaks off and is carried around blade

FIGURE 15.38

Broken saw tooth or chips of wood are a hazard to the saw operator.

which the cut is taking place. But if the workpiece should shift slightly, the exiting portion of the cut at the back of the blade will become misaligned, causing the edge of the material adjacent to the cut to contact the blade as it emerges from the table. This contact can result in a sudden and powerful upward thrust which causes the material to break contact with the table surface. Further misalignment becomes almost inevitable at this point, and the workpiece is grabbed firmly by the blade. If the workpiece is extremely thin or fragile, it will break up at this point, with thin portions or chips continuing with the blade down under the table. But a much more likely result is that the rigid material cannot follow the blade, and it is hurled at tangential velocity directly at the operator.

Both the spreader and the nonkickback fingers are intended to help prevent kickback. The spreader keeps the saw kerf open or spread apart in the completed portion of the cut so that the material will not contact the blade. The nonkickback fingers, or "dogs," are designed to arrest the kickback motion should it start to occur. The shape of the dog permits easy motion in the direction of the feed. A backward motion, however, causes the dog to grip the material and prevent a kickback.

Band Saws

It is essentially impossible to guard the point of operation in most band saw operations. However, the unused portion of the blade can feasibly be guarded. A sliding guard is used for this and is moved up or down to accommodate larger or smaller workpieces, respectively. This sliding guard, together with the wheel guards, permits the entire blade except for the working portion to be guarded.

Hand-Held Saws

Hand-held circular saws are subjected to a variation of the hazard of kickback, except that it is the *saw* that is kicked back instead of the material! Proper training and operator respect for the saw are important, as are a clean, sharp blade, a "dead-man" control, and a retractable guard for the lower portion of the blade. A dead-man control is simply a spring-loaded switch (button or trigger) that will immediately cut off power to the saw if the operator releases the switch.

The retractable guard for the lower portion of the blade on a hand-held circular saw is perhaps analogous to the lower blade guard on a radial saw and the hood guard on a table saw. However, on a hand-held circular saw, the retractable guard is much more important. If the operator locks the retractable guard open with a small wedge, as operators will sometimes do, the saw becomes very hazardous both before and after the cut. The blade is exposed, and direct damage or injury can result if the saw is dropped or set down on a surface.

Cutting aluminum with a hand-held circular saw can be a real problem. Hand-held circular saws are widely used to cut aluminum extrusions to length to manufacture windows, storm doors, shutters, and other architectural aluminum extrusion products. The problem is that the saw blade gets very hot, reaching the melting point of aluminum at around 1200°F; drops of molten aluminum start flying off the blade onto the guard, causing quite a mess. The aluminum later solidifies, causing the guard to stick and malfunction.

One prominent manufacturer of architectural aluminum extrusions wrestled with this problem for five years before turning to alternate means of protecting the operator. Reverse polarity brakes were installed on the saws, which caused the blade to come to an immediate halt as soon as the operator released the trigger. This removed most of the hazard, because it is while the blade coasts to a stop after use that the lower blade guard is most important. For added protection, the workers were provided with protective gloves and pads for the hands and wrists.

Chain Saws

Without doubt, the most hazardous hand-held saw of all is the chain saw. Binding of the blade can cause the saw to kickback and cause severe, perhaps fatal, injury to the operator. A dull or poorly lubricated chain can overheat and break, resulting in severe injury to the operator or to the other workers in the area. The Consumer Product Safety Commission is leading the effort to improve hand guards and minimize kickback.

Power Hacksaws

Power hacksaws are difficult to guard, more so than their cousins, the horizontal band saws. The band saw can be feasibly guarded with an adjustable guard along all portions of the blade except the portion actually performing the cut. Because of the reciprocating action of the power hacksaw, however, a much more complicated guard would be required to adjust back and forth *during* the cut every stroke. Guards, guardrails, or guarding by location would be appropriate for this hazard. Some modern power hacksaws are equipped with enclosures that contain the entire stroke of the reciprocating blade.

BELTS AND PULLEYS

Almost every industry has a wide variety of belts and pulleys and other driving systems for transmitting power from motors to machines. The hazards are well understood, and the technology is simple. Belts and pulleys represent a good target area for the safety and health manager to institute a low-cost program of in-house safety improvement.

It should be recognized that not all belts and pulleys are hazardous, and the standards acknowledge this fact by excluding certain small, slow-moving belts. The exclusions are intricate and are best represented by a decision diagram, as in Figure 15.39. A height of less than 7 feet off the floor or working platform is generally considered a working zone where personnel need protection from belts and other machine hazards.

Related to belts and pulleys are shaft couplings, such as are typically found between a pump and the motor that drives it. The preferred method of eliminating hazards with these couplings is to design them such that any bolts, nuts, and setscrews are used *parallel* to the shafting and are countersunk, as shown in Figure 15.40. If these fasteners do not extend beyond the edge of the flange as shown in the figure, it is unlikely that they will cause injury. The greatest hazard with exposed setscrew heads is that they will catch parts of loose clothing and then draw the worker into the machine. The setscrews and other projections are typically invisible on the rapidly spinning shaft or flange, adding to the hazard. The installation of U-type guards, wherever needed throughout the plant, is an objective that needs to be set by the safety and health manager. Once they are aware of the hazard, maintenance personnel can go about systematically taking measurements, having U-type guards fabricated in the sheet metal shop, and then installing these guards.

The possibility of safeguarding belts and pulleys by *location* should not be overlooked. Some belts and pulleys are located in a part of the machine that is protected from worker exposure. Some people believe that location is the answer to safeguarding the belt and pulley on motor-driven air compressors that operate intermittently. But because this equipment starts automatically, safeguarding by location may not be sufficient to eliminate the hazard.

One approach to guarding large air compressors is to place them in a room by themselves. The door should be kept locked. It is best that the room be kept small if heat dissipation methods permit so that it will not be used for storage or other purposes that will result in worker exposure. The safety of maintenance personnel who must enter the room to service the compressor should not be overlooked. Administrative procedures and training can be used to reduce hazards to these personnel.

It is important to call attention to the hazards of compressed air hoses used for cleaning. Compressed air hoses with nozzles are often used to spray chips away from machine areas. Safety standards specify that the air pressure used for such purposes must not exceed 30 psi. Most industrial compressed air systems operate at working pressures greater than 30 psi. Therefore, a reducer at the nozzle or a reducing nozzle is used to bring the air pressure within the specified maximum of 30 psi. Figure 15.41b shows one type of nozzle for this purpose.

Excessive air pressure from these nozzles can make flying chips hazardous. Even with proper air pressure, chip guarding and personal protective equipment are needed to protect the worker. If alternate means can be used to remove chips, it is usually in the

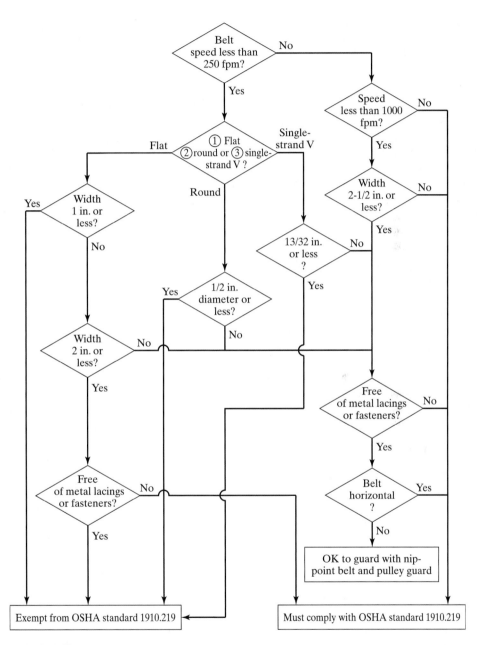

FIGURE 15.39

Decision diagram for OSHA belt guarding standard.

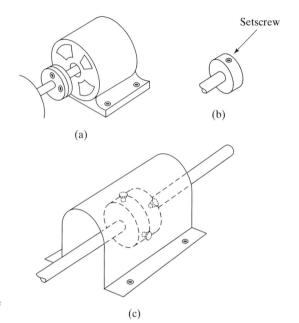

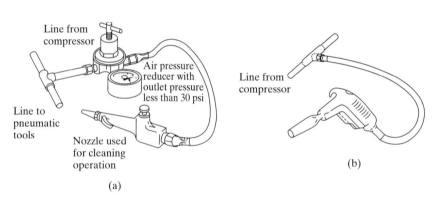

FIGURE 15.40

Safety with shaft couplings: (a) Screws or bolts are mounted parallel to the shafting and are countersunk; (b) setscrew is mounted in the periphery of the flange, but is countersunk and does not extend beyond the flange; (c) exposed setscrews present no hazard because the shaft coupling is covered by a U-shaped guard.

FIGURE 15.41

Two ways to comply with OSHA's requirement to reduce compressed air for cleaning to pressures less than 30 psi (a) pressure reducer in line; (b) special ventilating nozzle reduces pressure to less than 30 psi.

interest of safety to discontinue the use of air nozzles. Unfortunately, metal chips are very sharp and hazardous to handle, making the cleaning process somewhat of a problem.

The hazard of flying chips is fairly obvious, but most people do not realize that compressed air hoses used for cleaning can even present hazards of *fatalities*. There have been a few recorded cases of fatalities where horseplay has failed to consider the hazards of compressed air. There is no way that the human body can contain without serious damage internal overpressures of even 30 psi. Unfortunately, workers have rarely been trained to respect the lethal pressures presented by the ordinary and seemingly harmless compressed air nozzle used for cleaning.

Jacks

A frequent weekend accident is the familiar fatality when a "shade tree mechanic" is killed under an automobile that has fallen from the jack supporting it. A jack is essential for lifting, but the supported load is usually much more stable if transferred to secure blocks, removing the load from the jack. What is unsafe for the shade tree mechanic is also unsafe for the industrial worker as far as jacks are concerned.

As with a crane or a hoist chain, the temptation is to use a jack until failure, but the result of this policy will eventually be a catastrophic failure, an unacceptable alternative for the end of life for a jack. Therefore, the only other alternative is to inspect the jack at intervals throughout its life to watch for signs that the jack either needs repair or is worn to the danger point.

SUMMARY

Machine guarding is a term almost synonymous with industrial safety and is a high-priority item for safety and health managers. Although passed off by some health and safety professionals as old fashioned and nontechnical, machine guarding is actually a challenging task with new guarding technologies pressing the state of the art.

The most dangerous part of most machines is the point of operation, where the tool meets the workpiece. Unfortunately, it is also the most difficult part of the machine to guard in most cases. When guarding becomes impractical or infeasible, electromechanical devices have been devised for protecting the operator from the point of operation. Not to be overlooked are the indirect hazards at the point of operation such as flying chips or sparks.

Next in importance to the point of operation are belts, pulleys, and other power transmission apparatus. Usually more easily guarded than the point of operation, belts, pulleys, gears, shafts, and chains should receive plantwide attention. Guards often can be fabricated in-house with supervision of the safety and health manager. Remember to consider the possibility of guarding by location. Not to be confused with guarding by location is guarding by *distance*. Guarding by distance is intended for point of operation guarding for machines, such as press brakes, that fabricate large workpieces.

One of the most important and dangerous of production machines is the mechanical power press. Safety requirements for guarding the point of operation of presses can be quite technical and complicated. Many of the guarding methods specified for presses can be used as principles for guarding machines in general.

The hazards of radial saws are exemplary of those of other types of saws. Kickback is a hazard with most saws, and various mechanical devices in addition to training workers in the mechanism of kickback are viable remedies.

OSHA's abrasive wheel machinery standard is quite complicated, but the principal problems are short and simple—guarding the wheel, nut, and flange; adjustment of the workrests on grinding machines; and adjustment of the tongue guards. The safety and health manager should also give attention to such miscellaneous equipment as hand-held power tools, compressed-air equipment, and jacks. Significant among these is the requirement to reduce air pressure to 30 psi or less when the air is used for cleaning.

EXERCISES AND STUDY QUESTIONS

15.1 Explain the term *point of operation*.

15.2 Name several types of mechanical hazards on machines in general. Which is the most important from a safety standpoint?

15.3 Identify several examples of in-running nip points.

15.4 Name two ways of safeguarding a machine that require no physical guard or device at all. What is the difference between these two methods of safeguarding?

15.5 What is a lockout? How does it differ from an interlock?

15.6 What are the disadvantages of nylon mesh guards for fans?

15.7 The stopping time of the ram on a certain part-revolution press has been measured to be 0.333 second. At what minimum safety distance should a presence-sensing device be placed?

15.8 A popular mechanical power press has a full-revolution clutch and 14 engagement points on the flywheel, which rotates at 90 rpm. At what minimum distance from the point of operation on this press should a two-hand trip device be placed?

15.9 Name some reasons why a machine might have bolt holes in its feet.

15.10 Name several types of safeguarding devices for the point of operation.

15.11 What is the difference between an interlocked-barrier guard and a gate?

15.12 What is the difference between two-hand controls and two-hand trips?

15.13 What is the difference between type A and B gates?

15.14 What is the advantage of Allen-head screws over wing nuts for machine guards?

15.15 A guard has a maximum opening size of 3/4 inch. The guard openings are 6 inches from the danger zone. Does the opening size meet requirements?

15.16 What is an awareness barrier? What is a jig guard?

15.17 Explain the terms *full revolution* and *part revolution* as applied to press clutches. Which is safer?

15.18 What is muting of safeguarding devices, and when is it permitted?

15.19 What is the principal reason that galvanized wire mesh is a better material for constructing machine guards than is ordinary screen mesh? (*Hint*: The answer is not rust prevention.)

15.20 What is the biggest disadvantage of pullbacks?

15.21 What is the difference between pullbacks and restraints?

15.22 A part-revolution clutch press has a brake stop time of 0.37 second. At what minimum distance should two-hand controls be placed?

15.23 Where should a presence-sensing device be placed on the press of Exercise 15.22?

15.24 If the press of Exercise 15.22 had been a full-revolution press operating at 60 rpm and having four engagement points, at what distance should a two-hand trip device be placed?

15.25 Would a two-hand control offer any improvement in the press of Exercise 15.24? What about a presence-sensing device?

15.26 What are the three biggest problems with grinding machines? Why are they so important from a safety standpoint?

15.27 What single characteristic of the point of operation can exempt a mechanical power press from the requirement for guarding or safeguarding?

15.28 Compare hazards for table saws that are used as ripsaws versus those used as crosscut saws.

15.29 What is the ring test?

15.30 How can compressed air used for cleaning be dangerous? Under what circumstances is it permitted?

15.31 When do shaft couplings need no guards?

15.32 The expanded metal mesh in a machine guard has a maximum opening size of 3/8 inch. What minimum distance from the point of operation can this guard be legally placed?

15.33 Describe the nature of frequent OSHA citations involving lockout/tagout.

15.34 Explain how a machine designed to be started and stopped with push-button on/off switches can be modified to qualify for lockout.

15.35 What is the meaning of guarding by location? How does guarding by location differ from guarding by distance?

15.36 Explain the term *zero mechanical state*.

15.37 Explain the term *energy isolation device*.

15.38 The term *shield guard* applies to which type of machine hazard?

15.39 Which of the fail-safe principles discussed in Chapter 3 applies to the term *zero mechanical state*?

15.40 What is generally the best color for a point-of-operation guard? Explain why.

15.41 Comment on the suitability of hand-feeding tools or tongs for safeguarding the point of operation.

15.42 Name the types of machine guards that are inappropriate for hand feeding the point of operation.

15.43 Discuss the comparative advantages of die enclosure guards versus fixed-barrier guards.

15.44 Explain likely outcomes when punch press dies close on a misaligned workpiece.

15.45 One recognized method of safeguarding the point of operation can be made safe for hand feeding if facilitated with hand tools or tongs. Name this method of safeguarding.

15.46 Is it legal to use a type B gate on a full-revolution press? Is it advisable? What is the potential hazard in using a type B gate on a full-revolution press?

15.47 Discuss at least two advantages of a friction clutch drive on a punch press versus a positive mechanical drive engagement.

15.48 List and describe all mechanical press setups for which a brake monitor and control system are required by standards.

15.49 List and describe all mechanical press setups for which it is legal to use hand feeding.

15.50 **Design Case Study.** A full-revolution press has 100 rpm flywheel rotation. The press is equipped with two-hand palm buttons mounted on the machine at a distance of 20 inches from the point of operation. How many engagement points on the flywheel would be necessary to make this machine safe?

15.51 What optional feature of presence-sensing devices is permissible to give the presence-sensing device the same production advantage that the type B gate has over the type A gate? Explain the advantage.

15.52 Explain the difference between the terms "brake monitor" and "brake-stop-time measurement device."

15.53 Relate at least two advantages of using infrared instead of visible light as the medium for presence sensing devices.

15.54 What happens when the top-stop overtravel limit switch is tripped at the end of a press cycle?

15.55 The top-stop overtravel limit switch on a brake monitor for a mechanical power press is adjustable. What is the advantage of adjusting the overtravel high? What is the advantage of adjusting the overtravel low?

15.56 What characteristic of a hand-held circular saw makes the retractable blade guard even more important than it is on table saws or radial saws?

15.57 Describe two required features on a table saw that are intended to prevent kickback. Which one prevents kickback from starting and which one arrests the kickback motion if it does start?

15.58 Explain why large fixed-barrier guards can permit large distances between the guard and the point of operation and a coarser mesh for the guard material.

15.59 What is the difference between a power hacksaw and a band saw? Which is more difficult to guard? Why?

15.60 Explain the benefits of using a footswitch to trip a press in which the point of operation is protected by an infrared light screen.

15.61 Describe the disadvantages of awareness barriers in lieu of machine guards.

15.62 What design features of a presence-sensing system are permitted to improve system efficiency in the same way that type B gates improve workstation efficiency over type A gates?

15.63 Describe two advantages of using infrared light instead of visible light as the medium for a presence-sensing device.

15.64 Why should control pedestals, if used for punch presses, be rigidly fixed to the floor?

15.65 Comment on the relationship between the speed of a punch press and its safety. When is a slower machine usually better and when is a fast one better?

15.66 **Design Case Study.** A full-revolution press with two engagement points on the flywheel has two-hand palm buttons installed at a distance of 16 inches from the point of operation. Calculate the minimum flywheel speed that will permit this press to be in compliance with standards. Explain why a slower flywheel speed would be more dangerous.

15.67 **Design Case Study.** A workstation is being redesigned to increase its production rate to make it more competitive in an industry that competes in the global marketplace. The old workstation is a punch press operation, utilizing a full-revolution flywheel press employing a type A gate. By utilizing robot feeding, the production standard for this workstation is 600 units per hour. Describe three alternative designs for this workstation and explain why they should be capable of substantially increasing the production rate without violating safety standards. Qualify each alternative with a description of the disadvantages of each approach.

15.68 **Design Case Study.** In the press design of Exercise 15.50, suppose that the maximum available number of flywheel engagement points is 14. What modification of the mechanical drive would be needed to make the press setup described in Exercise 15.50 safe?

STANDARDS RESEARCH QUESTIONS

15.69 Review OSHA enforcement statistics to determine the frequency of citation for "general point-of-operation guarding." How does the OSHA standard for general point-of-operation guarding rank among all OSHA standards in terms of frequency of citation?

15.70 Review OSHA enforcement statistics to determine the frequency of citation for "point-of-operation guarding" on mechanical power presses.

15.71 Review OSHA enforcement statistics to determine the top three most frequently cited standards in the abrasive wheel machinery standard.

C H A P T E R   1 6

Welding

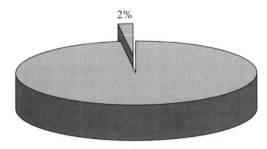

2%

Percentage of OSHA General Industry citations addressing this subject

After as broad a subject as machine guarding, it may seem ridiculously specific to address a field as narrow as welding. However, it may surprise some to learn that welding processes present some of the greatest hazards to both safety and health. In terms of breadth of hazard, this chapter encompasses even more than the chapter on machine guarding, and for that matter, more than any other chapter of this book.

The term *welding* is to be taken in a very broad sense to include gas welding, electric-arc welding, resistance welding, and even related processes such as soldering and brazing, which technically are not welding processes at all. Welding processes are so diverse, that before addressing the hazards of these processes, it is necessary to name them and to provide the necessary background in welding process terminology.

PROCESS TERMINOLOGY

The key to understanding welding hazards is to know how the process itself works, and unless safety and health managers have this knowledge, their credibility with their manufacturing and operations counterparts will be minimal. Everyone knows that welding requires that material melt or fuse to form a rigid joint. The first question to ask to determine the process is "What material melts?" If the melted material is of the parts to be joined themselves or of like filler material, the process is *welding*. If the material is some other material of lower melting temperature, the process is *brazing* or *soldering.* The breakpoint between brazing and soldering is 800°F (427°C), with brazing above and soldering below 800°F.

Since welding requires materials to be melted, heat is required, usually applied intensely, to meet the demands of high melting points of welding materials. The method of applying this intense heat usually identifies the process. Excluding the unusual and exotic processes, such as thermit and laser processes, the three basic categories of conventional welding are as follows:

- Gas welding
- Electric-arc welding
- Resistance welding

Gas welding is typified by the familiar oxyacetylene torch process, in which the very hot burning acetylene gas is made to burn even hotter by supplying pure oxygen to the flame. For welding materials with lower melting points, alternate and safer gases such as natural gas, propane, or MAPP gas (a trade name) can be used. These alternate gases are often used for brazing and soldering.

The safety and health manager should be careful not to confuse gas welding with some types of electric-arc welding that use an inert gas to facilitate the process. Indeed, some of these processes have names such as "gas metal arc welding" or "gas tungsten arc welding," but they are not gas welding. The telling feature of gas welding is that the gas must be used as a fuel for the process, not as an inert gas.

Even more diverse than the types of gas welding are the various types of *electric-arc welding.* Arc welding requires a small gap between electrodes, one of which is usually the workpiece itself. The intense heat is provided by the electric arc that forms between the electrodes. The process that typifies electric arc is *stick electrode* or shielded metal arc welding (SMAW),[1] shown in Figure 16.1. This highly portable and most popular operation is seen in welding structural steel for buildings, in repair of steel components, and in a wide variety of manufacturing processes. The *stick* is a piece of *welding rod* that is held by a gripper and is consumed by the process. The stick consists of a filler metal surrounded by a *flux,* a term to be explained later. Similar to SMAW welding is *flux-cored arc welding* (FCAW), in which the flux is on the inside of the rod, reminiscent of acid- or resin-core solder. Sometimes the welding process does not consume the electrode, a good example being the process commonly called TIG (tungsten inert gas) or GTAW (gas tungsten arc welding) shown in Figure 16.2. In a related process, GMAW (gas metal arc welding), the electric arc consumes a *flexible* electrode that is spooled on a reel and is continuously fed to the arc during welding.

The terms *flux* and *inert gas* have been used earlier and need some explanation. The extremely hot melting temperatures of steel and other metals make these metals very vulnerable to oxidation, which is harmful to the weld. Flux is typically a chemical compound that combines with impurities and with oxygen to prevent harmful oxidation of the hot metals. After combining with impurities while the flux is in the molten state, the resultant molten liquid is called *slag,* which later solidifies and must be removed from the finished weld. With some processes, one of the inert gases, such as

---

[1] American Welding Society recommended abbreviation.

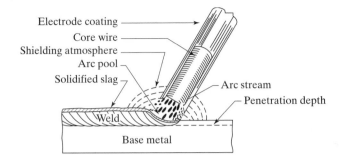

Electrode coating
Core wire
Shielding atmosphere
Arc pool
Solidified slag
Arc stream
Penetration depth
Weld
Base metal

FIGURE 16.1

Stick electrode or SMAW (shielded metal arc welding).

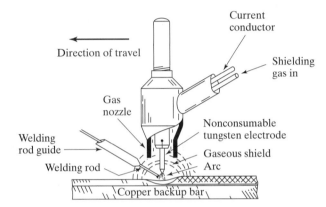

Direction of travel
Current conductor
Shielding gas in
Gas nozzle
Nonconsumable tungsten electrode
Welding rod guide
Gaseous shield
Welding rod
Arc
Copper backup bar

FIGURE 16.2

TIG (tungsten inert gas) welding or GTAW (gas tungsten arc welding).

argon or helium, is used for the same purpose. The inert gas displaces the air ambient to the weld and thus keeps harmful oxygen away from the hot metals. Unfortunately, this inert gas also sometimes keeps the oxygen away from the *welder*, an obviously undesirable characteristic that will be explored subsequently in the section on hazards mechanisms.

One other electric-arc process should be mentioned at this point: *submerged arc welding*, or SAW. In this process, the flux is granular and, as can be seen in Figure 16.3, the electric arc is hidden under the pile and puddle of granular and melted flux, respectively. This has great safety and health advantages, and submerged arc is growing in popularity. Automatic welding machines are often programmed to apply flux automatically and move the electrode over a long, straight path in the manufacture of large structural steel beams or plate girders. However, SAW welding in overhead positions is a problem because the granular flux will fall by gravity instead of covering the weld.

Resistance welding (Figure 16.4) is one of the least hazardous welding processes. It is widely used in mass-production manufacturing. However, resistance is generally restricted to relatively thin sheets of material. The concept of resistance welding is to pass electrical current *through* the material to be welded, enabling the heat so generated to melt the material. Physical pressure is also applied at the point of the weld. The attractive thing about resistance welding is that the melting generally occurs only where the mating surfaces meet. The outside and adjacent surfaces that are exposed to atmospheric contaminants harmful to the weld do not reach melting point, and damage

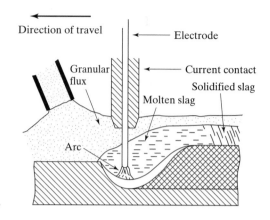

FIGURE 16.3

Submerged arc welding (SAW).

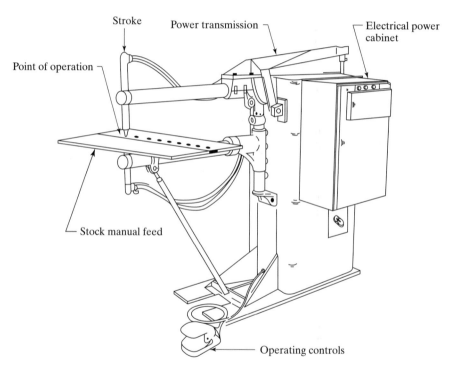

FIGURE 16.4

Resistance spot welder.

to the material is minimal. This fact also precludes the need for a flux or inert gas to complicate both the production process and the safety and health aspects.

Resistance spot welding (RSW) is widely used to join sheet metal coverings, housings, guards, and shields on such products as space heaters and grain bins. Another important industry for spot welding is the automobile industry. Seam welding (RSEW) is generally preferred over spot welding for watertight seals because a pair of rollers apply continuous pressure and a series of electrical pulses make, in effect, a seam of overlapping spot welds.

Some of the more unusual or exotic processes should be mentioned because, although they may rarely be used or seen by the safety and health manager, the nature of their hazards may be entirely different. The *thermit* method of welding (TW) employs a chemical reaction to produce the welding heat. Thermit is good for awkward applications or perhaps where the welding to be done is remote from convenient electrical power or gas sources. *Laser beam welding* (LBW) uses a concentrated laser light beam to generate the welding heat. Lasers pinpoint the weld so precisely that they are used for almost microscopic applications on tiny parts.

GAS WELDING HAZARDS

The myriad of different processes for welding should give some clue as to why the breadth of welding hazards is so great. We now examine these hazards, beginning with gas welding because it has given safety and health managers the most problems.

Acetylene Hazards

Gas welding cylinders are so familiar that it is difficult to keep in mind their devastating destructive power. Acetylene gas, the fuel gas for most gas welding, is so unstable that its pressurization in manifolds to pressures greater than 15 psig (30 psia) is prohibited. Contrast this low pressure with the familiar oxygen, nitrogen, or other ordinary compressed-gas cylinder, which contains pressure greater than 2000 psi.

There are tricks to avoid the instability hazards of acetylene gas; the most popular one is to dissolve the gas in a suitable solvent, usually acetone. Then the pressure can be raised to around 200 psi. As the acetylene gas is used from the acetylene cylinder, the pressure is lowered slightly, permitting a greater quantity of gas to bubble out of the solution, resulting in equilibrium at a pressure suitable for welding. In this fashion, a relatively large quantity of acetylene can be stored in a reasonably portable cylinder.

Figure 16.5 shows the inside of an acetylene cylinder. Most people do not know that the cylinder contains a solid absorbent filler material for the acetone–acetylene solution. The contents are more liquid than gas, and this gives rise to a hazard mechanism. Acetylene cylinders should be kept valve end up both while in storage and while

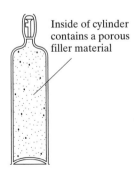

Inside of cylinder
contains a porous
filler material

FIGURE 16.5

Acetylene cylinder.

in use. There is no harm in tilting cylinders slightly, and in fact this is common practice in the use of hand trucks to handle cylinders connected for use. Good advice, however, is not to tip acetylene cylinders to an angle more than 45° from vertical.

If the cylinders are stored horizontally or valve end down, liquid acetone could enter the valve passages instead of the intended acetylene gas. Then later, when the valve is opened, the welder could get an unexpected flow of highly flammable liquid acetone instead of acetylene gas. Since the purpose of opening the valve is usually to ignite the torch with a sparking source, it is obvious that it would be easy to ignite the liquid acetone accidentally. A burning quantity of spilled acetone is difficult to control and is quite dangerous. The reader is referred to Chapter 11 to consider the flammability characteristics of acetone.

Another way to get liquid acetone through the cylinder valve is to use the cylinder when it is nearly empty. Acetone coming through the valve or leaking elsewhere is easy to detect. The principal active ingredient in nail-polish remover is acetone, the odor of which is familiar to most everyone.

Acetylene cylinders have been known to leak around the valve stem, causing what welders call "stem fires." Another place to check for leaks is from the plug in the bottom of the cylinder. In one incident, fortunately not a serious accident, the welder was perplexed by occasionally hearing a small explosion, audible but causing no damage. The mystery was finally solved when it was discovered that the explosions were occurring in the small concave area beneath the bottom of the cylinder. The little explosions were harmless, but imagine the hazard that could have accumulated if there had been no ignition, and the defective cylinder had been allowed to slowly release acetylene, as in overnight storage.

It is important to be able to turn off the fuel flow quickly in an emergency, especially when the fuel is acetylene. Some acetylene cylinder valves are designed to accept a special wrench, but ordinary hammers and wrenches are in general inappropriate for opening cylinder valves. The special wrench should be kept available for immediate use.

A carbide "miner's lamp" burns on acetylene by dropping lumps of calcium carbide into water. The resulting chemical reaction is as follows:

$$CaC_2 + 2H_2O \rightarrow Ca(OH)_2 + C_2H_2 \uparrow$$

Thus, acetylene gas slowly bubbles out of the water solution and fuels the lamp. The same chemical reaction can be used for generating acetylene gas for welding by means of an acetylene generator. This process avoids the hazards of large-scale storage of acetylene cylinders. But the process of storing acetylene in cylinders has been so perfected, so commercialized, and made relatively safe, that acetylene generators are rarely seen anymore. The cylinder supplier has a large acetylene generator somewhere, but this is of no concern to most safety and health managers.

Before leaving the subject of acetylene hazards, we should consider alternative fuel gases. If a safety and health manager really wants to win the approval of top management, he or she will offer a production innovation that will make the workplace safer, reduce the company's legal vulnerability, and cut production costs—*all at the same time!* It is difficult to do, but not impossible, and the selection of welding fuel gas presents a potential opportunity. It takes some "homework" and particular diligence to pursue the issue, but the rewards to the company and to everyone involved can be dramatic.

MAPP gas, natural gas, and propane were mentioned earlier as possible alternatives to acetylene. The first objection that the safety and health manager will hear to the idea of using these gases is that they do not burn hot enough. It is true that acetylene excels when a very hot flame is needed, but many industrial applications do not need temperatures as high as welders want for *some* applications. Alternate gases are certainly hot enough for brazing and soldering, as well as being hot enough for some welding applications.

One reason that halfhearted attempts to switch to alternate gases do not work is that welders attempt to switch gases while using the same torch tips they used with acetylene. Special torch tips may be the secret to making the new idea a success. Welders may also need to be taught how to adjust their torches to achieve a "neutral flame" with the proper burning characteristics.

Case Study 16.1 will illustrate how one safety and health manager had a positive economic impact on his company by making a suggestion that improved safety while reducing production costs.

CASE STUDY 16.1

REDUCING A HAZARD BY SUBSTITUTION

The safety and health manager of a manufacturing plant noticed that a production operation was using conventional gas welding equipment to perform a brazing operation. The conventional gas welding setup is a portable cart with a torch connected to oxygen and acetylene cylinders. However, in a production operation in which the process stays in one place within the plant, there is no need for the welding equipment to be on a portable cart and use fuel gases supplied in expensive portable bottles or tanks. Furthermore, since the operation was brazing, not welding, a much lower temperature was needed to do the job, which gave the safety and health manager the idea that perhaps dangerous and expensive acetylene gas, with its characteristic hot flame, might not be needed. The cheapest alternative welding fuel of all, natural gas piped into manifolds in the plant, was suggested as an alternative to acetylene. Unfortunately, public utility supplies of natural gas are at very low pressures, approximately 4 oz/in^2 and welders scoffed at the idea, saying that the natural gas pressures were too low to be effective in this production process. Not giving up easily, the safety and health manager negotiated an arrangement with the supplier utility to supply the natural gas to the plant at higher pressures than is normally used for other gas customers. The result was a very successful brazing operation using natural gas as a fuel substitute for the much more expensive acetylene gas. By following through with his idea, this safety and health manager gained prestige with the plant manager and throughout the plant by demonstrating that he was a problem solver with an eye for production cost and efficiency as well as for safety. At the same time, he succeeded in reducing the plant's exposure to OSHA citation for some of the most frequently cited OSHA standards ever.

Oxygen Cylinders

We have seen the hazards of highly unstable acetylene, and by comparison oxygen is much more stable; in fact, it is almost completely safe *if* it is kept away from fuel sources. But ironically, oxygen cylinders are more dangerous than acetylene cylinders. The reason for this danger is the extremely high pressure contained by the oxygen cylinder. Figure 16.6 depicts the familiar oxygen cylinder; note its resemblance to the shape of a bomb or a rocket. It is difficult to comprehend the energy that can be released by the sudden rupture of the valve on an oxygen cylinder containing 2000-psi pressure. There have been numerous accounts of cylinders becoming airborne and crashing into brick walls, demolishing the walls. If a cylinder valve ruptures while the cylinder is confined in a relatively small room, it may ricochet off walls until it kills anyone unfortunate enough to happen to be in the same room when the valve breaks. Consider the hazard of a heavy cylinder flying wildly around the room like a rapidly deflating balloon.

Workers often unwittingly drop oxygen cylinders onto the ground or bang them violently together. Oxygen cylinders are often seen standing alone and unsupported. Even though they are quite heavy, their small bases make them easy to tip over, with the resultant danger of knocking off the valve. The temptation to leave oxygen cylinders standing alone and unsecured needs to be dealt with in safety training sessions.

Another temptation with oxygen cylinders is that they seem to make perfect rollers for supporting and moving heavy items about. No matter whether the cylinders are full or "empty" (even a spent cylinder is not completely empty), use of cylinders as rollers or support can damage the cylinder and perhaps the valve. Furthermore, the large, heavy cylinders used as rollers present a problem of *control*. Once a heavy load begins to roll on a set of welding cylinders, it can even *run over* an unsuspecting victim.

A perennial problem is keeping track of the valve protection cap, which must be removed to use the cylinder, but which also must be screwed back into place when the cylinder is in storage. This cap protects the important valve from damage, and if the cylinder is ever moved about without it, there is a risk of the cylinder falling and knocking off the valve, with disastrous results.

When an oxygen cylinder is strapped to a wheeled cart or hand truck along with its companion acetylene cylinder and the regulator assembly is in place for welding, the

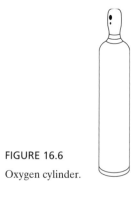

FIGURE 16.6

Oxygen cylinder.

cylinder is generally considered to be in operational status, not in storage. Therefore, industry practice says that the valve protection caps do not have to be in place in these situations.

By looking carefully at the valve protection cap in Figure 16.6, a somewhat odd-looking vertical slot can be seen; actually, there are two, but one is hidden in this view. These slots have a definite engineering purpose, but not the purpose most people think. If the valve comes off while the protection cap is screwed into place, the escaping gas will impact at high velocity on the closed top portion of the cap, tending to counter the force of the gas escaping at the valve. The slotted parts on the sides of the cap permit the gas to escape, but in directions exactly opposite to each other, balancing forces and leaving the cylinder relatively at rest.

Unfortunately, the existence of the slotted openings on the cap are an invitation to their misuse by the worker who is attempting to handle the cylinder. Cylinders are heavy and unwieldy, especially to the worker who has had to handle a lot of them in one day. Furthermore, in cold weather these cylinders have a tendency to become frozen to the ground, to a slab, or even to each other. The worker will want to find a means to break them apart from each other or from the slab to which they are frozen. The slotted opening on the cap seems to be an ideal place to insert a pry bar to obtain some leverage. But this is *not* the purpose of the slots, and misuse in this fashion is what can lead to a broken or damaged valve.

As if the extreme pressure hazards were not enough, oxygen presents additional hazards due to its chemical properties. As stated before, it is relatively stable in the absence of fuel sources, but the fire hazard of pure oxygen under pressure in the presence of a combustible substance is extraordinary. A substance as benign as ordinary grease can suddenly become explosively combustible in the presence of pure oxygen under pressure. A worker will often hold his or her hand over the valve opening when first opening the valve to test the cylinder. If the hands or gloves are greasy, an explosive combustion can follow in which a hand can easily be lost.

We have examined the separate hazards of acetylene and oxygen, but when oxygen and acetylene cylinders are stored together, the hazards are multiplied. There is always the possibility that one or more cylinders will leak. The reader will perhaps recall the earlier account describing small explosions from a leaking plug in the bottom of an acetylene cylinder. Acetylene is already highly flammable, and the presence of pure oxygen makes the situation about five times as serious. A noncombustible barrier at least 5 feet high must separate oxygen and acetylene cylinders, or they must be moved apart at least 20 feet.

Torches and Apparatus

Because of their vital role in safety, torches, manifolds, regulators, and related apparatus must be "approved," usually by a recognized testing laboratory, such as Underwriters' or Factory Mutual.

The familiar torch, illustrated in Figure 16.7, is a more sophisticated piece of engineering than most people realize. The torch is often taken to be simply a handy double tube-and-valve assembly for delivering both oxygen and fuel gas to the weld flame, but it

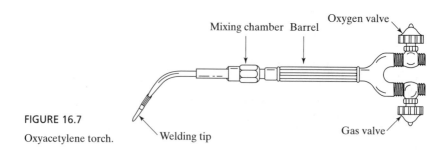

FIGURE 16.7

Oxyacetylene torch.

is more than that. Note in the figure that it has a *mixing chamber*. The torch is designed so that the mixing takes place at the right time and in the correct *total* volume. The welder controls the *proportion* of volumes of the mix by adjusting the torch valves for oxygen or acetylene, an oxidizing flame being oxygen rich and a reducing flame being fuel rich, but the total volume of the mixture is determined by the torch itself. The mixing chamber is mated to the various correct apertures for the approved torch tips, and this also is an important balance. If the balance is disturbed, flow rates may also be disturbed, and the flame may begin to travel back up the mixture stream and begin to burn *inside* the torch! This is really not all that uncommon; all welders know this phenomenon and commonly call it *flashback*. A popping or snapping sound is a warning that flashback is about to occur. Once the flashback begins, a distinctive humming sound can be heard. The heat being generated inside the torch will soon ruin it, and it also presents a safety hazard. The dangerous situation is alleviated by turning off both torch valves quickly.

Even with approved torches and tips, flashback can occur because of the deterioration of the equipment, especially the tips. The tips are close to the heat and naturally become brittle or burned and crack, or pieces may break off. If the tip is not replaced, flashback is likely.

Despite the importance and sophistication of the torch and tip, it is not uncommon to see the torch used as a hammer or chisel! This temptation arises due to the formation of a slag, a waste product of the flux mentioned earlier. This hard coating of slag generally covers the weld and sticks to it. To finish the job, the welder or helper must chip the brittle slag off the finished weld. The torch and tip are so handy for this purpose (take note of shape in Figure 16.7) that welders will often take the shortcut of using the torch as a chipping tool. This is a good way to ruin an expensive piece of apparatus and at the same time increase the likelihood of flashback due to a damaged torch or tip.

The torch assembly is expensive and may be owned personally by the welder. Even when the company owns the torch, the individual welder to whom it is assigned may rightfully be very possessive because of the importance of care of the apparatus. This can cause another safety problem. Welders may want to keep their torches in their locked toolboxes, but the danger here is that almost all toolboxes contain at least some grease or oil materials. Grease or oil on the torch is dangerous because of the oxygen hazards discussed earlier. Case Study 16.2 illustrates a related hazard, locking welding torches in personal lockers.

CASE STUDY 16.2

WELDING TORCH SAFETY

Welding torches are valuable pieces of equipment, and welders are reluctant to leave them unsecured at the end of a work day. Personal lockers for personal valuables are sometimes designed so that welding torches can be locked inside while the torches remain connected to the hoses that in turn remain connected to the welding fuel and oxygen cylinders used by the welder. In the state of Iowa, a welder once locked his torch inside his personal locker in this way. The welder closed his torch valves before leaving work, but he did not close the cylinder valves on the welding cart. The torch valves leaked a small amount of acetylene gas and oxygen, which overnight built up an explosive mixture in the confined space inside the locker. The next morning the welder opened his locker and was apparently smoking a cigarette at the time. The acetylene–oxygen mixture inside the locker ignited in a powerful explosion that decapitated the welder.

The next most abused piece of welding apparatus is the hose for delivering gas to the torch. To be practical, this hose must be flexible, and it is thus subject to the physical hazards of wear and tear and deterioration to which such materials are generally susceptible. Multiple hoses can easily become entangled, and because of this, welders commonly wrap the acetylene and oxygen hoses together with tape simply to keep them more orderly, but such taping practices may hide defects in the hose. It is a good idea to keep at least 8 of every 12 inches uncovered.

Manifolds are the rigid tubing networks that enable one or more cylinders to supply one or more torches. Manifold setups are sometimes found when a regular production operation requires long-term, regular use of welding gas. The principal purpose of welding manifolds is to increase gas welding volumes, but safety is also usually enhanced by the more permanent arrangement. An example manifold setup was described in Case Study 16.1.

Service Piping

Manifolds, described in the previous section, are not to be confused with *service piping*, an even more permanent arrangement. Some plants use so much welding gas that it becomes practical to pipe the gas to the workstation, which can present some problems. The piping for acetylene must be either steel or wrought iron because copper might react with the acetylene to produce copper acetylide, a dangerous explosive. A danger with oxygen piping is again the possibility of contact with oil or grease. Fittings and pipe must be checked before assembly and thoroughly cleaned if necessary. A solution of hot water and caustic soda or trisodium phosphate is suggested for this purpose. A modern solvent such as chlorothane (1,1,1-trichloroethylene) is recommended by some engineers today. Oxygen piping systems should be purged after assembly by blowing them out with oil-free nitrogen or oil-free carbon dioxide. Chlorothane should be used to be sure that every trace of oil is removed.

Flashback can occur in service pipe systems, too, and flashback protection devices are specified, as are check valves in appropriate positions. One design for flashback protection devices involves a simple water lock. But if the water freezes, the system will not work, so antifreeze protection is needed.

The things that can go wrong with a service piping system for welding are numerous and sometimes subtle. These problems involve inadequate emergency venting, improper joints, mistakes in installation in tunnels, and other items too numerous to mention here. The purpose of this book is to alert the safety and health manager to the potential problems accompanying these systems so that he or she can be sure that personnel obtain and follow the appropriate standards for installing these systems.

ARC WELDING HAZARDS

Arc welding is a more popular process and is in many ways even more hazardous than is gas welding, even with its more stormy safety record. This is one of the ironies of the subject. The major hazards of arc welding are health hazards, fires and explosions, eye (radiation) hazards, and confined space hazards, but they also appear to a lesser degree in gas welding and other welding. Therefore, these subjects are addressed in later sections. Before such arc welding hazards are covered, however, the equipment used in arc welding should be understood.

Equipment Design

Industrial arc welding manufacturers do all they can, by means of federal standards they have helped to write and in other ways, to promote their equipment to the exclusion of lesser and cheaper models that might compete. But there is a logic to their efforts other than the profit motive. Small and relatively inexpensive models of arc welding machines are available that operate off ordinary household 110-volt current. But there are physical disadvantages with welding that uses ordinary household current. Welding requires large amounts of electrical power in the form of low-voltage, high-amperage circuits. Since household circuits are rated for amperages insufficient for effective welding, the small household machines make up in voltage what they lack in amperage. The high-voltage hazards of these small welding machines are admittedly difficult to understand because the industrial welding machines are supplied by much higher voltages—typically 240 to 480 volts! The key to this paradox is that the industrial machines step down the high voltage to less than 80 volts while raising the amperage to effective levels. Chapter 17 will return to the subject of voltage and amperage and will perhaps add clarity to the welding machine problem.

Grounding

Even with machines operating at proper voltages, the welder or other personnel can receive an electrical shock from contact with the machine if something goes wrong. The protection for this is to be sure that the frame of the welding machine is properly grounded. Thus, if there is a dangerous short to the frame of the machine, the overcurrent protection mechanism on the circuit will be tripped, protecting personnel. Welding machine grounding needs to be strong, both physically and electrically, to meet the demands of the current that may be applied to it. This is an especially important consideration for portable machines.

Operation

In the case of welding equipment, safety and health training will pay dividends in longer service lives on the equipment, a point sometimes overlooked. Welding cable carries so much electrical current that it can overheat and damage the insulation. Coiling the cable, although convenient, contributes to this hazard. Coiled cable should be spread out before welding. Splices in the cable are not permitted within 10 feet of the electrode holder. The splices themselves must be properly insulated. Some judgment is required to determine when welding cables should be replaced. Certainly, damage to the extent that some conductors have bare spots is cause for replacement.

Care must be taken by the welder to prevent the wrong items from becoming a part of the welding circuit, either during welding or while the electrode holders are not in use. Voltage is not so much the danger as is heat produced by the potentially high amperages. Compressed gas tanks or cylinders must not be part of the electrical circuit, regardless of the flammability of their contents. The heat buildup caused by a high-amperage current through the conducting metal cylinder can cause pressure buildup in the cylinder that can exceed its design limits.

Some types of arc welding are safer than others, and they are gaining in popularity. This is especially true with respect to the hazards of fume generation and radiation, which will be discussed later.

RESISTANCE WELDING HAZARDS

The cleanest, most healthful, and probably safest form of welding is resistance welding. There are still hazards from electrical shock, but more important are the mechanical hazards surrounding the point of operation.

Shock Hazards

As in the spark coil of an automobile, many resistance welding machines build up electrical energy in a bank of capacitors for sudden release when the weld is made. The voltage may reach hundreds or even thousands of volts at peak. These voltages are not the empty variety seen in discharges of static electricity collected by walking on a thick carpet. The voltages may be at the same level, but the welding machine voltages carry with them the capacity to deliver a burning current. The capacitors that store this electrical energy should have interlocked doors and access panels. Not only must the inter-lock stop the power to the machine, it must also short-circuit all capacitors. Without this short-circuiting, the capacitors could deliver a lethal shock even with the power *disconnected*. The short-circuiting of welding machine capacitors when the machine is turned off is an example of the concept of "zero mechanical state," discussed in Chapter 15, and of the general fail-safe principle, discussed in Chapter 3.

Guarding

Spot and seam welding machines apply pressure to the materials when the weld is made. For spot welding machines, this pressure makes the machine analogous to a power press, and the operator can be injured from the mechanical hazards alone. The reader may want to refer back to Figure 16.4 to study the operation of the spot welder.

Seam welding machines are not like power presses, but they, too, have point-of-operation hazards. The nature of the hazard for seam welders is that the opposing rotation of the rollers produces a pair of in-running nip points both above and below the material being welded. The hazards of seam welders do not seem to be as pressing as those of spot welders because seam welding is more often a part of an automatic or mechanized production operation, and thus operator exposure is not as great.

FIRES AND EXPLOSIONS

Welding is one of the principal causes of industrial fires. Perhaps even more than for any other welding hazards, the safety and health manager can have an impact on this particular hazard because preventing welding fires is more of a procedural matter than anything else. This means that training becomes a very important element in the hazard prevention strategy. Fortunately for the safety and health manager, there is a wealth of audiovisual aids, literature materials, and case studies on the subject of welding fires.

To use just one case history as an example, one of the most devastating and tragic industrial accidents in the nation's history occurred in Arkansas in the 1960s. A welder's spark started a fire in a missile silo, and 53 workers trapped inside the silo were killed. This accident dramatically illustrates how welding adds to other hazards, such as confined workspaces, flammable and combustible materials, and lack of ventilation. Welding on old oil drums or pipes that have contained asphalt or other petroleum products has resulted in a large number of explosions and senseless fatalities.

Welding, a dangerous operation, and confined spaces, which are dangerous locations, often appear together. In Chapter 12, we addressed primarily the health hazards of working in confined spaces. When the job to be done in the confined space is welding, both health and safety hazards are compounded. The use of inert gas to protect the weld can cause oxygen deficiency in a confined space. In another scenario, the presence of oxygen and ignition sources from gas welding processes can aggravate the problem of fires and explosions when the welding is in a confined space. Welding is often a repair operation, and unfortunately the location of the necessary repair is in a confined space.

People do not seem to realize the ignition potentials of welding operations. Welding is not safe to watch directly because of the eye hazards, so unfortunately, except for welders themselves or their helpers, few people realize what kind of a fireworks display is really taking place. Some industrial movies are good for illustration of these fireworks. Sparks are flying everywhere—not just the benign variety as seen flying from a typical bench grinder, but visible chunks and spatters of red-hot molten metal that can burn a hole completely through heavy fabric, plastic containers, and cracks in floors. Welders are more likely to know the potential for fires generated by the arcs and sparks generated by the welding process. One would think that the welder would thus hesitate to weld in areas in which the welding sparks could cause fires. The subtlety lies in the fact that welding is often a short repair operation, and the temptation is to take a few chances because of the short duration of the hazard. This temptation often leads to the type of tragedy described in Case Study 16.3.

CASE STUDY 16.3

WELDING IN A CONFINED SPACE

An employee of a trailer service company entered a 8500-gallon cargo tank to weld a leak on the interior wall of the tanker. Despite the presence of strong fumes of lacquer thinner (the material previously carried in the tanker), the welder decided to proceed with the repairs even though the written company safety policy required the use of an explosion meter at that point. When he began welding, an explosion occurred. The employee was removed from the tank and taken to a nearby hospital, where he was declared dead by the attending physician (ref. Preamble).

Welding Permits

A permit system for confined spaces was discussed in Chapter 12. Even before federal standards required permits for entering certain confined spaces, the advantages of a permit system for welding operations were recognized. There is sometimes a tendency to ignore the short-term hazards of welding repair operations and proceed to get the job done quickly. The safety and health manager is wise to counter this natural tendency by instituting a system of requiring approval and written permits to do welding operations, even for quick repairs. There are so many special precautions to take that a signed checklist is a good idea. The responsible party is usually the supervisor of the area in which the welding is to take place, but in some instances the welder can make the necessary checks and sign the form. The safety and health manager's responsibility is to set up the permit system and ensure that it is executed properly by actually checking permits occasionally when welding is seen to be taking place in areas of potential hazard. Any responsible party will give the matter some conscientious attention before signing a welding permit form, especially if personnel have received safety training that exposes them to the devastating hazards of welding fires such as the one that killed 53 workers in Arkansas. Such training will certainly make the supervisor or welder think twice before signing the form.

Judgment should be exercised to attempt to set up a permit system that will be considered reasonable by welders and plant personnel alike. The key to this reasonableness will be to set up blanket permits or exemptions from the permit system in those areas of the plant in which fire hazards are minimal. Welding done in the welder's own welding shop, for instance, should be his or her responsibility, and for the safety and health manager to attempt to impose a permit system in the welding shop would obviously constitute unwise interference. For some plants, the entire plant might be reasonably safe from welding fires, and no permit system may be needed, except for confined spaces. Hazard identification surveys and advance planning for exactly when and where the permit system is really needed will go a long way toward the establishment of a reasonable system.

EYE PROTECTION

Eye protection comes under the topic of personal protective equipment (Chapter 12), but eye protection for welding operations is so important that this chapter on welding would not be complete without a section on this subject.

Note the careful reference in the preceding paragraph to welding opera*tions*, not welding opera*tors*. The welders themselves need protection, of course, but it is easy to forget about welding helpers and others in the area. Almost every welder has received an eyeburn at some time during his or her career and knows to be careful to wear eye protection as much to prevent pain and discomfort as to prevent long-range eye injury. But less experienced personnel may need more supervision and administrative controls to ensure protection.

A summary of minimum appropriate eye protection shades is shown in Figure 16.8. In selecting eye protection, one should remember that the darker shades (higher shade numbers) provide more protection than do the lighter (lower shade number) shades. The trade-off is visibility. Some shades are so dark that the user can hardly see the workpiece until the arc is turned on. To get the maximum protection, the darkest shade consistent with adequate visibility should be used. The various arc welding methods produce much more intense radiation and thus require higher shade numbers than does gas welding. The radiation from arc welding, except submerged arc methods, is so intense that a helmet is necessary to protect the entire face area from painful burns. Gas welding is typically done while wearing goggles.

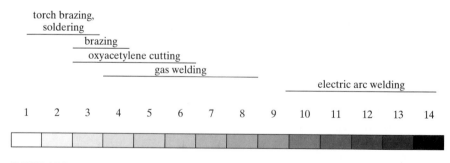

FIGURE 16.8

Summary of recommended shade numbers for various welding, cutting, brazing, and soldering operations (adapted from OSHA standard 1910 Subpart I, Appendix B).

PROTECTIVE CLOTHING

Proper clothing is serious business to the professional welder. Virtually every experienced arc welder has, some time in his or her career, sustained a "sunburn" from the ultraviolet rays produced by the welding arc. Such a welder does not need to be told to use protective clothing to cover all skin areas that would otherwise be exposed to the arc rays. But ultraviolet burn is only one of several hazards against which protective clothing is intended. Hot, burning sparks or small pieces of molten metal can fall into openings around shoes or between pieces of clothing. Even when the clothing is seamless or when openings are well shielded, a hot piece of molten metal can be trapped in a fold and burn its way through to the welder.

Leather is the traditional favorite material for welding gloves, aprons, and leggings because of its superior thermal protective qualities. Wool is also very durable. But Nomex and other synthetic materials are also becoming popular for welders' protective clothing. Cotton fabric is attacked by the radiation and soon disintegrates even if it escapes ignition by the sparks.

Hazards to the welder from falling sparks and weld metal are greatly increased when the welding must be done overhead. The welder is virtually taking a shower in welding sparks. All clothing openings must be carefully shielded, and even under the helmet, the welder's head must be protected from such misfortunes as a welding spark in the ear.

GASES AND FUMES

There are two extremes in degree of concern over welder breathing hazards. One extreme, call it position A, is taken generally by welders themselves, who often have no concern at all over chronic exposure to welding "smoke." Some welders even enjoy the smell of welding fumes in the air. The other extreme, position B, is the occasionally overzealous industrial hygienist who can find a hazard somewhere in almost all welding fume situations. Both extremes are only partially correct and can lead to dangerous errors in safety and health strategies.

The principal error in position A is that persons taking this extreme position are usually overlooking the long-term effects of chronic exposure. These persons tend to believe that if the welding smoke does not make them nauseated, dizzy, or give rise to some other acute symptom, the fumes are safe. From the principles covered in Chapters 1 and 9, the chronic exposures can actually be the most dangerous because of their adverse effects on worker health.

Position B exaggerates the effects of tiny exposures to dangerous contaminants. It is terrifying to realize that some welding releases phosgene gas, the same gas that has been used in chemical warfare. But the exposures are generally very low and can be controlled by appropriate procedures. In the final analysis, no epidemiological studies have shown welding to be an extremely dangerous occupation. From a health standpoint, welders do not have significantly shorter life spans than workers in general. Having considered this perspective, let us categorize the hazards of welding atmospheres and rationally examine what should be done about them.

Contaminant Categories

Figure 16.9 diagrams the major kinds of welding atmosphere contaminants: particulates and gases. The particulates are dust particles or even tinier smoke particles. Metal fumes in the welding atmosphere are tiny particles of metal that have been vaporized by the arc and then resolidified into particles as they cool. The gases may be either already present, as in inert shielding gases, or they may be chemical reaction products of the process.

The term *pneumoconioses* in Figure 16.9 was explained in Chapter 9; it is merely a general term that literally means "reactions to dust in the lungs." Everyone's lungs must deal with dust to some extent, and some welders' pneumoconioses are no more hazardous than would be caused by sweeping the floor. Some welding dusts are more hazardous, however, because they cause *fibrosis*, the building up of useless fibrous tissue in the lungs. The most harmful dusts are those in which the microscopic particles have the shape of fibers instead of more rounded particles. Asbestos and silica are examples of such dusts.

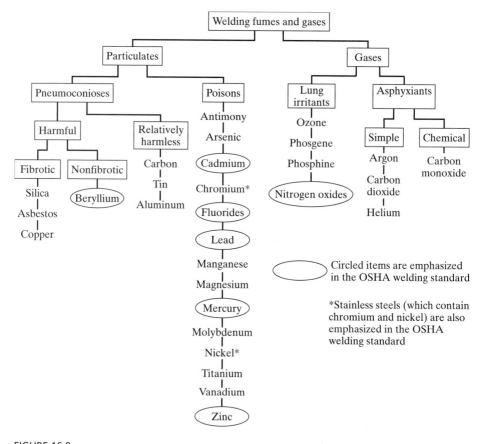

FIGURE 16.9

Classification of welding fumes and gases by hazard.

The "pulmonary irritants" are simpler in that they attack the lungs directly, whether the irritants be particulate or gas. The more insidious hazards, though, are from those particulates or gases that do not irritate the lungs directly, but through the lungs gain access to the rest of the body, where they act as systemic poisons.

It would be nice to have quantified tables of expected contaminant levels for various atmospheric constituents for various types of welding. There have been some experimental attempts at this (see refs. Fumes, Welding), but there are so many variables to be controlled that it is almost impossible to obtain reliable predictions of weld fume content. The best strategy is to be aware of potential hazardous contaminants and to know what conditions are most likely to produce these contaminants. Atmospheric sampling can then be used in suspect situations to establish whether contaminant levels are indeed excessive.

Hazard Potentials

The biggest contributor to atmospheric contaminants around welding is the coating or condition of the surfaces to be joined. It is true that welding on clean iron or ordinary construction steel produces fairly large concentrations of iron oxide fume, but fortunately siderosis, the pneumoconiosis resulting from iron oxide, is not really a very dangerous disease when it occurs alone. If the surface of the metal has a coating of material containing asbestos, however, this coating must be removed to prevent asbestos contamination of the air.

Even the act of cleaning the metal surfaces to be welded can result in secondary hazards. If chlorinated hydrocarbons, such as trichloroethylene, are used to clean the metal, these solvents must also be removed thoroughly before the welding takes place. The energy of the welding arc can cause decomposition of the solvent into dangerous phosgene gas.

Galvanized is a term referring to a zinc coating on the metal to prevent rust. Welding on galvanized steel needs extra caution and good ventilation because the welding arc can produce fumes of zinc or zinc oxide. Zinc is not as dangerous as its relative, lead, but it can cause a brief but uncomfortable "metal fume fever." Daily exposure results in a sort of immunity, but this immunity is lost in a few days, even in just a week-end away from exposure. The next Monday morning the nausea and chills are back again, causing this disease to be known as "Monday morning sickness," although admittedly there are other things about Mondays that often make workers "sick."

Plating metals are usually much more dangerous to weld than the iron or steel on which the plating is used. Cadmium is a plating metal for which welding fumes are considered very dangerous. This is one fume that has been known to be fatal in a single acute exposure. Even worse, acute exposures to cadmium usually do not display warning symptoms. Chronic exposures have been associated with emphysema and kidney impairment.

Stainless steel is one of the most dangerous materials to weld because of its high chromium content. Chromium trioxide is formed by the oxidation triggered by the welding heat and reacts with water to produce chromic acid. Ample sources of water can be found on human skin and the mucous membranes, resulting in chromic acid ulceration of these surfaces. Other chromic acid hazards were discussed in Chapter 12, where the phenomenon of chrome holes was discussed.

Welding in confined spaces complicates the atmospheric contamination problem. In confined spaces, the hazards of gases increase dramatically. Nitrogen and argon are inerting agents for the protection of the weld, but they are also simple asphyxiants to the welder. Another simple asphyxiant in welding atmospheres is carbon dioxide. Contrasted with the simple asphyxiants is the chemical asphyxiant carbon monoxide, also present to some extent in welding atmospheres, especially for gas welding.

Nitrogen is not as inert as argon or helium, mentioned earlier as inerting agents, although it is true that nitrogen is a relatively stable element. But nitrogen can be oxidized, especially in the extremes of welding temperatures, creating oxides that can be harmful. Since there are several oxides of nitrogen and they are somewhat difficult to isolate, industrial hygienists often refer to them as a group, naming them "NO_x." Nitrous oxide, N_2O, sometimes called "laughing gas," was once considered harmless and was even used as a dental anesthetic. Much more harmful are its dangerous cousins, nitric oxide (NO) and especially nitrogen dioxide (NO_2). According to Sax (ref. Sax), NO_x in concentrations of 60 to 150 ppm can have a delayed effect after the initial irritation of nose and throat. After breathing fresh air, the irritation goes away and the victim may feel all right. However, some 6 to 24 hours later, the following chain of symptoms can begin: tightness and burning sensation in the chest, shortness of breath, restlessness, air hunger, cyanosis, loss of consciousness, and, finally, death. Welding atmospheres are usually not this concentrated, but it should be noted that 100 ppm is only 0.01%.

Lead and mercury are well-known systemic poisons, and airborne fumes are the prime avenues for entry of these poisons into the body. Most welding does not involve these two metals. Soldering is used widely with lead alloys, but the low temperatures of soldering render the lead fumes relatively harmless.

Beryllium is a very useful alloy metal used in steel, copper, and aluminum. Unfortunately, the presence of the beryllium alloy in the material makes the metal very dangerous to weld. Since beryllium fume (particulate) hazards are both acute and chronic, most welders are wary of beryllium dangers.

Fluorine and fluorine compounds, usually fluorides, enter the welding atmosphere through welding flux or coverings. The popular shielded metal arc welding (SMAW) process is subject to hazards of fluorine compounds. The principal hazard is chronic, not acute, exposure, and long-term exposures cause abnormalities in the victim's bones. Other cleaning compounds and fluxes may also be hazardous, and personnel should check ingredients and heed manufacturer's instructions.

Before leaving the subject of welding gases and fumes, an important point is to be emphasized: None of the toxic materials or hazardous conditions described in this section is so dangerous as to prohibit welding. Welding atmospheres can be made safe by local or general exhaust ventilation or by personal protective equipment. The key is to recognize the potentially hazardous conditions, test atmospheres for excessive contaminant levels, and take corrective action if required.

SUMMARY

Welding represents a microcosm for the study of the entire field of occupational safety and health. It involves mechanical hazards, fire hazards, air-contamination hazards, personal protective equipment considerations, and almost every other subject addressed in this book. Welding processes are many and varied, and most safety and health managers know little about the technical aspects and terminology. A little study of the basics of welding, however, can open up opportunities for revision or substitution of processes that can enhance health and safety and at the same time improve efficiency and cut production costs. No other subject area seems to offer so much opportunity for safety and health managers.

EXERCISES AND STUDY QUESTIONS

16.1 What are the three basic categories of conventional welding? Which of the three is the cleanest and most healthful?

16.2 What distinguishes soldering and brazing from welding?

16.3 What distinguishes soldering from brazing?

16.4 What is the commonly used name for the most popular process of arc welding? What is the official AWS designation for this process?

16.5 Identify the following welding processes:

    (a) GTAW

    (b) GMAW

    (c) SAW

    (d) RSEW

    (e) RSW

16.6 Why should acetylene cylinders be stored valve end up?

16.7 Why are oxygen cylinders charged at so much higher pressures than acetylene cylinders?

16.8 In what ways are greasy gloves of particular hazard to the oxyacetylene welder?

16.9 Describe a way in which gas welding can sometimes be modified to cut production costs and at the same time avoid hazards.

16.10 Why are valve protection caps important to safety? Explain the purpose of the slots in the caps. How are the slots often misused?

16.11 Explain the phenomenon of flashback in welding operations.

16.12 How can taping welding hoses together to keep them orderly result in a hazard?

16.13 Which of the following materials should be avoided in pipes used for delivering acetylene to the work station: steel, wrought iron, or copper? Explain.

16.14 How can small arc welding machines that use ordinary household current be more dangerous than industrial arc welding machines.

16.15 Why should welding cables be uncoiled before using?

16.16 What is the principal hazard of allowing metal tanks to become a part of a welding circuit?

16.17 What arc welding process is gaining popularity because it is more healthful than other arc welding processes? What big disadvantage does it have?

16.18 What is the principal mechanical hazard of spot welders?

16.19 Give at least two reasons why people are psychologically inclined to risk the chance of welding fires.

16.20 Against what principal hazard are welding permits aimed?

16.21 Which of the following welding operations requires the most eye protection: SMAW, SAW, or RSEW? Explain.

16.22 What is the best natural material for welders' protective aprons, gloves, and leggings?

16.23 Name the pneumoconiosis resulting from exposure to iron oxide fume. Is it a severe hazard?

16.24 Why is it difficult to provide accurate tables of welding fume content?

16.25 What distinguishes welding fumes from toxic gases produced by the welding process?

16.26 Describe some workplace material characteristics that, when present, make welding fumes and gases more dangerous to the welder.

16.27 Consider the conditions narrated in Case Study 16.4 (obviously contrived), and describe apparent hazard mechanisms along with potential consequences.

CASE STUDY 16.4

A welder has constructed a manifolding arrangement for an assortment of oxygen and acetylene cylinders stored together lying on the floor. The manifold pressurizes the gaseous acetylene and depressurizes the oxygen to 50-psig for both. The room smells strongly of nail-polish remover. The welder is wearing greasy gloves and is chipping welding slag from the weld using his torch tip. The welding torch is connected to the manifold by two flexible hoses that are carefully wrapped together with duct tape, completely covering the hoses.

16.28 Nitrogen is a gas very widely used in welding operations, except gas welding. Explain.

16.29 In parts per million, calculate the approximate concentration of nitrogen in normal breathing air. How can nitrogen in breathing air be a hazard?

16.30 Acetylene is an unstable gas. Describe steps taken to make it safe to containerize and handle.

16.31 Discuss the principal hazards associated with oxygen cylinders used for welding.

16.32 Name some alternative fuels for acetylene in welding processes. In what way is acetylene superior to these fuels?

16.33 Explain why some people exaggerate the hazards of welding operations. Also explain why some people overly minimize the hazards of welding.

16.34 If items to be welded are cleaned with a solvent before welding, what dangerous gas may be generated if the solvent is not thoroughly removed before welding commences?

16.35 Describe the welders' illness "Monday morning sickness." What causes it?

16.36 Is exposure to welding flux an acute or chronic hazard? What adverse effect does welding flux have on the body?

RESEARCH EXERCISES

16.37 Examine the relative dangers of the various oxides of nitrogen, generically labeled NO_x. Which ones have both acute and chronic effects?

16.38 Look up records of industrial disasters involving welding.

16.39 Check the Internet home page for the American Welding Society. What sources of information there are valuable to the safety and health manager?

STANDARDS RESEARCH QUESTIONS

16.40 This chapter made a point about separating oxygen cylinders from acetylene cylinders in storage either by distance or by a noncombustible barrier at least 5 feet high. Search the OSHA standards for a provision dealing with this problem. Check the NCM database to determine whether this is a frequently cited standard.

16.41 Suppose your company is comparing processes to decide whether to use an arc welding process or a gas welding process. Do some research on the OSHA standards to compare these two types of welding. How do they compare with respect to the generation of OSHA citation activity?

Electrical Hazards

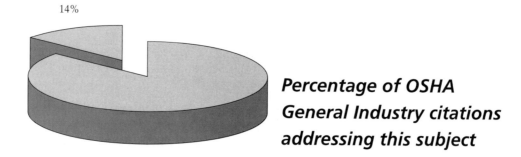

14%

Percentage of OSHA
General Industry citations
addressing this subject

Year after year, the National Center for Health Statistics reports approximately 500 to 1000 accidental electrocutions annually in the United States, with about 1 in 4 being industry and farm related. Everyone knows that electrical shock can be fatal, but the mechanism of the hazard is a mystery to most people. The mystery is due in large part to the fact that electricity is invisible. The use of electricity throughout our homes has led to a degree of complacency which is a factor in most electrocutions.

ELECTROCUTION HAZARDS

The first step toward safety from electrocution is to overcome the myth that "ordinary 110-volt circuits are safe." The truth is that ordinary 110-volt circuits can easily kill, and actually do kill, many more people than do 220- or 440-volt circuits, which nearly everyone respects. But the myth about 110 volts persists because almost everyone has sustained an electrical shock around the home or on the job without serious injury. An accident like this leads victims to the false conclusion that, although a 110-volt shock can be startling, it probably will not be fatal. Although they know that others have been killed by such shocks, they may somehow feel resistant or too strong to be seriously injured.

It is true that some persons are more resistant to electrocution hazards than others, but a far more important factor is the set of conditions surrounding the accident. Wet or damp locations are known to be hazardous, but even body perspiration can provide the dampness that can make electrical contact fatal. Another important condition is the point of contact. If current flow enters the body through the fingers and passes out through a contact at the elbow, no vital organs receive direct exposure. But if the flow is from a hand through the body to the feet, vital organs such as the heart, chest muscles, and diaphragm are affected, with possibly fatal results. Contact by the body torso to complete a circuit can also produce vital exposure to electrical current. Another factor can be the presence of wounds in the skin, which can result in a much higher current flow if contact is made where the skin is broken.

Physiological Effects

The central nervous system of our bodies is the conduit of signals between our brains and our muscles, including muscles of vital organs such as the heart and diaphragm. These signals are tiny electrical voltages that tell our muscles when to contract and when to relax. An external electric shock can send currents through the body that are many times greater than the tiny natural currents within our nervous systems. These larger currents can cramp or freeze muscles into a violent contraction—one that will not allow the victim to let go of the object contacted, or one that will stop breathing or stop the heart.

The heart is obviously our most important muscle. Its function is a rhythmic contraction and relaxation, which is timed by natural electrical pulses. The heart is thus very vulnerable to any pulsating electrical current. Common electric utility power supplies alternating current that cycles at a frequency of 60 hertz. It is ironic that 60 hertz is one of the most dangerous frequencies to which the heart can be exposed. This frequency tends to cause the heart to convulse weakly and irregularly at a rate too rapid to accomplish anything, a phenomenon known as *fibrillation*. Once fibrillation starts, death is almost a certainty, except that fibrillation has sometimes been stopped by controlled electric shocks to the heart muscle. The controlled electric shocks reestablish the heart's natural rhythms. Unfortunately, a defibrillation device is rarely available soon enough to save the life of an electrocution victim.

Stopped breathing from electric shock is due to cramped muscles responsible for respiration, such as the diaphragm and those controlling rib cage expansion. The first-aid remedy is artificial respiration, the same as for near drowning or other respiratory crises.

Just how much electrical current is fatal? There is no set answer to this question, but Figure 17.1 summarizes the opinions of several experts. The horizontal scale is logarithmic and is in units of milliamps or thousandths of an ampere. To put the chart in perspective, an ordinary table lamp with a 60-watt bulb draws about 500 milliamperes of current, far more than is needed to be fatal. An ordinary house circuit of 20 or 30 amperes will not trip the circuit breaker until there is a current flow of 20,000 to 30,000[1] milliamperes, respectively, about 100 to 1000 times as much as the lethal dose.

With such lethal potential available from an ordinary 110-volt house circuit, it would appear that almost no one could survive an electrical shock from such a circuit. But the body, especially the skin, has resistance that limits the flow of electric current

---

[1] 1 ampere = 1000 milliamperes.

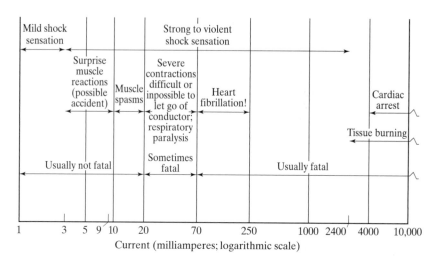

FIGURE 17.1

Effect of alternating electric current on the human body.

when exposed to 110-volt potential. To understand this resistance, some fundamentals of electricity may need review.

Ohm's Law

The basic law of electric circuits is *Ohm's law*, stated as

$$I = \frac{V}{R} \qquad (17.1)$$

where I = current in amperes

R = resistance in ohms

V = voltage in volts

The law can be rewritten as

$$V = IR \ \text{ or } \ R = \frac{V}{I}$$

Wattage is a measure of power and can be computed from known quantities of current and voltage or resistance as

$$W = V \times I \ \text{ and } \ W = I^2R \qquad (17.2)$$

where *W* is the power in watts.

Of principal concern are alternating-current (ac) circuits, which are the predominant type in both domestic and industrial use. Standard ac circuits cycle 60 times per second (in the United States and Canada), as shown in Figure 17.2. Alternating currents are more convenient to generate and distribute than are direct currents. But computations of current, resistance, and voltage using Ohm's law are somewhat awkward

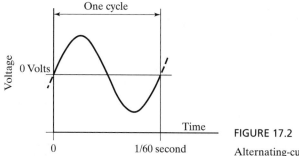

One cycle

Voltage

0 Volts

Time

0   1/60 second

FIGURE 17.2

Alternating-current voltage.

for ac circuits because the voltage varies from zero to positive, back to zero, to negative, and back to zero again every cycle. For convenience, an "effective" current for an ac circuit is computed as a value somewhat less than the current peaks. A direct current operating through a given load is found to generate as much heat as an alternating current that has peak currents 41.4% higher than the direct current. Thus, the ratio of effective current to peak current is computed as follows:

$$\frac{\text{Effective current}}{\text{Peak current}} = \frac{100\%}{100\% + 41.4\%} = 0.707 = 70.7\%$$

Effective voltages are computed by the same ratios as effective currents since they are related by Ohm's law. An ordinary 110-volt circuit then has an effective voltage of 110 volts, even though peaks of voltage over 150 volts occur every cycle.

The current drawn by an ordinary 60-watt lamp bulb can be computed by arranging Equation (17.2) as

$$I = \frac{W}{V} = \frac{60 \text{ watts}}{110 \text{ volts}} = 0.55 \text{ ampere}$$

Since the wire and other parts of the circuit would consume some power, a fair approximation to the current flow in the 60-watt table lamp is $\frac{1}{2}$ ampere, or 500 milliamperes, as stated earlier.

Returning now to the question of why more people are not killed by ordinary 110-volt circuits, we can use Ohm's law to determine how much the skin can limit the flow of electric current through our bodies. Human skin, if it is dry enough, is a good insulator and may have a resistance of 100,000 ohms or more. By using Ohm's law, a 110-volt exposure then would result in only a tiny current:

$$I = \frac{V}{R} = \frac{110 \text{ volts}}{100,000 \text{ ohms}} = 0.0011$$

$$= \text{approximately 1 milliampere}$$

From Figure 17.1, it can be seen that such a tiny current will probably not even be noticed. But add any perspiration or other moisture to the skin, and the resistance drops sharply. Due to perspiration alone, the skin resistance can be reduced 200 times, to a level of about 500 ohms with good contact with the electrical conductor. Once inside the body, the electrical resistance is very low and the current flows almost unimpeded. If the total resistance in the circuit is only 500 ohms, the current is calculated as

$$I = \frac{V}{R} = \frac{110 \text{ volts}}{500 \text{ ohms}} = 0.22 \text{ ampere}$$
$$= 220 \text{ milliamperes}$$

From Figure 17.1 it can be seen that an alternating current at this level passing through the body, including the heart, will most likely be fatal. Therefore, if you have ever received an electrical shock, and most of us have, you can be glad that you were not perspiring enough, or that you did not have a good enough contact, or that the path of the current bypassed the trunk of your body, or that you were poorly grounded, or that some other resistance impeded the current. Otherwise, you would have been killed by the ordinary 110-volt circuit, no matter how resistant you think you are to electrical shock. The principles and concepts of the electrocution hazards of ordinary house current are illustrated by Case Studies 17.1 and 17.2

CASE STUDY 17.1

A worker is using a hand-held circular saw to cut extruded aluminum strips in the manufacture of storm windows. He is holding the workpiece firmly in his left hand and holding the saw in his right hand. Aluminum is an excellent conductor of electricity, and the workpiece is making solid electrical contact with ground. In an accident that happens frequently, the worker accidentally saws the electrical cord in half. What are the probable consequences in the following three sets of circumstances:

- Case A: The tool is grounded through the third prong of the electrical plug.
- Case B: The tool is double insulated.
- Case C: The tool has a three-prong plug that is connected through an adapter to a two-hole wall socket; the tool is ungrounded.

Solution

Case A. Current will flow through the metal case of the saw handle into two paths, one through the grounding circuit, and the other through the worker's right hand, through his body, crossing through his torso and through his left hand into the well-grounded workpiece. Although the resistance through each of these paths might be relatively low, the resistance through the third prong grounding conductor should be the lower of the two, on the order of 2 to 3 ohms. A resistance as low as 3 ohms would immediately trip a 15- or 20-ampere circuit breaker, the type one would expect to find on such a circuit, as can be confirmed in the following calculation using Ohm's law:

$$I = \frac{V}{R} = \frac{110 \text{ volts}}{3 \text{ ohms}} = 36 + \text{amperes}$$

The current flow just calculated would be in addition to whatever a current might flow through the man's body and other paths to ground, including perhaps some flow through the tool itself before the accident completely severs the cord. The total current would therefore easily trip any reasonable circuit breaker and interrupt the flow of current, protecting the worker.

Case B. A double-insulated tool would have a nonconducting housing, resulting in no flow through the handle and the worker's body. The breaker would likely still be tripped as the metallic blade would make contact with both the hot wire and the well-grounded neutral. In addition, if the blade was cutting the aluminum workpiece at the time of the accident, another excellent path to ground would be through the metallic blade and into the grounded workpiece, causing an overcurrent to trip the breaker.

Case C. With no double insulation to protect the worker and no grounding conductor to trip the breaker, conditions might be present to cause the common accident to result in an electrocution. The well-grounded left hand of the worker would permit a substantial flow of current through his upper torso, the danger zone for heart and lung exposure. A reasonable value for the resistance in a well-grounded path through the worker's left hand and the aluminum workpiece would be 600 ohms. The current in such a grounding circuit would be calculated as follows:

$$I = \frac{V}{R} = \frac{110 \text{ volts}}{600 \text{ ohms}} = 0.183 \text{ ampere}$$

The current just calculated, only a small fraction of an ampere, would have no effect on a normal 15- to 20- ampere circuit breaker. However small a 183-milliampere current may be for breaking the circuit, it is a very large and dangerous current to flow through the worker's upper body. Such a current is shown by Figure 17.1 to represent a strong to violent, usually fatal, shock capable of producing heart fibrillation. The metallic blade of the saw might provide a good grounding path through the severed neutral or the well-grounded workpiece, accommodating an overcurrent that would trip the breaker and save the worker's life. But such a grounding through the metallic blade would be dependent upon chance; without such grounding the accident would likely be fatal.

CASE STUDY 17.2

A worker uses a trouble light suspended from the hood of an automobile while he repairs the engine. He leans across the fender of the car as he works so that his chest makes firm contact with the metallic fender, although that contact is resisted somewhat by a thin T-shirt he is wearing and slightly, by the paint on the fender of the automobile. The light, which has been through many years of severe usage, has developed a worn connection at the point at which the flexible cord is connected to the lamp socket. As the worker adjusts the light's position the worn connection results in accidental contact between the worker's index finger and the hot wire. Current passes through the man's finger and arm, continuing through multiple paths through his torso, most of it flowing to ground through his chest and the fender of the automobile and some through his feet and shoes. The contact between the hot wire and man's finger is only partial, and the electrical resistance of the skin in the man's finger at the point of contact is about 800 ohms. If this resistance represents about half the effective total resistance in the short circuit, how much current would flow through the man's torso? Would the circuit breaker, rated at 15 amperes, be tripped? Would the shock likely be fatal?

Solution

In this situation the current would flow through many parallel paths through the man's body, but for purposes of considering the total current flow due to the short circuit, one can consider the path to be equivalent to one effective path with a resistance of twice 800 ohms, or 1600 ohms. Using Ohm's law, we find that

$$I = \frac{V}{R} = \frac{110 \text{ volts}}{1600 \text{ ohms}} = 0.069 \text{ amphere}$$
$$= 69 \text{ milliamperes}$$

Such a current flow is much too small to trip the 15-ampere breaker, even if combined with the current flow through a 60-watt lighted lamp, which was calculated earlier in this chapter to be 0.55 ampere. If the entire 69- milliampere short circuit passes through the central part of the man's body, he is in critical danger of electrocution. Figure 17.1 reveals that 69 milliamperes is in the region of "sometimes fatal" and "respiratory paralysis." The victim may survive if an alert bystander is trained in cardiopulmonary resuscitation and applies artificial respiration, and if the victim is fortunate enough to avoid heart fibrillation.

Grounding

In the previous discussion, the term *grounded* was used. Just what does this electrical term mean? A requirement for electrical current to flow is that its path make a complete loop from the source of electrical power through the circuit and back again to the power source. We understand this loop as we connect a lantern bulb to the posts of a lantern battery, as shown in Figure 17.3. Disconnection of the circuit at any point in the complete loop stops the flow of current. This means that there must always be two conductors: one to carry the current to the device (usually called the "load") that uses it and another to carry the current from the load back to the electrical source. However, a trick makes the long trip back to the electrical source very simple in most applications of electrical power.

The earth for the most part is a fairly good conductor of electricity. Besides this, it is so massive that it is difficult for a human-made source of electricity to affect it much one way or another. Thus, no matter what we do on the surface of the earth, the earth

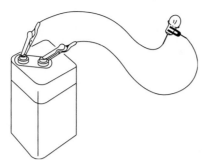

FIGURE 17.3

An electrical circuit makes a complete loop.

maintains a relatively even potential or charge. This means that if we drive two stakes firmly into the earth, even at great distances from each other, we may consider the resistance between them to be nil. The current flow may not be directly from one stake to the other because there are millions of electrical contacts to the earth at all times. Some of these contacts are positive and some are negative, but the total result is zero or earth potential. Thus, any electrical conductor driven into the earth immediately assumes the zero reference potential of the earth. This is a very convenient characteristic of the earth because it enables us to use it as one great common conductor back to the source of power. Figure 17.4 illustrates the use of the ground as a return conductor.

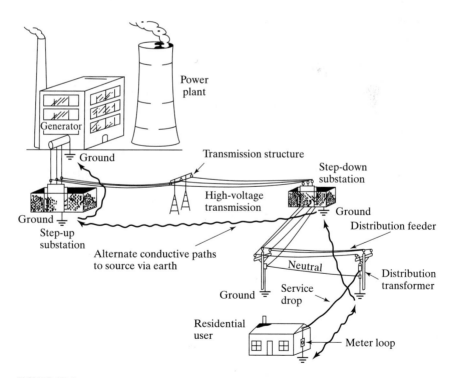

FIGURE 17.4

Alternative conductive paths to source through earth.

A careful examination of Figure 17.4 reveals that the power company provides a separate neutral conductor for the completion of the circuit back to the source. There are conditions that make dependence on the common potential of the earth somewhat unreliable. A very dry season, for instance, may make the surface of the earth lose its conductivity. This is especially a problem if the area is dry around the grounding conductor stake that has been driven into the ground. The neutral conductor then ensures the completion of the circuit regardless of conditions.

The use of ground in electrical circuits is so advantageous as to be considered indispensable. However, the very convenience and proximity of the ground everywhere present a hazard. If a person contacts an energized conductor and at the same time is in contact with the ground or some other object that has a conductive path to ground, that person completes the electrical circuit loop by passing electric current through his or her body. A major portion of the *National Electrical Code*®[2] is devoted to prevention of this hazard.

The principal way in which persons are protected from becoming a part of the path to ground is by insulation of conductors. In addition, exposed conductive surfaces are given a good connection to the ground, usually by means of the ground wire, so that opportunity for a person's body to be the path to ground is minimal. Paradoxically, in some rare instances, the *National Electrical Code*® takes exactly the opposite approach. For some systems, it makes more sense to *isolate* the entire structure from ground. If the structure is isolated, workers are protected by not being in contact with conductors that could connect them with ground.

Wiring

A typical 110-volt circuit has three wires: *hot, neutral*, and *ground*. Sometimes the neutral is called the "ground*ed*" conductor, in which case the ground is called the "ground*ing*" conductor. The purpose of the hot wire (usually a black insulated wire) is to provide contact between the power source and the device (load) that uses it. The neutral (usually a white insulated wire) completes the circuit by connecting the load with ground. Both the hot and the neutral normally carry the same amount of current, but the hot is at an effective voltage of 110 volts with respect to ground, whereas the neutral is at a voltage of nearly zero with respect to the ground.

The third wire is the ground wire and is usually either green or is simply a bare wire. The purpose of the ground wire is safety. If something goes wrong so that the hot wire makes contact with the equipment case or some other conductive part of the equipment, the current in its path to ground can bypass the load and take a shortcut, commonly called a *short*. Since the load is bypassed, the short is a very low-resistance path to ground and by Ohm's law draws a very high current. This high current in a properly protected circuit will almost immediately either "blow" a fuse or "trip" a circuit breaker, depending on the type of overcurrent protection provided in the circuit, and stop all flow of current in the circuit.

---

[2]The National Electrical Code®, commonly abbreviated NEC, is published regularly by the National Fire Protection Association (NFPA), 470 Atlantic Avenue, Boston, MA, 02201.

It is possible, of course, to have a short without a ground wire. The equipment may be naturally grounded by its location or installation, or the hot wire can somehow contact the neutral. Sometimes the short to ground is only partial because there is considerable resistance in the short path to ground. Such a short may go undetected because the short flow of current is of insufficient amperage to cause the total circuit current to trip the overcurrent protection in the circuit. In this case, current will continue to flow, and equipment loads will continue to operate in the presence of these shorts, or *ground faults*, as this type of short is sometimes called. Such ground faults can be especially dangerous on construction sites. This hazard is the basis for ground-fault circuit-interruptor (GFCI) devices on construction sites. The GFCI protection is in addition to overcurrent protection such as circuit breakers or fuses.

Figure 17.5 explains how a GFCI works. Whenever the current flow in the neutral is less than the current flow in the hot wire, a ground fault is indicated, and the current flow is stopped by a switch that breaks the entire circuit. Building codes have required the installation of GFCI circuits in residential bathrooms. GFCI electrical receptacles have small, red "RESET" buttons, as shown in Figure 17.6. One difficulty with GFCIs is that some leakages to ground are almost impossible to prevent, especially when conditions are wet or extension cords are extremely long. This causes the GFCI to trip even when no hazard exists, a condition known in the construction industry as "nuisance tripping." An alternative to GFCIs is for the employer to test, inspect, and keep records of the condition of equipment grounding conductors.

One misconception about shorts is the idea that a good fuse or circuit breaker is sufficient to stop the flow of a dangerous short through a person's body. A reexamination of Figure 17.1 shows that a person will almost certainly be killed by exposure to a current that would not blow even the smallest popular household fuses (i.e., 15- or 20-ampere fuses). A fuse or breaker rated at 15 amperes will handle up to 15,000 milliamperes before blowing, several times as great as the fatal current shown in Figure 17.1. The beauty of the third wire or grounding wire is that it provides a very low-resistance, high-current short to ground, which will trip the fuse or breaker immediately, before *other* short paths to ground (such as through a person's body) can do their damage.

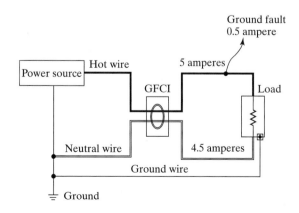

FIGURE 17.5

Ground-fault circuit interruptor (GFCI). The 0.5-ampere fault to ground causes a current flow imbalance between the hot and the neutral. This imbalance triggers the GFCI to break the circuit.

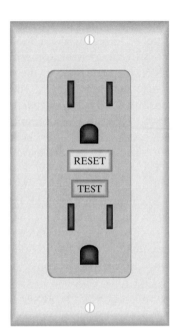

FIGURE 17.6

An electrical receptacle equipped with a GFCI.

Double Insulation

Unfortunately, less than half of the electric hand tools in actual use are properly grounded. Studies of equipment returned to the factory for repair have shown that a large number of units have been altered so that the grounding system is no longer intact. A common alteration is to cut off the third prong of the plug so that it can be plugged into an old two-wire receptacle. To counter this practice, the use of "double-insulated" tools is permitted in lieu of equipment grounding. A second covering of insulation gives an extra measure of protection to the operator of double-insulated tools in case of a short to the equipment case.

Most double-insulated tools have a plastic, nonconductive housing, but this is not a fully reliable indication that the tool is double insulated. The second covering of insulation must be applied according to precise specifications before the tool can qualify to receive the designation "double insulated." Qualifying tools have the manufacturer's mark "double insulated" or a square within a square ▢ to indicate double insulation.

Miswiring Dangers

In the original wiring job, electricians sometimes make mistakes or use slipshod practices that increase hazards. One of these practices is to "jump" (connect) the ground wire to the neutral wire. Actually, this is a trick that will work, and usually no one will be the wiser, but the practice does cause hazards. Figure 17.7 shows how a circuit is correctly wired, revealing that both the neutral and grounding wire are connected directly to ground. So in Figure 17.8, in which it is shown that the ground is jumped to the neutral, a third wire is not used in the wiring system.

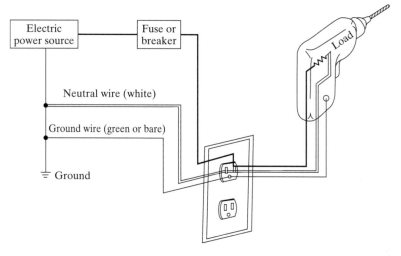

FIGURE 17.7

Correctly wired 110-volt circuit.

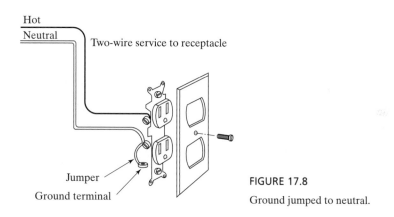

FIGURE 17.8

Ground jumped to neutral.

The main hazard in jumping the ground to neutral is that it can create low voltages on exposed parts of equipment. The equipment case or housing is connected to the ground wire. Since no current normally flows through the ground wire, it serves as an excellent method of keeping the voltage on the equipment case close to zero with respect to ground. But the neutral does carry considerable current. By using Ohm's law, it can be determined that this large current on the neutral may cause it to have a low voltage with respect to ground, especially if the neutral wire must travel a long distance back to the ground at the meter. If the circuit is carrying a current of 20 amperes and the resistance of the neutral wire is $\frac{1}{2}$ ohm, the voltage on the equipment case is calculated to be

$$V = IR = 20 \times \tfrac{1}{2} = 10 \text{ volts.}$$

This is a low voltage, but is theoretically capable of producing a fatal current through a person's body if conditions are just right (or rather, wrong). The real hazard, though, is that a loose or corroded connection somewhere in the neutral circuit would increase its resistance, perhaps to 4 or 5 ohms, causing the voltage to increase several times.

Another common wiring mistake is *reversed polarity*, which simply means that the hot and neutral wires are reversed. This is another subtle problem because most equipment will operate perfectly well with reversed polarity. One hazard of reversed polarity is that the designated leads (black lead, hot; white lead, neutral) become reversed, and the confusion could bring on an accident to an unsuspecting technician. Another hazard is that a short to ground between the switch and the load could cause the equipment to run indefinitely, independent of whether the switch is on or off. (See Figure 17.9.) Finally, bulb sockets can become hazardous when the polarity is reversed. In Figure 17.10(a), a correctly wired socket shows the screw threads to be neutral. But in a reversed-polarity socket, as shown in Figure 17.10(b), the exposed screw threads become hot, and the button, which is naturally more protected at the bottom of the socket, becomes neutral.

Perhaps the most common wiring mistake of all is to fail to connect the ground terminal to a ground wire, a condition known as *open ground*, or *ground not continuous*. This is another error that can easily go unnoticed because equipment attached to

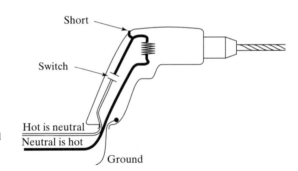

FIGURE 17.9

Reversed polarity. A short in the position indicated in the circuit for this drill will cause the drill to operate continuously, independent of the switch.

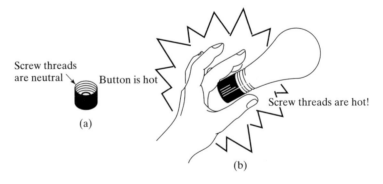

FIGURE 17.10

Hazards of reversed polarity in lamp socket: (a) correctly wired socket; (b) reversed-polarity socket.

circuits miswired in this way will usually operate normally. But if an accidental short to the equipment case occurs, the worker is in danger of electrocution.

The three cases of miswiring discussed in this chapter are not the only mistakes that can be made in wiring electrical circuits; they are not even the most hazardous. But because they permit electrical circuits to "work normally," they go unnoticed by uninformed users of equipment attached to such circuits. Because the errors are not immediately crippling to the function, they are frequently committed. Some simple checks with inexpensive testers can easily demonstrate the problems. These testers will be examined later.

FIRE HAZARDS

Most people think of electrocution when they think of electrical safety, but electrical codes have as much to do with fire hazards as they do with electrocution. Many systems, such as fuses or circuit breakers, protect against both fire and electrocution, but their primary function is fire prevention.

Wire Fires

One of the most common causes of electrical fires is wires that become overheated because they conduct too much current. Wire diameters (gauges) must be properly sized to handle the expected current load, and overcurrent protection (fuses or breakers) must ensure that these loads are not exceeded. Substitution of fuses with copper pennies is a common method of defeating the overcurrent protection so that the circuit will handle larger loads. If no fuse is present to burn in two, the wire itself may act as the next weakest link. If the wire becomes hot enough to burn in two, any contact with combustible material along the wire run is likely to produce a fire.

Arcs and Sparks

Whenever two conductors make a physical contact to complete a circuit, a tiny (or not so tiny) electric arc jumps the air gap just prior to contact. This arc may be so small as to be undetectable, but it is usually hot enough to ignite explosive vapors or dusts within their dangerous concentration ranges.

When the electric arc is an instantaneous discharge of a statically charged object, it is sometimes called a *spark*. Such sparks are capable of igniting an explosive mixture, the spark plug of an automobile engine being ample testimony. Sparks are prevented by electrically connecting, or "bonding," two objects that may be of different static charge. This is especially important when pouring flammable liquids from one container to another.

The arc that occurs when an ordinary electrical circuit is completed is virtually impossible to prevent. This means that switches, lights, receptacles, motors, and almost any electrical device, even telephones, are a source of ignition to hazardous concentrations of explosive vapors or dusts. Chapter 11 discussed the ranges of explosive vapors and defined the lower and upper explosive limits (LEL and UEL, respectively). Since the arc is impossible to prevent, some means must be used to separate it from the hazardous concentrations in the air. This is done by using wire, conduit, or equipment that is either vapor tight or strong enough to contain and prevent the propagation of an explosion inside the conduit or equipment. This is an expensive undertaking, and it is tempting to take shortcuts. The *National Electrical Code*® has a strict code for electrical wiring and equipment designed for hazardous locations. The safety and health manager should be able to identify the operations or hazardous locations within the plant that require special wiring and electrical equipment. Accordingly, this identification scheme is discussed next.

Hazardous Locations

One of the most difficult tasks in the field of industrial safety is the definition of various industrial locations that require special wiring and equipment to prevent explosions. Industrial processes are so diverse as to defy a general definition. In addition, the ignition mechanisms are different for different materials. For instance, the hazard of heat buildup on electric equipment housings and bearings coated with ignitable dusts is altogether different from the hazard of spark ignition of explosive vapors derived from flammable liquids. Difficult as the problem is, it must be dealt with because some industrial locations are just too dangerous to allow exposure to electric ignition sources.

The *National Electrical Code*® meticulously defines various conditions to classify hazardous locations roughly into six categories. Within these classifications are various groups that identify the substance group that is causing the hazard.

The major classification is according to the physical type of dangerous material present in the air and is designated *Class*. The next classification is called *Division* and relates to the extent of the hazard by considering the relative frequency with which the process releases hazardous materials into the air. The criteria for "Division" are subjective, not quantitative, except around paint spray areas. This subjectivity introduces problematic gray areas.

Figure 17.11 is a decision chart that attempts to simplify the complicated process of the classification of hazardous locations. The chart is approximate only because strict definition would require enumeration of pages of exceptions and conditions, many of which would overlap. The thing to remember is that the "Class" is the material and the "Division" is the extent of the hazard. Thus, one can state that Division 1 locations are more hazardous than Division 2 locations, but cannot state absolutely that Class I locations are more hazardous than Class II or III locations.

Since classification is so complex, industry often relies on examples common in similar industries to decide whether a location is Division 1 or 2 or not of sufficient hazard to classify at all. Examples of some common hazardous locations are listed in Table 17.1.

Equipment qualifying for hazardous locations usually qualifies for both Division 1 and 2. When in doubt, most industries want to install equipment approved for Division 1 in order to be prepared for the worst. Most violations of code are not for selection of Division 2 equipment when Division 1 should have been selected; most violations are from the use of thin-walled conduit and conventional electrical equipment in Division 1 or 2 locations.

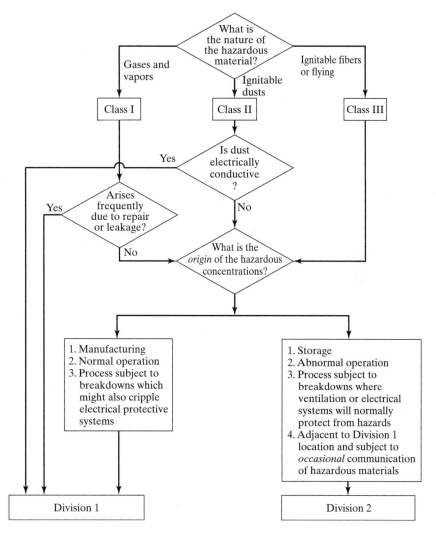

FIGURE 17.11

Decision chart for classifying hazardous locations that are dangerous from the standpoint of ignition of materials in the air.

"Approval" of electrical equipment for use in hazardous locations means that the manufacturer's design has been tested and approved by a recognized testing laboratory such as Underwriters' or Factory Mutual. A label must be on the equipment designating its classification if that equipment is to be used in hazardous locations.

Figure 17.12 shows several examples of *explosion-proof equipment* approved for Class I, Division 1 locations. Explosion-proof conduit looks more like pipe than conventional thin-walled conduit, which looks more like tubing. Explosion-proof junction boxes are castings, as opposed to conventional formed sheet metal boxes. The complicated

TABLE 17.1 Classification of Common Hazards.

Description	Classification
Paint spray areas (flammable paint)	Class I, Division 1
Areas adjacent to, but outside of, paint spray booth	Class I, Division 2
Areas of open tanks or vats of volatile flammable solvents	Class I, Division 2
Storage areas for flammable liquids	Class I, Division 2
Inside refrigerators containing open or easily ruptured containers of volatile flammable liquids	Class I, Division 1
Gas generator rooms	Class I, Division 1
Grain mills or processors	Class II, Division 1
Grain storage areas	Class II, Division 2
Coal pulverizing areas	Class II, Division 1
Powdered magnesium mill	Class II, Division 1
Parts of cotton gin locations	Class III, Division 1
Excelsior storage areas	Class III, Division 2
Closed piping for flammable liquids (piping has no valves, checks, meters, or similar equipment)	Not classified

structures for telephones and even light switches make obvious the fact that explosion-proof equipment costs several times as much as conventional equipment.

In Division 1 locations, it is recognized that there is no way to ensure that the vapors will be kept out of the conduit and equipment. During maintenance, installation, or other open periods, vapors will enter the system. Therefore, the electrical equipment designer assumes this inescapable reality and designs the Division 1 equipment to withstand an internal explosion and cool the explosion gases as they escape before they can ignite the entire area in a devastating explosion.

By contrast, Division 2 electrical equipment enjoys some isolation from dangerous explosive vapors *most* of the time. Therefore, if the Division 2 equipment can be properly sealed with gaskets to make it vapor tight, it will be safe. Class I, Division 2 equipment is characterized as vapor tight, whereas Class I, Division 1 equipment is *not* vapor tight but is explosion proof. Of course, if equipment is classified as Class I, Division 1 (explosion proof), it is also acceptable for use in Class I, Division 2 areas even though the equipment is not vapor tight.

The foregoing comparison of Division 1 and Division 2 equipment and locations acknowledged the generalization that any equipment approved for Division 1 locations is also acceptable for Division 2 locations *of the same classification*. However, it should be noted here that equipment approved for Class I locations is not necessarily approved for Class II or III. The hazard mechanisms of ignitable Class II dusts or Class III fibers are somewhat different from the hazards of Class I vapors. Dusts and fibers can settle on warm equipment, insulating it from necessary heat dissipation during operation. Such insulation can cause a substantial heat buildup on the equipment that can result in a smoldering dust ignition and subsequent explosion.

A common error when selecting electrical receptacles for hazardous locations is to mistake weatherproof electrical outlets for approved equipment. Ordinary weatherproof outlets, as shown in Figure 17.13, are not approved for any type of hazardous locations in either Division 1 or 2. The spring-loaded cover protects the receptacle from

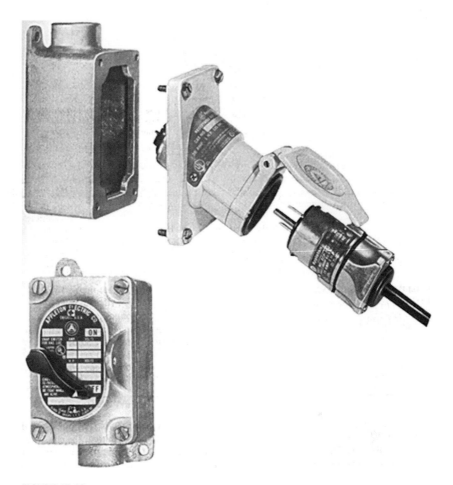

FIGURE 17.12

Explosion-proof electrical equipment approved for Class I, Division I hazardous locations. Note heavy-duty, machined components. (a) Electrical outlet plug and receptacle; (b) wall switches (*Source:* Courtesy of Appleton Electric Co.).

FIGURE 17.13

Ordinary weatherproof outlets not approved for hazardous locations.

weather when the receptacle is not in use, but when a plug is inserted into the receptacle, the receptacle is exposed as much as any conventional one.

The safety and health manager may be baffled when reading equipment labels to find that equipment is classified and labeled by Class and *Group*, rather than Class and *Division*. Like the Class designation, the Group designation identifies the type of

TABLE 17.2   Groups and Classes.

Class	Group	Description	Examples	Industries
I	A		Acetylene	Welding fuel generators
	B	Highly flammable gases and some liquids	Hydrogen	Chemicals and plastics
	C	Highly flammable chemicals	Ethyl ether, hydrogen sulfide	Hospitals, chemical plants
	D	Flammable fuels, chemicals	Gasoline	Refineries, chemical plants, paint spray areas
II	E	Metal dusts	Magnesium	Chemical plants
	F	Carbon, coke, coal dust	Carbon black	Mines, steel mills, power plants
	G	Grain dusts	Flour, starch	Grain mills and elevators

material present in the atmosphere, but the Group classification is more detailed. Four of the Groups belong to Class I and three to Class II and are summarized in Table 17.2.

A typical classification label will state "approved for Class I, Groups A and B," omitting the "Division" designation. When the "Division" is omitted, the classification is invariably acceptable for both Division 1 and 2 locations. If the equipment is merely vapor tight and approved only for Division 2, the "Division" designation should be displayed on the label. Class I equipment is usually either approved for Groups C and D or is approved for all four Class I Groups: A, B, C, and D.

TEST EQUIPMENT

Pursuant to general electrical safety, there are a few inexpensive items of test equipment that the safety and health manager should have access to for occasional in-house inspections. Brief descriptions of these items follow.

Circuit Tester

A circuit tester (Figure 17.14) simply has two wire leads connected by a small lamp bulb, usually neon. Whenever one of the leads touches a hot wire and the other lead touches a grounded conductor, completing the circuit, the bulb glows. The tester works only for a given voltage range, but most are able to handle both 110- and 220-volt circuits. The circuit tester is somewhat of a safety device in itself, enabling maintenance workers to be sure that power is turned off before touching hot wires. The circuit tester can also be used to determine whether various exposed parts of machines or conductors are *live* or *energized*.

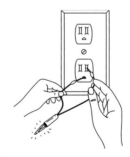

FIGURE 17.14

Circuit tester.

Receptacle Wiring Tester

One of the simplest and most widely used test devices is the receptacle wiring tester, which can be carried around in one's pocket. (See Figure 17.15). No battery or cord is required. The user simply plugs the device into any standard 110-volt outlet and interprets the arrangement of indicator lights to determine whether the receptacle is wired improperly. Note carefully the use of the word *improperly* instead of the word *properly* in the preceding sentence. Only certain errors will be detected, and a "correct" indication of the lights means only that those certain errors were not found. The receptacle could still be wired incorrectly.

Figure 17.16 reveals that the receptacle wiring tester is nothing more than three simple circuit testers combined into one tester. The tabular portion of the figure shows types of wiring errors that can be interpreted from the indicator lights. One of the most common wiring errors—"ground jumped to neutral"—is shown by the tester to be wired correctly.

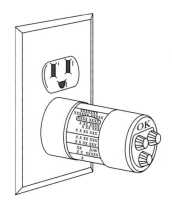

FIGURE 17.15

Receptacle wiring tester.

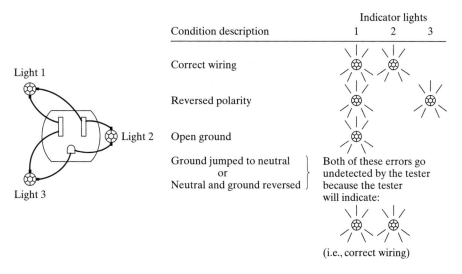

Condition description	Indicator lights		
	1	2	3
Correct wiring	⊗	⊗	
Reversed polarity	⊗		⊗
Open ground	⊗		
Ground jumped to neutral or Neutral and ground reversed	Both of these errors go undetected by the tester because the tester will indicate:		
	⊗	⊗	
	(i.e., correct wiring)		

Light 1

Light 2

Light 3

FIGURE 17.16

Schematic of receptacle wiring tester.

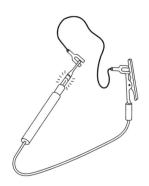

FIGURE 17.17

Continuity tester

Continuity Tester

It is sometimes useful to check a dead circuit simply to see whether all the connections are complete or whether a break in a conductor has occurred. Figure 17.17 illustrates a simple continuity tester, which is similar to a circuit tester except that a small battery is included to provide power for the light. The continuity tester also has much less electrical resistance than the circuit tester because it is used only on dead circuits. If the continuity tester is used on a live 110-volt circuit, it will immediately blow a fuse or throw a breaker in the circuit.

An important application for a continuity tester is to check the path to ground. One terminal of the tester can be connected to the exposed equipment case of a machine in question and the other terminal to a known-grounded object. If the lamp lights, the machine is known to be grounded. If the lamp does not light, there is a break somewhere in the path to ground.

FREQUENT VIOLATIONS

Having been versed in the principles of electrical hazards, the safety and health manager should know what to look for when making an inspection. But for review, the remainder of this chapter will describe frequently cited violations of the *National Electrical Code*®.

Grounding of Portable Tools and Appliances

It should come as no surprise that the most frequently cited electrical code violation is for grounding. This is the code cited for ungrounded electric drills, sanders, saws, and other portable hand-held equipment. Nonportable equipment, such as refrigerators, freezers, and air conditioners, must also be grounded. Special attention should be given to the use of portable electric tools used in damp or wet locations and, of course, to plugs that have the ground pin broken off.

Exposed Live Parts

Almost as frequent as ungrounded portable equipment is the existence of exposed live parts. If the actual equipment conductors cannot be insulated and terminals covered, locked rooms should be used to prevent exposure to workers. One of the most frequent observations of exposed live parts is sloppy electrical installations in which covers are left off junction boxes or receptacle cover plates are missing. A switch box, fuse, or breaker box in which the door is open also constitutes "exposed live parts."

Improper Use of Flexible Cords

Makeshift or temporary electrical installations are prohibited as are substitutions of flexible cords for fixed wiring. Obvious instances are flexible cords run through holes in the walls, ceilings, floors, doorways, or windows.

Marking of Disconnects

This is an easy thing to correct. The disconnect box or switch panel for motors and appliances must be identified so that equipment can be quickly and confidently disconnected. Also, the origins of branch circuits, such as in the breaker box, must be labeled to indicate their purposes. If the location or arrangement of the disconnect makes it obvious which machines or circuits it controls, marking may not be necessary. Very few of the disconnect marking violations are designated as "serious."

Connection of Plugs to Cords

Here is another extremely simple item. When repairing a plug to an extension cord, appliance cord, or any other flexible cord, be sure to tie a knot in the cord or do something that will prevent a pull on the cord from being transmitted directly to joints or terminal screws. This is just a basic principle of electrical maintenance.

SUMMARY

This chapter began with some sobering thoughts of what electricity, even in small amounts, can do to the human body. The greatest hazard is with 110-volt circuits, not 220-volt or higher circuits. This is because of the popularity of 110-volt circuits and the complacency of the people who use them. Besides electrocution hazards, electricity also presents fire hazards, not to speak of burns and other electrical exposure hazards.

Simple testers can demonstrate quickly some of the frequent wiring mistakes that might otherwise go undetected in normal operations. But some of these testers are too simple and overlook common errors such as "ground jumped to neutral."

Providing for electrical equipment in industrial processes that produce flammable vapors, dusts, or fibers is a difficult and expensive task. The definitions and codes for hazardous locations with explosive atmospheres are also complicated and tricky. The safety and health manager is advised to look for required markings on electrical equipment installed in hazardous locations to be sure that the equipment is approved for the "Class" and "Division" of the location in which it has been installed.

Violations of electrical code are often for conditions that are easy to correct. The largest single item to remember is grounding—grounding of both portable and fixed equipment and the circuits that serve them. Of less frequency, but of great seriousness, are the OSHA citations for unapproved conduit and electrical equipment used in explosive atmospheres of flammable vapors or dusts.

EXERCISES AND STUDY QUESTIONS

17.1 What are the two principal hazards of electricity?

17.2 Approximately how many people die each year of electrocution in the United States?

17.3 Compare 110-, 220-, and 440-volt hazards of electricity.

17.4 At approximately what current levels through the heart and lungs does the sustained flow of electricity become fatal?

17.5 What part does grounding play in an electrical circuit? Why do safety regulations require equipment to be grounded?

17.6 Compare the functions of hot, neutral, and ground wires.

17.7 What is a GFCI, and where is it used?

17.8 Explain the term *double insulation*.

17.9 What are the hazards of reversed polarity?

17.10 Compare electrical "arcs" and "sparks."

17.11 Explain the difference between Class I, II, and III hazardous locations.

17.12 What do the terms *Division* and *Group* mean in the classification of hazardous locations?

17.13 Explain the difference between a continuity tester and a circuit tester.

17.14 Name some of the most frequent violations of electrical code.

17.15 A certain string of Christmas tree lights has eight bulbs, each rated at 5 watts. If there is no ground fault, how much current flows through the hot wire at the receptacle plug? How much flows through the neutral?

17.16 In Exercise 17.15, if all the current flowing through the hot wire would pass through a person's heart muscle, would the current likely be fatal?

17.17 A maintenance worker wires a 110-volt electric lamp without disconnecting the circuit. With no load energized in the circuit, the worker accidentally contacts the bare hot wire. What will probably happen? Under what conditions would the worker likely be killed?

17.18 In Exercise 17.17, suppose the wire that was contacted was the neutral wire. What would probably happen? Are there conditions under which the worker would be killed?

17.19 In Exercise 17.17, suppose the wire that was contacted was the equipment ground wire. What would probably happen?

17.20 Portable "trouble lights" used in automobile repair have been involved in many fatalities. Explain probable hazard mechanisms.

17.21 Explain the difference between a GFCI and an ordinary circuit breaker.

17.22 Explain why it is so difficult to detect the wiring defect "ground jumped to neutral."

17.23 A worker was electrocuted when he drilled through a wall and the drill bit contacted an electrical wire inside the wall by cutting through its insulation. The ground pin of the electric plug for the drill was broken off. Describe several ways in which this fatality could have been prevented.

17.24 The broken-off ground pin represented a code violation in the fatality case described in Exercise 17.23. How did this contribute to the hazard? What would probably have happened had the ground pin been intact?

17.25 It can be said that "grounding" with respect to electricity can be both good and bad. Explain the "good" and the "bad" about electrical grounding.

17.26 Explain the distinction between the terms "grounded conductor" and "grounding conductor."

17.27 For each of the following situations, explain the nature of the hazard and describe probable outcomes:

(a) A worker brushes his pant leg across a 120-volt terminal strip with screw terminals exposed. Every other screw is hot, with neutral screws intervening.

(b) A worker's boot contacts a 440-volt bus bar. A ground path occurs through the worker's foot and the nails in the sole of the boot to the concrete floor. The resistance of the ground path is 10,000 ohms.

(c) A worker wires a 220-volt circuit "hot" using a wood-handle screwdriver. The screwdriver slips, and the shaft of the screwdriver makes a direct short across the hot terminal and the adjacent neutral.

(d) A worker changes a 120-volt wall receptacle in a garage while standing on a concrete floor. The circuit is energized, but the worker is careful not to touch both the hot wire and the neutral wire at the same time.

17.28 A right-handed worker is repairing a 120-volt lamp socket using a screwdriver with an insulated handle. He is wearing a short-sleeve shirt, and his bare right arm is braced against a water pipe. A bare hot wire contacts the metal housing of the socket assembly the worker is holding in his hand while he is using the screwdriver in his other hand. A ground path occurs with total resistance of 600 ohms. Calculate the current flow and describe its probable path. Will the breaker likely be tripped? Is there a risk of electrocution? If so, what factors contribute to the risk?

17.29 Describe the phenomenon of fibrillation and its probable effect.

17.30 What characteristic about electric utility power aggravates the hazard of fibrillation?

17.31 Calculate the peak voltage for an effective line voltage of 240 volts ac.

17.32 An ac circuit has peak voltages of ±80 volts. Calculate the effective voltage.

17.33 An ac circuit has peak voltages of ±170 volts. Calculate the *effective* current flow in this circuit if the total circuit load operates at 60 watts of power. Calculate the peak current flow.

17.34 Explain why it is so important for the resistance to be low in the "third-wire" path to ground. (*Hint:* "Electricity follows the path of least resistance" is not the correct answer to this question.)

17.35 Explain how a "shorted-out" electric tool may continue to operate. There are at least two different sets of conditions that will lead to this phenomenon.

17.36 Explain why the common wiring defect "open ground" easily goes undetected.

17.37 Is it okay to use Division 1 approved equipment in Division 2 hazardous locations? Why or why not? Is it okay to use Class I approved equipment in Class II hazardous locations? Why or why not?

17.38 **Design Case Study.** In an actual case history of an electrocution fatality, Figure 17.18(a) illustrates the way the chuck key for a hand-held electric drill was conveniently attached to the power cord so that it would always be readily available for operator use in changing bits. The fatality occurred when the twisted wire wore through the cord insulation after continued use. Figure 17.18(b) illustrates a much safer way to connect the chuck key. What other factors probably contributed to this fatality and how could it have been prevented? Suppose that the ground pin had been broken off the plug for this drill. How would this code violation have affected this fatality?

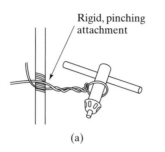 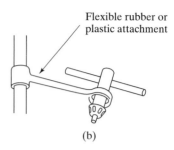

(a)   (b)

FIGURE 17.18

Cause of electrocution fatality: chuck key secured to cord by twisted wire tie.
(a) Unsafe, improvised attachment of chuck key. Wire or tape will eventually
damage cord insulation. (b) Safer method of attaching chuck key.

17.39 **Design Case Study.** An architectural design and engineering firm seeks advice regarding
federal safety standards for wiring for a new process being set up to manufacture dyestuff.
The process uses chlorobenzene and releases ignitable concentrations in the vicinity of the
process equipment. Specify the appropriate Class, Division, and Group classification for
wiring and electrical equipment located in this area.

17.40 **Design Case Study.** In the case study of Exercise 17.39, suppose an alternate supplier of
process equipment proposes a completely closed system in which releases would only
occur during repair or in event of a leak. Such occurrences arise frequently out of a need
to regularly clean the in-feed mechanism. Would the closed system have an impact on
safety and on the prescribed wiring system?

RESEARCH EXERCISES

17.41 Use the Internet to find out who publishes the *National Electrical Code*. What other aids
for electrical safety are available from this organization?

17.42 Check recent statistics to determine the number of electrocutions annually in the United
States. What percentage are "on the job"? Is the annual number of electrocutions increasing or decreasing?

17.43 OSHA citation frequencies change somewhat from year to year. Check enforcement statistics to determine the top five most frequently cited electrical standards. Are they the
same as the ones identified in this chapter?

STANDARDS RESEARCH QUESTIONS

17.44 This chapter concludes those related to the General Industry standards. If the reader has
been studying the chapters of this book sequentially, then at this point he or she should
have a great deal of familiarity with the OSHA General Industry standards and the NCM
database tool for examining inspection statistics for these standards. In previous chapters,
this text has assigned specific questions for "STANDARDS RESEARCH QUESTIONS."
In this chapter, however, rather than address specific questions for research related to
electrical standards, it is suggested that the reader go "browsing" in the OSHA General

Industry electrical standards, using the OSHA website. Try to relate the standards to topics that were discussed in this chapter. Try to pick out particular provisions in the OSHA standard that you reason would be frequently cited. Check whether your theories are correct by querying the NCM database for your selected provisions to see whether they are indeed frequently cited. Try to pick out standards that OSHA would cite as "serious" violations in many cases. Check the NCM database to determine whether your intuitions about these standards are correct.

CHAPTER 18

Construction

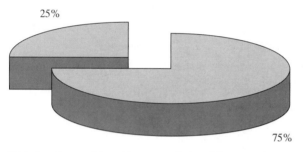

25%

75%

*Percentage of OSHA
citations addressing
this subject*

This book would not be complete without a chapter on the construction industry and its relationship to the field of safety and health. Why is this industry singled out, whereas the book has otherwise taken a general approach to the subject? There are two principal reasons. One is that the construction industry has long been recognized as more hazardous than most. Workplace installations are temporary, and economics dictate a different approach to such facilities as guardrails and stairways. The very nature of the work dictates that risks be taken that are not necessary in general industry. This added risk alone would ensure that OSHA would take a special interest in construction, but there is another reason for OSHA's emphasis that is probably more pervasive than the risk factor. The Occupational Safety and Health Act of 1970 was preceded by the Construction Safety Act, Public Law 91–54. The result of this sequence is that federal standards for construction safety were already in place at the time of passage of the more general OSHA law. The OSHA law provided for adoption of "existing federal standards" as national consensus standards, bypassing the lengthy promulgation procedures required for new standards. Thus, a set of rather comprehensive standards for construction (Part 29 CFR 1926) has been maintained separately from the General Industry (Part 29 CFR 1910) standards. The construction standard is a classic example of "vertical" standards, as compared with OSHA's general approach of adopting "horizontal" standards, as was covered in Chapter 4. The safety and health manager for a construction company should look first to the construction vertical standards, but should be aware that if OSHA cannot find a construction standard that covers a given hazard found at a construction site, the compliance officer is free to turn to the general industry standards to write a citation. This places a new dimension on the problem of bringing a construction industry into compliance with standards.

This chapter is primarily for safety and health managers of construction companies, but even the general industry safety and health manager should find the information herein useful. Almost every industry has occasional remodeling or expansion projects that result in construction. Even if the company contracts out the construction project, what goes on is still important to the company because its own employees can become exposed to hazards created by the contract construction personnel. Such hazards can be both physical and legal.

GENERAL FACILITIES

Lighting

Curiously, there are standards for adequate lighting on construction sites, specifying minimum illumination intensities for various areas, whereas general industry has no such general[1] table of intensities. The reason perhaps has to do with trip hazards and pitfalls (in the literal sense of the word) common to construction sites, hazards that are intensified by poor illumination. The minimum illumination standard of 5 footcandles is really quite low for general construction area lighting, and the standard drops even lower, to 3 footcandles for concrete placement, excavation and waste areas, accessways, active storage areas, loading platforms, and refueling and field maintenance areas. Lighting requirements are higher for most construction shops and indoor areas.

Materials Handling and Storage

The real structural test of most buildings is during their own construction. The heaviest load a floor will probably ever support is the stacked materials used during its own construction. Planning is needed to prevent overloading and possible collapse. All nails should be withdrawn from used lumber before it is stacked.

Rigging equipment for material handling such as chain, rope, and wire rope is, unfortunately, often used until failure. If failure does occur on a construction site, the safety and health manager must be prepared to explain why, because rigging equipment is required to be inspected prior to use on *each shift*.

Disposal of scrap material requires vigilance throughout the construction phase. It is difficult to concentrate on scrap and waste removal when working on a tight project completion schedule, but haphazard scrap accumulation is not only unsafe, it also slows progress. Enclosed chutes are needed for dropping materials from distances of greater than 20 feet. In addition, some scrap material, such as asbestos, may require special protection.

---

[1]Although OSHA does not have a general illumination standard for appropriate levels of lighting, there is a somewhat general table in the standard for Hazardous Waste Operations and Emergency Response, as was discussed in Chapter 7.

PERSONAL PROTECTIVE EQUIPMENT

Construction requirements for personal protective equipment are similar to those for general industry, the principal difference being one of emphasis.

Hard Hats

At the top of the list is head protection; indeed, the hard hat is a symbol of the construction industry. So obvious is the absence of a hard hat in a construction work crew that the hard-hat rule can be a source of embarrassment to worker and manager alike. Only general conditions are laid out to describe *when* hard hats are needed. This lays the decision in the safety and health manager's lap, and as was pointed out in Chapter 12, discretion is advised.

Hearing Protection

It may surprise some readers that hearing protection is a concern for *construction*, but construction work often involves damaging levels of noise. Consider the noise levels and durations of exposures of the compressed-air "jackhammer," for example.

Eye and Face Protection

The greatest concern for construction workers' eyes is for mechanical injury, as from using structural steel riveters, grinders, powder-actuated tools, woodworking tools, concrete nozzles, and other spark- and chip-producing equipment. Surprisingly, construction workers can even be exposed to lasers used as tools for checking steel girder alignment and deflection in bridges and buildings.

Fall Protection

On the bases of both fatalities and injuries, falls are probably the greatest hazard in construction work. Where general industry has a permanent wall, construction may have only a guardrail. Where general industry has a permanent guardrail, construction may have a temporary guardrail or perhaps will have no protective structure at all. Where general industry has a permanent stairway, construction may have a temporary ladder. General industry fixed ladders may have cages or ladder safety devices; construction ladders often have no such safety devices.

Personal protective equipment is the answer to many construction industry fall hazards, simply because protection by other means may be awkward or even impossible. Body harnesses and lanyards tied to lifelines are essential to the safety of construction workers subject to fall hazards.

One mistake made in selecting fall protection equipment is to improvise with ordinary leather belts and ropes. An ordinary leather belt and hardware will not satisfy the specified 4000-pound tensile test for fall protection belt hardware. No one weighs 4000 pounds, but what matters in a fall is the shock load, which can be several times as great as the ordinary dead weight. A falling 200-pound person can therefore result in $\frac{1}{2}$ to 1 ton of force on the fall protection system. Using a safety factor of approximately 4, it is easy to see why the standard specifies a 4000-pound tensile load limit.

The lanyard is that part of the fall protection system that attaches to the body harness on one end and the lifeline or structure on the other. The lanyard must have a nominal breaking strength of 5400 pounds. The applicable standard specifies "$\frac{1}{2}$-inch nylon or equivalent." Beware of substituting materials of equivalent breaking strength to $\frac{1}{2}$-inch nylon. Tensile or breaking strength is not the only consideration in selecting a lanyard. A certain resiliency or elasticity exists with artificial fiber ropes that lessens the shock load when arresting a fall.

An important point with safety lanyards is that they must not be too long. The standard specifies "a maximum length to provide for a fall of no greater than 6 feet." The rationale is that there is no point in breaking a fall with the lanyard if the worker has already fallen so far that the shock of the rope will be lethal. However, the standard is often misunderstood on this point. Note carefully that the wording just quoted does not limit the lanyard length to 6 feet. Figure 18.1 clarifies this point.

One difficulty with limiting the length of lanyards is that they restrict the movement of the worker. Therefore, as with other systems that are designed for protection of the individual, the worker finds a way to get around the system. A common practice in the construction industry is to use "cheater cables," a sort of "extension cord" for a safety lanyard. The worker has a sense of security because he or she is still attached to the lifeline by means of a lanyard plus the cheater. But this sense of security is somewhat misleading, because an accidental fall might be of too great a distance, resulting in a fatal shock load to the worker—even if the worker does not hit the ground or lower platform level.

One difficulty is attempting to attach a lanyard to a vertical lifeline, especially when the lanyard must be adjusted upward or downward on the lifeline, such as in use with a scaffold. It is necessary for the attachment to slide easily when this is intended, but the attachment must lock and hold if the worker falls. There are mechanical devices, but an easy-to-tie knot for this purpose is the triple rolling hitch shown in Figure 18.2.

Fall protection is usually considered from the standpoint of height, but working over or near water presents a different hazard. Even the best of swimmers will have

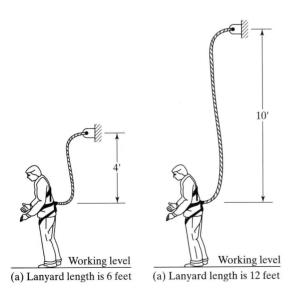

(a) Lanyard length is 6 feet   (a) Lanyard length is 12 feet

FIGURE 18.1

Maximum lanyard length must provide for a fall of no greater than 6 feet.

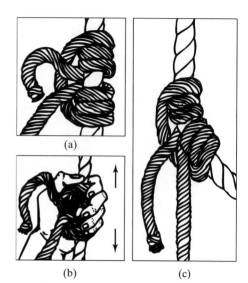

FIGURE 18.2

(a) Triple rolling hitch tied with free end of lanyard; (b) easily raised and lowered slipping on cable or lifeline; (c) when the lanyard is pulled tight, as in a fall, the triple rolling hitch will not slip on the lifeline.

(a)

(b)   (c)

difficulty with a fall into water if he or she is fully dressed and perhaps burdened by tools, equipment, or materials such as rivets or bolts. If the water is rather cold, the danger of hypothermia increases the drowning hazard.

Drowning hazards are taken very seriously in federal standards, which require

1. Life jackets or buoyant work vests *and*
2. Ring buoys every 200 feet *and*
3. A lifesaving skiff

whenever employees work over or near water and a danger of drowning exists.

FIRE PROTECTION

From a property-loss standpoint, fires are more dangerous after a building is completed, but to protect construction workers, fire hazards must also be controlled *during* construction. Construction sites have somewhat more latitude in distributing fire extinguishers than do general industries. Even an ordinary garden hose may be used in place of fire extinguishers on construction sites. But there are so many restrictions on such use of a garden hose that most safety and health managers will regret having considered this alternative.

The biggest problem with fire prevention during construction is the handling of flammable liquids. For ordinary flammable liquids such as gasoline, quantities handled must be no more than 1 gallon unless approved metal safety cans are used. Approved metal safety cans must be used even for quantities up to 1 gallon unless the flammable liquid is used out of its *original* container. Containers should be kept off stairways and away from exits and aisles.

TOOLS

It has been well publicized that mushroomed heads on chisels, wedges, and other impact tools are unsafe. The hazard is that a sliver of metal can break off and cause a severe eye injury, even total loss of sight. Another problem with hand tools is defective handles, in particular loose hammer heads.

Construction sites may employ pneumatic tools, such as jackhammers, staplers, or nailers. Pneumatic tools need to be secured to the hose by some positive means to prevent accidental disconnection. Hoses larger than $\frac{1}{2}$ inch inside diameter need a pressure-reducer device to prevent whip action in case of hose failure. Figure 18.3 is a diagram of a pressure-reducer device for this purpose. It is a simple in-line device, usually placed between the hose and the compressor. Unfortunately, the device reduces the overall capacity of the system. Operating at full capacity, as from several tools operating at once, the pressure downstream from the device becomes so low that the device (a spring-loaded valve) begins to close as if the downstream line had ruptured. This shuts off or nearly shuts off the supply of air to the tools, making the system useless at that capacity. The result is that the worker removes the device from the line, and often as not, it is soon lost. This is a sore point with many construction safety and health managers.

Some construction applications are being found for hydraulically operated tools, especially in the public-utility construction field. Hydraulic tools operate under the same principle as pneumatic tools but use liquids instead of air as the medium. Fluid pressures in these tools can reach 3000 psi gauge, approximately 20 times the maximum pressure achievable with pneumatic tools. Such tremendous pressures give hydraulic tools much greater power than pneumatic tools, but a hazard can exist if operating pressure limits of the equipment are exceeded. Under the criterion of occupational noise, however, hydraulic tools have a safety and health advantage over pneumatic tools. Additional hazards of hydraulic fluids are electrical conductivity and fire. These hazards tend to conflict in that the more fire resistant fluids are electrical conductors. When working in construction and alteration of electric utility transmission and distribution systems, the hazard of electrical conductivity is more serious than the hazard of fire. The hydraulic fluids used for the insulated sections of derrick trucks, aerial lifts, and hydraulic tools that are used on or around energized lines and equipment for power transmission and distribution are required to be of the insulating type. The fluids for hydraulic tools used in other applications are required to be fire resistant.

Powder-actuated tools carry an explosive charge to provide the driving force. The applications of these tools are increasing because they are both fast and effective. Driving fasteners into concrete, masonry, or steel demands large, accurately placed impact forces. Powder-actuated tools are able to provide these forces in a convenient way,

FIGURE 18.3

In-line device for preventing whip action in event of pneumatic hose failure.

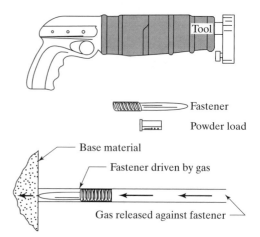

FIGURE 18.4

Powder-actuated fastening tool (*Source*: NIOSH, ref. 121).

speeding up production on construction projects. But together with this speed, force, and convenience come safety hazards.

A powder-actuated tool looks and operates very much like a handgun, as can be seen in Figure 18.4. Even the powder cartridges look like bullets for a gun. In powder-actuated tools, however, the projectile is separate from the cartridge, as shown in Figure 18.4.

In some ways, powder-actuated tools are even more dangerous than handguns. Powder-actuated tools are capable of handling a variety of cartridges with a wide range of power ratings. These cartridges must be selected with care by a knowledgeable person. Insufficient power will fail to do the job, but too much power may drive the fastener completely through the material and kill a coworker in another room. (This has actually happened.) To protect against such hazards and for convenience, cartridges are color coded for easy identification. Metal-cased cartridges are available in a range of 12 different power ratings, as shown in Table 18.1. It may be necessary to back the material with a substance that will prevent the fastener from passing completely through.

TABLE 18.1   Color Identification for Cased Power Loads for Powder-Actuated Tools

Power level	Case color	Load color	
1	Brass	Gray	Lightest
2	Brass	Brown	
3	Brass	Green	
4	Brass	Yellow	
5	Brass	Red	
6	Brass	Purple	
7	Nickel	Gray	
8	Nickel	Brown	
9	Nickel	Green	
10	Nickel	Yellow	
11	Nickel	Red	
12	Nickel	Purple	Heaviest

Knowledge and judgment are also required in avoiding very hard or brittle materials, such as cast iron, glazed tile, surface-hardened steel, glass block, live rock, face brick, or hollow tile. If the material is spalled or cracked by an unsatisfactory previous fastening, the new fastening must be driven elsewhere. Fasteners driven too close to the edge of the material can cause explosive chipping of the material at the edge. Obviously, eye protection is always needed when using powder-actuated tools.

ELECTRICAL

Construction workers are often in close contact with ground and frequently work in wet locations under adverse conditions. Electrocution ranks with falls near the top of the list of causes of fatalities among construction workers.

A principal requirement on construction sites is that all 15- and 20-ampere outlets have either ground-fault circuit-interruptor (GFCI) protection or a program of equipment ground-conductor assurance, including inspection, testing, and recordkeeping. The safety and health manager in a construction company is faced with a decision between the two alternatives, and the paragraphs that follow are intended to assist in making that decision.

The purpose and operating principle of the GFCI were discussed in Chapter 17. The hazards of electrical shock by faults to ground are greater on construction sites than in the general industrial workplace. Because of the increased hazards, the *National Electrical Code*® specified GFCIs for construction, but not for general industry, although they would be of value in any electric circuit serving hand-held appliances or tools. Chapter 17 illustrated how GFCIs work and showed a residential-type receptacle equipped with a GFCI.

It seems that almost every safety device has its drawbacks, and the GFCI is no exception. The GFCI closely monitors any difference in current flow between the ground and neutral conductors, as low as fractions of a milliampere. But there are ways in which these currents can be unbalanced when no hazard exists. Tiny amounts of current leakage to ground occur for quite innocent reasons. Damp or weakened insulation might produce tiny currents in various locations. Even an extension cord that is too long can create a condition of capacitance between the conductor and the ground, resulting in a tiny leak. Although none of these conditions of itself is a significant hazard, the cumulative effect is one that may be great enough to trip the GFCI, shutting down the entire circuit. This is so-called *nuisance tripping* and has made the GFCI a controversial issue in the construction industry.

Chapter 17 mentioned a permissible alternative to the GFCI: the careful maintenance of the grounding conductors of electrical equipment. Such maintenance includes regular inspections and records of these inspections. The idea behind the "assured equipment grounding-conductor program" is that if an electrical tool shorts to the case or handle, the third-wire grounding system will shunt the current to quickly throw the circuit breaker. A good grounding conductor can therefore provide protection similar to the GFCI.

The assured equipment grounding-conductor program is appealing to many construction companies because they can avoid buying the GFCI equipment. They can also avoid the nuisance tripping of the GFCIs discussed earlier. But although the costs are less tangible, they nevertheless exist with the grounding assurance alternative. It takes instruments and time to test the grounding conductors, and the business of recordkeeping always involves intangible costs. An economic impact analysis of the two alternatives has been made that estimated a cost of compliance of $87.5 million for purchase, installation, and first-year maintenance of GFCIs. The study estimated for the alternative assured equipment grounding-conductor program a similar cost of $36 to $43.8 million. Inflation changes absolute annual cost estimates, but the relative difference between the costs of the two alternatives suggests that the assured equipment grounding-conductor program is cheaper.

Temporary lighting is an electrical problem on construction sites more than in general industry. Often seen are ordinary incandescent bulbs suspended from electrical cords. Cords and lights that are suspended in this way must actually be *designed* for this purpose; all electric cords for temporary lighting must be heavy duty, and insulation must be maintained in a safe condition. To prevent accidental contact, bulbs must be guarded, unless the construction of the reflector is such that the bulbs are deeply recessed.

A construction site is a profusion of temporary conditions, and electrical cords or extension cords strung about the area are a common sight. Unfortunately, the area is also visited by heavy-duty vehicles such as excavation equipment, heavily loaded trucks, and very heavy concrete delivery trucks. The situation is too hazardous to permit electrical cords to pass through work areas unless covered or elevated to protect them from hazardous damage. No splices are permitted in flexible cord unless properly molded or vulcanized.

LADDERS AND SCAFFOLDS

The "care and use" provisions for ladders are important for construction ladders, just as they are for ladders used in general industry, a topic that was covered in Chapter 7. Construction ladders have some differences, however, that have caused some problems.

Job-Made Ladders

Construction companies often make their own ladders, and such ladders are not illegal if made properly. The first requirement is to determine how many persons will be needing the ladder. If simultaneous two-way traffic is anticipated, a conventional ladder will not work, and a double-cleat ladder, as shown in Figure 18.5, should be used. In fact, if the ladder is the only means of access or exit from a working area for 25 or more employees, the double-cleat ladder is *mandatory*, unless two ladders are provided.

The biggest mistake made in building job-made or homemade ladders is to fail to inset the cleats into the side rails. (See Figure 18.6.) It is much more trouble to inset the cleats or use filler blocks than to simply nail the cleats in place, but the security and stability of the cleats are increased many times by this additional effort to make the ladder safe.

Scaffolds

The subject of scaffolds can become somewhat technical, and these technical details can be very important. The safety and health manager will find it useful, and in some

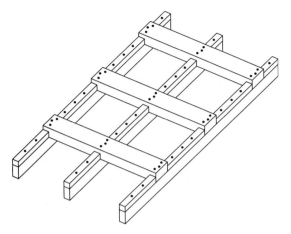

FIGURE 18.5

Double-cleat ladder for simultaneous two-way traffic

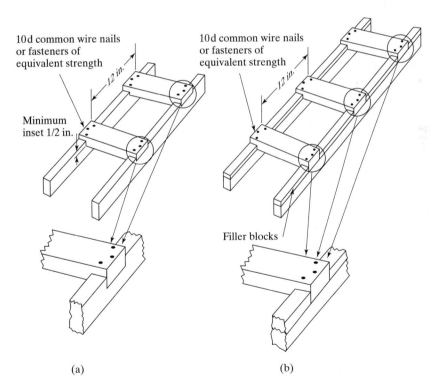

FIGURE 18.6

Two acceptable ways of setting cleats in construction job-made ladders: (a) cleats inset into side rails; (b) cleats braced by filler blocks.

cases imperative, to obtain the services of a registered professional engineer. This is one area in which the credential as well as the knowledge can be quite useful.

One of the technical aspects of scaffolds is safety factor. Design safety factor for scaffolds and their components is a factor of four. This factor increases to a factor of six for the suspension ropes supporting the suspended type of scaffolding. The application of counterbalances, tie-downs, footings, and the allowance for wind loading all can be quite technical, and an engineering evaluation is advisable.

The safety and health manager may be frustrated by the many bewildering names of scaffolds listed in applicable construction standards. But the majority of the scaffolds with unfamiliar names, such as "window-jack scaffolds," "outrigger scaffolds," and "chicken ladders," are rarely seen. The most popular scaffolds are the following types:

- Welded frame (or "bedstead" scaffolds)
- Manually propelled mobile scaffolds (on casters)
- Two-point suspension (or "swinging" scaffolds)
- Tube and coupler scaffolds

Some scaffolds, such as tube and coupler scaffolds and welded frame scaffolds, are supported by structure on the ground and must have sound footings. If the ground slopes, scaffold jacks may be necessary to assure that the footings are level. Certain engineered "cribbing" may be acceptable, but such unstable objects as barrels, boxes, loose brick, or concrete blocks are criticized. Concrete blocks are a real problem because most people feel that they are very strong and rigid. But scaffolds direct very highly concentrated loads on their relatively tiny feet, and such loads can break through a molded concrete block. Once a scaffold support breaks through its footing, a major shift can occur aloft, and the consequences can be extremely serious. Such a mishap is most likely to occur at the worst time (such as when personnel are on the scaffold).

For scaffolds suspended from above, such as from two-point suspension (swinging) scaffolds, the security of the attachment on the roof is of obvious importance. Since rooftops vary in structure and design, an engineer is very useful in ensuring a safe anchor point. Cornice hooks are designed to hook over the edge, not into it. In addition, tiebacks are needed as a secondary means of support. Sometimes a sound structure for the tieback is not available on the rooftop, and the only solution is to cross the entire roof and go down to the ground on the other side of the building. Tying a scaffold to an ordinary vent pipe on the roof is asking for trouble.

Safety belts and lifelines for personnel on suspended swinging scaffolds must be tied to the building, not to the scaffold. Thus, if the scaffold falls, the personnel can still be saved. For guidance on the attachment of the lanyard to the lifeline, refer back to the earlier discussion of fall protection.

The floor of the scaffold is also important. Loose planks can be especially hazardous if the overhang beyond the support is insufficient for security. But too much overhang can also be dangerous because a worker might step beyond the support and cause the plank to tip like a seesaw. Figure 18.7 illustrates minimums and maximums for plank overhang. Figure 18.8 illustrates the minimum for overlap, unless planks are secured from movement.

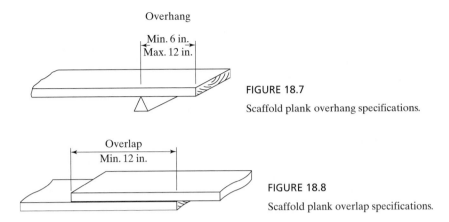

FIGURE 18.7

Scaffold plank overhang specifications.

FIGURE 18.8

Scaffold plank overlap specifications.

FLOORS AND STAIRWAYS

As in general industry, the standard for guarding of open-sided floors and platforms is one of the most frequently cited standards in construction. Curiously, though, instead of setting the vertical fall distance at 4 feet as in general industry, the specification for construction is *6 feet*. Thus, an open-sided floor or platform 6 feet or more above the adjacent floor level is required to be guarded by a standard railing. Runways 4 feet high or more must be guarded.

During construction of buildings with stairways, construction employees use the new stairways during finishing operations for the building. If the building's stairways are properly designed (see Chapter 7), the safety and health manager for the construction company has little to worry about.

A pitfall needs mentioning on the subject of use of new building stairways during construction. Many stairways and landings today are of steel construction with hollow pan-type treads that are filled on-site with concrete or other materials. Contractors often save until last the job of pouring the treads. Meanwhile, during building construction, workers are walking on the unfilled treads and are subject to the trip hazard created by the steel nosing along the leading edge of the riser. Of course, exposure to the hazard is unavoidable during the actual construction of the stairway itself. But after the stairways have been installed, lumber or other temporary material can be used to fill the hollow space and eliminate the trip hazard during completion of the building.

CRANES AND HOISTS

Cranes, hoists, and other material- or personnel-handling equipment are essential tools of the construction industry. Chapter 14 investigated the hazards of these machines in detail. This chapter discusses the subject from the viewpoint of the construction industry, the most important user of these machines.

One of the hazards addressed in Chapter 14 was two-blocking. Although two-blocking can be a hazard with any crane or hoist, most of the fatalities resulting from this hazard have occurred in the construction industry. A crane can be two-blocked in many ways. Hoisting the load, extending the boom, or even *lowering* the boom on a crane with a stationary winch mounted to the rear of the boom hinge can lead to two-blocking. It is difficult for the crane operator to avoid all of the ways in which a crane can be two-blocked, and consequently fatalities have occurred even when experienced crane operators are at the controls. In 1973, the ANSI standard[2] for mobile hydraulic cranes incorporated a requirement for a "two-blocking damage prevention feature" on telescoping boom cranes with less than 60 feet of extended boom. Safety and health managers should beware of the temptation to purchase old equipment that might not be built to safe standards. And for equipment that has been purchased, the feasibility for retrofit should be considered.

On construction sites, the simultaneous execution of many parts of the overall project means that personnel will often be working around or near a crane in operation. The crane operator will avoid moving the bucket or other load over personnel and will attempt to keep the cab from striking personnel, but it is impossible for the crane operator to watch all moving parts at all times. Particularly hazardous is the rear of the cab, which on many models swings outside the crawler or other substructure when the cab and boom rotate, as shown in Figure 18.9. This motion generally occurs on every cycle of a crane's operation and is a constant threat to personnel on the ground. The hazard is a very serious one, and accidents carry a high likelihood of fatality. The employee may be either struck or crushed between the cab and some other object, such as a building wall, stack of materials, or another vehicle.

Another serious hazard with construction cranes is the possibility of contact with exposed live overhead utility lines. Crane boom contact with high-voltage transmission

FIGURE 18.9

Rear of crane cab is a hazardous area due to swing radius (*Source*: courtesy of the Construction Safety Association of Ontario).

---

[2]ANSI B30.15–1973.

lines results in fatalities every year. The higher the voltage, the greater must be the clearance between the crane and the electric transmission line, to prevent arcing. Federal standards recognize this physical reality by setting up formulas for calculating the minimum clearance required for various voltages. Recognizing that during transit it may be more difficult to maintain clearances, OSHA standards are a little more lenient for cranes in transit. Further complicating the problem is that requirements for cranes in general industry are different from requirements in construction, resulting in a very complicated set of requirements. These requirements are summarized in Figure 18.10.

A common worker practice is "riding the headache ball." Figure 18.11 identifies the headache ball as the ball-shaped weight used to keep a necessary tension on the wire rope when the hook is not loaded. It is possible for workers to stand on this ball and take a ride, using the crane as an elevator, a practice that usually horrifies passers-by. Riding the headache ball is not explicitly addressed in the OSHA standards, but is generally considered a dangerous practice. OSHA can turn to the General Duty Clause to cite dangerous practices likely to cause death or serious physical harm. Also, lack of fall protection can be cited in cases where persons are riding the headache ball without fall protection. The practice is so highly visible to the public from the street that it easily can trigger an OSHA inspection. If persons are to be elevated using a crane, the recommended practice is to use a lift cage attached to the crane hook.

Hammerhead tower cranes are large structures that take advantage of counterweights on the end of the jib opposite the work. (See Figure 18.12.) This crane is often used for large building construction. Sometimes construction workers will have duties

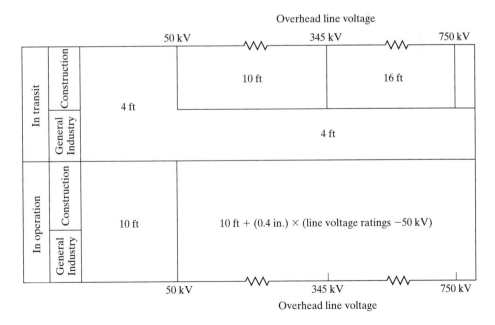

FIGURE 18.10

Clearance for cranes from electric transmission lines.

FIGURE 18.11

Headache ball. These balls range in weight from less than 100 pounds to 1 ton or more. The weight of the ball overcomes the friction of the sheaves for the running rope of the crane. The headache ball should not be confused with the much heavier wrecking ball.

FIGURE 18.12

Hammerhead tower crane.

on the horizontal jib of a hammerhead tower crane. This presents a fatal fall hazard, and guardrails or safety belts, lanyards, and lifelines are needed to protect the worker.

Helicopters are sometimes used in construction for such operations as the placement of a steeple. Although the use of helicopters in construction is rare, the nature of the hazards of the operation is somewhat strange and therefore deserves some mention. Ordinary tag lines, used with all cranes to control the load from below, can be a hazard to helicopters because the lines can be drawn up into the rotors, resulting in a tragedy. Tag lines must be of a length that will not permit their being drawn up into the rotors.

Cargo hooks are another problem with helicopters used as cranes. With ordinary cranes, the sole concern is that the hook will hold and not release at the wrong time. With helicopters, there is this concern and the additional concern that the hook might *not* release at the right time. Cargo hooks for helicopter cranes need to have an emergency mechanical control to release the load in case the electrical release fails.

Another unusual effect that can represent a hazard is the generation of a static electrical charge on the load. This charge is developed by the friction of the air on the rotor and other moving parts. To deal with this hazard, a grounding device can be used to dissipate the charge before ground personnel touch the load, or protective rubber gloves can be worn. Once the load actually touches down, the static charge is generally dissipated through the load itself directly into the ground.

An indirect hazard that can develop from the use of helicopters is fire on the ground. The rotating blades generate such a wind on the ground that it is considered unsafe to have open fires in the path of a low-flying helicopter.

Material and Personnel Hoists

Temporary external elevators are often used on construction sites to move workers and materials and must be designed, maintained, and used properly to avoid a serious hazard. The seriousness of this hazard is quite personal to me because my brother narrowly escaped a fatal accident on such a hoist when two of five members of his engineering inspection section were killed when a personnel hoist failed in an oil refinery. Two others were totally and permanently disabled, and the fifth, my brother, was uninjured because he had been asked to remain in the office that fateful day to complete an engineering drawing.

The design requirements for personnel hoists and material hoists are different, and one of the principal safety factors is maintaining the distinction between the two hoists during use. Material hoists must be conspicuously marked "No riders allowed." On the other hand, it is permissible to move material on a personnel hoist provided that rated capacities are not exceeded.

Latched gates are needed to guard the full width of the landing entrance for both material and personnel hoists. In the case of personnel hoists, an electrical interlock must not allow movement of the hoist when the door or gate is open. Furthermore, on personnel hoists, the hoistway doors or gates must have mechanical locks that are accessible only to persons in the car.

Aerial Lifts

An alternative to scaffolds, ladders, and hoists is needed for high and awkward locations on a construction site. The use of vehicle-mounted boom platforms or aerial "buckets" is becoming increasingly popular. The boom is usually *articulating* (capable of bending in the middle) or is *hydraulically extensible* (telescoping) or both.

The biggest problem with aerial lifts is not their construction, but the way they are used. Anyone in an aerial bucket needs to recognize the difference between his or her perch and terra firma. The floor of the bucket is the only place to stand—not on a ladder or plank carried aloft. Sitting on the bucket's edge is also dangerous. Even if the worker *does* stand properly in the bucket, a body belt with lanyard tied to the boom or bucket is needed to protect against a hazard such as a surprise encounter with a tree limb or a dip in terrain that can toss the operator from the basket.

HEAVY VEHICLES AND EQUIPMENT

Next to falls and electrocutions, more construction fatalities involve vehicles, tractors, and earthmoving equipment than any other hazard source. The fatality hazard is both to drivers of the equipment and to their coworkers. Vehicle *rollovers* are the principal cause of driver fatalities, whereas vehicle *runovers* are the problem for coworkers. Not to be excluded, however, is a significant number of fatalities from repairing tires for these vehicles.

ROPS

The acronym ROPS (rhymes with "hops") represents the term *rollover protective structures* and is a major change in construction vehicle design brought about by federal safety standards. The purpose of the ROPS, illustrated in Figure 18.13, is to protect the operator from serious injury or death in the event the vehicle rolls over. The following kinds of construction equipment require ROPS:

- Rubber-tired, self-propelled scrapers
- Rubber-tired, front-end loaders
- Rubber-tired dozers
- Wheel-type agricultural and industrial tractors
- Crawler tractors
- Crawler-type loaders
- Motor graders

Exempted are sideboom pipelaying tractors.

To be effective, the ROPS system must be able to withstand tremendous shock loads, which increase as the weight of the vehicle increases. The standards are quite specific regarding the structural tests to which ROPS systems must be subjected in order to qualify. For all wheel-type agricultural and industrial tractors used in construction, either a laboratory test or a field test is required to determine whether performance requirements are met. The laboratory test may be either static or dynamic. In

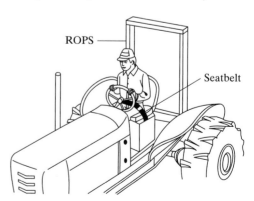

FIGURE 18.13

Rollover protective structures for construction vehicles prevent fatalities when used with seatbelts.

the static test, the stationary tractor chassis is gradually loaded while strain is measured by deflection instruments. The required input energy is a function of tractor weight, which in turn is required to be a function of tractor horsepower. In other words, the tractor's gross weight cannot be lightened below rated horsepower limits in order to meet ROPS tests.

The dynamic test, an alternative to the static test, uses a 2-ton pendulum that provides an impact load on the rear and side of the ROPS in successive tests. The height from which the pendulum is dropped is dependent on the calculated tractor weight based on horsepower, as in the static test just described. Deflection limits must not be exceeded.

If the field test is used, the tractor is actually rolled over, both rearward and sideways, as both types of accidents can easily occur. If the actual weight of the tractor is less than specified for its horsepower, ballast must be added for the tests.

For all ROPS tests, it is a good idea to remove protective glass and weather shields, which would probably be destroyed during the test. If there is any question whether such shields may absorb some of the energy, thereby assisting ROPS to pass the test, the shields *must* be removed.

Having read thus far, one can see that retrofitting an old tractor to meet current requirements for ROPS is not an easy task. The safety and health manager is cautioned against taking the tractor to a local welder and requesting a ROPS system to be fabricated. Unless the tractor is actually going to be subjected to the standard ROPS test, the frame should be of a design identical to a frame actually tested for the model tractor in question. Very old or rare model tractors are obviously a problem. If a qualified ROPS system is removed for any reason, it must be remounted with bolts or welding of equal or better quality than those required for the original. The ROPS must be permanently labeled with manufacturer's or fabricator's name and address and the machine make, model, or series number that the structure is designed to fit. This labeling requirement is a sobering thought for the welder or fabricator who is asked to retrofit an old tractor with a ROPS system. All told, it is easy to see why most companies dump the old equipment on the used-equipment market, and many are shipped to foreign countries that do not require ROPS.

After the safety and health manager has ensured that all appropriate construction equipment has been equipped with ROPS systems, the next task is to ensure that the equipment is used properly. Essential to the effectiveness of the ROPS system is that the operator wear a seat belt. If the operator is thrown out of the vehicle, the ROPS will afford no protection at all and may actually contribute to a fatality. Passengers are another hazard unless the vehicle is equipped with seat belts for passengers. Hitchhiking on heavy equipment at construction sites is a dangerous practice.

Runover Protection

Most of the balance of fatalities with heavy construction equipment is due to personnel being run over by the equipment. Confrontation of this major fatality category has two main thrusts: operator visibility and pedestrian awareness.

Operator visibility as good as the visibility in a private automobile is simply not feasible for a huge piece of earthmoving machinery. It is no wonder that runovers occur frequently on construction sites. The operator needs all the help affordable, but ironically, some of the poorest windshield conditions occur on construction equipment.

In the morning, the operator, foreman, and everyone concerned is anxious to get equipment rolling, but if the morning is cold, defrosting and defogging are essential. Often, the defrosting or defogging equipment is ineffective. Dirty or cracked windshields are also a common sight in the harsh environment of the construction job.

The second link in the hazard-prevention chain for runovers is the horn used by the operator to warn personnel when visibility *is* good enough to notice the endangered worker on the ground. Personnel are generally distributed all over a construction site, and the operator needs a good operable horn to warn them when they are dangerously close.

Besides the ordinary horn, many construction vehicles also need "backup alarms." Earthmoving equipment and construction vehicles that have an obstructed view to the rear need these backup alarms if they are used in reverse gear. The term *obstructed view* may be somewhat vague, but most safety and health professionals are taking the position that earthmoving machines of all types need these backup alarms. Too many people have been killed by machines backing over them to take this requirement lightly. This is said although it is acknowledged that the steady beep-beep-beep of the backup alarms can be very monotonous on a construction site and perhaps even lead to a certain complacency on the part of the personnel endangered. An alternative to the beeping backup alarms is the use of an observer standing behind the machine to alert others every time the machine backs up. This is expensive, though, and has the disadvantage of being an *administrative* or *work-practice control*, instead of the preferred *engineering control* represented by the backup alarm.

Dump Trucks

One more hazard with construction vehicles and equipment needs emphasis. Dump trucks can cause a terrible accident if the raised dump body falls while the driver or some other worker has crawled into the exposed area for maintenance or inspection work. Sometimes all that holds the dump body aloft is the pressure in a hydraulic line, which can be suddenly lost due to any of a variety of failure modes. For this reason, the safety of the maintenance or inspection worker inside the exposed area demands that the truck be equipped with some positive means of support, permanently attached and capable of being locked into position.

TRENCHING AND EXCAVATIONS

A major cause of construction fatalities is the sudden collapse of the wall of a trench or excavation. It is difficult to imagine the drama of digging for a coworker who has been literally buried alive in such a cave-in. Before becoming involved in the field of occupational safety and health, I coincidentally became an eyewitness to such a drama in Tempe, Arizona. The trench was located directly below a public stairway landing on which I happened to be standing, and thus my vantage point was directly over the scene of the cave-in. The impression was unforgettable, and its memory would motivate anyone to try to prevent such accidents in the future, a point supporting the principles of hazard avoidance set forth in Chapter 3. Recognizing the seriousness of this hazard, OSHA has undertaken several special-emphasis programs on trenching and excavation cave-ins. Joseph Dear, in the mid-1990s (ref. Dear), cited a dramatic result of such emphasis in the State of Indiana where trenching and excavation fatalities decreased from six per year to one per year after conducting special trenching programs.

All trenches are excavations, but not all excavations are trenches. Trenches are narrow, deep excavations; the depth is greater than the width, but the width is no greater than 15 feet, according to the standard definition. A trench is more confined and generally more dangerous than other excavations, especially because both walls can collapse, trapping the worker. However, the walls of a trench are easier to shore than the walls of an excavation. Both are dangerous if over 5 feet deep and will easily snuff out the life of anyone who stands in the path of a collapsing wall. The hazard is not simply one of suffocation. A cave-in generally represents tons of falling earth, which can crush the body and lungs of the worker even if the face and breathing passages are left clear.

The *angle of repose* is defined as the greatest angle above the horizontal plane at which a material will lie without sliding. The angle naturally varies with the material, and approximate angles are shown in Figure 18.14. The science of soil slides is not exact, and the uncertainty thwarts attempts to control the hazard. It is difficult to say whether a particular soil type is "typical" or "compacted angular gravels," or something between the two. The specifications for trench shoring are more detailed, as can be seen in Table 18.2.

Adding to the uncertainty of the cave-in hazard are certain hazard-increasing factors, such as

- Rainstorms, which soften the earth and promote slides
- Vibrations from heavy equipment or street traffic nearby
- Previous disturbances of the soil, as from previous construction or other excavations
- Alternate freezing and thawing of the soil
- Large static loads, as from nearby building foundations or stacked material

Although judgment is required in deciding whether to employ shoring, the hazard is so serious that it is wise to adopt a conservative policy, well clear of the marginal

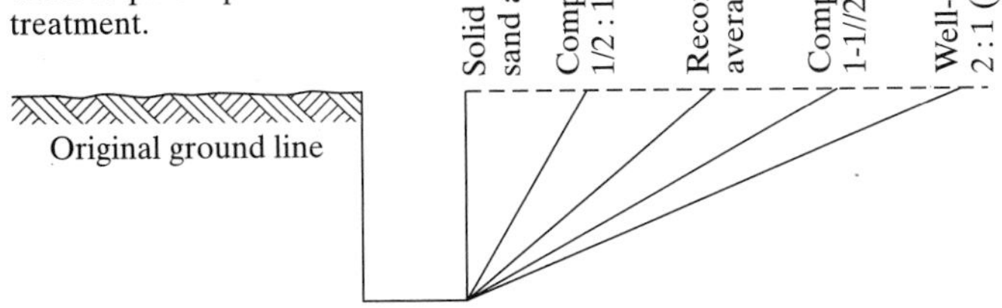

Note: Clays, silts, loams or nonhomogenous soils require shoring and bracing. The presence of ground water requires special treatment.

Original ground line

Solid rock, shale, or cemented sand and gravels (90")
Compacted angular gravels 1/2 : 1 (63" 26')
Recommended slope for average soils 1 : 1 (45")
Compacted sharp sand 1-1/2 : (33" 41')
Well-rounded loose sand 2 : 1 (26"/34')

FIGURE 18.14

Approximate angle of repose for sloping the sides of excavations.

TABLE 18.2 Trench Shoring—Minimum Requirements

Depth of trench (ft)	Kind or condition of earth	Uprights Minimum dimension (in.)	Uprights Maximum spacing (ft)	Stringers Minimum dimension (in.)	Stringers Maximum spacing (ft)	Cross braces[a] (in.) Width of trench (ft) Up to 3	3–6	6–9	9–12	12–15	Maximum spacing (ft) Vertical	Horizontal
5–10	Hard, compact	3 × 4 or 2 × 6	6			2 × 6	4 × 4	4 × 6	6 × 6	6 × 8	4	6
	Likely to crack	3 × 4 or 2 × 6	3			2 × 6	4 × 4	4 × 6	6 × 6	6 × 8	4	6
	Soft, sandy, or filled	3 × 4 or 2 × 6	Close sheeting	4 × 6	4	4 × 4	4 × 6	6 × 6	6 × 8	8 × 8	4	6
	Hydrostatic pressure	3 × 4 or 2 × 6	Close sheeting	6 × 8	4	4 × 4	4 × 6	6 × 6	6 × 8	8 × 8	4	6
10–15	Hard	3 × 4 or 2 × 6	4	4 × 6	4	4 × 4	4 × 6	6 × 6	6 × 8	8 × 8	4	6
	Likely to crack	3 × 4 or 2 × 6	2	4 × 6	4	4 × 4	4 × 6	6 × 6	6 × 8	8 × 8		6
	Soft, sandy, or filled	3 × 4 or 2 × 6	Close sheeting	4 × 6	4	4 × 6	6 × 6	6 × 8	8 × 8	8 × 10	4	6
	Hydrostatic pressure	3 × 6	Close sheeting	8 × 10	4	4 × 6	6 × 6	6 × 8	8 × 8	8 × 10	4	6
15–20	All kinds or conditions	3 × 6	Close sheeting	4 × 12	4	4 × 12	6 × 8	8 × 8	8 × 10	10 × 10	4	6
Over 20	All kinds or conditions	3 × 6	Close sheeting	6 × 8	4	4 × 12	8 × 8	8 × 10	10 × 10	10 × 12	4	6

[a]Trench jacks may be used in lieu of, or in combination with, cross braces. Shoring is not required in solid rock, hard shale, or hard slag. Where desirable, steel sheet piling and bracing of equal strength may be substituted for wood.

Source: Code of Federal Regulations 29 CFR 1926.652.

area where a cave-in might or might not occur. One thing is certain: *After* the cave-in *does* occur, and there is a fatality, the OSHA officer will come to the scene, and everyone (including the OSHA officer) will conclude that the shoring or cave-in protection was insufficient.

A trench shoring system is shown in Figure 18.15. The trench jacks may be either screw-type or hydraulically operated. They need to be secured to prevent falling or sliding if they loosen as the sidewalls adjust slightly. Care must be taken to level the jacks and also to be sure that they are below the plane of the surface of the surrounding earth. A trench jack placed too high can be subjected to bending stresses, as shown in Figure 18.16. This can damage or even ruin the trench jack. The shoring system should be removed slowly and carefully. It is sometimes necessary to remove braces or jacks by means of ropes from above after everyone has cleared the trench.

A cave-in is not the only hazard to working in trenches and excavations. All workers in the ditch are subject to the hazards of falling rocks, tools, timbers, or pipes. Head protection is needed for workers below, and good housekeeping is important along the edges of the excavation.

Even when personnel are not inside the excavation and machines do all the work, another hazard presents itself. Utility lines are often broken, resulting in fire, explosion, or inhalation hazards. Such accidents are not only dangerous, but are always costly and require additional coordination with the utility company—coordination that would have been better handled in advance of the excavation. The safety and health

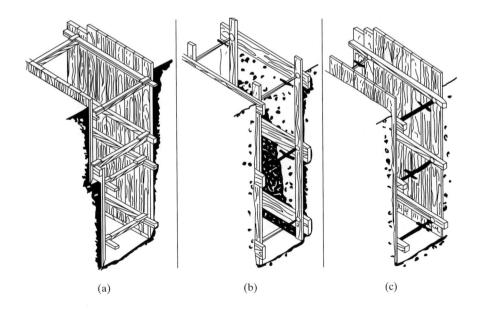

(a) (b) (c)

FIGURE 18.15

Trench shoring system: (a) bracing used with two lengths of sheet piling; (b) bracing with screw jacks, hard soil; (c) screw jacks used with complete sheet piling (*Source*: courtesy of the National Safety Council, Chicago; used with permission).

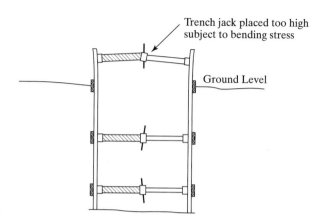

Trench jack placed too high subject to bending stress

Ground Level

FIGURE 18.16

Improperly placed trench jack.

manager should install a procedure that ensures that someone stops and checks with utility companies before proceeding "blind" into an excavation project. Excavation contractors are generally familiar with the information system for dye marking underground utilities. Signs often post warnings: "DO NOT DIG; BURIED UTILITIES." Most communities have a toll-free number accessing a free service for dye marking underground utilities to warn excavation contractors before proceeding.

Signs of an imminent cave-in, rupture of a utility line, dangerous accumulations of toxic gases in deep excavations, or other possible emergency situations dictate the need for quick and easy exit. A ladder, steps, or other adequate means of exit should be located so as to require no more than 25 feet of lateral travel for escape.

It was stated at the beginning of this section that OSHA has undertaken several "special-emphasis" programs for enforcement of trenching and excavation hazards. Just how much "special emphasis" does OSHA place on this type of hazard? Some insight is provided in a 1997 news release (ref. Fleming, 1997) in which OSHA stated that in the previous 5-year period the agency and state agencies having enforcement jurisdiction had conducted a total of 9400 trenching inspections. Of these, almost 200 inspections were investigations of fatal accidents. Trenching and excavations remain as one of the principal hazards of the construction industry, and it is likely that OSHA will continue its emphasis well into the 21st century.

CONCRETE WORK

Perhaps the most dramatic industrial accident in history was the partial collapse of a vertical wall of concrete. The year was 1978, and the location was Willow Island, West Virginia, where a huge cooling tower was under construction for a nuclear power plant. The continuously poured concrete walls of the structure supported scaffolds for workers 170 feet above the ground. At the time of the accident the "green" concrete wall was insufficiently cured to accept the load. The wall failed, dropping the scaffold; 51 workers fell to their deaths.

The pressure to keep a construction project on schedule is antagonistic to the careful curing of concrete. But the consequences of rushing the job are serious even if no one is injured. The memory of the tragic Willow Island accident serves to remind concrete project managers of these consequences. Concrete curing time is a function of both time and temperature. About 70°F is ideal, and temperatures either too hot or too cold can delay curing.

Even before the concrete is poured there are hazards to the placement of the reinforcing steel *rebars*. For vertical structures, rebars need guys or other support to prevent collapse. Another hazard is the protruding points of exposed vertical rebars. It may seem farfetched, but workers have actually been *impaled* by falling on these bars. The most serious hazards are for workers on ladders when these ladders are placed over protruding rebars. But even at floor or ground level, a trip and fall on an exposed rebar can be fatal. In one case, a worker stumbled at ground level and fell on an exposed rebar, which impaled his neck and ruptured his jugular vein. The life of the worker was saved by the quick action of a trained first-aid person. The answer to the exposed rebar problem is to bend down the ends, cover them with plywood, or wrap them with canvas until ready for pouring.

Concrete forms need carefully designed shoring systems to prevent collapse together with hazards reminiscent of the excavation cave-in hazards discussed earlier. The hydrostatic pressure of wet concrete can be very great just after it is poured. Then, at a time when the forms' stress is greatest, vibrating equipment is often applied to ensure even distribution, adding to the stresses on the forms. Overdesign is necessary to prevent the hazard of forms "kickout."

Concrete is often poured by buckets handled by cranes. Unfortunately, vibrator crews have to work with the freshly poured concrete in close proximity to the moving bucket. The pouring strategy should be to keep the vibrator crews away from the overhead path of the bucket. Riding the concrete bucket is prohibited.

STEEL ERECTION

Who has not marveled at the daring of the highrise steelworker "walking the beam" hundreds of feet up the steel superstructure of a new building under construction? Perhaps this work will always be dangerous, but the hazard has been mollified somewhat by requiring safety nets to be installed whenever the fall distance exceeds two stories or 25 feet. An alternative is to use scaffolds or temporary floors. Case Study 18.1 is a classic example of the benefits of using safety nets in steel erection.

CASE STUDY 18.1

SAFETY NETS ON THE GOLDEN GATE BRIDGE

The construction of San Francisco's Golden Gate Bridge in the 1930s was a dramatic engineering feat made even more perilous by the wind, rain, and ocean tides. Many workers lost their lives due to falls into the treacherous water. After the death toll reached, 23 workers refused to continue without increased safety protection. Work was resumed after safety nets were installed. An additional 10 workers fell from the bridge, but all 10 were saved by the safety nets (ref. Avers).

A safety railing around the perimeters of temporary floors is now required for tier buildings and other multifloored structures. During structural steel assembly, however, the use of $\frac{1}{2}$-inch wire rope approximately 42 inches high is permitted for the safety railing. To be effective, the wire rope should be checked frequently to be sure that it remains taut.

To maintain structural integrity on the way up, the permanent floors should follow the structural steel as the work progresses. The general rule is no more than eight stories between the erection floor and the uppermost permanent floor. No more than four floors or 48 feet of unfinished bolting or welding is permitted above the foundation or uppermost permanently secured floor.

Structural steel erection sites are subject to constant hazards of falling objects. Rivets, bolts, and drift pins are required to be stored in secured containers, and if all else fails, there is the hard hat to protect the worker below. For some objects at the steel erection site, the hard hat is no protection, however. I am reminded of an instance in Chicago in which a falling steel beam flattened a parked car from left taillight to right headlight. The car was parked next to a sidewalk along which I walked to and from a meeting. The accident occurred during the meeting; luckily, no one was injured.

DEMOLITION

Some would say that the subject of demolition does not belong in this chapter, but demolition and construction are actually closely related. If the construction site is not clear, demolition of previous structures may be the first step in construction of a new building. Many of the tools and equipment, such as cranes and bulldozers, are the same.

People identify skill, knowledge, and quality as important to construction jobs, but most people do not think of demolition as requiring skill and knowledge. But often the engineering expertise required of a demolition job far exceeds the engineering for the original construction. Buildings to be demolished have often been previously damaged by fire or may have been condemned for some serious reason, such as structural damage. Required for every demolition operation is a written report of an engineering survey conducted in advance.

A demolition operation begins with manual operations, such as disassembly of salvage items, and then proceeds to material teardown and dumping to street level. Dangers exist in the debris dumping operation. The area below needs protection if it is outside the walls of the structure. Well-designed chutes capable of withstanding impact loads, together with substantial discharge gates, are needed to control the dropping material. One hazard is that personnel can fall down the chute while dumping debris. A substantial guardrail about 42 inches high is needed to protect against this hazard. A toeboard or bumper is also needed, if wheelbarrows are used, to prevent losing the wheelbarrow down the chute.

Once light teardown operations are completed, heavier demolition equipment such as cranes with wrecking balls are used. Most walls are unstable without lateral support, so they should not be allowed to stand alone at heights greater than one story. No unstable standing wall should be left at the end of a shift.

One sensational demolition technique that is gaining in popularity is controlled explosive demolition, illustrated in Figure 18.17. In this method, carefully engineered

FIGURE 18.17

Building collapses under controlled explosive demolition (*Source*: courtesy of Jim Wolfe Photo, Tulsa, Oklahoma).

explosive charges are detonated to precipitate a catastrophic failure of the building structure, resulting in an immediate and total collapse. The operation has been carried out successfully on many downtown buildings in U.S. cities, typically triggered at the quiet time of dawn on Sunday mornings. Although the operation is dramatic and seems dangerous, it is really quite safe and avoids many of the hazards of a slow tear-down process.

EXPLOSIVE BLASTING

Demolition is only one application for blasting; the construction industry has others. The preparation of roadway cuts is the most important. The chief concern with construction blasting is the safe handling, storage, and transportation of the explosives themselves. The reader might want to review some of the concepts of explosives handling covered in Chapter 11.

Almost everyone has witnessed the familiar warning to "turn off two-way radio" when in a blasting area. The chance is remote, but an electrically fired blasting cap could be detonated by a small stray current induced by a radio transmitter. Lightning is even more of a hazard, and all blasting operations should cease when an electrical storm is near. Radar, nearby power lines, and even dust storms can also be sources of stray currents.

Good visibility reduces many hazards, and explosive blasting hazards are in this category. Aboveground blasting should be conducted in the daytime only. Black powder blasting has been replaced by safer modern methods, and black powder blasting is now prohibited in construction.

The transportation of explosive materials is subject to Department of Transportation (DOT) regulations familiar to suppliers and most construction operators who use explosives. Signs saying **EXPLOSIVES** in large (4-inch) red letters are required on all four sides of the vehicle. Blasting caps should be transported in a separate vehicle from the vehicle transporting other explosives, and both should be transported separate from other cargoes.

Vehicles for explosives need a good fire extinguisher rated at least 10-ABC on board. It would be foolhardy to attempt to control a fire in the cargo compartment of a vehicle transporting explosives. However, most vehicle fires begin in the engine compartment or outside but adjacent to the vehicle. Such fires can sometimes be controlled with a good fire extinguisher wielded by a trained operator, thereby averting a major explosives catastrophe.

ELECTRIC UTILITIES

A specialized type of construction is the erection and modification of electric transmission and distribution lines and equipment. The efficient transmission of usable levels of electrical power necessitates very high voltages. The rules for handling high voltages are quite different from the rules for handling ordinary household and industrial and commercial voltages. For instance, with ordinary voltages, a danger is the contact with exposed live parts. With high voltages, it can be dangerous even to approach the *vicinity* of live parts, as is reflected in Table 18.3 based on the OSHA standard. For voltages in the kilovolt range, the atmosphere may not be an effective insulator, and arcing becomes a hazard. Therefore, safety distances must be maintained. Of course, the distances shown in Table 18.3 apply safety factors. The actual physical arcing distances are much smaller, but there is an element of uncertainty due to such factors as humidity and barometric pressure. Furthermore, the electric-utility lineman may not be able to estimate precisely his distance from the high-voltage line or equipment, making safety factors essential.

Personal protective equipment for high-voltage work takes on a new dimension, that of *degree* of protection. Ordinary insulators on tools, protective gloves, and other insulating equipment that are effective insulators for ordinary applications might completely break down in high-voltage exposures. The whole business of working with energized high-voltage lines is a strange world to the uninitiated—a world fraught with

TABLE 18.3   Minimum Clearance Distances for Live-Line Bare-Hand Work (Alternating Current)

Voltage range (phase to phase) (kV)	Distance (feet and inches) for maximum voltage	
	Phase to ground	Phase to phase
2.1–15	2–0	2–0
15.1–35	2–4	2–4
35.1–46	2–6	2–6
46.1–72.5	3–0	3–0
72.6–121	3–4	4–6
138–45	3–6	5–0
161–169	3–8	5–6
230–242	5–0	8–4
345–362	7–0[a]	13–4[a]
500–552	11–0[a]	20–0[a]
700–765	15–0[a]	31–0[a]

[a]For 345–362, 500–552, and 700–765 kV, the minimum clearance distance may be reduced provided that the distances are not made less than the shortest distance between the energized part and a grounded surface.

Source: Code of Federal Regulations 29 CFR 1926.955.

curious physical effects. The electric-utility industry offers a prime example of an industry in which training in awareness and understanding of hazards is the key to a safe workplace, echoing the principles set forth in Chapter 3.

SUMMARY

The construction industry deserves special consideration because it is so dangerous and also because OSHA has watched construction more closely than general industries. The safety and health manager for construction jobs should remember that the principal task is to avoid *fatalities*. The top five categories of fatalities in the construction industry are

- Falls
- Electrocutions
- Vehicle rollover
- Personnel runover by vehicle
- Excavation cave-ins

If the safety and health manager keeps these fatality categories in mind, it will help to place overall efforts in proper perspective on the construction site.

This complete chapter has represented somewhat a summary of the entire book. The construction industry displays virtually every hazard presented by general industry, but in construction, the hazard is usually worse. Compounding the problem is the transitory nature of the problems that are encountered. It is difficult to pursue costly safeguarding procedures such as trench shoring when the exposure to the hazard will consist of only a few days or even hours. Construction schedules are always demanding for several reasons. High-stakes investments are on the line, costly interruptions to facilities and street traffic are often present, and unplanned chance events are always popping up to ensure that the construction project manager will always struggle to stay on schedule. In this environment, there will always be room for improvements to the safety and health program. What is true for construction is also true of general industry, although perhaps to a lesser degree. The reader should recognize the challenge that this reality represents.

It is hoped that this book has cast some light on the challenges that face the safety and health manager in today's industrial and regulatory environment and some insights for dealing with these challenges. The field is certain to present new challenges in the coming years. Each new challenge ushers in new opportunities for safety and health managers to have an impact on the lives of their co-workers and on the health and financial well-being of their companies.

EXERCISES AND STUDY QUESTIONS

18.1 What is the minimum illumination level permitted for general construction areas? In what areas may lighting be reduced to 3 foot candles?

18.2 How are lasers used in construction?

18.3 What tensile strength is specified for safety belt hardware? Give two reasons that the specification is so much higher than the weight of any human being.

18.4 What is a safety belt lanyard? What nominal breaking strength is specified for lanyards?

18.5 What is a triple rolling hitch?

18.6 Compare hydraulic versus pneumatic power tools from a safety and health standpoint.

18.7 How are construction cranes particularly hazardous to persons on the ground?

18.8 Why are helicopter hooks (for load attachment) more complicated than those for ordinary construction cranes?

18.9 What does the acronym *ROPS* represent?

18.10 What are the two principal strategies for preventing personnel from being run over by construction equipment?

18.11 What is the difference between a trench and an excavation?

18.12 How can trench jacks be damaged by improper placements?

18.13 What is a rebar? Why is it dangerous?

18.14 When must steel erection workers be protected against falls with safety nets?

18.15 What type of safety rail construction is permitted during steel erection?

18.16 Why is an engineering survey required prior to demolition of a building?

18.17 Is it a good idea to carry a fire extinguisher aboard a truck that transports explosives? Why or why not?

18.18 A crawler-type construction crane is operating near a 550-kilovolt power line. What is the minimum distance the boom should approach the line?

18.19 The crawler-type construction crane in Exercise 18.18 finishes its job and travels to the next job, passing under the same 550-kilovolt power line where the power line crosses over a city street at a different location in the neighborhood. What minimum distance is specified for this situation?

18.20 Suppose in Exercise 18.18 that the crawler crane had been employed in general industry instead of construction. What minimum distance would be permitted in general industry?

18.21 For the crane of Exercise 18.20, what is the minimum clearance from the 550-kV power line if the crane is in transit?

18.22 A painter is standing on a work platform that is 27 feet above ground level. For fall protection, the worker's safety belt is attached to a 12-foot safety line, which serves as a lanyard and is securely fastened to the structure at a point 40 feet above ground level. Does this arrangement violate standards for fall protection? Explain.

18.23 A convenient and secure attachment point for fall protection lines is located 35 feet above ground level on the exterior wall of a building. A 20-foot safety line, which can be used as a belt lanyard, is available for connection to this attachment point. What is the lowest and highest level at which a work platform can be safely positioned for workers to be protected by a 20-foot belt lanyard fastened to this attachment point? State any assumptions necessary to your solution.

18.24 Explain the hazard of mushroomed heads on chisels.

18.25 What provisions are specified by federal standards to protect workers from drowning?

18.26 Explain the coding system for identifying the power level for cased power loads for powder-actuated tools.

18.27 To comply with standards, how often should material handling rigging on a construction site be checked?

18.28 What is the standard method of disposing of scrap materials from higher than ground level during construction? What special precaution is required when the material to be discarded is at a height greater than 20 feet?

18.29 What is the principal fire hazard on construction sites?

18.30 What personnel hazard arises when a pneumatic power hose is severed? What device is specified by federal standards to deal with this hazard? What problems have been encountered in the field with the use of this device?

18.31 Describe several ways in which a crane can be maneuvered into a two-blocking situation. Prove your point using diagrams to clarify your reasoning.

18.32 What hazard takes priority in the selection of the type of hydraulic fluid to be used for general construction tools? How does this priority change when the hydraulic tools are used in the public utility industry?

18.33 Compare the hazards of handguns and powder-actuated tools.

18.34 Explain the controversy behind the requirement for GFCIs in the construction industry.

18.35 What is a cornice hook? What additional precautions are needed when cornice hooks are used?

18.36 What is the principal hazard of using concrete blocks for scaffold "cribbing"?

18.37 What are hollow, pan-type treads, and what hazard do they represent?

18.38 **Case Study.** A construction crane is resting with no load, and the headache ball is close to, but not touching, the nose of the crane boom, which is in an almost erect position. The running rope of the crane is mounted on a stationary winch to the rear of the boom hinge. While the operator is in the process of lowering the boom, the wire rope breaks and the headache ball and hook assembly falls to the ground. Explain the most likely cause of this accident. Use a diagram to show how the hazard would develop. What precaution could have prevented it?

18.39 How many lives were lost due to falls during the construction of the Golden Gate Bridge? How many additional falls occurred after the erection of safety nets? How many of these additional falls were fatalities?

18.40 At what time in a building's life is the floor likely to be subjected to its heaviest load?

18.41 What is "two-blocking," and under what conditions do ANSI standards require "two-blocking damage prevention features?"

18.42 How often must rigging for construction material handling be inspected for damage or wear?

18.43 What characteristic of artificial fiber ropes besides strength and weight make them especially favorable for selection as lanyard ropes?

18.44 What is "spalling" and what has it to do with the need for eye protection?

18.45 Is it okay to splice flexible extension cords on construction sites? Explain.

18.46 What is a "headache ball" and what is its purpose?

18.47 What is the principal advantage of a hammerhead tower crane over a conventional crawler crane?

18.48 Cargo hooks for helicopters used in construction have an additional hazard that is not present for cargo hooks for construction cranes. What is this additional hazard?

18.49 Is it okay to use a material hoist for personnel? What about vice versa? Explain.

18.50 Explain the terms *articulating* and *hydraulically extensible* with respect to vehicle-mounted boom platforms.

18.51 Identify an example "engineering control" versus an example "administrative or work-practice control" for the hazard of personnel being run over by heavy construction equipment.

18.52 Explain the concept of "angle of repose" with respect to construction hazards.

18.53 Identify the two types of extraordinary stress on concrete forms that creates a need for overdesign to prevent forms kickout.

18.54 Identify the top five fatality hazards to construction workers.

RESEARCH EXERCISES

18.55 Construction and agriculture share many common hazards. Examine available data on agriculture hazards. Specifically, attempt to identify the single most common fatal farm accident. How many fatalities are estimated to occur each year due to this cause?

18.56 Less than 10 years after the Willow Island, West Virginia, tragedy that took the lives of 51 construction workers, another accident of the same type occurred. Use the Internet or other sources to look up details of this second tragedy of the same type resulting in the highest number of construction fatalities since Willow Island.

18.57 **Research Case Study.** In an actual accident,[3] a construction worker was installing steel channel for a roof of a new building. He was using a fall protection system, by attaching his standard 6-foot lanyard to a line that ultimately connected to a rigid structure. However, the attachment line was an extra 10 feet of line looped between the lanyard and the rigid attachment. Since the 10-foot line was looped, its working length was half that: 5 feet. The worker was standing on a 10″ beam, which was 20 feet off the concrete slab floor. This beam served as the rigid attachment for the worker's lifeline system. The worker lost his balance and accidentally fell. How far do you estimate that he fell? (*Hint: The answer is not 11 feet.*)

---

[3]The source of the information surrounding this accident has been withheld upon request.

Bibliography

Accident Facts. Itasca IL: National Safety Council, 1993.

AFI 91–202. *The Air Force Mishap Prevention Program*. U.S. Air Force, 1991.

AFI 91–204. *Safety Investigations and Reports*. U.S. Air Force, 1995.

Arc Welding and Gas Welding and Cutting: Safety and Health (NIOSH 78–138). Cincinnati, OH: U.S. Department of Health, Education, and Welfare (NIOSH), 1978.

Arkansas Assigned Risk Rates for 2002, National Council on Compensation Insurance, 2002.

Arkansas Workers' Compensation Laws and Rules of the Commission (rev. ed.). Little Rock, AR: Workers' Compensation Commission, June 1986.

Asfahl, C. Ray, ed., *OSHA Standards Digest*, Fayetteville, AR: New Century Media, 1995.

Auto and Home Supply Stores: Health and Safety Guide (NIOSH 76–113). Cincinnati, OH: U.S. Department of Health, Education, and Welfare (NIOSH), 1975.

Avers, Laura, "A Bridge Builder's Worst Nightmare," *Ohio Monitor*, July–August 1993, Vol. 65, No. 3, p. 16.

"Back Belts," *NIOSH FACT*, www.cdc.gov/niosh/backfs.html, August 31, 2002.

Barciela, Susana, "Working Column," *Miami Herald*, March 21, 1994.

Barnett, Ralph L., "Foot Controls: Riding the Pedal," *Safety Brief*, July 1997, Vol. 12, No. 4.

Bischoff, Kenneth B., and Robert J. Lutz, "Pharmacokinetics and Risk Assessment," course description, *Continuing Education*. New York: American Institute of Chemical Engineers, 1992, p. 45.

Bland, Jay, ed., *The Welding Environment*. Miami, FL: American Welding Society, 1973.

Bloodborne Pathogens Final Standard: Summary of Key Provisions, Fact Sheet OSHA 92–46. Washington, DC: U.S. Department of Labor, August 5, 1992.

"Brake Monitoring," *Manufacturing Engineering*, February 1976.

Briscoe, G. J., *Risk Management Guide*, DOE 76–45, SSDC–11, revision 1. Idaho Falls, ID: EG&G Idaho, Inc., 1982.

Chaffin, D., and K. Park, "Biomechanical Evaluation of Two Methods of Manual Load Lifting," *AIIE Transactions*, June 1974, Vol. 6, No. 2.

Chapnik, Elissa-Beth, and Clifford M. Gross, "Evaluation, Office Improvements Can Reduce VDT Operator Problems," *Occupational Health and Safety*, July 1987, Vol. 56, No. 7.

Chemistry Laboratory Safety Library (5th ed.). Boston: National Fire Protection Association, 1975.

Christensen, Herbert E., et al., eds., *Registry of Toxic Effects of Chemical Substances* (1975 ed.). Rockville, MD: U.S. Department of Health, Education, and Welfare (NIOSH), 1975.

Christensen, Herbert E., et al., eds., *Suspected Carcinogens*. Rockville, MD: U.S. Department of Health, Education, and Welfare (NIOSH), 1975.

"Cleaning up the environment," *U.S. News and World Report*, March 25, 1991, p. 45.

"Color Detector Tubes and Direct Reading Gas Detection Instruments," *SKC Bulletin 9206*, June 1, 1992.

A Common Goal. Little Rock, AR: Arkansas Department of Labor, 1975.

"Computer Chips and Miscarriages" column. *Occupational Hazards*, December 1992, Vol. 54, No. 12.

Concepts and Techniques of Machine Safeguarding (OSHA 3067). Washington, DC: U.S. Department of Labor, Occupational Safety and Health Administration, 1980.

Concrete Products and Industry: Health and Safety Guide (NIOSH 75–163). Cincinnati, OH: U.S. Department of Health, Education, and Welfare (NIOSH), 1975.

Construction and Related Machinery Manufacturers: Health and Safety Guide (NIOSH 78–103). Cincinnati, OH: U.S. Department of Health, Education, and Welfare (NIOSH), 1977.

Construction Industry OSHA Safety and Health Standards (29 CFR 1926/1910: OSHA 2207) (rev.). Washington, DC: Department of Labor, Occupational Safety and Health Administration, 1979.

Construction vs. Manufacturing 1974–94, U.S. Department of Labor Washington, DC: Department of Labor, 1996.

Cost of Government Regulation Study. Chicago: Arthur Andersen, 1979.

Covan, John M., *Electric Hazards Control Manual*, Little Rock, AR: Arkansas Department of Labor, 1977.

Cranes: A Guide to Good Work Practices for Operators (NIOSH 78–192). Cincinnati, OH: U.S. Department of Health, Education, and Welfare (NIOSH), 1978.

Crites, Thomas R., "Reconsidering the Costs and Benefits of a Formal Safety Program," *Professional Safety*, December 1995, Vol. 40, No. 12, p. 28.

Cross, Rich, "EPA's Tank Rule: A Compliance Nightmare for Motor Carriers," *Commerical Carrier Journal*, November 1988.

Dear, Joseph A., "The Rate of Injuries and Illnesses in Construction." Speech to the Northwest Indiana Business Roundtable, March 23, 1995.

DeGroff, Lola, "Deadline to Submit Comments on OSHA Draft Guidelines on Night Retail Violence Extended Again" Release USDL: 96–437. Washington, DC: *U.S. Department of Labor News*, October 17, 1996.

Dickie, D. E., *Crane Handbook*. Toronto: Construction Safety Association of Ontario, 1975.

Douglass, Cynthia, "Indoor Air Quality Act of 1991," Congressional Testimony. Washington, DC: U.S. House of Representatives, Committee on Science, Space, and Technology, Subcommittee on Environment, February 7, 1992.

Draeger Tube Handbook. Pittsburgh, PA: National Draeger, 1992.

Drug Testing Monitor (brochure). Washington, DC: Traffic World, 1989.

Eckhardt, Robert, "The Safety Professional in the Corporate Social Structure," *Professional Safety*, May 1993, Vol. 38, No. 5, p. 31.

Engineering Control of Welding Fumes (NIOSH 75–115). Cincinnati, OH: U.S. Department of Health, Education, and Welfare (NIOSH), 1975.

Fabricated Structural Metal Products Industry: Health and Safety Guide (NIOSH 78–100). Cincinnati, OH: U.S. Department of Health, Education, and Welfare (NIOSH), 1977.

Fatal Facts, No. 36. Washington, DC: U.S. Department of Labor, 1988.

Federal Register, OSHA Standard 29 CFR 1910.146, with preamble, 58 FR 4549, January 14, 1993.

Feldman, Marye C., and James B. Bramson, "What Is the Cost of Compliance," *Journal of the American Dental Association*, June 1994, Vol. 125, p. 682.

Felsenthal, Edward, "Out of Hand: Is It an Epidemic or Largely a Fad? The Debate Over Repetitive-Stress Injury Heats Up." *Wall Street Journal*, February 18, 1994, Vol. 223, No. 35.

Fiberglass Layup and Sprayup (NIOSH 76–158). Cincinnati, OH: U.S. Department of Health, Education, and Welfare (NIOSH), 1976.

Fire Protection Handbook (15th ed.). Quincy, MA: National Fire Protection Agency, 1981.

Flashpoint Index of Trade Name Liquids (9th ed.). Boston: National Fire Protection Association, 1978.

Fleming, Susan Hall, "Florida Excavation Firm Involved in Fatal Trench Collapse Faces $448,000 OSHA Penalty, Third Highest for Trenching," Release USDL: 97–147. Washington, DC: *U.S. Department of Labor News*, April 30, 1997.

Fleming, Susan Hall, "Secretary of Labor Reich Announces Violence Prevention Guidelines for Health Care and Social Service Workers," Release USDL: 96–99, Washington, DC: *U.S. Department of Labor News*, March 14, 1996.

Foremanship and Accident Prevention in Industry. Boston, MA: American Mutual Liability Insurance Company, 1943.

Foulke, Edwin G., Jr., "Contested Cases Get a Fair Shake," *Safety and Health*, September 1992, Vol. 148, No. 3, p. 68.

Foundries: Health and Safety Guide (NIOSH 76–124). Cincinnati, OH: U.S. Department of Health, Education, and Welfare (NIOSH), 1976.

Fox, Stephen, "Manhole Cover Industry Hurt by Industry Imports," (an Associated Press interview with Jim Pinkerton, Pinkerton Foundry, Lodi, CA), *Northwest Arkansas Times*, 1981.

"Fraud: It doesn't pay to cheat," *BWC Focus Magazine*, Autumn, 1997, Vol. 1 No. 1.

Friend, Mark A., "What Can Responders Learn from Sept. 11?" *Responder Safety*, a quarterly supplement to *Occupational Hazards*, 2002, Vol. 1, No. 1.

Fumes and Gases in the Welding Environment. Miami, FL: American Welding Society, 1979.

Garg, A., D. B. Chaffin, and G. D. Herrin, "Prediction of Metabolic Rates for Manual Materials Handling Jobs," *American Industrial Hygiene Association Journal*, 1978, Vol. 39, pp. 661–674.

General Industry OSHA Safety and Health Standards (29 CFR 1910, OSHA 2206) (rev.). Washington, DC: U.S. Department of Labor, Occupational Safety and Health Administration, 1989.

Glantz, Stanton A., "Health Hazards of Secondhand Smoke," *Trial*, June 1, 1991, Vol. 29, No. 6, p. 36.

Gonzales, Claire, "EEOC Sues Exxon for Disability Act Violation," U.S. Equal Employment Opportunity Commission News Release, June 28, 1995.

Grain Mills: Health and Safety Guide (NIOSH 75–144). Cincinnati, OH: U.S. Department of Health, Education, and Welfare (NIOSH), 1975.

Greene, Warner C., "AIDS and the Immune Systems," *Scientific American*, September 1, 1993, Vol. 269, No. 3, p. 98.

Griffin, Mark F., "A Review of the Effectiveness of OSHA's Safety Enforcement Policy," Master's thesis, University of Arkansas, Fayetteville, AR, 1993.

Grimaldi, John V., and Rollin H. Simonds, *Safety Management* (3d ed.). Homewood, IL: Richard D. Irwin, 1975.

Hamilton, Robert W., "The Role of Nongovernmental Standards in the Development of Mandatory Federal Standards Affecting Safety or Health," *Texas Law Review*, November 1977, Vol. 56, No. 8.

Hammer, Willie, *Occupational Safety Management and Engineering* (2d ed.). Englewood Cliffs, NJ: Prentice-Hall, 1981.

Handbook of Organic Industrial Solvents, Technical Guide 6 (5th ed.). Chicago: Alliance of American Insurers, 1980.

Hawley, Gessner G. (revision author), *The Condensed Chemical Dictionary* (9th ed.). New York: Van Nostrand Reinhold, 1975.

"Hazardous Employer Program" (Amended Rule 2) Little Rock, AR: Arkansas Workers' Compensation Commission, 1997.

Health and Safety Guide for Public Warehousing. Cincinnati, OH: U.S.Department of Health, Education, and Welfare (NIOSH), 1978.

Heinrich, H. W., *Industrial Accident Prevention* (4th ed.). New York: McGraw-Hill, 1959.

Hotels and Motels: Health and Safety Guide (NIOSH 76–112). Cincinnati, OH: U.S. Department of Health, Education, and Welfare (NIOSH), 1975.

Houston Chronicle, January 10, 1984.

Hyatt, E. C., et al., "Effect of Facial Hair on Worker Performance," *American Industrial Hygiene Association Journal*, April 1973.

Imre, John, unpublished doctoral dissertation, Michigan State University, East Lansing, MI, 1974.

The Industrial Environment: It's Evaluation and Control. Cincinnati, OH: U.S. Department of Health, Education, and Welfare (NIOSH), 1973.

Industrial Noise Control Manual (NIOSH 79–117) (rev. ed.). Cincinnati, OH: U.S.Department of Health, Education, and Welfare (NIOSH), 1978.

Industrial Ventilation (15th ed.). Lansing, MI: Committee on Industrial Ventilation, American Conference of Governmental Industrial Hygienists, 1978.

Injury Facts, 2002 Edition. Itasca, IL: National Safety Council, 2002.

Kamp, John, *Job Candidate Profile Technical Manual* (2nd ed.). St. Paul, MN: St.Paul Fire and Marine Insurance Company, 1991.

Konz, Stephan, and Steven Johnson, *Work Design Industrial Ergonomics* (6th ed.). Scottsdale, AZ: Holcomb Hathaway Publishers, June 2003.

Kroemer, Karl, Henrike Kroemer, and Katrin Kroemer-Elbert, *Ergonomics* (2d ed.). Upper Saddle River, NJ: Prentice Hall, 2001.

LaBar, Gregg, "Hamlet, NC: Home to a National Tragedy," *Occupational Hazards*, September 1992, Vol. 54, No. 9, p. 29.

LaBar, Gregg, "Substituting Safer Materials," *Occupational Hazards*, November 1997, Vol. 59, No. 11, pp. 49–51.

LaBar, Gregg, "Testing the Limits of Industrial Hygiene," *Occupational Hazards*, May 1993, Vol. 55, No. 5, p. 56.

Lapedes, Daniel N., ed. *Dictionary of Scientific and Technical Terms* (2d ed.). New York: McGraw-Hill, 1978.

Lastowka, James A., "OSHA's Process Safety Standard: Key Lessons from the First Five Years," *Occupational Hazards*, July 1997, Vol. 59, No. 7, p.45.

Lee, Cynthia, and Sheila Hall, "Region V: Partnership and Teamwork for Success," *Job Safety and Health Quarterly*, Winter 1995, Vol. 6, No. 2, p. 10.

Lithographic Printing Industry: Employee Health and Safety (NIOSH 77–223). Cincinnati, OH: U.S. Department of Health, Education, and Welfare (NIOSH), 1977.

Loewer, Otto J., Thomas C. Bridges, and Ray A. Bucklin, *On-farm Drying and Storage*. St. Joseph, MI: American Society of Agricultural Engineers, 1994.

Lorenzi, Neal, "From Research to Reality," *Professional Safety*, August 1995, Vol. 40, No. 8, p. 16.

Lovett, John N., *Industrial Noise Control Manual*, Little Rock, AR: Arkansas Department of Labor, 1976.

Machine Guarding: Assesment of Need (NIOSH 75–173). Cincinnati, OH: U.S. Department of Health, Education, and Welfare (NIOSH), 1975.

Main, Jeremy, "The Big Cleanup Gets It Wrong," *Fortune*, May 20, 1991, Vol. 123, No. 10, p. 95.

Management of Work-Related Musculoskeletal Disorders, ANSI Accredited Standards Committee Z365 working draft. Itasca, IL: National Safety Council, August 6, 2002.

Manual of Respiratory Protection against Radioactive Materials (NUREG-0041). Washington, DC: U.S. Atomic Energy Commission, Directorate of Regulatory Standards, 1974.

Manufacturers of Paints and Allied Products: Health and Safety Guide (NIOSH 75–179). Cincinnati, OH: U.S. Department of Health, Education, and Welfare (NIOSH), 1975.

McClay, Robert E., "Toward a More Universal Model of Loss Incident Causation," *Professional Safety*, January 1989, Vol. 34, No. 1.

McElroy, Frank E., ed., "Administration and Programs," *Accident Prevention Manual for Industrial Operations* (10th ed.). Chicago: National Safety Council, 1992.

McElroy, Frank E., ed., "Engineering and Technology" *Accident Prevention Manual for Industrial Operations* (10th ed.). Chicago: National Safety Council, 1992.

Metal Stamping Operations, Health and Safety Guide (NIOSH 75–174). Cincinnati, OH: U.S. Department of Health, Education, and Welfare (NIOSH), 1975.

Method of Recording and Measuring Work Injury Experience (ANSI Z16.1–1967). New York: American National Standards Institute, 1967.

Moore, Larry R., "Preventing Homicide and Acts of Violence in the Workplace," *Professional Safety*, July 1997, Vol. 42, No. 7, p. 20.

Most Frequently Asked Questions Concerning the Bloodborne Pathogens Standard, OSHA Administrative Memo Information Booklet, Washington, DC: U.S. Department of Labor, Occupational Safety and Health Administration, February 1, 1993.

Mukherjee, Sougata, "OSHA's Cost to Business Tops $33 Billion," *Baltimore Business Journal*, November 1, 1996, Vol. 14, No. 24, p. 27.

National Electrical Code® (NEC). Boston: National Fire Protection Association, 1971 and 1981.

National Safety Council, Safety Training Institute program announcement, Itasca, IL: National Safety Council, October, 1995.

Nemeth, John C., "Risk: Dollars and Trust," *Environmental Spectrum*, 1991, Vol. 8, No. 1.

NFPA Inspection Manual (3d ed.). Boston: National Fire Protection Association, 1970.

NIOSH Certified Equipment: Cumulative Supplement (NIOSH 77–195). Morgantown, WV: U.S. Department of Health, Education, and Welfare (NIOSH), 1977.

Nonmandatory Compliance Guidelines for Hazard Assessment and Personal Protective Equipment Selection, OSHA standard 29 CFR–1910, Subpart I, App. B.

Occupational Diseases: A Guide to Their Recognition. Cincinnati, OH: U.S. Department of Health, Education, and Welfare (NIOSH), 1951.

Occupational Exposure Sampling Manual (NIOSH 77–173). Cincinnati, OH: U.S. Department of Health, Education, and Welfare (NIOSH), 1977.

Occupational Injuries and Illnesses in the United States by Industry. Washington, DC: U.S. Department of Labor, Bureau of Labor Statistics, 1985.

Occupational Safety and Health Cases, Vols. 1–9. Washington, DC: Bureau of National Affairs, 1974–1982.

Occupational Safety and Health Directory (NIOSH 80–124). Cincinnati, OH: U.S. Department of Health and Human Services (NIOSH), 1980.

Occupational Safety and Health Reporter (weekly periodical), Vols. 1–11. Washington, DC: Bureau of National Affairs, Inc., 1971–1982.

Occupational Safety and Health in Vocational Education (NIOSH 79–125). Cincinnati, OH: U.S. Department of Health and Human Services (NIOSH), 1979.

Olishifski, Julian B., and Frank E. McElroy, eds. *Fundamentals of Industrial Hygiene*, Chicago: National Safety Council, 1971.

"OSHA," column, *Occupational Hazards*, December 1992, Vol. 54, No. 12, p. 11.

"OSHA," column, *Occupational Hazards*, September 1993, Vol. 55, No. 9, p. 26.

OSHA Consultation Service, Fact Sheet No. OSHA 97–04, Washington, DC: U.S. Department of Labor, January 1, 1997.

"OSHA Proposes Fines for Ergonomics-Related Injuries," *Material Handling Engineering*, February 1989, Vol. 44, No. 2, p. 42.

OSHA Requests Comments On Proposed Improvements to Twenty-three Health Standards, OSHA Trade News Release, Washington, DC: U.S. Department of Labor, October 30, 2002.

OSHA's Role at the World Trade Center Emergency Project, OSHA News Release, Washington, DC: U.S. Department of Labor, September 21, 2002.

Ott, Wayne R., and John W. Roberts, "Everyday Exposure to Toxic Pollutants," *Scientific American*, February 1998, Vol. 278, No. 2, p. 86.

Overhead and Gantry Cranes (ANSI B.30.2.0–1976). New York: American National Standards Institute, 1976.

Parker, Sandra C., C. Ray Asfahl, and Susan Johnsen, "An Industrial Chemical Hazards Database with Natural Language Interface: An Application of Artificial Intelligence," *Proceedings, Tenth Annual Conference for Computers and Industrial Engineering*, Dallas, TX, 1988.

Peterson, Donald R., and David B. Thomas, *Fundamentals of Epidemiology*. Lexington, MA: Lexington Books, 1978.

Plumbing, Heating and Air Conditioning Contractors: Health and Safety Guide (NIOSH 76–127). Cincinnati, OH: U.S. Department of Health, Education, and Welfare (NIOSH), 1975.

Power Actuated Fastening Tools (NIOSH 78–178A). Cincinnati, OH: U.S. Department of Health, Education, and Welfare (NIOSH), 1978.

"Preamble to the OSHA Lockout/Tagout Standard," *Federal Register*, September 1, 1989, Vol. 54, No. 169.

A Prescription for Battery Workers (NIOSH 76–153). Cincinnati, OH: U.S. Department of Health, Education, and Welfare (NIOSH), 1976.

"Psychological Tests & Workplace Violence," *The Florida Bar Journal*, March 1994.

Public Law 91–596 (Williams–Steiger Occupational Safety and Health Act of 1970), U.S. Congress, December 19, 1970.

Public Law 101–336 (Americans with Disabilities Act of 1990), U.S. Congress, July 26, 1990.

Ramachandran, Kumar, "An Analysis Tool for Building Design Based on the Simulation Modeling of Human Behavior in Buildings During Emergencies/Fires," proposal for a doctoral dissertation, Department of Industrial Engineering, University of Arkansas, Fayetteville, AR, August 1997.

"Rapid Rater," Independent Insurance Agents of Arkansas, May 1, 1991.

Recordkeeping Requirements under the Occupational Safety and Health Act of 1970. Washington, DC: U.S. Department of Labor, Occupational Safety and Health Administration, 1978.

Registry of Toxic Effects of Chemical Substances. Cincinnati, OH: U.S. Department of Health, Education, and Welfare (NIOSH), 1976.

Rekus, John F., "OSHA's Lockout Tagout Standard Mandates Control of Energy Sources," *Occupational Health and Safety*, Vol. 59, No. 11, p. 108.

Respiratory Protection: An Employer's Manual (NIOSH 78–193A). Cincinnati, OH: U.S. Department of Health, Education, and Welfare (NIOSH), 1978.

Respiratory Protection: A Guide for the Employee (NIOSH 78–193B). Cincinnati, OH: U.S. Department of Health, Education, and Welfare (NIOSH), 1978.

Rigging Manual. Toronto: Construction Safety Association of Ontario, 1975.

Ritzel, Dale O., and Rodney G. Allen, "Validity of the Basic Principle of Safety Management or Loss Control," *Professional Safety*, February 1996, Vol. 41, No. 2, p. 24.

Ryan, Joseph P., "Power Press Safeguarding: A Human Factors Perspective," *Professional Safety*, August 1987, Vol. 32, No. 8, pp. 23–26.

Safety Requirements for Woodworking Machinery (ANSI 01.1–1975). New York: American National Standards Institute, 1975.

Safety Standard for Monorail Systems and Underhung Cranes (ANSI B30.11–1973). New York: American National Standards Institute, 1973.

Sample Bloodborne Pathogens Exposure Control Plan. Philadelphia: OSHA Philadelphia Regional Office of Technical Support, March 25, 1992.

Saulter, Gilbert J., Regional Administrator, OSHA Region VI, in a speech entitled "OSHA Update," Little Rock, AR, July 27, 1988.

Sax, N. Irving, *Dangerous Properties of Industrial Materials* (5th ed.). New York: Van Nostrand Reinhold, 1975.

Scannell, Gerard F., "OSHA's Efforts to Protect Workers from Indoor Air Pollution," Congressional Testimony. Washington, DC: Committee on Energy and Commerce, Subcommittee on Environment, April 10, 1991.

Schuster, Eric S., "Understanding the ADA," *Association Management*, April 1, 1991, Vol. 43, No. 4, p. 51.

Sensidyne Product Literature, *The First Truly Simple Precision Gas Detector System.* Largo, FL: Sensidyne, 1984.

Sign and Advertising Display Manufacturers: Health and Safety Guide (NIOSH 76–126). Cincinnati, OH: U.S. Department of Health, Education, and Welfare (NIOSH), 1976.

Smith, Harry C., "BLEVE Can Blow You to Oblivion," *Ohio Monitor*, October 1979, Vol. 52, No. 10, p. 16.

"Standard Code of Practise for Safety of Machinery," British Standard BS5304:1988.

Threshold Limit Values (TLV booklet). Cincinnati, OH: American Conference of Governmental Industrial Hygienists. September 1993–1994.

Title III Fact Sheet. Washington, DC: U.S. Environmental Protection Agency, 1987. U.S. Government Printing Office Document 1987–718–872–1302/1280.

"The Tobacco Settlement, Statements & Information" Washington, DC: The American Cancer Society, June 20, 1997, Internet address: www.cancer.org/tobacco/tralertl.html

Tyson, Pat, speech at the 18th Annual Employment Law & Legislative Conference, sponsored by the Society for Human Resource Management, Washington, DC, 2001.

Wallace, L.A., "Human Exposure to Environmental Pollutants: A Decade of Experience," *Clinical and Experimental Allergy*, 1995, Vol. 25, No. 1, pp. 4–9.

The Welding Environment. Miami, FL: American Welding Society, 1973.

Wells, A. Judson, "Deadly Smoke," *Occupational Health and Safety*, September 1989.

Work-Related Musculoskeletal Disorders (WMSDs), draft ANSI standard Z365, August 6, 2002.

Wilkinson, Bruce S., "Substance Abuse Programs," *Proceedings of the American Society of Safety Engineers Annual Professional Development Conference and Exposition*, Baltimore, MD, June 14–17, 1987.

Wooden Furniture Manufacturing: Health and Safety Guide (NIOSH 75–167). Cincinnati, OH: U.S. Department of Health, Education, and Welfare (NIOSH), 1975.

Worker Exposure to AIDS and Hepatitis B. Washington, DC: U.S. Department of Labor, Occupational Safety and Health Administration, 1987.

Work Injury and Illness Rates. Chicago: National Safety Council, 1981.

Work Practices Guide for Manual Lifting (NIOSH 81–122). Cincinnati, OH: U.S. Department of Health and Human Services (NIOSH), 1981.

"Workplace Smoking Rule Moves Too Slowly for ASH," *Keller's Industrial Safety Report*, May 1997, Vol. 7, No. 5.

Zumar, Tony, "Worker's Comp Safety Division Update," presentation in Little Rock, AR, September 16, 1993.

APPENDIX A

OSHA Permissible Exposure Limits

APPENDIX A.1: GENERAL TABLE OF PELs

LIMITS FOR AIR CONTAMINANTS

The permissible exposure levels (PELs) are 8-hour time-weighted averages (TWAs) unless otherwise noted; a (C) designation denotes a ceiling limit. Concentrations are to be determined from breathing-zone air samples.

Substance	CAS No.[a]	permissible exposure level ppm[b]	mg/m[3c]	Skin
Acetaldehyde	75-07-0	200	360	
Acetic acid	64-19-7	10	25	
Acetic anhydride	108-24-7	5	20	
Acetone	67-64-1	1000	2400	
Acetonitrile	75-05-8	40	70	
2-Acetylaminofluorine; see 1910.1014	53-96-3			
Acetylene dichloride; see 1,2-Dichloroethylene				
Acetylene tetrabromide	79-27-6	1	14	
Acrolein	107-02-8	0.1	0.25	
Acrylamide	79-06-1		0.3	X
Acrylonitrile; see 1910.1045	107-13-1			
Aldrin	309-00-2		0.25	X
Allyl alcohol	107-18-6	2	5	X
Allyl chloride	107-05-1	1	3	
Allyl glycidyl ether (AGE)	106-92-3	(C) 10	(C) 45	
Allyl propyl disulfide	2179-59-1	2	12	
α-Alumina	1344-28-1			
Total dust			15	
Respirable fraction			5	
Aluminum metal (as Al)	7429-90-5			
Total dust			15	
Respirable fraction			5	

Substance	CAS No.[a]	permissible exposure level ppm[b]	mg/m[3c]	Skin
4-Aminodiphenyl; see 1910.1011	92-67-1			
2-Aminoethanol; see Ethanolamine				
2-Aminopyridine	504-29-0	0.5	2	
Ammonia	7664-41-7	50	35	
Ammonium sulfamate	7773-06-0			
Total dust			15	
Respirable fraction			5	
n-Amyl acetate	628-63-7	100	525	
sec-Amyl acetate	626-38-0	125	650	
Aniline and homologs	62-53-3	5	19	X
Anisidine (*o*-, *p*-isomers)	29191-52-4		0.5	X
Antimony and compounds (as Sb)	7440-36-0		0.5	
ANTU (*α* = naphthylthiourea)	86-88-4		0.3	
Arsenic, inorganic compounds (as As); see 1910.1018	7440-38-2		10 μg/m^3	
Arsenic, organic compounds (as As)	7440-38-2		0.5	
Arsine	7784-42-1	0.05	0.2	
Asbestos; see 1910.1001 for construction, see 1926.1101	varies with compound			
Azinphos-methyl	86-50-0		0.2	X
Barium, soluble compounds (as Ba)	7440-39-3		0.5	
Barium sulfate	7727-43-7			
Total dust			15	
Respirable fraction			5	
Benomyl	17804-35-2			
Total dust			15	
Respirable fraction			5	
Benzene; See 1910.1028 see Appendix A.2 for the limits applicable in the operations or sectors excluded in 1910.1028[d]	71-43-2			
Benzidine; see 1910.1010	92-87-5			
Benzo (*a*) pyrene; see Coal tar pitch volatiles				
p-Benzoquinone; see Quinone				
Benzoyl peroxide	94-36-0		5	
Benzyl chloride	100-44-7	1	5	
Beryllium and beryllium compounds (as Be)	7440-41-7		see Appendix A.2	
Biphenyl; see Diphenyl				
Bismuth telluride				
Undoped	1304-82-1			
Total dust			15	
Respirable fraction			5	
Boron oxide	1303-86-2			
Total dust			15	
Boron trifluoride	7637-07-2	(C) 1	(C) 3	
Bromine	7726-95-6	0.1	0.7	

Substance	CAS No.[a]	permissible exposure level ppm[b]	mg/m³[c]	Skin
Bromoform	75-25-2	0.5	5	X
Butadiene (1,3-Butadiene)	106-99-0	1–5 ppm STEL		
see 1910.1051;1910.19				
Butanethiol; see Butyl mercaptan.				
2-Butanone	78-93-3	200	590	
(methyl ethyl ketone)				
2-Butoxyethanol	111-76-2	50	240	X
n-Butyl-acetate	123-86-4	150	710	
sec-Butyl acetate	105-46-4	200	950	
tert-Butyl-acetate	540-88-5	200	950	
n-Butyl alcohol	71-36-3	100	300	
sec-Butyl alcohol	78-92-2	150	450	
tert-Butyl alcohol	75-65-0	100	300	
Butylamine	109-73-9	(C) 5	(C) 15	X
tert-Butyl chromate [as CrO₃]	1189-85-1		(C) 0.1	X
n-Butyl glycidyl ether (BGE)	2426-08-6	50	270	
Butyl mercaptan	109-79-5	10	35	
p-tert-Butyltoluene	98-51-1	10	60	
Cadmium (as Cd); see 1910.1027	7440-43-9			
Calcium carbonate	1317-65-3			
Total dust			15	
Respirable fraction			5	
Calcium hydroxide	1305-62-0			
Total dust			15	
Respirable fraction			5	
Calcium oxide	1305-78-8		5	
Calcium silicate	1344-95-2			
Total dust			15	
Respirable fraction			5	
Calcium sulfate	7778-18-9			
Total dust			15	
Respirable fraction			5	
Camphor, synthetic	76-22-2		2	
Carbaryl (Sevin)	63-25-2		5	
Carbon black	1333-86-4		3.5	
Carbon dioxide	124-38-9	5000	9000	
Carbon disulfide	75-15-0		see Appendix A.2	
Carbon monoxide	630-08-0	50	55	
Carbon tetrachloride	56-23-5		see Appendix A.2	
Cellulose	9004-34-6			
Total dust			15	
Respirable fraction			5	
Chlordane	57-74-9		0.5	X
Chlorinated camphene	8001-35-2		0.5	X
Chlorinated diphenyl oxide	55720-99-5		0.5	
Chlorine	7782-50-5	(C) 1	(C) 3	
Chlorine dioxide	10049-04-4	0.1	0.3	
Chlorine trifluoride	7790-91-2	(C) 0.1	(C) 0.4	
Chloroacetaldehyde	107-20-0	(C) 1	(C) 3	
α-Chloroacetophenone	532-27-4	0.05	0.3	
(phenacyl chloride)				

| Substance | CAS No.[a] | permissible exposure level | | Skin |
		ppm[b]	mg/m³[c]	
Chlorobenzene	108-90-7	75	350	
o-Chlorobenzylidene malononitrile	2698-41-1	0.05	0.4	
Chlorobromomethane	74-97-5	200	1050	
2-Chloro-1,3-butadiene; see β-Chloroprene				
Chlorodiphenyl (42% Chlorine) (PCB)	53469-21-9		1	X
Chlorodiphenyl (54% Chlorine) (PCB)	11097-69-1		0.5	X
1-Chloro-2,3-epoxypropane; see Epichlorohydrin				
2-Chloroethanol; see Ethylene chlorohydrin				
Chloroethylene; see Vinyl chloride				
Chloroform (trichloromethane)	67-66-3	(C) 50	(C) 240	
bis (Chloromethyl) ether; see 1910.1008	542-88-1			
Chloromethyl methyl ether; see 1910.1006	107-30-2			
1-Chloro-1-nitropropane	600-25-9	20	100	
Chloropicrin	76-06-2	0.1	0.7	
β-Chloroprene	126-99-8	25	90	X
2-Chloro-6 (trichloromethyl) pyridine	1929-82-4			
Total dust			15	
Respirable fraction			5	
Chromic acid and chromates [as CrO (III)]	varies with compound		see Appendix A.2	
Chromium (II) compounds (as Cr)	7440-47-3		0.5	
Chromium (III) compounds (as Cr)	7440-47-3		0.5	
Chromium metal and insoluble salts (as Cr)	7440-47-3		1	
Chrysene; see Coal tar pitch volatiles				
Clopidol	2971-90-6			
Total dust			15	
Respirable fraction			5	
Coal dust (less than 5% SiO2), respirable fraction			see Appendix A.3	
Coal dust (greater than or equal to 5% SiO2), respirable fraction			see Appendix A.3	
Coal tar pitch volatiles (benzene soluble fraction), anthracene, BaP, phenanthrene, acridine, chrysene, pyrene	65966-93-2		0.2	
Cobalt metal, dust, and fume (as Co)	7440-48-4		0.1	
Coke oven emissions; see 1910.1029			150 $\mu g/m^3$	

Substance	CAS No.[a]	permissible exposure level		Skin
		ppm[b]	mg/m[3c]	
Copper	7440-50-8			
Fume (as Cu)			0.1	
Dusts and mists (as Cu)			1	
Cotton dust[e]; see 1910.1043			200–750 μg/m[3]	
in cotton gin; see 1910.1046				
Crag herbicide (Sesone)	136-78-7			
Total dust			15	
Respirable fraction			5	
Cresol, all isomers	1319-77-3	5	22	X
Crotonaldehyde	123-73-9	2	6	
	4170-30-3			
Cumene	98-82-8	50	245	X
Cyanides (as CN)	varies with compound		5	
Cyclohexane	110-82-7	300	1050	
Cyclohexanol	108-93-0	50	200	
Cyclohexanone	108-94-1	50	200	
Cyclohexene	110-83-8	300	1015	
Cyclopentadiene	542-92-7	75	200	
2,4-D (Dichlorophen oxyacetic acid)	94-75-7		10	
Decaborane	17702-41-9	0.05	0.3	X
Demeton (Systox)	8065-48-3		0.1	X
Diacetone alcohol (4-Hydroxy-4-methyl-2-pentanone)	123-42-2	50	240	
1,2-Diaminoethane; see Ethylenediamine				
Diazomethane	334-88-3	0.2	0.4	
Diborane	19287-45-7	0.1	0.1	
1,2-Dibromo-3-chloropropane (DBCP); see 1910.1044	96-12-8	1 ppb (parts per billion)		
1,2-Dibromoethane; see Ethylene dibromide				
Dibutyl phosphate	107-66-4	1	5	
Dibutyl phthalate	84-74-2		5	
o-Dichlorobenzene	95-50-1	(C) 50	(C) 300	
p-Dichlorobenzene	106-46-7	75	450	
3,3'-Dichlorobenzidine; see 1910.1007	91-94-1			
Dichlorodifluoromethane	75-71-8	1000	4950	
1,3-Dichloro-5, 5-dimethyl hydantoin	118-52-5		0.2	
Dichlorodiphenyltri-chloroethane (DDT)	50-29-3		1	X
1,1-Dichloroethane	75-34-3	100	400	
1,2-Dichloroethane; see Ethylene dichloride				
1,2-Dichloroethylene	540-59-0	200	790	
Dichloroethyl ether	111-44-4	(C) 15	(C) 90	X
Dichloromethane; see Methylene chloride				

Substance	CAS No.[a]	permissible exposure level ppm[b]	mg/m³[c]	Skin
Dichloromonofluoromethane	75-43-4	1000	4200	
1,1-Dichloro-1-nitroethane	594-72-9	(C) 10	(C) 60	
1,2-Dichloropropane; see Propylene dichloride				
Dichlorotetrafluoroethane	76-14-2	1000	7000	
Dichlorvos (DDVP)	62-73-7		1	X
Dicyclopentadienyl iron	102-54-5			
Total dust			15	
Respirable fraction			5	
Dieldrin	60-57-1		0.25	X
Diethylamine	109-89-7	25	75	
2-Diethylaminoethanol	100-37-8	10	50	X
Diethyl ether; see Ethyl ether				
Difluorodibromomethane	75-61-6	100	860	
Diglycidyl ether (DGE)	2238-07-5	(C) 0.5	(C) 2.8	
Dihydroxybenzene; see Hydroquinone				
Diisobutyl ketone	108-83-8	50	290	
Diisopropylamine	108-18-9	5	20	X
4-Dimethylaminoazobenzene; see 1910.1015	60-11-7			
Dimethoxymethane; see Methylal				
Dimethyl acetamide	127-19-5	10	35	X
Dimethylamine	124-40-3	10	18	
Dimethylaminobenzene; see Xylidine				
Dimethylaniline (*N,N*-Dimethylaniline)	121-69-7	5	25	X
Dimethylbenzene; see Xylene				
Dimethyl-1,2-dibromo-2, 2-dichloroethylphosphate	300-76-5		3	
Dimethylformamide	68-12-2	10	30	X
2,6-Dimethyl-4-heptanone; see Diisobutyl ketone				
1,1-Dimethylhydrazine	57-14-7	0.5	1	X
Dimethylphthalate	131-11-3		5	
Dimethyl sulfate	77-78-1	1	5	X
Dinitrobenzene (all isomers)			1	X
ortho	528-29-0			
meta	99-65-0			
para	100-25-4			
Dinitro-*o*-cresol	534-52-1		0.2	X
Dinitrotoluene	25321-14-6		1.5	X
Dioxane (Diethylene dioxide)	123-91-1	100	360	X
Diphenyl (biphenyl)	92-52-4	0.2	1	
Diphenylmethane diisocyanate; see Methylene bisphenyl isocyanate				
Dipropylene glycol methyl ether	34590-94-8	100	600	X
Di-*sec* octyl phthalate (Di-(2-ethylhexyl) phthalate)	117-81-7		5	

Substance	CAS No.[a]	permissible exposure level		Skin
		ppm[b]	mg/m³[c]	
Emery	12415-34-8			
Total dust			15	
Respirable fraction			5	
Endrin	72-20-8		0.1	X
Epichlorohydrin	106-89-8	5	19	X
EPN	2104-64-5		0.5	X
1,2-Epoxypropane; see				
Propylene oxide				
2,3-Epoxy-1-propanol; see Glycidol				
Ethanethiol; see Ethyl mercaptan				
Ethanolamine	141-43-5	3	6	
2-Ethoxyethanol (cellosolve)	110-80-5	200	740	X
2-Ethoxyethyl acetate	111-15-9	100	540	X
(cellosolve acetate)				
Ethyl acetate	141-78-6	400	1400	
Ethyl acrylate	140-88-5	25	100	X
Ethyl alcohol (ethanol)	64-17-5	1000	1900	
Ethylamine	75-04-7	10	18	
Ethyl amyl ketone	541-85-5	25	130	
(5-Methyl-3-heptanone)				
Ethyl benzene	100-41-4	100	435	
Ethyl bromide	74-96-4	200	890	
Ethyl butyl ketone (3-heptanone)	106-35-4	50	230	
Ethyl chloride	75-00-3	1000	2600	
Ethyl ether	60-29-7	400	1200	
Ethyl formate	109-94-4	100	300	
Ethyl mercaptan	75-08-1	(C) 10	(C) 25	
Ethyl silicate	78-10-4	100	850	
Ethylene chlorohydrin	107-07-3	5	16	X
Ethylenediamine	107-15-3	10	25	
Ethylene dibromide	106-93-4	see Appendix A.2		
Ethylene dichloride	107-06-2	see Appendix A.2		
(1,2-Dichloroethane)				
Ethylene glycol dinitrate	628-96-6	(C) 0.2	(C) 1	X
Ethylene glycol methyl acetate;				
see Methyl cellosolve acetate				
Ethyleneimine see 1910.1012	151-56-4			
Ethylene oxide; see 1910.1047	75-21-8	1		
Ethylidene chloride;				
see 1,1-Dichlorethane				
N-Ethylmorpholine	100-74-3	20	94	X
Ferbam	14484-64-1			
Total dust			15	
Ferrovanadium dust	12604-58-9		1	
Fluorides (as F)		see Appendix A.2	2.5	
Fluorine	7782-41-4	0.1	0.2	
Fluorotrichloromethane	75-69-4	1000	5600	
(trichlorofluoromethane)				
Formaldehyde; see 1910.1048	50-00-0	1		
Formic acid	64-18-6	5	9	
Furfural	98-01-1	5	20	X

Substance	CAS No.[a]	permissible exposure level ppm[b]	mg/m[3c]	Skin
Furfuryl alcohol	98-00-0	50	200	
Glycerin (mist)	56-81-5			
Total dust			15	
Respirable fraction			5	
Glycidol	556-52-5	50	150	
Glycol monoethyl ether; see 2-Ethoxyethanol				
Grain dust (oat, wheat barley)			10	
Graphite, natural respirable dust	7782-42-5		see Appendix A.3	
Graphite, synthetic				
Total dust			15	
Respirable fraction			5	
Guthion; see Azinphos methyl				
Gypsum	13397-24-5			
Total dust			15	
Respirable fraction			5	
Hafnium	7440-58-6		0.5	
Heptachlor	76-44-8		0.5	X
Heptane (n-Heptane)	142-82-5	500	2000	
Hexachloroethane	67-72-1	1	10	X
Hexachloronaphthalene	1335-87-1		0.2	X
n-Hexane	110-54-3	500	1800	
2-Hexanone (Methyl n-butyl ketone)	591-78-6	100	410	
Hexone (Methyl isobutyl ketone)	108-10-1	100	410	
sec-Hexyl acetate	108-84-9	50	300	
Hydrazine	302-01-2	1	1.3	X
Hydrogen bromide	10035-10-6	3	10	
Hydrogen chloride	7647-01-0	(C) 5	(C) 7	
Hydrogen cyanide	74-90-8	10	11	X
Hydrogen fluoride (as F)	7664-39-3		see Appendix A.2	
Hydrogen peroxide	7722-84-1	1	1.4	
Hydrogen selenide (as Se)	7783-07-5	0.05	0.2	
Hydrogen sulfide	7783-06-4		see Appendix A.2	
Hydroquinone	123-31-9		2	
Iodine	7553-56-2	(C) 0.1	(C) 1	
Iron oxide fume	1309-37-1		10	
Isomyl acetate	123-92-2	100	525	
Isomyl alcohol (primary and secondary)	123-51-3	100	360	
Isobutyl acetate	110-19-0	150	700	
Isobutyl alcohol	78-83-1	100	300	
Isophorone	78-59-1	25	140	
Isopropyl acetate	108-21-4	250	950	
Isopropyl alcohol	67-63-0	400	980	
Isopropylamine	75-31-0	5	12	
Isopropyl ether	108-20-3	500	2100	
Isopropyl glycidyl ether (IGE)	4016-14-2	50	240	
Kaolin	1332-58-7			
Total dust			15	
Respirable fraction			5	

| Substance | CAS No.[a] | permissible exposure level | | Skin |
		ppm[b]	mg/m[3c]	
Ketene	463-51-4	0.5	0.9	
Lead inorganic (as Pb); see 1910.1025 (for General Industry) or 1926.62 (for Construction)	7439-92-1		50 μg/m^3	
Limestone	1317-65-3			
Total dust			15	
Respirable fraction			5	
Lindane	58-89-9		0.5	X
Lithium hydride	7580-67-8		0.025	
LPG (Liquefied petroleum gas)	68476-85-7	1000	1800	
Magnesite	546-93-0			
Total dust			15	
Respirable fraction			5	
Magnesium oxide fume	1309-48-4			
Total particulate			15	
Malathion	121-75-5			
Total dust			15	X
Maleic anhydride	108-31-6	0.25	1	
Manganese compounds (as Mn)	7439-96-5		(C) 5	
Manganese fume (as Mn)	7439-96-5		(C) 5	
Marble	1317-65-3			
Total dust			15	
Respirable fraction			5	
Mercury (aryl and inorganic) (as Hg)	7439-97-6		see Appendix A.2	
Mercury (organo) alkyl compounds (as Hg)	7439-97-6		see Appendix A.2	
Mercury (vapor) (as Hg)	7439-97-6		see Appendix A.2	
Mesityl oxide	141-79-7	25	100	
Methanethiol; see Methyl mercaptan				
Methoxychlor	72-43-5			
Total dust			15	
2-Methoxyethanol; (methyl cellosolve)	109-86-4	25	80	X
2-Methoxyethyl acetate (methyl cellosolve acetate)	110-49-6	25	120	X
Methyl acetate	79-20-9	200	610	
Methyl acetylene (propyne)	74-99-7	1000	1650	
Methyl acetylene propadiene mixture (MAPP)		1000	1800	
Methyl acrylate	96-33-3	10	35	X
Methylal (dimethoxymethane)	109-87-5	1000	3100	
Methyl alcohol	67-56-1	200	260	
Methylamine	74-89-5	10	12	
Methyl amyl alcohol; see Methyl isobutyl carbinol				
Methyl n-amyl ketone	110-43-0	100	465	
Methyl bromide	74-83-9	(C) 20	(C) 80	X
Methyl butyl ketone; see 2-Hexanone				

| Substance | CAS No.[a] | permissible exposure level | | Skin |
		ppm[b]	mg/m[3c]	
Methyl cellosolve; see 2-Methoxyethanol				
Methyl cellosolve acetate; see 2-Methoxyethyl acetate				
Methyl chloride	74-87-3		see Appendix A.2	
Methyl chloroform (1,1,1-Trichloroethane)	71-55-6	350	1900	
Methylcyclohexane	108-87-2	500	2000	
Methylcyclohexanol	25639-42-3	100	470	
o-Methylcyclohexanone	583-60-8	100	460	X
Methylene chloride	75-09-2		see Appendix A.2	
Methyl ethyl ketone (MEK); see 2-Butanone				
Methyl formate	107-31-3	100	250	
Methyl hydrazine (monomethyl hydrazine)	60-34-4	(C) 0.2	(C) 0.35	X
Methyl iodide	74-88-4	5	28	X
Methyl isoamyl ketone	110-12-3	100	475	
Methyl isobutyl carbinol	108-11-2	25	100	X
Methyl isobutyl ketone; see Hexone				
Methyl isocyanate	624-83-9	0.02	0.05	X
Methyl mercaptan	74-93-1	(C) 10	(C) 20	
Methyl methacrylate	80-62-6	100	410	
Methyl propyl ketone; see 2-Pentanone				
α-Methyl styrene	98-83-9	(C) 100	(C) 480	
Methylene bisphenyl isocyanate (MDI)	101-68-8	(C) 0.02	(C) 0.2	
Mica; see Silicates				
Molybdenum (as Mo)	7439-98-7			
Soluble compounds			5	
Insoluble compounds				
Total dust			15	
Monomethyl aniline	100-61-8	2	9	X
Monomethyl hydrazine; see Methyl hydrazine				
Morpholine	110-91-8	20	70	X
Naphtha (coal tar)	8030-30-6	100	400	
Naphthalene	91-20-3	10	50	
α-Naphthylamine; see 1910.1004	134-32-7			
β-Naphthylamine; see 1910.1009	91-59-8			
Nickel carbonyl (as Ni)	13463-39-3	0.001	0.007	
Nickel, metal and insoluble compounds (as Ni)	7440-02-0		1	
Nickel, soluble compounds (as Ni)	7440-02-0		1	
Nicotine	54-11-5		0.5	X
Nitric acid	7697-37-2	2	5	
Nitric oxide	10102-43-9	25	30	

Substance	CAS No.[a]	permissible exposure level ppm[b]	mg/m[3c]	Skin
p-Nitroaniline	100-01-6	1	6	X
Nitrobenzene	98-95-3	1	5	X
p-Nitrochlorobenzene	100-00-5		1	X
4-Nitrobiphenyl; see 1910.1003	92-93-3			
Nitroethane	79-24-3	100	310	
Nitrogen dioxide	10102-44-0	(C) 5	(C) 9	
Nitrogen trifluoride	7783-54-2	10	29	
Nitroglycerin	55-63-0	(C) 0.2	(C) 2	X
Nitromethane	75-52-5	100	250	
1-Nitropropane	108-03-2	25	90	
2-Nitropropane	79-46-9	25	90	
N-Nitrosodimethylamine; see 1910.1016				
Nitrotoluene (all isomers)		5	30	X
o-isomer	88-72-2			
m-isomer	99-08-1			
p-isomer	99-99-0			
Nitrotrichloromethane; see Chloropicrin				
Octachloronaphthalene	2234-13-1		0.1	X
Octane	111-65-9	500	2350	
Oil mist, mineral	8012-95-1		5	
Osmium tetroxide (as Os)	20816-12-0		0.002	
Oxalic acid	144-62-7		1	
Oxygen difluoride	7783-41-7	0.05	0.1	
Ozone	10028-15-6	0.1	0.2	
Paraquat, respirable dust	4685-14-7 1910-42-5 2074-50-2		0.5	X
Parathion	56-38-2		0.1	X
Particulates not otherwise regulated (PNOR)[f]				
Total dust			15	
Respirable fraction			5	
PCB; see Chlorodiphenyl (42% and 54% chlorine)				
Pentaborane	19624-22-7	0.005	0.01	
Pentachloronaphthalene	1321-64-8		0.5	X
Pentachlorophenol	87-86-5		0.5	X
Pentaerythritol	115-77-5			
Total dust			15	
Respirable fraction			5	
Pentane	109-66-0	1000	2950	
2-Pentanone (Methyl propyl ketone)	107-87-9	200	700	
Perchloroethylene (tetrachloroethylene)	127-18-4		see Appendix A.2	
Perchloromethyl mercaptan	594-42-3	0.1	0.8	
Perchloryl fluoride	7616-94-6	3	13.5	
Perlite	93763-70-3			
Total dust			15	
Respirable fraction			5	

| Substance | CAS No.[a] | permissible exposure level | | Skin |
		ppm[b]	mg/m[3c]	
Petroleum distillates (naphtha) (rubber solvent)		500	2000	
Phenol	108-95-2	5	19	X
p-Phenylene diamine	106-50-3		0.1	X
Phenyl ether, vapor	101-84-8	1	7	
Phenyl ether-biphenyl mixture, vapor		1	7	
Phenylethylene; see Styrene				
Phenyl glycidyl ether (PGE)	122-60-1	10	60	
Phenylhydrazine	100-63-0	5	22	X
Phosdrin (Mevinphos)	7786-34-7		0.1	X
Phosgene (carbonyl chloride)	75-44-5	0.1	0.4	
Phosphine	7803-51-2	0.3	0.4	
Phosphoric acid	7664-38-2		1	
Phosphorus (yellow)	7723-14-0		0.1	
Phosphorus pentachloride	10026-13-8		1	
Phosphorus pentasulfide	1314-80-3		1	
Phosphorus trichloride	7719-12-2	0.5	3	
Phthalic anhydride	85-44-9	2	12	
Picloram	1918-02-1			
Total dust			15	
Respirable fraction			5	
Picric acid	88-89-1		0.1	X
Pindone (2-Pivalyl-1,3-indandione)	83-26-1		0.1	
Plaster of paris	26499-65-0			
Total dust			15	
Respirable fraction			5	
Platinum (as Pt)	7440-06-4			
Metal				
Soluble Salts			0.002	
Portland cement	65997-15-1			
Total dust			15	
Respirable fraction			5	
Propane	74-98-6	1000	1800	
beta-Propriolactone; see 1910.1013	57-57-8			
n-Propyl acetate	109-60-4	200	840	
n-Propyl alcohol	71-23-8	200	500	
n-Propyl nitrate	627-13-4	25	110	
Propylene dichloride	78-87-5	75	350	
Propylene imine	75-55-8	2	5	X
Propylene oxide	75-56-9	100	240	
Propyne; see Methyl acetylene				
Pyrethrum	8003-34-7		5	
Pyridine	110-86-1	5	15	
Quinone	106-51-4	0.1	0.4	
RDX: see Cyclonite				
Rhodium (as Rh), metal fume and insoluble compounds	7440-16-6		0.1	
Rhodium (as Rh), soluble compounds	7440-16-6		0.001	
Ronnel	299-84-3		15	
Rotenone	83-79-4		5	

| Substance | CAS No.[a] | permissible exposure level | | Skin |
		ppm[b]	mg/m[3c]	
Rouge				
Total dust			15	
Respirable fraction			5	
Selenium compounds (as Se)	7782-49-2		0.2	
Selenium hexafluoride (as Se)	7783-79-1	0.05	0.4	
Silica, amorphous, diatomaceous earth, containing less than 1% crystalline silica	61790-53-2		see Appendix A.3	
Silica, amorphous, precipitated and gel	112926-00-8		see Appendix A.3	
Silica, crystalline cristobalite, respirable dust	14464-46-1		see Appendix A.3	
Silica, crystalline quartz, respirable dust	14808-60-7		see Appendix A.3	
Silica, crystalline tripoli (as quartz), respirable dust	1317-95-9		see Appendix A.3	
Silica, crystalline tridymite, respirable dust	15468-32-3		see Appendix A.3	
Silica, fused, respirable dust	60676-86-0		see Appendix A.3	
Silicates (less than 1% crystalline silica)				
Mica (respirable dust)	12001-26-2		see Appendix A.3	
Soapstone, total dust			see Appendix A.3	
Soapstone, respirable dust			see Appendix A.3	
Talc (containing asbestos): use asbestos limit: see 29 CFR 1910.1001			see Appendix A.3	
Talc (containing no asbestos), respirable dust	14807-96-6		see Appendix A.3	
Tremolite, asbestiform; see 1910.1001				
Silicon	7440-21-3			
Total dust			15	
Respirable fraction			5	
Silicon carbide	409-21-2			
Total dust			15	
Respirable fraction			5	
Silver, metal and soluble compounds (as Ag)	7440-22-4		0.01	
Soapstone; see Silicates				
Sodium fluoroacetate	62-74-8		0.05	X
Sodium hydroxide	1310-73-2		2	
Starch	9005-25-8			
Total dust			15	
Respirable fraction			5	
Stibine	7803-52-3	0.1	0.5	
Stoddard solvent	8052-41-3	500	2900	
Strychnine	57-24-9		0.15	
Styrene	100-42-5		see Appendix A.2	
Sucrose	57-50-1			
Total dust			15	
Respirable fraction			5	

Substance	CAS No.[a]	permissible exposure level ppm[b]	mg/m[3c]	Skin
Sulfur dioxide	7446-09-5	5	13	
Sulfur hexafluoride	2551-62-4	1000	6000	
Sulfuric acid	7664-93-9		1	
Sulfur monochloride	10025-67-9	1	6	
Sulfur pentafluoride	5714-22-7	0.025	0.25	
Sulfuryl fluoride	2699-79-8	5	20	
Systox; see Demeton 2,4,5-T (2,4,5-trichlorophenoxyacetic acid)	93-76-5		10	
Talc; see Silicates				
Tantalum, metal and oxide dust	7440-25-7		5	
TEDP (Sulfotep)	3689-24-5		0.2	X
Tellurium and compounds (as Te)	13494-80-9		0.1	
Tellurium hexafluoride (as Te)	7783-80-4	0.02	0.2	
Temephos	3383-96-8			
Total dust			15	
Respirable fraction			5	
TEPP (Tetraethyl pyrophosphate)	107-49-3		0.05	X
Terphenylis	26140-60-3	(C) 1	(C) 9	
1,1,1,2-Tetrachloro-2, 2-difluoroethane	76-11-9	500	4170	
1,1,2,2-Tetrachloro-1, 2-difluoroethane	76-12-0	500	4170	
1,1,2,2-Tetrachloroethane	79-34-5	5	35	X
Tetrachoroethylene; see Perchloroethylene				
Tetrachloromethane; see Carbon tetrachloride				
Tetrachloronaphthalene	1335-88-2		2	X
Tetraethyl lead (as Pb)	78-00-2		0.075	X
Tetrahydrofuran	109-99-9	200	590	
Tetramethyl lead, (as Pb)	75-74-1		0.075	X
Tetramethyl succinonitrile	3333-52-6	0.5	3	X
Tetranitromethane	509-14-8	1	8	
Tetryl (2,4,6-Trinitro-phenylmethylnitramine)	479-45-8		1.5	X
Thallium, soluble compounds (as TI)	7440-28-0		0.1	X
4,4'-Thiobis (6-tert, Butyl-m-cresol)	96-69-5			
Total dust			15	
Respirable fraction			5	
Thiram	137-26-8		5	
Tin, inorganic compounds (except oxides) (as Sn)	7440-31-5		2	
Tin, organic compounds (as Sn)	7440-31-5		0.1	
Titanium dioxide	13463-67-7			
Total dust			15	
Toluene	108-88-3		see Appendix A.2	
Toluene-2,4-diisocyanate (TDI)	584-84-9	(C) 0.02	(C) 0.14	
o-Toluidine	95-53-4	5	22	X

| Substance | CAS No.[a] | permissible exposure level | | Skin |
		ppm[b]	mg/m³[c]	
Toxaphene; see				
Chlorinated camphene				
Tremolite; see Silicates				
Tributyl phosphate	126-73-8		5	
1,1,1-Trichloroethane;				
see Methyl chloroform				
1,1,2-Trichloroethane	79-00-5	10	45	X
Trichloroethylene	79-01-6		see Appendix A.2	
Trichloromethane;				
see Chloroform				
Trichloronaphthalene	1321-65-9		5	X
1,2,3-Trichloropropane	96-18-4	50	300	
1,1,2-Trichloro-1,2,	76-13-1	1000	7600	
2-trifluoroethane				
Triethylamine	121-44-8	25	100	
Trifluorobromomethane	75-63-8	1000	6100	
2,4,6-Trinitrophenyl;				
see Picric acid				
2,4,6-Trinitrophenylmethyl				
nitramine; see Tetryl				
2,4,6-Trinitrotoluene (TNT)	118-96-7		1.5	X
Triorthocresyl phosphate	78-30-8		0.1	
Triphenyl phosphate	115-86-6		3	
Turpentine	8006-64-2	100	560	
Uranium (as U)	7440-61-1			
Soluble compounds			0.05	
Insoluble compounds			0.05	
Vanadium	1314-62-1			
Respirable dust (as V_2O_5)			(C) 0.5	
Fume (as V_2O_5)			(C) 0.1	
Vegetable oil mist				
Total dust			15	
Respirable fraction			5	
Vinyl benzene; see Styrene				
Vinyl chloride;	75-01-4			
see 1910.1017				
Vinyl cyanide; see Acrylonitrile				
Vinyl toluene	25013-15-4	100	480	
Warfarin	81-81-2		0.1	
Xylenes (o-, m-, p-isomers)	1330-20-7	100	435	
Xylidine	1300-73-8	5	25	X
Yttrium	7440-65-5		1	
Zinc chloride fume	7646-85-7		1	
Zinc oxide fume	1314-13-2		5	
Zinc oxide	1314-13-2			
Total dust			15	
Respirable fraction			5	
Zinc stearate	557-05-1			
Total dust			15	
Respirable fraction			5	
Zirconium compounds (as Zr)	7440-67-7		5	

[a]The CAS number is for information only. Enforcement is based on the substance name. For an entry covering more than one metal compound measured as the metal, the CAS number for the metal is given, not CAS numbers for the individual compounds.

[b]Parts of vapor or gas per million parts of contaminated air by volume at 25°C and 760 torr.

[c]Milligrams of substance per cubic meter of air. When entry is in this column only, the value is exact; when listed with a ppm entry, it is approximate.

[d]The final benzene standard in 1910.1028 applies to all occupational exposures to benzene except in some circumstances the distribution and sale of fuels, sealed containers and pipelines, coke production, oil and gas drilling and production, natural gas processing, and the percentage exclusion for liquid mixtures; for the excepted subsegments, the benzene limits in Appendix A.2 apply. See 1910.1028 for specific circumstances.

[e]This 8-hour TWA applies to respirable dust as measured by a vertical elutriator cotton dust sampler or equivalent instrument. The time-weighted average applies to the cotton waste-processing operations of waste recycling (sorting, blending, cleaning, and willowing) and garnetting. See also 1910.1043 for cotton dust limits applicable to other sectors.

[f]All inert or nuisance dusts, whether mineral, inorganic, or organic, not listed specifically by substance name are covered by the "Particulates Not Otherwise Regulated" (PNOR) limit, which is the same as the inert or nuisance dust limit of Appendix A.3.

Source: Code of Federal Regulations 29 CFR 1910.1000. 54 FR 36767, Sept. 5, 1989; 54 FR 41244, Oct. 6, 1989; 55 FR 3724, Feb. 5, 1990; 55 FR 12819, Apr. 6, 1990; 55 FR 19259, May 9, 1990; 55 FR 46950, Nov. 8, 1990; 57 FR 29204, July 1, 1992; 57 FR 42388, Sept. 14, 1992; 58 FR 35340, June 30, 1993.; 61 FR 56746, Nov. 4, 1996.

APPENDIX A.2: PELs AND STELS FOR SOME SPECIALIZED MATERIALS

Substance	8-hr time-weighted average	Acceptable ceiling concentration	Acceptable maximum peak above the acceptable ceiling concentration for an 8-hr shift	
			Concentration	Maximum duration
Benzene[a] (Z37.40-1969)	10 ppm	25 ppm	50 ppm	10 min
Beryllium and beryllium compounds (Z37.29-1970)	2 $\mu g/m^3$	5 $\mu g/m^3$	25 $\mu g/m^3$	30 min
Cadmium dust[b] (Z37.5-1970)	0.2 mg/m^3	0.6 mg/m^3		
Cadmium fume[b] (Z37.5-1970)	0.1 mg/m^3	0.3 mg/m^3		
Carbon disulfide (Z37.3-1968)	20 ppm	30 ppm	100 ppm	30 min
Carbon tetrachloride (Z37.17-1967)	10 ppm	25 ppm	200 ppm	5 min in any 4 hr
Chromic acid and chromates (Z37-7-1971)		1 mg/10 m^3		
Ethylene dibromide (Z37.31-1970)	20 ppm	30 ppm	50 ppm	5 min
Ethylene dichloride (Z37.21-1969)	50 ppm	100 ppm	200 ppm	5 min in any 3 hr
Fluoride as dust (Z37.28-1969)	2.5 mg/m^3			
Formaldehyde: see 1910.1048				
Hydrogen fluoride (Z37.28-1969)	3 ppm			

Hydrogen sulfide (Z37.2-1966)		20 ppm	50 ppm	10 min once only, if no other measurable exposure occurs
Mercury (Z37.8-1971)		1 mg/10 m^3		
Methyl chloride (Z37.18-1969)	100 ppm	200 ppm	300 ppm	5 min in any 3 hr
Methylene chloride see 1910.1052	25 ppm		125 ppm	15 min
Organo (alkyl) mercury (Z37.30-1969)	0.01 mg/m^3	0.04 mg/m^3		
Styrene (Z37.15-1969)	100 ppm	200 ppm	600 ppm	5 min in any 3 hr
Tetrachloroethylene (Z37.22-1967)	100 ppm	200 ppm	300 ppm	5 min in any 3 hr
Toluene (Z37.12-1967)	200 ppm	300 ppm	500 ppm	10 min
Trichloroethylene (Z37.19-1967)	100 ppm	200 ppm	300 ppm	5 min in any 2 hr

[a]This standard applies to the industry segments exempt from the 1-ppm 8-hour TWA and 5-ppm STEL of the benzene standard at 1910.1028.

[b]This standard applies to any operations or sectors for which the cadmium standard, 1910.1027, is stayed or otherwise not in effect.

Source: Code of Federal Regulations 29 CFR 1910.1000. 57 FR 42388, Sept. 14, 1992; 58 FR 21781, Apr. 23, 1993; 58 FR 35340, June 30, 1993; 62 FR 1493, Jan. 10, 1997.

APPENDIX A.3: PELs FOR MINERAL DUSTS

Substance	mppcf[a]	mg/m^3
Silica		
Crystalline		
Quartz (respirable)	$\frac{250^{b}}{\%SiO_2 + 5}$	$\frac{10\ mg\backslash m^{3d}}{\%SiO_2 + 2}$
Quartz (total dust)		$\frac{80\ mg\backslash m^{3}}{\%SiO_2 + 2}$
Cristobalite: Use $\frac{1}{2}$ the value calculated from the count or mass formulas for quartz		
Tridymite: Use $\frac{1}{2}$ the value calculated from the formulas for quartz		
Amorphous, including natural diatomaceous earth	20	
Silicates (less than 1% crystalline silica)		
Mica	20	
Soapstone	20	
Talc (not containing asbestos)	20[c]	
Talc (containing asbestos). Use asbestos limit		
Tremolite, asbestiform (see 29 CFR 1910.1001)		
Portland cement	50	
Graphite (natural)	15	

Substance	mppcf[a]	mg/m^3
Coal Dust[e]		
Respirable fraction less than 5% SiO$_2$	$\dfrac{2.4\ \text{mg}\backslash\text{m}^3}{\%\,\text{SiO}_2\ +\ 2}$	
Respirable fraction greater than 5% SiO$_2$	$\dfrac{10\ \text{mg}\backslash\text{m}^3}{\%\,\text{SiO}_2\ +\ 2}$	
Inert or nuisance dust[d]		
Respirable fraction	15	5 mg/m^3
Total dust	50	15 mg/m^3

Note: Conversion factors—mppcf $\times$ 35.3 = million particles per cubic meter = particles per cc.

[a]Millions of particles per cubic foot of air, based on impinger samples counted by light-field techniques.

[b]The percentage of crystalline silica in the formula is the amount determined from airborne samples, except in those instances in which other methods have been shown to be applicable.

[c]Containing less than 1% quartz; if 1% quartz or more, use the quartz limit.

[d]All inert or nuisance dusts, whether mineral, inorganic, or organic, not listed specifically by substance name are covered by this limit, which is the same as the "Particulates not Otherwise Regulated" (PNOR) limit in the table of Appendix A.1.

[e]Both concentration and percent quartz for the application of this limit are to be determined from the fraction passing a size selector with the following characteristics:

Aerodynamic diameter (unit density sphere)	Percent passing selector
2	90
2.5	75
3.5	50
5.0	25
10	0

The measurements under this note refer to the use of an AEC (now NRC) instrument. The respirable fraction of coal dust is determined with an MRE: the figure corresponding to that of 2.4 mg/m^3 in the table for coal dust is 4.5 mg/m^3.

APPENDIX B

Medical Treatment

- Treatment of INFECTION
- Application of ANTISEPTICS during second or subsequent visit of medical personnel
- Treatment of SECOND-OR THIRD-DEGREE BURN(S)
- Application of SUTURES (stitches)
- Removal of FOREIGN BODIES EMBEDDED IN EYE
- Removal of FOREIGN BODIES from wound if procedure is COMPLICATED because of depth of embedment, size, or location
- Use of PRESCRIPTION MEDICATIONS (except a single dose administered on first visit for minor injury or discomfort)
- Use of hot or cold SOAKING THERAPY during second or subsequent visit to medical personnel
- Application of hot or cold COMPRESS(ES) during second or subsequent visit to medical personnel
- CUTTING AWAY DEAD SKIN (surgical debridement)
- Application of HEAT THERAPY during second or subsequent visit to medical personnel
- Use of WHIRLPOOL BATH THERAPY during second or subsequent visit to medical personnel
- POSITIVE X-RAY DIAGNOSIS (fractures, broken bones, etc.)
- ADMISSION TO A HOSPITAL or equivalent medical facility for treatment

Source: Ref. Recordkeeping.

First-Aid Treatment

- Application of ANTISEPTICS during first visit to medical personnel
- Treatment of FIRST-DEGREE BURN(S)
- Application of BANDAGE(S) during first visit to medical personnel
- Application of BUTTERFLY ADHESIVE DRESSING(S) or STERISTRIP(S) in lieu of sutures.*
- Use of ELASTIC BANDAGE(S) during first visit to medical personnel
- Removal of FOREIGN BODIES NOT EMBEDDED IN EYE if only irrigation is required
- Removal of FOREIGN BODIES from wound if procedure is UNCOMPLICAT-ED, and is, for example, by tweezers or other simple technique
- Use of NONPRESCRIPTION MEDICATION and administration of single dose of PRESCRIPTION MEDICATION on first visit for minor injury or discomfort
- SOAKING THERAPY on initial visit to medical personnel or removal of bandages by SOAKING
- Application of hot or cold COMPRESS(ES) during first visit to medical personnel
- Application of OINTMENTS to abrasions to prevent drying or cracking
- Application of HEAT THERAPY during first visit to medical personnel
- Use of WHIRLPOOL BATH THERAPY during first visit to medical personnel
- NEGATIVE X-RAY DIAGNOSIS
- OBSERVATION of injury during visit to medical personnel

The following procedure, by itself, is not considered medical treatment:

- Administration of TETANUS SHOT(S) or BOOSTER(S). However, these shots are often given in conjunction with the more serious injuries; consequently, injuries requiring tetanus shots may be recordable for other reasons.

Source: Ref. Recordkeeping

---

*OSHA Recordkeeping standard change

Classification of Medical Treatment

OCCUPATIONAL SKIN DISEASES OR DISORDERS

Examples: Contact dermatitis, eczema, or rash caused by primary irritants and sensitizers or poisonous plants; oil acne; chrome ulcers; chemical burns or inflammations, etc.

DUST DISEASES OF THE LUNGS (PNEUMOCONIOSES)

Examples: Silicosis, asbestosis and other asbestos-related diseases, coal worker's pneumoconiosis, byssinosis, siderosis, and pneumoconioses.

RESPIRATORY CONDITIONS DUE TO TOXIC AGENTS

Examples: Pneumonitis, pharyngitis, rhinitis or acute congestion due to chemicals, dusts, gases, or fumes; farmer's lung, etc.

POISONING (SYSTEMIC EFFECT OF TOXIC MATERIALS)

Examples: Poisoning by lead, mercury, cadmium, arsenic, or other metals; poisoning by carbon monoxide, hydrogen sulfide, or other gases; poisoning by benzol, carbon tetrachloride, or other organic solvents; poisoning by insecticide sprays such as parathion or lead arsenate; poisoning by other chemicals such as formaldehyde, plastics, or resins, etc.

DISORDERS DUE TO PHYSICAL AGENTS (OTHER THAN TOXIC MATERIALS)

Examples: Heatstroke, sunstroke, heat exhaustion, and other effects of environmental heat; freezing, frostbite, and effects of exposure to low temperatures; caisson disease; effects of ionizing radiation (isotopes, X rays, radium); effects of nonionizing radiation (welding flash, ultraviolet rays, microwaves, sunburn), etc.

DISORDERS ASSOCIATED WITH REPEATED TRAUMA

Examples: Noise-induced hearing loss; synovitis, tenosynovitis, and bursitis; Raynaud's phenomena; and other conditions due to repeated motion, vibration, or pressure.

ALL OTHER OCCUPATIONAL ILLNESSES

Examples: Anthrax, brucellosis, infectious hepatitis, malignant and benign tumors, food poisoning, histoplasmosis, coccidioidomycosis, etc.

Highly Hazardous Chemicals, Toxics, and Reactives

This appendix contains a list of toxic and reactive highly hazardous chemicals that present a potential for a catastrophic event at or above the threshold quantity. List is current as of October 10, 2002. OSHA standard 29 CFR 1910. 119, APP.A.

Chemical name	CAS[a]	TQ[b]
Acetaldehyde	75-07-0	2500
Acrolein (2-propenal)	107-02-8	150
Acrylyl chloride	814-68-6	250
Alkylaluminums	Varies	5000
Allyl chloride	107-05-1	1000
Allylamine	107-11-9	1000
Ammonia, anhydrous	7664-41-7	10,000
Ammonia solutions (greater than 44% ammonia by weight)	7664-41-7	15,000
Ammonium perchlorate	7790-98-9	7500
Ammonium permanganate	7787-36-2	7500
Arsine (also called arsenic hydride)	7784-42-1	100
Bis (chloromethyl) ether	542-88-1	100
Boron trichloride	10294-34-5	2500
Boron trifluoride	7637-07-2	250
Bromine	7726-95-6	1500
Bromine chloride	13863-41-7	1500
Bromine pentafluoride	7789-30-2	2500
Bromine trifluoride	7787-71-5	15,000
3-Bromopropyne (also called propargyl bromide)	106-96-7	100
Butyl hydroperoxide (tertiary)	75-91-2	5000
Butyl perbenzoate (tertiary)	614-45-9	7500
Carbonyl chloride (see Phosgene)	75-44-5	100
Carbonyl fluoride	353-50-4	2500
Cellulose nitrate (concentration greater than 12.6% nitrogen)	9004-70-0	2500
Chlorine	7782-50-5	1500
Chlorine dioxide	10049-04-4	1000
Chlorine pentrafluoride	13637-63-3	1000
Chlorine trifluoride	7790-91-2	1000

Chemical name	CAS[a]	TQ[b]
Chlorodiethylaluminum (also called diethylaluminum chloride)	96-10-6	5000
1-Chloro-2,4-dinitrobenzene	97-00-7	5000
Chloromethyl methyl ether	107-30-2	500
Chloropicrin	76-06-2	500
Chloropicrin and methyl bromide mixture	None	1500
Chloropicrin and methyl chloride mixture	None	1500
Cumene hydroperoxide	80-15-9	5000
Cyanogen	460-19-5	2500
Cyanogen chloride	506-77-4	500
Cyanuric fluoride	675-14-9	100
Diacetyl peroxide (concentration greater than 70%)	110-22-5	5000
Diazomethane	334-88-3	500
Dibenzoyl peroxide	94-36-0	7500
Diborane	19287-45-7	100
Dibutyl peroxide (tertiary)	110-05-4	5000
Dichloroacetylene	7572-29-4	250
Dichlorosilane	4109-96-0	2500
Diethylzinc	557-20-0	10,000
Diisopropyl peroxydicarbonate	105-64-6	7500
Dilauroyl peroxide	105-74-8	7500
Dimethyldichlorosilane	75-78-5	1000
1,1-Dimethylhydrazine	57-14-7	1000
Dimethylamine, anhydrous	124-40-3	2500
2,4-Dinitroaniline	97-02-9	5000
Ethyl methyl ketone peroxide (also methyl ethyl ketone peroxide; concentration greater than 60%)	1338-23-4	5000
Ethyl nitrite	109-95-5	5000
Ethylamine	75-04-7	7500
Ethylene fluorohydrin	371-62-0	100
Ethylene oxide	75-21-8	5000
Ethyleneimine	151-56-4	1000
Fluorine	7782-41-4	1000
Formaldehyde (formalin)	50-00-0	1000
Furan	110-00-9	500
Hexafluoroacetone	684-16-2	5000
Hydrochloric acid, anhydrous	7647-01-0	5000
Hydrofluoric acid, anhydrous	7664-39-3	1000
Hydrogen bromide	10035-10-6	5000
Hydrogen chloride	7647-01-0	5000
Hydrogen cyanide, anhydrous	74-90-8	1000
Hydrogen fluoride	7664-39-3	1000
Hydrogen peroxide (52% by weight or greater)	7722-84-1	7500
Hydrogen selenide	7783-07-5	150
Hydrogen sulfide	7783-06-4	1500
Hydroxylamine	7803-49-8	2500
Iron, pentacarbonyl	13463-40-6	250
Isopropylamine	75-31-0	5000
Ketene	463-51-4	100
Methacrylaldehyde	78-85-3	1000
Methacryloyl chloride	920-46-7	150
Methacryloyloxyethyl isocyanate	30674-80-7	100
Methyl acrylonitrile	126-98-7	250
Methylamine, anhydrous	74-89-5	1000

Chemical name	CAS[a]	TQ[b]
Methyl bromide	74-83-9	2500
Methyl chloride	74-87-3	15,000
Methyl chloroformate	79-22-1	500
Methyl ethyl ketone peroxide (concentration greater than 60%)	1338-23-4	5000
Methyl fluoroacetate	453-18-9	100
Methyl fluorosulfate	421-20-5	100
Methyl hydrazine	60-34-4	100
Methyl iodide	74-88-4	7500
Methyl isocyanate	624-83-9	250
Methyl mercaptan	74-93-1	5000
Methyl vinyl ketone	79-84-4	100
Methyltrichlorosilane	75-79-6	500
Nickel carbonyl (nickel tetracarbonyl)	13463-39-3	150
Nitric acid (94.5% by weight or greater)	7697-37-2	500
Nitric oxide	10102-43-9	250
Nitroaniline (paranitroaniline)	100-01-6	5000
Nitrogen dioxide	10102-44-0	250
Nitrogen oxides (NO; NO_2; N_2O_4; N_2O_3)	10102-44-0	250
Nitrogen tetroxide (also called nitrogen peroxide)	10544-72-6	250
Nitrogen trifluoride	7783-54-2	5000
Nitrogen trioxide	10544-73-7	250
Nitromethane	75-52-5	2500
Oleum (65 to 80% by weight; also called fuming sulfuric acid)	8014-94-7	1000
Osmium tetroxide	20816-12-0	100
Oxygen difluoride (fluorine monoxide)	7783-41-7	100
Ozone	10028-15-6	100
Pentaborane	19624-22-7	100
Peracetic acid (concentration greater 60% acetic acid; also called peroxyacetic acid)	79-21-0	1000
Perchloric acid (concentration greater than 60% by weight)	7601-90-3	5000
Perchloromethyl mercaptan	594-42-3	150
Perchloryl fluoride	7616-94-6	5000
Peroxyacetic acid (concentration greater than 60% acetic acid; also called peracetic acid)	79-21-0	1000
Phosgene (also called carbonyl chloride)	75-44-5	100
Phosphine (hydrogen phosphide)	7803-51-2	100
Phosphorus oxychloride (also called phosphoryl chloride)	10025-87-3	1000
Phosphorus trichloride	7719-12-2	1000
Phosphoryl chloride (also called phosphorus oxychloride)	10025-87-3	1000
Propargyl bromide	106-96-7	100
Propyl nitrate	627-3-4	2500
Sarin	107-44-8	100
Selenium hexafluoride	7783-79-1	1000
Stibine (antimony hydride)	7803-52-3	500
Sulfur dioxide (liquid)	7446-09-5	1000
Sulfur pentafluoride	5714-22-7	250
Sulfur tetrafluoride	7783-60-0	250
Sulfur trioxide (also called sulfuric anhydride)	7446-11-9	1000
Sulfuric anhydride (also called sulfur trioxide)	7446-11-9	1000
Tellurium hexafluoride	7783-80-4	250
Tetrafluoroethylene	116-14-3	5000
Tetrafluorohydrazine	10036-47-2	5000
Tetramethyl lead	75-74-1	1000

Chemical name	CAS[a]	TQ[b]
Thionyl chloride	7719-09-7	250
Trichloro (chloromethyl) silane	1558-25-4	100
Trichloro (dichlorophenyl) silane	27137-85-5	2500
Trichlorosilane	10025-78-2	5000
Trifluorochloroethylene	79-38-9	10,000
Trimethyoxysilane	2487-90-3	1500

[a] Chemical Abstract service number.

[b] Threshold quantity in pounds (amount necessary to be covered by this standard). OSHA std 1910.119, Appendix A. 57 FR 7847, Mar. 4, 1992.

Standard Industrial Classification [SIC] Code

PRINCIPAL MANUFACTURING CATEGORIES

SIC Code	Description
20	Food and Kindred Products
21	Tobacco Manufacturers
22	Textile Mill Products
23	Apparel and Other Finished Products Made from Fabrics and Other Similar Materials
24	Lumber and Wood Products, Except Furniture
25	Furniture and Fixtures
26	Paper and Allied Products
27	Printing, Publishing, and Allied Industries
28	Chemicals and Allied Products
29	Petroleum Refining and Related Industries
30	Rubber and Miscellaneous Plastics Products
31	Leather and Leather Products
32	Stone, Clay, Glass, and Concrete Products
33	Primary Metal Industries
34	Fabricated Metal Products, Except Machinery and Transportation Equipment
35	Machinery, Except Electrical
36	Electrical and Electronic Machinery, Equipment and Supplies
37	Transportation Equipment
38	Measuring, Analyzing, and Controlling Instruments; Photographic, Medical, and Optical Goods; Watches and Clocks
39	Miscellaneous Manufacturing Industries

PARTIALLY EXEMPT INDUSTRIES

Employers classified in the following SIC Codes are not required to keep general OSHA injury and illness records unless asked by OSHA, the Bureau of Labor Statistics (BLS), or an authorized state agency. Fatalities or workplace incidents that result in the hospitalization of three or more employees must still be reported.

SIC Code	Industry Description	SIC Code	Industry Description
525	Hardware Stores	725	Shoe Repair and Shoe-shine Parlors
542	Meat and Fish Markets	726	Funeral Service and Crematories
544	Candy, Nut, and Confectionery Stores	729	Miscellaneous Personal Services
545	Dairy Products Stores	731	Advertising Services
546	Retail Bakeries	732	Credit Reporting and Collection Services
549	Miscellaneous Food Stores	733	Mailing, Reproduction, and Stenography Services
551	New- and Used-Car Dealers	737	Computer and Data Processing Services
552	Used-Car Dealers	738	Miscellaneous Business Services
554	Gasoline Service Stations	764	Reupholstery and Furniture Repair
557	Motorcycle Dealers	78	Motion Pictures
56	Apparel and Accessory Stores	791	Dance Studios, Schools, and Halls
573	Radio, Television, and Computer Stores	792	Producers, Orchestras, Entertainers
58	Eating and Drinking Places	793	Bowling Centers
591	Drug Stores and Proprietary Stores	801	Offices and Clinics of Medical Doctors
592	Liquor Stores	802	Offices and Clinics of Dentists
594	Miscellaneous Shopping Goods Stores	803	Offices of Osteopathic Physicians
599	Retail Stores not elsewhere classified	804	Offices of Other Health Practitioners
60	Depository Inst (banks and savings inst)	807	Medical and Dental Laboratories
61	Nondepository Institutions (credit inst)	809	Health and Allied Services not elsewhere class
62	Security and Commodity Brokers	81	Legal Services
63	Insurance Carriers	82	Educational Services (schools, colleges, university libraries)
64	Insurance Agents, Brokers, and Services	832	Individual and Family Service
653	Real Estate Agents and Managers	835	Child Day-care Services
654	Title Abstract Offices	839	Social Services not elsewhere classified
67	Holding and Other Investment Offices	841	Museums and Art Galleries
772	Photographic Portrait and Studios	86	Membership Organizations
723	Beauty Shops	87	Engineering Accounting Research, Management, Related, Services
724	Barber Shops	899	Services not elsewhere classified

States Having Federally Approved[1] State Plans for Occupational Safety and Health Standards and Enforcement

Alaska
Arizona
California
Connecticut[2]
Hawaii
Indiana
Iowa
Kentucky
Maryland
Michigan
Minnesota
Nevada
New Jersey[2]
New Mexico

New York[2]
North Carolina
Oregon
South Carolina
Tennessee
Utah
Vermont
Virginia
Washington
Wyoming
Also, Puerto Rico and the
    Virgin Islands

---

[1]Approved by authority of Section 18.b of Public Law 91–596. The list is current as of November 2002. The most recent change was in 2001.

[2]The Connecticut, New Jersey, and New York plans cover the public sector only (state, local government, and municipal employees only). The federal OSHA agency covers enforcement for private sector employees in these three states.

Glossary

ACGIH American Conference of Governmental Industrial Hygienists.

ADA Americans with Disabilities Act (of 1990).

Aggravating factor In the study of loss incident causation, a circumstance that makes the outcome of a loss incident more severe.

AIDS Acquired immune deficiency syndrome. A terminal illness or condition resulting from exposure to the HIV virus. (*See also* HIV.)

AIHA American Industrial Hygiene Association.

AL Action level.

amp Ampere.

ANSI American National Standards Institute.

Anti tiedown Means of preventing a control panel from being tied down and thus rendering the machine unsafe.

Anti two-block device Mechanism for preventing a crane hook block from being drawn up to the point at which it contacts the boom point.

API American Petroleum Institute.

ASME American Society of Mechanical Engineers.

Asphyxiants Substances that prevent oxygen from reaching the body cells.

ASSE American Society of Safety Engineers.

ASTM American Society for Testing and Materials.

Barrier guard In machine guarding terminology, a rigid partition on a machine to prevent any part of the operator's or any other's body from entering the danger zone. (Compare with Shield guard.)

BCSP Board of Certified Safety Professionals

BLEVE Boiling liquid expanding vapor explosion.

Block Mechanical assembly containing one or more freely rotating pulleys.

Block and Bleed *See* Double block and bleed.

Block and tackle Assembly consisting of (usually two) blocks reeved together to achieve mechanical advantage.

Bloodborne pathogens Term used in the OSHA standards to refer principally to the HIV and HBV viruses.

Boom Long pivoting structure or arm on a crane.

Brake monitor Device that monitors stopping time or ram overtravel every time the clutch disengages on a part-revolution mechanical press.

Bridge crane Industrial crane that accesses loads by means of a bridge that travels on parallel overhead rails and by means of a trolley that travels back and forth on the bridge.

Bridgeplate Support surface for the transition between dock and a vehicle being loaded; dockboard.

C Ceiling value; maximum acceptable exposure concentration.

Carcinogen a substance that causes or is suspected of causing cancer.

CAS Chemical Abstracts number, a reference list for chemical substances.

CERCLA Comprehensive Environmental Response, Compensation, and Liability Act.

CIH Certified Industrial Hygienist.

Combustible Having a flashpoint higher than 100°F.

CPSC Consumer Product Safety Commission.

CSP Certified Safety Professional.

dB Decibel(s).

dBA Decibel, by the A-weighted scale.

Detector tube Small (usually glass) tube installed in-line in a flexible tube connected to an air-sampling pump. The tube contains a material that changes color in quantifiable layers to provide a measure of the concentration of the particular air contaminant being tested.

Distal cause Secondary or indirect cause of a loss incident, such as a deficient management policy. Distal causes create and shape proximal causes.

Dockboard Support surface for the transition between a dock and a vehicle being loaded; a bridgeplate.

Dosimeter Instrument worn on the person to collect a cumulative measure of exposure over a specified time period; used especially for TWAs. (*See also* TWA.)

Double block and bleed Method of positive isolation of fluids in which two successive valves are closed and between the two closed valves a small "bleeder" valve is opened to relieve pressure that might build up if one of the closed valves were to leak.

EEOC Equal Employment Opportunity Commission.

Egregious violation Glaring or flagrant safety violation that invokes high penalties from OSHA.

Energy isolation device Device for positively locking off a source of energy to a machine or other equipment.

Engulfment Capture and sinking of a person in a dry granular solid material that has fluidlike qualities.

EPA   Environmental Protection Agency.

Epidemiology   Statistical study of populations of victims of various diseases or disorders.

Ergonomics   Study of human capability in relation to the work environment.

Experience rating   Insurance rating based on a company's claim history.

Extra-hazardous employer   In some states, a Workers' Compensation Commission designation given to certain establishments in which the rate of lost workday injuries and illnesses exceed certain defined averages for the industry. The designation results in extra duties for the employer and follow-up consultation to determine whether the firm has complied with the extra duties.

Fail-safe principles   Principles of engineering design that consider the consequences of component failure within the system.

Fault-tree analysis   Logic diagram used to analyze probabilities associated with various causes and their adverse effects.

FCAW   Flux core arc welding.

Fibrillation   Rapid, irregular, convulsions of the heart; the condition is induced by electric shock, especially when the shock is from alternating current.

Flammable   Having a flashpoint lower than 100°F.

Flashback   In welding terminology, the phenomenon that occurs when an oxygen/fuel gas flame begins to propagate back into the mixing chamber of the welding torch or welding gas manifold.

Flux   In soldering and welding, a material that melts along with the metal and is used to facilitate the process by combining with impurities and preventing oxidation.

FMEA   Failure modes and effects analysis.

Full revolution   In presswork terminology, a type of mechanical press drive in which the ram is tripped by a positive engagement of the flywheel and cannot be disengaged midstroke.

Fume   Tiny particles of resolidified vapors of substances that are normally solids. Metal fumes are often encountered in welding operations.

Gate   In presswork terminology, a temporary barrier that closes to protect the operator during the dangerous part of the machine cycle. Type A closes at the beginning of the cycle and stays closed for the entire cycle. Type B closes at the beginning of the cycle, then reopens during the less dangerous portion of the stroke (e.g., when the ram starts back up and the dies begin to reopen).

GFCI   Ground-fault circuit interrupter.

GMAW   Gas metal arc welding (also known as MIG).

Gross load   Total load being supported by the system, including both the payload and the equipment being used to move it. In the case of a hoist mechanism, gross load equals the weight of the payload plus the weight of the load block.

Ground   Object or conductor that has zero voltage potential from ground potential; in electrical wiring terminology, "ground" usually refers to the noncurrent-carrying conductor that is intended to ground unwanted voltage, sometimes called the ground*ing* conductor. Sometimes "ground" is used to refer to the current-carrying neutral conductor; sometimes called the ground*ed* conductor.

GTAW   Gas tungsten arc welding (also known as TIG).

HBV   Hepatitis B virus.

HIV   Human immunodeficiency virus.

Hot   In electrical wiring terminology, the current-carrying conductor that has a significant voltage potential from ground potential.

Hydrostatic test   Periodic test performed on fire extinguishers to verify the integrity of the shell to contain adequate pressures.

Hz   Hertz (cycles per second).

IBP   Initial boiling point.

ICHD   Industrial chemical hazards database.

IDL   Immediately dangerous to life.

IDLH   Immediately dangerous to life or health.

Impairment   A physical or mental handicap that qualifies under the Americans with Disabilities Act (ADA) for protecting the worker from discrimination in employment.

Incidence rate   Rate of job-related injuries and illnesses, including fatalities, lost-work-day cases, number of lost workdays, specific-hazard incidence, lost-workday-injury rate.

Interlock   Switch, usually electrical, that turns off power to a machine whenever a guard is removed or a gate opened or some other safety device is disabled.

Irreversibility   *See* Point of irreversibility.

Irritants   Substances that inflame surfaces of parts of the body by their corrosive action.

JCP   Job candidate profile; a screening test used in job placement testing.

Kickback   Accidental engagement between the material and a rotating tool that causes the material to be thrown back at the operator, or alternatively, if the material is fixed, causes the tool, such as a hand-held circular saw, to be jerked out of control.

Lay   In wire rope terminology, the length, measured along the core of a wire rope, required for one strand to make one complete revolution (twist) about the core.

LBW   Laser beam welding.

LEL   Lower explosive limit (for flammable vapors); the percent concentration in air below which the mixture is too lean to ignite.

LFL   Lower flammable limit (for flammable vapors); the lowest concentration of a material that will propagate a flame. The LFL is usually expressed as a percent by volume of the material in air. The term is used somewhat interchangeably with LEL and is applied to the control of fires around dip tanks.

Load block   Pulley assembly to which the load is attached.

Lockout   Method of assuring that machines being serviced are not turned back on prematurely. The power switch or control box is locked out by the maintenance person performing the service. Only the maintenance person performing the service has the key to the lock. Multiple maintenance personnel each have their own locks so that all locks must be removed before the machine can be returned to service. Lockout is considered more reliable than "tagout."

Lost workday   According to OSHA definition and current terminology, a "lost" workday includes both days away from work and days in which the worker is temporarily transferred to another job within the firm because of a workplace injury or illness.

LPG   Liquefied petroleum gas.

LWDI   Lost-workday incidence rate (for injuries only, not illnesses); a standard occupational safety and health index for comparing industry safety records. The LWDI was formerly used by OSHA Compliance Officers to determine whether a particular firm should receive a full inspection. Later the LWDI became an index for nationwide decision making in determining industry (SIC) priorities for inspection.

ma   Milliampere.

MAC   Maximum acceptable ceiling.

MAPP gas   Commercially available welding fuel gas used as a somewhat safer substitute for acetylene when lower welding temperatures are acceptable for the process.

Mechanical advantage   Favorable ratio of output force to the required input force for a mechanism.

MIG   Metal inert gas (welding); also known as GMAW.

Mitigating factor   In the study of loss-incident causation, a circumstance that makes the outcome of a loss incident less severe.

MSDS   Material safety data sheet.

MSHA   Mine Safety and Health Administration.

Mutagens   Substances that affect chromosomes and are thus a hazard to the species.

NEC   *National Electrical Code®*.

Neutral   In electrical wiring terminology, the current-carrying conductor that is at ground or near-ground potential; sometimes called the grounded conductor.

NFPA   National Fire Protection Association.

NIOSH   National Institute for Occupational Safety and Health.

Nip point   Position at which moving parts meet and can draw clothing or body parts into the machine. Examples are points at which gears mesh, chains engage sprockets, or belts engage pulleys.

Nomex   Commercially available material used in clothing and personal protective equipment for welders.

Nominal   Generally, a specification or dimension before safety factor has been applied. (*See also* Rated.)

NOx   Oxides of nitrogen (e.g., NO, NO_2, N_2O).

NSC   National Safety Council.

OSHA   Occupational Safety and Health Administration.

Part revolution   In presswork terminology, a type of mechanical press drive in which the flywheel is equipped with a friction clutch and can be disengaged midstroke and a brake applied to stop the ram.

Parts of rope   Mechanical advantage provided by block and tackle; number of lines supporting the load block.

Payload   Weight of the load exclusive of the vehicle, material-handling device, or other equipment being used to move the load.

PE   Registered professional engineer.

PEL   Permissible exposure limit.

Pharmacokinetics   Study of the absorption, disposition, metabolism, and elimination of chemicals in the body.

Plugging   Using reverse torque as a brake to stop the travel of a crane hoist.

Point of irreversibility   In the study of loss-incident causation, the point in the causal diagram that, if reached, will definitely result in a loss incident.

Power rail   Fixed electrical conductor for delivering continuous power to a moving device.

PPE   Personal protective equipment.

Proximal cause   Direct hazard; a primary, immediate cause of a loss incident.

Pullback   Method of safeguarding the point of operation of a press by a mechanical linkage between the action of the press ram and a set of wristlets worn by the operator. As the ram descends, the linkage pulls the operator's hands away from the danger zone.

Rail clamp   In industrial crane terminology, a device for clamping a crane bridge to the rail on which it is traveling; the clamp is used to prevent drifting of the bridge under wind loading.

Rated   Describes a specification or dimension for which a safety factor has already been applied. (*See also* Nominal.)

RCRA   Resource Conservation and Recovery Act.

Reeving   Threading of ropes through pulleys.

Risk assessment code (RAC)   A U.S. Air Force classification system for accidents that is based on both mishap severity and mishap probability of occurrence.

Roadmap approach   A method of filing documents in which a central file tells where within the plant to find various other required documents.

ROPS   Rollover protective structures.

RSEW   Resistance seam welding.

RSW   Resistance spot welding.

SARA   Superfund Amendments and Reauthorization Act.

SAW   Submerged arc welding.

SCBA   Self-contained breathing apparatus.

Sheave   Pulley used in a hoisting arrangement, especially when several pulleys are used together to achieve mechanical advantage.

Shield guard   In machine guarding terminology, a barrier to protect against flying chips or sparks flying out of a machine, (*Compare with* Barrier guard.)

SIC   Standard industrial classification.

Slag   In welding, the nonmetallic molten, then solid, residue that consists of solidified flux combined with impurities.

Sling   In material-handling terminology, the wire rope, chain, or other connector (not to be confused with the hoist rope itself) used to attach a load to the crane, hoist, helicopter, or other lifting device.

SLM   Sound-level meter.

SMAW   Shielded metal arc welding (also known as stick electrode welding).

Sphere of control   In a loss-incident causation diagram, the region prior to the point of irreversibility.

Spreader   In power saw terminology, a device to maintain separation between the two walls of a cut to assure that the saw blade will not accidentally reengage the material that has already been cut and thus result in a kickback.

STEL   Short-term exposure limit.

Strand   In wire rope terminology, any of several bundles of wires that are twisted about the core of the wire rope.

Sweep   Obsolete (no longer legally recognized) method of safeguarding a press. Sweeps are mechanically linked to the press ram mechanism and strike and brush away the operator's arms or other objects in the area of the entrance to the danger zone.

Switch loading   Filling a tank with a material different from that which it contained previously. Switch loading usually refers to the switching of flammable/combustible fuels to be transported in a tank truck.

TAG   Popular test method for determining flashpoint.

Tagout   Method of assuring that machines being serviced are not turned back on prematurely. A tag placed on the power switch warns operators and others not to turn on the machine until the tag is removed by the person performing the service. (*See also* Lockout.)

Teratogens   Substances that have harmful effects on fetuses.

TIG   Tungsten inert gas (welding); also known as GTAW.

TLV   Threshold limit value.

Tongue gap   On a wheel-type grinding machine, the clearance between the upper opening (tongue) guard and the grinding wheel. The maximum legal clearance is $\frac{1}{4}$ inch.

Tongue guard   Adjustable plate attached to the upper edge of the opening that exposes the wheel on a grinding machine. The purpose is to contain flying fragments in the event of wheel failure.

TOSCA   Toxic Substances Control Act.

Toxicology   Study of the nature and effects of poisons.

Tribology   The study of the mechanisms and phenomena of friction; in safety, especially applies to the study of trips and falls.

Trip bar   Bar that can be quickly and conveniently accessed by the operator or others nearby to deactivate or disengage a machine in an emergency. Also called a triprod.

Trip wire   Flexible cable that can be pulled or otherwise quickly and conveniently accessed by the operator or others nearby to deactivate or disengage a machine in an emergency.

Trolley   In industrial crane terminology, the assembly that travels back and forth on top of the bridge on an overhead bridge crane. The trolley carries the hoist mechanism.

TW   Thermit welding.

TWA   Time-weighted average.

Two-blocking   Hazardous, accidental contact between blocks of sheaves that occurs when the hoist rope is drawn too far. The condition will quickly break the hoist rope, causing the load block and the load (if any) to fall.

UEL   Upper explosive limit (for flammable vapors); percent concentration in air above which the mixture is too rich to ignite.

Vapors   Gases of substances that are normally liquid or solid, usually liquid.

Workers' compensation   Statutory compensation levels to be paid by the employer for various injuries that may be incurred by the worker.

Zero mechanical state   State of a machine after any residual sources of energy have been relieved or restrained to render them harmless, after the machine has been turned off.

Index